Rights to Plant Genetic Resources and Traditional Knowledge

Basic Issues and Perspectives

Rights to Plant Genetic Resources and Traditional Knowledge

Basic Issues and Perspectives

Edited by

Susette Biber-Klemm

World Trade Institute, University of Berne, Switzerland

and

Thomas Cottier

World Trade Institute, University of Berne, Switzerland

Associate Editor

Danuta Szymura Berglas

World Trade Institute, University of Berne, Switzerland

www.cabi.org

on behalf of the Swiss Agency for Development and Cooperation and the World Trade Institute

CABI is a trading name of CAB International

CABI Head Office
Nosworthy Way
Wallingford
Oxfordshire OX10 8DE
UK

Tel: +44 (0)1491 832111
Fax: +44 (0)1491 833508
E-mail: cabi@cabi.org
Website: www.cabi.org

CABI North American Office
875 Massachusetts Avenue
7th Floor
Cambridge, MA 02139
USA

Tel: +1 617 395 4056
Fax: +1 617 354 6875
E-mail: cabi-nao@cabi.org

A catalogue record for this book is available from the British Library, London, UK.

Library of Congress Cataloging-in-Publication Data

Rights to plant genetic resources and traditional knowledge : basic issues and perspectives / edited by Susette Biber-Klemm and Thomas Cottier.
p. cm.
Includes bibliographical references and index.
ISBN-13: 978-0-85199-033-0 (alk. paper)
ISBN-10: 0-85199-033-9 (alk. paper)
1. Plant varieties--Patents. 2. Genes--Patents. 3 Seeds--Patents. 4. Biological diversity conservation. I. Biber-Klemm, Susette. II. Cottier, Thomas. III. Title.

K3876.R54 2005
346.04´86--dc22

2005011727

ISBN-13: 978 0 85199 033 0
ISBN-10: 0 85199 033 9

Typeset by MRM Graphics Ltd, Winslow, Bucks.
Printed and bound in the UK by Biddles Ltd, King's Lynn.

Contents

Contributors **xv**
Preface and Acknowledgements **xvii**
Abbreviations **xix**

Introduction **xxiii**

I.1 A New Regulatory Field xxiii
I.2 Goals and Basic Assumptions xxiv
I.3 The Current Legal Status of Plant Genetic Resources xxv
I.4 Policy Measures Submitted xxvii
I.4.1 A multiple approach required xxvii
I.4.2 The protection of assignable traditional knowledge (Traditional Intellectual Property Rights) xxviii
I.4.3 Reform and novel uses of existing intellectual property rights: patents and plant breeders' rights xxviii
I.4.4 The protection of non-assignable traditional knowledge and plant genetic resources for food and agriculture xxix
I.4.5 Support of trade in products resulting from the use of traditional plant genetic resources for food and agriculture and traditional knowledge xxix
I.4.6 Tariff measures facilitating the export of traditional knowledge-related products xxx
I.5 Conclusions xxxii

PART I GENERAL FRAMEWORK **1**

1 Problems and Goals **3**
Susette Biber-Klemm and Danuta Szymura Berglas

1.1 Biodiversity and Traditional Knowledge: Factual Background and Problems 3
1.1.1 Introduction 3
1.1.2 Biodiversity 4
Terminology 4
Characteristics of plant genetic resources 6
Distribution and origin of biological diversity 6

The value of biological diversity 7
The loss of biological diversity 10
Conclusions 14
1.1.3 Traditional knowledge 15
Introduction 15
Terminology and characteristics 16
Value of traditional knowledge 21
The loss of traditional knowledge 23
Conclusions 26
1.1.4 Conservation and sustainable use of biodiversity and traditional knowledge: the interface with trade 27
Introduction 27
Economic failure as reason for the loss of biodiversity and traditional knowledge 29
Supporting conservation and sustainable use of biodiversity and traditional knowledge: financial aspects 30
The creation of incentives: the Convention on Biodiversity discussion 31
Conclusions 34
1.2 Goals of Instruments and Measures to Protect Traditional Knowledge and Plant Genetic Resources: Basic Principles and Ethical Reflections 34
1.2.1 Introduction 34
In general 34
Goals as laid down in international legal instruments concerning genetic resources and traditional knowledge 34
1.2.2 Equity and fairness 35
Introduction 35
Legal sources in the biodiversity context: conventions and soft law 35
Equity 36
Fairness 39
Conclusions 40
1.2.3 Sustainability 41
General: term and interpretation 41
The particulars of biological diversity and traditional knowledge with a view to sustainable use and development 44
1.2.4 Food security and basic health care 45
In general: theoretical background 45
Food security 46
Health care, medicinal care, traditional medicine 49
1.2.5 Conclusions 50
Bibliography 51

2 The Current Law of Plant Genetic Resources and Traditional Knowledge **56**
Susette Biber-Klemm, Thomas Cottier, Philippe Cullet and Danuta Szymura Berglas

2.1 Introduction 56
2.2 Permanent Sovereignty over Plant Genetic Resources 57
2.3 International Agreements and Institutions 59
2.3.1 The Convention on Biological Diversity 59
2.3.2 The International Treaty on Plant Genetic Resources for Food and Agriculture 63
2.3.3 The law and policy in the World Trade Organization 68

From GATT to the WTO 68
From tariffs to farm support levels and intellectual property protection 69
International dispute settlement and enforcement 70
The Agreement on Agriculture 72
The Agreement on the Application of Sanitary and Phytosanitary Measures 72
The Agreement on Trade-related Aspects of Intellectual Property Rights 73
WTO law and genetic resources 80
2.3.4 The International Union for the Protection of New Varieties of Plants (UPOV) 80
2.4 Selected Regional and National Legal Frameworks 82
2.4.1 Developing countries 82
The African continent: the OAU Model Legislation 82
Andean region 84
India 90
2.4.2 Developed countries 93
Europe 93
USA 95
Australia 96
Bibliography 110

3 Intellectual Property Rights, Plant Genetic Resources and Traditional Knowledge 112
Philippe Cullet, Christophe Germann, Andrea Nascimento Müller and Gloria Pasadilla

3.1 Introduction 112
3.1.1 Overview 112
3.1.2 Forms of intellectual property protection 113
3.1.3 The functions of intellectual property rights 114
3.2 The Protection of TK Related to Plant Genetic Resources 115
3.2.1 Definitions 115
3.2.2 Patent protection 116
Introduction 116
The rationale of patent protection 119
The concepts of invention and the protection of plant varieties 121
Protective criteria 125
Scope and limitations of patent protection 127
Preliminary conclusions on patent protection for holders of traditional knowledge 128
3.2.3 Plant breeders' rights 130
Background 130
What are PBRs? 131
Content and limitation of PBRs 131
3.2.4 *Sui generis* protection systems 132
Rationale of *sui generis* protection 132
Country examples of *sui generis* protection 133
3.3 Impacts on Existing PGR, Landraces and TK 134
3.3.1 Case studies 135
The turmeric patent 135
The neem patents 136
The kava case 137
3.3.2 Proposed solutions 137
Biodiversity registers 137

Benefit-sharing 138
Local innovation 139
3.3.3 Problem analysis 140
3.3.4 Impacts on PGR diversity 141
3.3.5 Impacts on TK 142
3.3.6 Tension between IPR and competition 142
Contract arrangement among sequential innovators 143
Some basic principles for IPR and competition policy 145
3.4 Conclusions 151
Bibliography 151

PART II TOWARDS *SUI GENERIS* RIGHTS **155**

4 Origin and Allocation of Traditional Knowledge and Landraces **157**

Part 1: Origin and Allocation of Traditional Knowledge and Traditional PGRFA: Basic Questions **157**
Susette Biber-Klemm

4.1 Introduction 157
4.2 Origin and Allocation of Traditional Knowledge 158
4.2.1 Introduction 158
Definition and relevant characteristics 158
Shortcomings of the formal IP system 159
4.2.2 Types of traditional knowledge 160
4.2.3 Degree of publicity: the 'public domain' discussion 161
4.2.4 The question of ownership and origin 161
Origin 161
4.2.5 Summary 162
4.3 The Concept of Origin of PGRFA 162
4.3.1 Background and terminology 162
Plant genetic resources for food and agriculture 163
In situ on-farm conservation 164
4.3.2 Origin 165
The scientific understanding of origin 165
Country of origin according to the CBD 165
Country of origin in the International Treaty on PGRFA 166
Discussion 166
Conclusions 168
4.3.3 A functional interpretation of origin 168
In general 168
The criterion of distinctness 169
Uniformity and identifiability 170
The element of time 171
4.3.4 Conclusion 171
4.3.5 Questions for further research 171

Part 2: Farmers, Landraces and Property Rights: Challenges to Allocating *Sui Generis* Intellectual Property Rights to Communities over their Varieties **173**
M. Halewood, J.J. Cherfas, J.M.M. Engels, Th. Hazekamp, T. Hodgkin and J. Robinson

4.4 Introduction 173

4.5 Historical Evolution of the Concept of Landrace 174
4.5.1 From nature to nurture: the anthropologization of the landrace 174
Case study 4.1: Farmers' key role in selecting and perpetuating maize landrace traits 176
4.5.2 Details of farmers' innovation: improving landraces' contribution to livelihoods is not necessarily linked to promoting their distinctness, stability and uniformity 177
4.6 How the Interplay of Plant Reproduction Systems, Human Uses and the Environment Affect the Incidence and Duration of Distinctions between Landraces 180
4.6.1 In-breeders 180
Case study 4.2: Gene flow between in-breeders; a case study of sorghum 182
4.6.2 Out-breeders 182
4.6.3 Clones 183
4.7 Assessing Potential Intellectual Property Policy Responses 185
4.7.1 Challenges 185
4.7.2 Hypothetical case 190
4.7.3 Cautionary note regarding perverse incentives 191
4.7.4 Converse situations: when *sui generis* protections could be feasible 192
4.8 Mining Data for Connections between Communities and *Ex Situ* Collections 193
4.8.1 Data selection 194
4.8.2 Results 194
4.8.3 Data-mining conclusions 195
4.8.4 Collecting forms, ethnobotanical descriptors 195
4.9 Assessing Possible Policy Responses to the Database Analyses 197
4.10 Conclusion 198
Bibliography 199

5 A New Generation of IPR for the Protection of Traditional Knowledge in PGR for Food, Agricultural and Pharmaceutical Uses 203
Thomas Cottier and Marion Panizzon

5.1 Introduction 203
5.2 The Strategic Case for Intellectual Property Protection 204
5.2.1 The legitimacy and rebalancing of the IPR system 204
5.2.2 The need to clarify the connection between traditional knowledge and existing IPR 205
5.2.3 Competing interests and conflicting legal concepts 205
The structure of the seed industry 206
PGRFA in the public domain 206
Linkages to global public goods 206
5.3 Current State of the Law and Efforts Made to Date 207
5.3.1 International instruments 207
Convention on Biological Diversity (CBD) and benefit sharing 207
Plant Variety Protection (UPOV) 207
International Treaty on Plant Genetic Resources for Food and Agriculture (ITPGRFA) 208
The WTO Agreement on Trade-Related Aspects of Intellectual Property Rights (TRIPS Agreement) and the Paris Convention for the Protection of Industrial Property 208
World Intellectual Property Organization (WIPO) 210

5.3.2 Recommendations by international bodies and national legislation 214
Model laws 214
Domestic legislation 214
5.3.3 Conclusions 215
5.4 Basic Objections to IPR Protection 216
5.4.1 Innovation *vs* traditions 216
5.4.2 Assignable resources and communitarian rights 217
5.4.3 Free flow and crop exchange 219
5.4.4 Destruction of traditional ways of life 219
5.4.5 Conclusion 220
5.5 A New IP Right in Traditional Knowledge 220
5.5.1 Subject matter 220
5.5.2 Right holders 221
5.5.3 Conditions, content and scope of rights 222
Intellectual activity 223
The object and content of the right 223
Rights conferred 225
5.5.4 Duration of rights 227
5.6 The Creation and Registration of Rights 228
5.6.1 Registration requirements 228
5.6.2 Opposition procedures and judicial review 230
5.7 The Interface with Other IP Rights 231
5.7.1 The complementary function of geographical indications 231
5.7.2 Relation to plant breeders' rights 232
5.8 The Fora for Future Negotiations 233
5.9 Conclusions 233
Bibliography 235

6 Flanking Policies in National and International Law 239
Susette Biber-Klemm, Philippe Cullet, Christophe Germann, Andrea Nascimento Müller and Jonathan Curci

6.1 Introduction 239
6.2 Existing Intellectual Property Rights: Avenues for Further Development 239
6.2.1 Novel uses of the patent model 240
Prior art, disclosure and prior informed consent 240
Joint inventorship 244
Traditional knowledge and existing conditions for patentability 245
6.2.2 Adaptation of the conditions for granting of plant breeders' rights 247
6.2.3 Geographical indications 248
Geographical indications and trademarks 249
Further development of geographical indications 250
Geographical indications and traditional knowledge 250
Case study: Potential benefits of geographical indications: the kava case in the South Pacific 253
6.3 Management and Enforcement of IPR and Traditional PGR: Institutional Design 253
6.3.1 Documentation and registration 254
Introduction and questions 254
Definitions and terminology 256
Current discussions 257
Different emphasis: examples 259

Registration as a marketing tool? 263
6.3.2 Collecting societies 264
General considerations 264
Collecting societies as a possible model 266
Domaine Public Payant 268
Further considerations 268
6.3.3 Clearing house mechanisms 269
What is a clearing house mechanism? 270
CHMs in the realm of TK management: the CBD CHM and the WIPO Platform 270
CHMs to facilitate trade in TK: examples 272
Analysis 275
Conclusions 277
Bibliography 278

PART III OPTIONS FOR COLLECTIVE AND TRADE POLICY MEASURES **281**

7 New Collective Policies **283**
Susette Biber-Klemm, Philippe Cullet and Katharina Kummer Peiry

7.1 Introduction 283
7.2 Farmers' Rights 283
7.2.1 Farmers' rights under the International Undertaking 284
7.2.2 Farmers' rights under the International Treaty on PGRFA 285
7.2.3 Farmers' rights as a means to protect traditional knowledge 286
7.3 Reviewing Access Legislation 288
7.3.1 Introduction 288
7.3.2 Legal background 289
In general 289
The regulation of the CBD 290
The regulation of ABS by the ITPGRFA 291
Conclusions and discussion 294
7.3.3 The ABS system at the interface with markets: issues to consider 295
Introduction 295
Value of PGRFA and TK and the creation of markets 296
Marketing and access legislation 297
Control on the side of the users of the information 298
The designation and integration of holders of TK, including capacity-building 299
7.3.4 Conclusions 300
7.4 Financial Mechanisms for Compensation of Non-assignable Traditional Plant Genetic Resources and Traditional Knowledge 301
7.4.1 The role of a future financial mechanism in the context of access to genetic resources 301
Aims and functions of a future financial mechanism 301
Elements of a future financial mechanism 302
7.4.2 Analysis of existing mechanisms as possible models 304
Financial mechanisms aiming at the provision of assistance to developing and transitional countries ('assistance funds') 304
Financial mechanisms forming part of a treaty system on civil liability for environmental damage caused by potentially hazardous activities ('compensation funds') 309

Mechanisms to provide incentives for implementation under the Kyoto Protocol 313
Relevant work in the framework of the FAO 317
7.4.3 Options for a financial mechanism in the context of access to genetic resources 318
Structure, organization and legal personality 318
Generation of funds: contributors 319
Disbursement of funds: claimants 320
7.4.4 Conclusions and recommendations 320
Bibliography 321

8 International Trade Regulation for Plant Genetic Resources and Related Products 324

Part 1: The Impact of Agricultural Subsidies 324
Susette Biber-Klemm and Michael Burkard

8.1 Introduction 324
8.1.1 General questions 324
8.1.2 Marrakesh Agreements: cross-cutting issues and recent developments 325
Relevance of topics: environment and development 325
Doha Declaration and Doha Work Programme 325
Recent developments: the July package 326
8.2 The Agreement of Agriculture and the Maintenance of TK and PGRFA 327
8.2.1 Introduction 327
8.2.2 Legal background 327
The AoA in general 327
Coverage 328
8.2.3 Reform programme: mandates and current stage 329
The negotiation mandates 329
Doha Declaration and Doha Work Programme: agriculture 330
8.3 The Structure of Agriculture in Developing Countries 333
8.3.1 Terminology 333
8.3.2 Economics of subsistence farming 333
8.3.3 Rural–urban migration 335
8.3.4 Structures of agriculture 337
8.3.5 Agriculture and genetic erosion in PGRFA 338
8.4 The Impact of Agricultural Export Subsidies and Food Aid on Agrobiodiversity 339
8.4.1 Introduction 339
8.4.2 Economics of agricultural export subsidies and food aid 340
Economics of agricultural export subsidies 340
Economic effects of food aid 341
Economic effects of a removal of agricultural export subsidies 341
8.4.3 Impact of agricultural export subsidies and food aid on agrobiodiversity: the PLA model 342
8.4.4 Impact of food aid on agrobiodiversity 344
8.4.5 Conclusions 344
8.4.6 Summary and outlook 346
Agricultural export subsidies 346
Food aid 346
Outlook: an integrated approach towards global multifunctionality 346

8.5 Direct Payments 347
8.5.1 Introduction 347
8.5.2 WTO regulation of domestic support: overview 348
8.5.3 The 'Green Box' measures 348
8.5.4 The development programmes 349
8.5.5 Assessment in view of support of agrobiodiversity 350
8.5.6 The debates 351
Critique of the present state in general 351
The multifunctionality *vs* food security debate 352
8.5.7 Discussion and conclusions 354

Part 2: Enhancing Market Access **357**
Thomas Cottier and Marion Panizzon

8.7 Introduction 357
8.8 Market Access for Developing Countries in PGR Products 361
8.8.1 Tariffication of trade barriers and the prospects of tariff reductions 361
8.8.2 Impact on traditional knowledge-based products 364
8.9 The Prospects of Differential Tariffs for Traditional Knowledge-based Products 364
8.9.1 Special and differential treatment 364
8.9.2 The case of special and differential treatment for TK-based products 365
Recourse to the General System of Preferences (GSP) 366
Adjusting tariff schedules and the Harmonized System 368
Limiting recourse to trade remedies against TK-based products 372
Labelling requirements 374
8.10 TK and the TBT and the SPS Agreements 375
8.10.1 Agreement on Technical Barriers to Trade (TBT Agreement) 375
Product standards 375
Product Process Methods (PPMs) 376
8.10.2 The Agreement on the Application of Sanitary and Phytosanitary Measures (SPS Agreement) 377
Product standards 377
Product Process Methods (PPMs) 378
8.11 The Need for Joint Standard-setting Operations 379
8.12 Non-discrimination Principles of GATT 1994 and Traditional Knowledge 380
8.12.1 Introduction 380
8.12.2 Legal issues 381
8.13 The Impact of MFN 382
8.14 The Impact of National Treatment 382
8.14.1 The narrow definition of likeness for purposes of taxation 383
8.14.2 The broader definition for regulatory purposes 384
8.14.3 Assessment 384
8.15 Product Process Methods (PPMs) in GATT 1994 385
8.15.1 The state of affairs 385
8.15.2 Towards a new deal on PPMs 387
8.16 Conclusions 387
Bibliography 388

Index **393**

Contributors

Biber-Klemm, S., *Faculty of Law, University of Basel, Maiengasse 51, CH 4056 Basel, Switzerland.*

Burkard, M., *World Trade Institute, University of Berne, Hallerstrasse 6, CH 3012 Berne, Switzerland.*

Cherfas, J.J., *International Plant Genetic Resources Institute (IPGRI), Via dei Tre Denari 472/a, 00057 Maccarese (Fiumicino), Rome, Italy.*

Cottier, T., *World Trade Institute, University of Berne, Hallerstrasse 6/8, CH 3012 Berne, Switzerland.*

Cullet, P., *School of Law, School of Oriental and African Studies, University of London, Thornhaugh Street, London WC1 0XG, UK.*

Curci, J.C., *Brigham Young University, 620N. 1066E. Provo, Utah, USA.*

Engels, J.M.M., *International Plant Genetic Resources Institute (IPGRI), Via dei Tre Denari 472/a, 00057 Maccarese (Fiumicino), Rome, Italy.*

Germann, C., *Institute of European and International Economic Law, University of Berne, Hallerstrasse 6, CH 3012 Berne, Switzerland.*

Halewood, M., *International Plant Genetic Resources Institute (IPGRI), Via dei Tre Denari 472/a, 00057 Maccarese (Fiumicino), Rome, Italy.*

Hazekamp, Th., *International Plant Genetic Resources Institute (IPGRI), Via dei Tre Denari 472/a, 00057 Maccarese (Fiumicino), Rome, Italy.*

Hodgkin, T., *International Plant Genetic Resources Institute (IPGRI), Via dei Tre Denari 472/a, 00057 Maccarese (Fiumicino), Rome, Italy.*

Kummer Peiry, K., *Kummer EcoConsult, Chemin de la Forêt 4, CH 1752 Villars-sur-Glane, Switzerland.*

Nascimento Müller, A., *Legal Advisor, Rue de Lyon 48, CH 1203 Geneva, Switzerland.*

Panizzon, M., *World Trade Institute, University of Berne, Hallerstrasse 6, CH 3012 Berne, Switzerland.*

Pasadilla, G.O., *Philippine Institute for Development Studies, NEDA Building, Amorsolo Street, Makati City, Philippines.*

Robinson, J., *International Plant Genetic Resources Institute (IPGRI), Via dei Tre Denari 472/a, 00057 Maccarese (Fiumicino), Rome, Italy. Permanent address: Tick-aho, Joroisniemenkehatie, 79600 Joroinen, Finland.*

Szymura Berglas, D., *Faculty of Law, University of Basel, Maiengasse 51, CH 4056 Basel, Switzerland.*

Preface and Acknowledgements

Current instruments in international agricultural, environmental and trade law do not respond sufficiently to the needs of effectively preserving biodiversity and improving conditions of subsistence farming for large parts of the globe's rural population. The same is true domestically in most countries. This book collects a series of studies undertaken under the auspices of the World Trade Institute at the University of Berne on behalf of the Swiss Agency for Development and Cooperation (SDC), Federal Department of Foreign Affairs. Upon assessing the current situation, the authors submit a number of policy proposals for future international negotiations in different fora, in particular within the multilateral trading system of the World Trade Organization (WTO), but also relating to other institutions such as the Food and Agriculture Organization (FAO), Consultative Group on International Agricultural Research (CGIAR), United Nations Environment Programme (UNEP) and World Intellectual Property Organization (WIPO). These proposals seek to reinforce economic incentives to making available traditional knowledge, to conserve it through use, and thereby enhance the livelihoods of farming and indigenous communities and reverse the decline of biodiversity, upon which, in return, long-term food security is based. Research was limited to traditional knowledge as it relates to the sustainable use and conservation of plant genetic resources, mainly for the purposes of agriculture and health care. It encompasses the knowledge of farming communities around the world, but also traditional knowledge of minority cultures (indigenous peoples). However, it does not include traditional knowledge relating to cultural diversity in a broad sense, such as folklore, music, dance and other cultural activities.

The proposals submitted are based upon a fundamental distinction between assignable and non-assignable traditional knowledge in plant genetic resources. The former allows creating *sui generis* intellectual property, so called TIP rights (Traditional Intellectual Property rights). Building upon the concept of unfair competition, such rights will enhance legal security and trade relating to plant genetic information. These improvements are crucial to rebalance the current legal system of protecting information in intellectual property which ignores the needs of traditional and informal sectors. The latter, non-assignable knowledge, needs to be supported by publicly funded policies. However, both areas will benefit from trade policy measures seeking to enhance market access and removing trade barriers for products based upon traditional knowledge. The suggestions made throughout the chapters of this book are based upon the conviction that these measures are crucial for the promotion of shared long-term interests, both by industrialized and developing countries alike. The topic is by no means limited to issues of intellectual property which have dominated the debate so far.

The editors and authors are indebted to the Swiss Development Corporation for initiating and funding the project. The project, undertaken by lawyers, social and natural scientists, was an inspiring and rewarding experience. We are in particular grateful to Paul Egger, Martin Sommer, Christina Grieder and Katharina Jenny from the SDC. They subsequently accompanied work in progress during the past 5 years. We are grateful to the International Plant Genetic Resources Institute (IPGRI) for cooperating in the research effort. We are grateful to the Members of the Consultative Board for their advice: Klaus Ammann, Botanical Institute of the University of Berne; Martin A. Girsberger, Swiss Federal Institute of Intellectual Property, Berne; Christina Grieder, Swiss Agency for Cooperation and Development (SDC), Berne; Katharina Jenny, Swiss Agency for Cooperation and Development (SDC), Berne; Martin Krebs, Federal Ministry of Foreign Affairs (EDA), Berne; Alwin Kopse, Federal Office for Agriculture (BLW) and State Secretariat for Economic Affairs (SECO), Berne; François Pythoud, Swiss Agency for the Environment, Forests and Landscape (BUWAL), Berne; Jürg Schneider, Swiss Agency for the Environment, Forests and Landscape (BUWAL), Berne; and Andreas Werthmüller, State Secretariat for Economic Affairs (SECO), Berne. We are grateful to all those who accompanied the work and commented upon drafts in two consecutive workshops held at the World Trade Institute (WTI), 24–26 November 2002 and 27–29 April 2003, as well as upon final drafts: Shakeel T. Bhatti, WIPO, Geneva; Carlos A. Primo Braga, The World Bank, Washington; Didier Chambovey, State Secretariat for Economic Affairs (SECO), Berne; Carlos Correa, University of Bueno Aires; Sabine Demangue, Max Planck Institute for Intellectual Property, Münich; Lothar Ehring, WTO, Geneva; Graham Dutfield, Queen Mary College, University of London; Christian Haeberli, Federal Office for Agriculture (BLW), Berne; Promilla Kapoor-Vijay, GB Pant University, Pant Nagar, Uttaranchal, Himalayas; Anatole Krattiger, Cornell University, New York; Lucretia Landmann, State Secretariat for Economic Affairs (SECO), Berne; Magdalena Lesjak, Swiss Agency for Development and Cooperation (SDC), Berne; Anne-Laure Mangard, Federal Office for Agriculture (BLW), Berne; Dieter Meier, Suissimage, Berne; Nicola Notaro, European Commission, Brussels; Dwijen Rangnekar, University College London; Jerome Reichman, Duke University, Durham, USA; Renato Salazar, Project for the Enhancement of Diversity of Plant Genetic Resources of Asia (PEDIGREA), Philippines; Aphrodite Smagadi, European University Institute, Florence; Walter Smolders, Syngenta AG, Basel; Marie Wollheim, Swiss Federal Institute of Intellectual Property, Berne; and Xiaoping Wu; WTO, Geneva. We are indebted to Tim Hardwick of CABI and his professional support in preparing the publication. All views expressed, as well as omissions and errors, remain the sole responsibility of the authors and editors of the volume.

Susette Biber-Klemm, Thomas Cottier, Danuta Szymura Berglas
Berne, March 2005

Abbreviations

AAUs	assigned amount units
ABS	access and benefit sharing
AIDS	Acquired Immune Deficiency Syndrome
AIJ	activities implemented jointly
AMS	aggregate measure of support
AoA	Agreement on Agriculture
ATC	Agreement on Textiles and Clothing (WTO)
BCH	Biosafety Clearing House
Cairns Group	group of states with a highly competitive agricultural sector
CBD	Convention on Biological Diversity
CDM	clean development mechanism
CERs	certified emission reductions
CFCs	chlorofluorocarbons
CGIAR	Consultative Group on International Agricultural Research
CGRFA	Commission on Genetic Resources for Food and Agriculture of the FAO
CHM	clearing house mechanism
CIEL	Centre for International Environmental Law
CITES	Convention on International Trade in Endangered Species of Wild Fauna and Flora
CLC	Civil Liability Convention, 1992
CLR	compensatory liability regime
CoA	Committee on Agriculture (WTO)
COP	Conference of the Parties
COP/MOP	Conference of the Parties serving as the Meeting of the Parties
CRISTAL	Contract Regarding an Interim Supplement to Tanker Liability for Oil Pollution, 1971
CSIR	Council of Scientific and Industrial Research (India)
CSIRO	Commonwealth Scientific and Industrial Research Organization (Australia)
CTE	Committee on Trade and Environment (WTO)
DCs	developing country or countries
DDA	Doha Development Agenda
DDAGTF	Doha Development Agenda Global Trust Fund

DNA	deoxyribonucleic acid
DSB	Dispute Settlement Body of the WTO
DUS	distinctness, uniformity and stability
EC	European Community
ECJ	European Court of Justice
EGS	environmentally friendly goods and services
EPBC Act	The Environment Protection and Biodiversity Conservation (EPBC) Act 1999 (Commonwealth)
EPC	European Patent Convention, 5 October 1973
EPO	European Patent Office
EPP	environmentally preferable products
ERUs	emission reduction units
EU	European Union
FAO	Food and Agriculture Organization
FC	Fund Convention, 1992
Flex Mex	Flexible Mechanisms of the Kyoto Protocol to the Climate Convention
FR	farmers' rights
FUDM	Farmer's Unit of Diversity Management
Fund Convention	Convention on the Establishment of an International Fund for Compensation of Oil Pollution Damage, 1971
GATS	General Agreement on Trade in Services
GATT	General Agreement on Tariffs and Trade
GDP	gross domestic product
GEF	The Global Environmental Facility, established in 1991 and restructured in 1994
GEF 'the instrument'	Instrument for the Establishment of the Restructured Global Environment Facility, 1994
GI	geographical indications
GIAN	Grassroots Innovation Augmentation Network, Gujarat, India
GMOs	genetically modified organisms
GPA	Global Plan of Action for the Conservation and Sustainable Utilization of Plant Genetic Resources for Food and Agriculture
GRAIN	Genetic Resources International
GSP	general system of preferences
HNS	hazardous and noxious substances
HS	harmonized system
HVVy	highly variable varieties
HYVs	high-yielding varieties
IARCs	International Agricultural Research Centres
IBCR	Institute for Biodiversity Conservation and Research (Ethiopia)
ICC	Intergovernmental Panel on Climate Change
ICCAT	International Commission for the Conservation of Atlantic Tunas
ICRISAT	International Crops Research Institute for the Semi-arid Tropics
ICTSD	International Centre for Trade and Development
IEA	International Environmental Agreements
IFAD	International Fund for Agricultural Development
IGC WIPO	Intergovernmental Committee on Intellectual Property and Genetic Resources, Traditional Knowledge and Folklore
ILM	International Legal Material
ILO	International Labour Organization
IMF	International Monetary Fund

IMO	International Maritime Organization
INBio	Costa Rican National Biodiversity Institute
INDECOPI	National Intellectual Property Institute (Instituto Nacional de Defensa de la Competencia y de la Protección de la Propiedad Intelectual)
IOPC	International Oil Pollution Compensation Funds
IP	intellectual property
IPGRI	International Plant Genetic Resources Institute
IPR	intellectual property right/s
IRRI	International Rice Research Institute
ITPGRFA	International Treaty on Plant Genetic Resources for Food and Agriculture
ITTA	International Tropical Timber Agreement
IU (or IUPGR)	International Undertaking on Plant Genetic Resources
IUCN	World Conservation Union
JI	joint implementation
LDCs	least developed countries
LMOs	living modified organisms
MEAs	Multinational Environmental Protection Agreements
MFN	most-favoured nation principle
MLS	multilateral system
MNC	multinational corporation
MOP	Meeting of the Parties
MTA	material transfer agreement
NBA	National Biodiversity Authority (India)
NFIDCs	Net Food-importing Developing Countries
NGO	non-governmental organization
NIF	Indian National Innovation Foundation
NTBs	non-trade barriers
NTCs	non-trade concerns
OAU	Organization of African Unity
OECD	Organization for Economic Cooperation and Development
PBR	plant breeders' right
PCT	Patent Cooperation Treaty
PGM	plant genetic material
PGR	plant genetic resources
PGRFA	plant genetic resources for food and agriculture
PIC	prior informed consent
PIIPA	Public Interest Intellectual Property Advisors
PLA	potential loss of agrobiodiversity/agriculture
POPs	persistent organic pollutants
PPM	process and production method
PSE	producer support estimate
PTO	Patent and Trademark Office
PVPA	Plant Variety Protection Act
R&D	research and development
RAFI	Rural Advancement Foundation International (now ETC Group)
RMUs	removal units
SACD	Société des auteurs et compositeurs dramatiques, 1776
SAP	Structural Adjustment Programme
SBSTTA	Subsidiary Body on Scientific, Technical and Technological Advice

SCM Agreement	Agreement on Subsidies and Countervailing Measures
SDC	Swiss Agency for Development and Cooperation
SDR	Special Drawing Rights defined by the International Monetary Fund
SDT	special and differential treatment
SGRP	System-wide Program on Genetic Resources
SINGER	(CGIAR) System-wide Information Network for Genetic Resources
SPGRC	SADC (Southern African Development Community) Plant Genetic Resources Centre
SPS	sanitary and phytosanitary measures
SPS Agreement	Agreement on Sanitary and Phytosanitary Measures
SRISTI	Society for Research and Initiatives for Sustainable Technologies and Institutions
SSG	Special Safeguard (under Article 5 AoA)
SSM	Special Safeguard Mechanism
TBGRI	Tropical Botanic Garden and Research Institute
TBRK	traditional biodiversity-related knowledge
TBT	technical barriers to trade
TBT Agreement	Agreement on Technical Barriers to Trade
TEK	traditional ecological knowledge; indigenous or aboriginal knowledge, traditional biodiversity-related knowledge (TBRK) or related knowledge
TIP rights	Traditional Intellectual Property rights
TK	traditional knowledge
TOVALOP	Tanker Owners Voluntary Agreement concerning Liability for Oil Pollution, 1968
TRIPS	Agreement on Trade-related Aspects of Intellectual Property Rights
TRQ	Tariff Rate Quota
TRR	traditional resource right
UN	United Nations
UNCTAD	United Nations Commission for Trade and Development
UNDP	United Nations Development Programme
UNEP	United Nations Environment Programme
UNESCO	United Nations Educational, Scientific, and Cultural Organization
UPOV	*Union internationale pour la protection des obtentions vegetales*: The International Union for the Protection of New Varieties of Plants
URM	usual marketing requirement
USPTO	United States Patent and Trademarks Office
VCLT	Vienna Convention on the Law of Treaties
VERs	voluntary export restraints
WHO	World Health Organization
WIPO	World Intellectual Property Organization
WTI	World Trade Institute
WTO	World Trade Organization
WWF	World Wide Fund for Nature

Introduction

Thomas Cottier

I.1 A New Regulatory Field

For centuries, plant genetic resources have provided the basis of civilization and agriculture, of life, health and prosperity around the globe. Knowledge and experience have been passed from generation to generation by custom. Whereas seeds in themselves were the primary objects of earlier trade, dissemination and exchange, more recently plant genetic resources have also come to be addressed in terms of rights to genetic information and their appropriation. Different forms of intellectual property have emerged and altered traditional patterns of exchange and dissemination. The Green Revolution in the 1960s greatly enhanced productivity and formalized the sector. The recent advent of biotechnology has added new dimensions and challenges. Customary patterns of breeding and conservation, still alive in many quarters of the world, are now faced with the impact of modern technology. A new interface is called for – another generation of legal foundations – well beyond national jurisdictions. A new topic has emerged in international law and policy. It is far from settled.

Since the adoption of the Convention on Biological Diversity (CBD) in 1992, the law of plant genetic resources (PGR) and the legal status of traditional knowledge (TK) has attracted increasing attention in international fora, governments, non-governmental organizations and academic research. A number of reasons account for this development: the alarming loss of biodiversity in plant genetic resources, and the imbalance between protected plant varieties and genetically engineered products on the one hand, and traditional crops and landraces in the public domain on the other hand. In international law, we witnessed the advent of the Agreement on Trade-related Aspects of Intellectual Property Rights (TRIPS Agreement) within the World Trade Organization (WTO) in 1995 and the International Treaty on Plant Genetic Resources for Food and Agriculture in (ITPGRFA) 2002. All this has stimulated the debate in recent years. The 2001 Doha Agenda of the WTO explicitly endorsed the issue of traditional knowledge as a subject for further work. What, some years ago, was a concern limited to ecological aspects of preserving biodiversity, has now moved to centre stage and affects the legitimacy of the multilateral trading system, and intellectual property in particular, and its interface with agricultural and environmental policies. We expect that the subject matter will further gain importance, and policy answers are a pressing need. What is required is a future regulatory system for plant genetic resources and traditional knowledge which serves the goals of equity and fairness in allocating resources among nations and within nations. It will contribute to enhanced legitimacy of the overall international trading system.

The matter before us is of great and puzzling complexity. Not only does it entail problems relating to different regulatory areas, both domestically and internationally, but it also touches upon the jurisditions of different governmental departments and international organitions, and upon a host of diverging interests, ultimately affecting, and relating to, global food security and welfare. It covers new ground due to the advent of biotechnology. Results will need to be achieved step by step, in different fora. It is for this reason that coherent goals need to be defined, and policy measures suggested in different areas need to be well coordinated and mutually consistent. This book seeks to contribute to this effort. A brief introduction may state the main propositions and encourage the reader to turn to the various chapters of the book. Assigned to different teams of authors within the project, they set forth different angles of the problem. As they stand in their own right, they partly overlap and may address an issue from different perspectives.

I.2 Goals and Basic Assumptions

Policy measures suggested in different fields developed and described in the various chapters of this book seek to contribute to and realize a number of shared essential goals. These goals will be set forth and reiterated throughout the different chapters. Before turning to policy measures and implementation – the crux of the matter – it would seem useful to recall the goals briefly at the outset in terms of underlying assumptions, agreed among the team of researchers.

Law and policies are to support long-term global food security, distributive justice and intergenerational equity within, and among, countries. They are not limited to improving conditions in developing countries, but ought to form part of a future global plant genetic resources policy in general, and agricultural policy in particular. They affect all countries equally and should entail potential for agriculture in developing and developed countries alike. They should form part of a policy that seeks to overcome the current divide between interests in the North and the South, and should be seen in the context of redefining public goods, public domain and property rights.

The future legal regime and status of plant genetic resources and traditional knowledge in international law should contribute to halting the loss of diversity in PGR and to reverse the current trend. Thus, it provides incentives to conserve the diversity of PGR and maintain traditional knowledge through use, and brings about recognition and validation of such use. It assists in generating income and thus makes a contribution to overcoming subsistence farming and improving the livelihoods of local communities in developing countries. Similarly, it creates additional sources of income and market opportunities for traditional farming in industrialized countries. A future legal regime respects the communal traditions and builds upon them. It creates instruments for participation of the local population in the relevant decision processes.

The future legal status of plant genetic resources and traditional knowledge also responds to the needs of the research-based industries, in particular for the purpose of genetic engineering. Defining appropriate levels of patent and plant variety protection must take these needs into account. In return, it brings about a better balance between biological resources appropriated by intellectual property and those currently in the public domain. In particular, it defines the relationship between industrial research and development, as well as breeding processes and traditional informal innovation, in such a manner that all contributions, past and present, are respected and obtain adequate rewards. Equitable treatment of traditionally bred crop varieties and traditional knowledge in the system of intellectual property rights therefore cannot be achieved by fundamentally reducing patent or intellectual property rights (IPR) protection, in general terms, as many advocate today. Rather, it needs to build upon rebalancing the entire system of intellectual property rights by enhancing compensatory mechanisms and rights.

The allocation of enhanced and balanced rights both for genetically engineered and traditional crops seeks to reduce dependence on full and exclusive public financing of the international crop genetic resources system by international agencies and governments. At the same time, it takes into account the need for a regime of open and unimpaired exchange of genetic materials under the International Treaty on Plant Genetic Resources for Food and Agriculture (ITPGRFA). It also takes into account the rules and principles of the WTO, and of international trade regulation, in general, and of other related fields of international law. Efforts to realize the goals set out above include reform of these instruments. While intellectual property is a key issue before us, the regulation of plant genetic resources is far from limited to this field. We seek realization of valuational goals in all the pertinent regulatory fields.

I.3 The Current Legal Status of Plant Genetic Resources

The current legal status of plant genetic resources (PGR) and traditional knowledge (TK) is far from achieving these objectives. The deficiencies are well known. Both areas operate under the doctrine of permanent sovereignty over natural resources, but there is a sharp divide between the formal and the informal sectors. While the formal sector of research-based breeding is fully subject to appropriation and enjoys the benefit of intellectual property rights, the informal sector is based upon customary practices, predominant in developing countries, and forms part of the public domain. Resources and related information are, by and large, freely available for appropriation in terms of IPR.

This constellation does not provide sufficient incentives to preserve traditional ways of innovation and propagation. Traditionally bred crops are rapidly replaced by high-yield and efficient, but largely uniform, varieties bought on the international market. The seed industry is highly concentrated, as are the crops made available by industry. Tools to offer compensation for recourse and use of traditional crops in biotechnology under the Convention on Biodiversity are limited to contractual arrangements under the Bonn Guidelines. They are essentially defined by negotiating power, rather than by rights and entitlement. There are no effective provisions and obligations as to transfer of technology in return for using traditionally generated information. Efforts to bring about benefit-sharing in the national laws of developing countries by means of access legislation impair research and are limited in their effect to domestic jurisdiction. Due to the territorial limitation of national law, there are no remedies available when biological resources and/or associated traditional knowledge are being deployed on third-country markets.

International instruments do not provide any rights, and the concept of farmers' rights has remained an empty shell. The benefits foreseen in the ITPGRFA, in particular the benefits of the Multilateral System on Facilitated Access and Benefit Sharing, are not designed in a way to be beneficial for individual, small-scale farmers. Support measures are known under the term of 'farmers' rights'. Yet, the notion is misleading as these measures, according to the minimal standard for their realization on the national level laid down in the ITPGRFA, are by no means individualized and specified, but defined as measures of general support. They fail to mirror plant variety rights and fall short of effective protection of traditional knowledge related to PGRFA, or the individual right to participate equitably in benefit-sharing and to participate in decision-making on matters related to the conservation and sustainable use of PGRFA.

International policies supporting the preservation of biodiversity through use therefore all depend on public policies and thus rely heavily on funding by the international community. The funding depends on recurring budgetary allocation and is not sufficiently secured and sustainable. Moreover, the multilateral trading system still largely ignores concerns relating to biodiversity in general, and to the diversity of crop species in particular. It has not addressed the problem in a thorough manner. In fact, extensive export subsidies still contribute to the destruction of traditional ways of production and markets. Products stemming from the use of traditional knowledge are treated like any other, and do not benefit from enhanced and privileged market access. The current system of intellectual property rights fails to address the concerns of traditional, non-industrial farming and holders of traditional knowledge. It leaves traditional knowledge without any protection, except for untested principles of unfair competition and the protection of undisclosed information. Unregistered traditional knowledge as prior art is difficult to prove, while industrial products, which make use of traditional knowledge, may benefit from full patent protection. It is true that products based on such knowledge may benefit from protection under trademarks and geographical indications; however, the latter is still limited to unfair competition and does not provide a sound basis for enhanced marketing efforts. In general, the current system of intellectual property protection does not sufficiently take into account the potential and needs of the informal agricultural sector upon which a majority of people in the developing world depends. It is only of long-term benefit to most developing countries. It does not yield immediate results and its legitimacy and usefulness are increasingly questioned. Tariffs and sanitary and phytosanitary measures do not differentiate between conventional products and those based upon traditional knowledge. Again, no incentives exist to stimulate and encourage the use of traditional knowledge and its rich diversity. As the law stands, the public good values of biodiversity are not sufficiently served.

The current legal status of plant genetic resources and traditional knowledge and its institutional implementation also reflects the fragmentation of the international system and the lack of coherence and global governance. A number of fora deal with the matter without close cooperation. The matter of traditional knowledge (in a broad sense) is addressed by the United Nations Educational, Scientific and Cultural Organization (UNESCO) and, with a view to health care, by the World Health Organization (WHO). From the perspective of intellectual property, the World Intellectual Property Organization (WIPO) is dealing with the issue. Traditional knowledge is addressed by the United Nations Environment Programme (UNEP) in the context of the CBD, and by the United Nations Commission for Trade and Development (UNCTAD) in the context of trade and development. With a view to agriculture and the International Treaty, it is dealt with by the Food and Agriculture Organization (FAO) and the Consultative Group on International Agricultural Research (CGIAR) and its various institutions, such as the International Plant Genetic Resources Institute (IPGRI). Finally, it will eventually be addressed by the WTO in the context of TRIPS and possibly other trade agreements. Efforts to support use and conservation of traditional PGRFA and technologies are further supported by multilateral and bilateral development programmes.

Since awareness of traditional knowledge is a recent issue, there is still a profound lack of evidence as to the practical relevance and potential of these qualities in different countries. The potential of using traditional knowledge in the context of industrial research and development (R&D) processes, in particular in biochemistry, pharmacology, and, with a view to PGRFA in biotechnology, is recognized, but examples are still limited. Data are not fully available as to the localization and origin of plant genetic resources. There is also little knowledge in relation to speciality products that might benefit from protection as geographical indications.

Finally, it is difficult to identify common directions or consensus in contemporary international policy making.

Issues relating to plant genetic resources, despite increasing awareness of the acuteness of the problem and perhaps an emerging consensus as to core objectives as depicted above, are highly controversial. Discussions are confused and still lack conceptual underpinnings. There are divergences as to the choice of fora. There are profound differences of opinion as to whether the future status first needs to develop in national law before efforts on the international level can be realistically made. A number of countries have introduced legislative measures to regulate access and benefit-sharing, partly in combination with *sui generis* protection of plant varieties under Article 27.3(b) of the TRIPS Agreement. India has enacted extensive legislation; other examples involve Peru. The Organization of African Unity (OAU) has adopted a model legislation. Australia, as a bio-diverse industrialized country, has dealt extensively with TK-related issues, mainly in case law. All these efforts are characterized by the lack of uniform and internationally coordinated efforts.

I.4 Policy Measures Submitted

I.4.1 A multiple approach required

Based upon the analysis of the current legal framework and issues, policy measures are submitted for discussion throughout the chapters of this book. They were elaborated and discussed by the research team of the World Trade Institute (WTI), and commentators provided valuable remarks on previous drafts. A brief introduction and roadmap shows that these proposals and suggestions do not seek single-handed answers. While they stress the importance of reinforcing rights to plant genetic resources, mainly by means of protecting traditional rights in the context of intellectual property, they equally emphasize the need to define coherent policies in financing conservation and use, as well as in trade regulation and in the further concretization of the CBD and the ITPGRFA. The future law of plant genetic resources is multi-dimensional. Without such qualities, the broad objectives discussed above cannot be achieved. We thus recommend the pursuit of a number of policies and initiatives. We offer a host of measures to be pursued in different international fora and on the level of national law, both in developing and industrialized countries.

The chapters of this book recommend adopting a pragmatic approach and using tools as they fit, independently of their legal nature. It is wrong to assume that public goods are exclusively achieved by publicly funded policies while private law instruments are limited to pursue appropriation and private interests. It is the combination of the two which will bring about the effective promotion of the public policy goals and objectives defined. Research shows that goals and objectives relating to TK and PGR need to be addressed by different legal instruments at both the international and national level. These measures will be based partly upon public law, and partly upon private law, such as IPRs and licensing agreements. However, all measures need to serve the public goods of biodiversity, equity, prosperity and advances in research.

Research efforts described in this book conclude that effective protection of TK and PGR, and in particular PGRFA, depends primarily on enhanced efforts on the international level. The current framework, in particular the access and benefit-sharing regime of the CBD, leaves the main efforts at the national level. Some hold strong views that the protection of TK needs, first of all, to adopt a bottom-up approach and to develop national legislation, and it is only on this basis that it may induce changes in the international framework. We respectfully disagree. National legislation based upon the CBD, or Article 27.3(b) of the TRIPS Agreement allowing for *sui generis* systems of protection of TK and PGRFA, may effectively regulate access, but are devoid of means of enforcing rights abroad, in particular on major consumer markets. Self-standing domestic regulations, as they currently exist in a number of countries, are meaningful for large jurisdictions such as India or China, which have an extended pharmaceutical and food industry of their own. Smaller countries, dependent on foreign markets, will need to rely on internationally agreed disciplines. The task ahead is to provide for appropriate frameworks to generate instruments at the national level in a coherent way and to assure their implementation in the context of international trade.

Effective protection, whatever form it will take, requires obligations incurred by industrialized countries in the context of their own policies. These obligations will not be assumed without addressing them on the international level as part of an overall concept and balanced package of rights and obligations. Moreover, we fear that the current panoply of diverging concepts and measures will render the achievement of international agreement in the field more difficult over time. Our recommendations therefore mainly focus on the level of international law development. We see great potential in using IT technology in bringing about an effective global system in international law for the management of PGR and future rights. Short of such action, the decline of biodiversity cannot be effectively reversed.

I.4.2 The protection of assignable traditional knowledge (Traditional Intellectual Property Rights)

Our research diagnoses the lack of clear allocation of rights and obligations as one of the main obstacles to bringing about effective protection, validation and exchange of TK and traditional PGRFA. For example, accessions to gene banks have declined partly due to lack of legal certainty relating to ownership of information. We submit that in the era of biotechnology and enhanced patent or plant variety protection law, the creation of legal rights to protect 'informally' created information will create a more level playing field, and bring about equity and legal security for trading informational values.

Whatever the precise nature of such rights, they are related to intellectual property protection which deals, in its different forms, with entitlement to information and knowledge. Seeking to develop such rights therefore may be termed intellectual property rights. It amounts to a new generation of IPR. We consider TK to amount to traditional intellectual property of rural communities or indigenous people, and therefore suggest using the term Traditional Intellectual Property Rights (TIP Rights). Other terms could be used, such as *sui generis* rights, or even farmers' rights. We avoid the latter term due to a different and non-substantial meaning given to it in the context of PGR. Developing TIP Rights is in the interest not only of the potential rightholders, but also in the long-term interests of the seed industry and consumers to preserve biodiversity in general, and agrobiodiversity in particular. These rights will enable partners to do business in a fair and legally secured manner. The creation of proper rights to PGR, in particular PGRFA, depends, in the first place, upon whether such resources can be assigned to particular regions and/or communities.

The assignability of TK regarding utilization of PGR, as, for example, in the context of pharmacology, veterinary or human medicine, plant insecticides and other technologies, seems to be more readily feasible, as localization is possible in many cases. Pertinent information is increasingly collected in data banks relating to proving prior art, and it could also be used to establish positive rights as they identify the origin and the holder of the information.

Developing foundations, definition and scope of such rights, as well as ways and means to implement them, form an important part of this book.

I.4.3 Reform and novel uses of existing intellectual property rights: patents and plant breeders' rights

Creating new TIP Rights on the international level is clearly ambitious and will take some time. Additional policy options relating to intellectual property may be introduced at an earlier stage and will continue to play an important role as flanking policies to TIP Rights, once these have been created. These policies relate exclusively to the patenting of new products based upon TK and enhanced protection of geographical indications. Both avenues are advanced and are being addressed in the relevant fora.

The patent and plant variety protection system could be adapted to provide a better framework within which some elements of TK could be protected. This book discusses a number of suggestions and options. These proposals relate to prior art, prior informed consent, conditions of patentability, joint ventures, petty patents and a reform of plant variety protection laws, as well as to the adaptation and use of existing institutional devices for the management and implementation of such rights. They provide a second mainstay of proposals made in this book.

I.4.4 The protection of non-assignable traditional knowledge and plant genetic resources for food and agriculture

Research has found that a large number of traditional PGRFA and/or TK will not be assignable to particular communities or regions. They are largely used by farmers throughout countries and continents as they spread over time. Here, TK is widely known in a specific country or region and cannot be localized. The information is in the public domain and bound to stay there. As such PGRFA form an important part of agrobiodiversity, and appropriate policy instruments to support conservation and use need to be employed and further developed.

Neither funding of the CBD nor the CGIAR and the International Treaty for PGRFA is assured, and continues to depend on annual budgetary decisions of donor countries. It would seem important to seek structures that allow for long-term and sustainable support of conservation and use of biodiverse PGRFA. Based on the study of different options used in the field of environmental programmes and institutions, the book recommends that a future financial mechanism that could substitute financial compensation of holders of traditional knowledge in cases where traditional or *sui generis* intellectual property rights cannot be allocated, would need to be based upon a proper international legal instrument. The required infrastructure would comprise a supreme body comprising all member states, an executive body and a secretariat. In this respect, there are a number of precedents in international environmental law that could serve as models. These models provide a third layer of proposals made.

I.4.5 Support of trade in products resulting from the use of traditional plant genetic resources for food and agriculture and traditional knowledge

The potential of trade policy and regulation other than the field of IPR in the narrow sense (mainly patents and geographical indications) has so far been largely ignored in international discussions. There is a significant potential for the promotion of traditional PGRFA and the development of markets for TK-based products. The book recommends considering these options and seeking to achieve a comprehensive policy for TK-based products which transcends the current debate in WTO limited to IPR. Options are not limited to the multilateral trading system. Support and enhanced market access may also be sought on the basis of unilateral measures and bilateral or regional agreements. These measures suggest a fourth mainstay of proposals upon which further action should be taken. They essentially focus on the idea of promoting niche products.

Instruments informing about specific qualities, ways of production or places of origin are a means for the marketing of niche products. They answer to specific demands of an increasing number of consumers, mainly in the industrialized world, who prefer to opt for products which are of a high quality, produced in an environmentally sound way and traded under socially fair conditions. We submit that efforts to market TK-based products should begin at home. These are efforts of a practical nature and do not entail international negotiations. They should be supported by corresponding donor programmes for the promotion of TK-based products. Additional international legal instruments may eventually support them in developing international markets. The following instruments could be made available.

Geographical indications (GIs) protect products, the quality of which is related to a particular geographical location, from unfair competition. They provide an important supplementary means to protect TK. They are of interest in the present context for the following reasons: GIs are collective in scope and are therefore well adapted to reflect the collective nature of much of TK. GIs limit the protection to a specific area rather than to a specific rights holder. GIs do not impose any novelty criterion and can specifically be used to protect traditional products as long as the particular characteristics of these products can be attributed to a specific geographical origin. GIs do not relate to one specific method of production of a given product and allow different production methods to be covered under a given indication and production method changes over time. GIs do not imply monopoly control over the knowledge that is embedded in the protected indication, as the knowledge remains in the public domain. Finally, GIs cannot be transferred outside their region of protection.

The protection of GIs in the field of TK and PGRFA for foodstuffs is currently limited to protection from unfair competition. Efforts to expand such protection are under discussion at the WTO. Developing countries are increasingly aware of this potential. It is recommended that efforts in bilateral programmes to assist in identifying TK-based products that will benefit from current and extended protection of GIs to be supported. They cannot replace the creation of *sui generis* TIP Rights protection information, but provide important complementary flanking support.

Trademark protection may be equally used for the protection of TK-based products, in particular in the form of certified marks that allow a group of producers to participate in the mark. Unlike GIs, there is no need to have the product relate to a particular region. These instruments are available under the TRIPS Agreement and do not need amendment in the present context.

In the framework of the WTO legislation, we do not see any limitation to the possibility of voluntarily and positively labelling TK-based products, describing the origin, methods of production, quality and context in a fair and transparent manner. The problem with labelling will be one of excluding producers in a particular region as they may not qualify for the label. In this sense, voluntary labelling may lead to effects similar to the requirements of the sanitary and phytosanitary measures mentioned below. Labelling may have exclusive, and thus protectionist, effects. It will be necessary to design international rules on labelling that minimize such effects, and support programmes actively identifying niche products supporting biodiversity.

I.4.6 Tariff measures facilitating the export of traditional knowledge-related products

Given the high average level of agricultural tariffs, tariff policy can potentially make an important contribution in facilitating market access for TK-based products. While products from least developed countries are likely to benefit from zero tariffs after the completion of the Doha Agenda ('all but arms' initiative), the same is not true for developing countries in general. We therefore recommend granting preferential tariffs for specific TK-based products supporting biodiversity, in order to compensate for what often amounts to less efficient and more expensive methods of biodiverse production. Under current practices, members enjoy some leeway as to tariff classification, and they are entitled to introduce subcategories to this effect for TK-based products, but are limited by the requirements for treating like products alike.

We therefore recommend creating legal

foundations of differential treatment in assuring that, in the first place, adequate categories can be created under the Harmonized System (HS). Alternatively, it is conceivable to bring about comparable effects by expanding unilaterally the General System of Preferences (GSP) to TK-based products from developing countries. We suggest combining these efforts with the protection of GIs. By doing so, eligible products can be defined clearly.

Upon importation, Members are free to give preference to imports of foreign products, and nothing would impair unilateral measures to prefer foreign TK-based products alike, for example in terms of taxation. The problem lies with giving preference to domestic TK-based products over imports, beyond tariff protection. This may be a problem for developing countries that wish to give privileges to their own TK production on the domestic market. Discussions therefore need to focus not only on defining appropriate tariff levels in developing countries for agricultural products, but also on defining likeness and unlikeness between conventional and TK-based products, which serve the additional purpose of biodiversity conservation and crop diversification and long-term food security. At this stage, it is unclear whether exceptions under Article XX(g) of the General Agreement on Tariffs and Trade (GATT) could be invoked. The problem is related to the debate on production and process methods (PPM), which should be extended from mainly environmental considerations to aspects of long-term food security. The problem addresses a core issue of GATT law, and plant genetic resources should be taken into account when it is discussed.

Finally, imports of TK-based products on world markets may be substantially impaired by strict phytosanitary measures. The general standards of the Codex Alimentarius are shaped with a view to developed markets, and may not sufficiently take into account considerations of TK-based biodiverse production. Many products from developing countries may face import restrictions which render marketing of TK-based products difficult. At the same time, it is important to develop and provide assistance to developing countries with a view to meeting the current standards in the production of TK-based products. While consumers in developed country markets show increasing interest in specialities and TK-based and ethnic foodstuffs, they are clearly entitled to be sure that the products are safe. We recommend examining to what extent additional programmes could be envisaged to address this issue.

Evidence shows that export subsidies relating to conventional crops are a major factor in destroying biodiverse agriculture in developing countries. Successful conservation and use of PGRFA in developing countries depends on reduction and elimination of export subsidies. We recommend supporting these efforts in the WTO from the point of view of strengthening the potential of TK-based products in developing countries. Export subsidies should be eliminated to the extent that they distort the competitiveness of TK-based local products. We also recommend strengthening disciplines on food aid in order to avoid similarly adverse effects under such programmes. They should be limited to true emergencies and no longer serve to dispose of surplus production in a systematic manner.

The WTO Agreement on Agriculture provides adequate instruments to support financial measures in support of TK-based PGRFA. The so-called 'Green Box' allows for direct and non-production-related payments and de-coupled income in support of farmers working with biodiverse crops (in traditional, diverse farming systems). The scope of measures contained in Annex II of the Agreement is sufficient to cover measures relating to research and conservation. The same is true for supporting marketing and promotional services. We recommend stressing the importance of Green Box measures in developing countries under the heading of sustainable development. We also recommend clarifying that international funding meeting these qualifications must not be challenged as trade distorting under WTO rules.

The same criteria apply to tax reductions for farmers producing TK-based products. The Agreement on Agriculture allows operating tax reductions to the same extent that it allows subsidization. Subject to their constitutional law, WTO members therefore could operate tax reduction or elimination schemes in support of TK-based methods of production. We recommend considering these options in the context of conditionalities of future aid and support programmes.

I.5 Conclusions

Law and policy of TK, PGR and PGRFA is at the heart of the interface of agricultural, environmental, trade and development policies. Its current foundations do not sufficiently take into account the need to promote and protect biodiversity, distributive justice and intergenerational equity. Existing legal tools are shaped in accordance with the needs of developed countries, but fail to take into account those of developing countries. By doing so, they also fail to take into account long-term interests in conserving global biodiversity and thus long-term food security. Future policies and law need to build on that common interest to preserve a viable nutritional base in PGRFA and therefore also to protect and value traditional knowledge around the world. We submit that this common interest will serve as a driving force to move the issue to the centre stage of trade policy. It offers the potential to overcome the current divide of interest between North and South, and to build a key part of global agricultural and environmental policies. We believe, in particular, that the idea of *sui generis* TIP Rights for TK, and other adjustments in the field of IPR, is a viable option which is in the long-term interest of both developing and industrialized countries alike. We also believe that there is substantial potential in international regulations for creating more favourable conditions for marketing of non-assignable TK-based products, in particular by combining GI protection and preferential tariff treatment, by reviewing sanitary and phytosanitary measures, supporting quality control and operating targeting measures of agricultural support with a view to preserving biodiversity and enhancing global food security.

Part I

General Framework

1 Problems and Goals

Susette Biber-Klemm and Danuta Szymura Berglas
World Trade Institute, University of Berne, and Faculty of Law, University of Basel, Maiengasse 51, CH 4056 Basel, Switzerland

1.1 Biodiversity and Traditional Knowledge: Factual Background and Problems[1]

1.1.1 Introduction

In past decades, insight into the significance of biological diversity and the value of genetic and biochemical information has increased significantly. On the ecosystem level, new scientific findings have furthered the understanding of ecosystem functions and the crucial role of biodiversity in assuring the regulatory mechanisms of ecosystems, which provide the basis for economic activities such as agriculture (Barbier, 1997; Costanza *et al.*, 1997).

On the level of genes, achievements in biotechnology and the progress made in research technology have brought about profound changes in the agricultural and pharmaceutical industries and opened up new fields of application of plant and animal genetic resources. Regarding plant genetic resources for food and agriculture (PGRFA), there is increasing recognition of the importance of genetic diversity for maintaining food security in a changing world. Concurrently, the role of indigenous and other local communities in environmental management and development is being better acknowledged: their knowledge and traditional practices have been crucial for the conservation and sustainable use of natural resources.

At the same time, there is growing awareness of the accelerated loss of biodiversity and associated traditional knowledge, and of the threat of this process to the survival of humanity. This leads to greater efforts to conserve biodiversity, which is reflected in the political, institutional and legal field, at the international and national level and within civil society itself.

This process led to the conclusion of the Convention on Biological Diversity (CBD).[2] The CBD reflects the scientific state of the art in integrating, in its definition, the concepts of genetic diversity and genetic resources. It is the first international convention on biological resources to link the conservation of biodiversity with its use, and so too, economic issues and questions of politics in trade and technology. Also, for the first time in international law, conservation issues are combined with the need to develop economically poor states, taking

[1] We thank J. Schneider and G. Dutfield for their comments on earlier versions of this section.
[2] Convention on Biological Diversity, adopted 5 June 1992, reprinted in 31 ILM 818 (1992).

account of (North–South) equity and justice. An important element in this context is the acknowledgement of state sovereignty over natural resources, thus deciding against the principle of common heritage. The discussion of legal means for the protection of traditional knowledge at the interface with international trade is to be seen in this context in the course of the following deliberations.

However, before being able to discuss possible solutions, the underlying problems must be analysed clearly. Therefore, the characteristics of biodiversity and traditional knowledge, their value and the reasons for their loss, will first be explored, taking into consideration both 'wild' and domesticated plant biodiversity and associated knowledge. It is submitted that the respective value and loss of biodiversity and traditional knowledge are interrelated, but otherwise follow different patterns. As this is important for the design of legal instruments to regulate and protect them, the basic issues will be analysed separately.

1.1.2 Biodiversity

Terminology

BIODIVERSITY IN GENERAL

The term 'biodiversity' emerged in the 1980s as a result of the changing understanding of the role of diversity in natural processes. Whereas before, insights were based on a static understanding of the balance of nature that could be disturbed by the loss of a species, now diversity became a symbol and guarantor for the dynamics of nature. It was Edward O. Wilson who initiated, with his discourse on biological diversity, *The Diversity of Life*, a new direction of the environmental debate (Wilson, 1992). He understood 'biodiversity' as an 'umbrella concept' of the biological sciences, thus aiming at a holistic and dynamic understanding of nature (see Stettler, 2002).

So, 'biodiversity' is, in fact, an abstract concept. Glowka *et al.* (1994, p. 16) point out that biodiversity is the variability of life in all forms, levels and combinations. Accordingly, the term does not mean the sum of all ecosystems, species and genetic material, but rather represents the variability within and among them. It is, therefore, an *attribute* of life, in contrast with 'biological resources', which are the tangible biotic components of ecosystems (Gaston, 1996, pp. 1–7).

The CBD is based upon a broad understanding of the notion of biological diversity. 'Biological diversity' (or 'biodiversity') is defined by Article 2(1) as being 'the variability among living organisms from all sources including, *inter alia*, terrestrial, marine and other aquatic ecosystems and the ecological complexes of which they are part; this includes diversity within species, between species and of ecosystems'. Thus, the definition includes three levels of diversity: the genetic, species and ecosystem levels.

The notions of 'biological resources' and 'genetic resources' are defined by the CBD with reference to their value and use. 'Biological resources' include 'genetic resources, organisms or parts thereof, populations, or any other biotic component of ecosystems with actual or potential use or value for humanity' (Article 2.2). 'Genetic resources', in turn, are defined as 'genetic material, i.e. any material of plant, animal, microbial or other origin, containing functional units of heredity, which has an actual or potential value' (Article 2.7).

It is important to note that biological or genetic resources, as viewed by the CBD, include genetic information found in both wild species of animals and plants (including organisms such as microbes) and in domesticated plant varieties and animal species. Further, the notion encompasses resources found in *in situ* conditions, meaning in their natural surroundings, as well as in stored *ex situ* collections, such as in gene banks or botanical gardens.

The essays in this book focus on the genetic resources of plants, including both wild and domesticated varieties. This is understood to include the biochemical information as contained in medicinal plants. Biological resources, for instance, raw materials traded for industrial uses or for consumption, are of indirect relevance at most. As the characteristics of domesticated plant varieties are important in view of the question of rights to traditional knowledge, the following discussion considers this aspect in some detail.

AGROBIODIVERSITY AND PLANT GENETIC RESOURCES FOR FOOD AND AGRICULTURE

Biodiversity in the context of agriculture is dealt with in both the framework of the CBD and the Food and Agriculture Organization (FAO). A Resolution by the Conference of the Parties of the CBD[3] delegated the issue of plant genetic resources for food and agriculture (PGRFA) to the process to revise the International Undertaking on Plant Genetic Resources, which at that time took place in the framework of the FAO's Global System for the Conservation of PGRFA. This revision process resulted in the conclusion of the International Treaty on Plant Genetic Resources for Food and Agriculture (ITPGRFA) in November 2001, which entered into force 29 June 2004 (for details see Chapter 2).

Agrobiodiversity: the Convention on Biodiversity parlance. The CBD follow-up process in matters of agricultural biodiversity conservation and sustainable use focuses on the 'ecosystem approach'. Accordingly, the term 'agrobiodiversity' is understood in a broad sense. It includes all components of biological diversity of relevance to food and agriculture, as well as all components of biological diversity that constitute the agroecosystem. It encompasses the 'variety and variability of animals, plants and micro-organisms, at the genetic, species and ecosystem levels, which are necessary to sustain key functions of the agroecosystem, its structure and processes' (UNEP/CBD/SBSTTA/5/10).

Plant genetic resources for food and agriculture (PGRFA). Article 2 of the International Treaty defines PGRFA as 'any genetic material of plant origin of actual or potential value for food and agriculture'. The term is understood in a broad sense, including all plant material valuable for the entirety of the agricultural ecosystem. Hence, not only cultivated but also wild plants are included (for details see Chapters 4 and 8).

Landraces. Landraces, also termed primitive cultivars, farmers' varieties or traditionally bred PGRFA are of specific interest in the context of traditional knowledge. The term designates local crop varieties that have been bred on-farm by local farmers or farming families, adapting them to the specific conditions of the local ecosystem, climate and diseases and pests found there, and to their own needs and tastes. Typically, their nature is the product of a continuous process of evolutionary change over generations of development by farmers. This process includes hybridization within and between populations of wild, weedy and cultivated plants; competition among genotypes; natural and conscious selection at the local level; and exchange of different genotypes among farmers and farms (Brush, 1994). Thus, the maintenance and evolution of landraces are dependent on the continuous active human management of the entire agroecosystem and of a system of open exchange (see Chapter 4).

The specific characteristics of agrobiodiversity. It is important to be aware of the fact that agrobiodiversity in all its dimensions, such as the maintenance of agroecosystems and the further evolution of PGRFA, is influenced by, and dependent on, human intervention. The Global Plan of Action (GPA)[4] points out the fact that, unlike most natural biodiversity, PGRFA require continuous active human

---

[3] Resolution 3, adopted 22 May 1992, on the Interrelationship between the Convention on Biological Diversity and the Promotion of Sustainable Agriculture.

[4] Global Plan of Action for the Conservation and Sustainable Utilization of Plant Genetic Resources for Food and Agriculture. Adopted by the International Technical Conference on Plant Genetic Resources, Leipzig, Germany, 17–23 June 1996.

management. Humanity, in turn, is itself dependent on agricultural biodiversity. Therefore, agrobiodiversity has not only a biological nature, but also a prominent social dimension.

Characteristics of plant genetic resources

Plant genetic resources have certain qualities not found in non-living resources such as mineral resources, water and air, which are important in our context.

First, biological resources have a so-called 'dual character'. Seeds, plants and domesticated animals are, on the one hand, goods that are produced for consumption as food, timber, textiles and industrial products, and are traded internationally as such. On the other hand, these same goods are of interest – as genetic resources – for the genetic information that they contain. In the case of plant genetic resources, the information is contained in each seed and, through the use of modern biotechnology, access to this information can be gained on the basis of analysing only a small quantity of the resource.

Second, biological resources are self-propagating. This has two consequences:

1. Theoretically, they can be exploited without being damaged, as in principle we are dealing with a renewable resource. However, once a species or a variety becomes endangered, it may fail to recuperate and become lost forever. Hence, biological resources also have an absolutely non-renewable character.
2. The self-propagating characteristic of biological resources often means that a specific resource may be exploited independently of its place of origin.

These characteristics mean that it is not possible for the original owner or holder of the resource to maintain control over the use of the genetic information found within them. Genetic resources have a ubiquitous character, which is specific to information of any kind, meaning that once it has been revealed, it becomes independent of its original source.

Distribution and origin of biological diversity

Biodiversity is distributed unevenly over the globe. Generally speaking, there is more diversity in warmer and wetter climates than in cooler and drier ones (McNeely *et al.*, 1990). Seventy per cent of the world's biodiversity is found in only 12 'mega-diverse' countries – Colombia, Ecuador, Peru, Brazil, Zaire, Madagascar, China, India, Malaysia, Indonesia, Australia and Mexico – which, with the exception of Australia, are all developing, non-Western nations. A wealth of biodiversity is also found in many other countries; for instance, South Africa contains the most biological diversity in plant species (Wilson, 1992, p. 260).[5]

The definition of the origin of either domesticated or non-domesticated plant genetic resources is important in the context of the implementation of the CBD concept of access to genetic resources and sharing the benefits of their use, and of the International Treaty's Multilateral System of Access and Benefit Sharing, which accord national governments the authority to determine access to genetic resources (see Article 15 CBD and Article 10 ITPGRFA).

The CBD, on the basis of sovereignty of states over their natural resources, takes the 'country' as a starting point for defining the origin of genetic resources. Accordingly, '*the country of origin of genetic resources*' is defined as the country that possesses the genetic resources in *in situ* conditions (Article 2.4). *In situ* conditions for non-domesticated species are defined as 'conditions where genetic resources exist within ecosystems and natural habitats' (Article 2.12). *In situ* conditions for domesticated and cultivated species[6] in turn are '*the*

[5] As noted by Wilson (1992, p. 260), it is an alarming fact that (in general) 'the richest nations preside over the smallest and least interesting biotas, while the poorest nations, burdened by exploding populations and little scientific knowledge, are stewards of the largest'.

[6] Defined as 'species in which the evolutionary process has been influenced by humans to meet their needs' (Article 2 CBD).

surroundings where they have developed their distinctive properties'.[7]

The International Treaty takes the 'centre of origin' and the 'centre of crop diversity' as references referring to the scientific rather than political connecting points for the definition of origin. It defines 'centre of origin' as 'a geographical area where a plant species, either domesticated or wild, first developed its distinctive properties', and 'centre of crop diversity' as 'the geographic area containing a high level of genetic diversity for crop species in *in-situ* conditions' (Articles 2.8 and 2.9).[8]

The value of biological diversity

The discussion of the value of biodiversity is made in the context of providing a case for its conservation. It appears in different disciplines – philosophy, biology, economics – and is thus described differently.[9] In 'Western' philosophies two main approaches can be distinguished: the anthropocentric or utilitarian approach and the ecocentric approach.

Under an anthropocentric approach, the value of biodiversity is explained by its contribution to human welfare and/or happiness. Thus, arguments for the preservation of biodiversity include its aesthetic value, and the moral responsibility of humanity to preserve natural resources, entailing the concept of human stewardship of the environment. Biodiversity is valued by science as an indicator of the sustainable use of resources. Furthermore, economic theories refer to the so-called direct, indirect and optional value of biodiversity.

The ecocentric approach is concerned with the intrinsic value of biodiversity, meaning its value independent from its contribution to human welfare.

In the North–South context it is important to take account of the philosophical approach of indigenous peoples, which is holistic in nature.

As the focus of this study is on legal and economic issues, the following analysis emphasizes the economic line of argument. Economists distinguish between: (i) the direct or actual use value; (ii) the indirect use value; and (iii) the long-term or option use value.

DIRECT OR ACTUAL USE VALUE

Direct use value refers to those elements of biodiversity that can be consumed directly, traded or used as an input to commercial activities, such as plants or animals that are collected or hunted for food, clothing, energy or shelter. Thus, direct use is just as important for indigenous, local and farming communities, who depend on it for their everyday survival, as for the industrial use of natural resources. There is no agreement on the actual financial value of biodiversity. Estimations vary according to the methods applied for their calculation. Direct use values can normally be realized by individuals or companies, and are thus *privately appropriable*. This includes using certain areas for activities, such as sightseeing or tourism (OECD, 1999, p. 29). The most important direct use values in our context are the use as food and as medicines and pharmaceuticals.

Plants are the most important source of food for all living things. About 50,000 varieties of edible plants have been discovered to date, based upon 10,000 years of settled agriculture, yet a mere 15 food crops provide 90% of the world's food energy intake (including sorghum/millet, barley, potato, sweet potato/yam, soybean and sugarcane), and just three of them – rice, maize and wheat – are the staple foods of 4 billion people (UNFPA, 2001, p. 17, citing World Bank, 1996; Henne, 1998, p. 68).

---

[7] *In situ* conditions for wild species are defined as the ecosystem where they exist in natural surroundings such as ecosystems and habitats (Glowka *et al.*, 1994, p. 22). Thus the definition differs for wild and for domesticated or cultivated species.

[8] See the discussion of the term 'origin' in Chapter 4.

[9] Compare, for example, the overview by Kunin and Lawton (1996) and the overview of the philosophical approaches in Krebs (1999).

On the one hand, the variety of PGRFA is important in guaranteeing food security in subsistence farming systems in developing countries. On the other hand, PGRFA are sources for tailoring food production in order to meet present and future challenges.[10] The variety of PGRFA forms the basis for reinvigorating cultivars in order to give them greater resistance against disease and insects, as well as for introducing new yield-enhancing traits such as increased tolerance to drought or saline soils.

As for the use of plant genetic resources for medicines and pharmaceuticals, the World Health Organization (WHO) estimates that 80% of the people in developing countries rely on traditional medicine for their primary health-care needs, and that about 85% of traditional medicine involves the use of plant extracts (Farnsworth, 1988, p. 91). This means that about 4 billion people rely on plants as an indispensable source of medicine. In China, traditional medicine is used by 80% of the people, based on 4600 species of plant; in India, 8000 species of plant are used for medicinal purposes (Henne, 1998, p. 79).

It is said that most of the world's most effective pharmaceutical products have been discovered from compounds derived from some natural product (Henne, 1998, citing Albers-Schönberg, 1996, p. 77; UNFPA, 2001, p. 17). Examples include medicines against malaria (the substance quinine comes from a tree, the properties of which were known by the Incas); childhood leukaemia (*Catharanthus roseus*, a plant which is found in Madagascar); and antibiotics, such as the one based upon the mould *Cephalosporium*. In the period 1983–1994, 78% of new antibacterial medicines and 61% of new anticancer medicines in the USA had natural origins. About one-third of all medicines in Germany have a plant basis, and about one-quarter of all medicines in the USA also stem from plants (Henne, 1998, p. 72).

These medicines generate an impressive turnover. The two natural substances vincristine, used against childhood cancer, and vinblastine, used against Hodgkin's disease, alone are said to generate US$100 million per year (Henne, 1998, p. 77). Ten Kate and Laird (2000, p. 334) analysed the sales of the 25 best-selling drugs worldwide in 1997. They concluded that 42% of the sales were made up of biological, natural products or entities derived from natural products, with a total value of US$17.5 billion.

According to Farnsworth (1988, p. 93), about 119 pure chemical substances extracted from fewer than 90 species of plant are used in modern medicine throughout the world today. Given that there are more than 250,000 species of plants on Earth, one can only imagine how many more useful compounds may reasonably be predicted to exist, as yet undiscovered. A study by Mendelssohn and Balick (1995, p. 223) estimates the worth of as yet undiscovered pharmaceutical substances from tropical rainforests to be between US$147 and US$900 billion.

Wild species are used around the world to meet basic requirements for human survival and socio-economic development. At one end of the spectrum are the people living in subsistence systems, who depend on the supplies offered by plant and animal species found in their surroundings. At the other end is the utilization of wild plant and animal species products in industrial production processes; for example the production and consumption of timber, the consumptive use of fish, wildlife hunting and non-timber plant products (Freese, 1998).

Biological diversity of wild plants and of PGRFA is a basic asset for the survival of people living in subsistence economies in developing countries, and is also valuable for use in the context of industrial research and development (R&D) and formal plant breeding.

However, wild (and domesticated) animal and plant species are also important

[10] See, for instance, the Leipzig Declaration on conservation and sustainable utilization of plant genetic resources for food and agriculture. Adopted by the International Technical Conference on Plant Genetic Resources, Leipzig, Germany, 17–23 June 1996, para. 3.

assets for consumptive use in the context of industrial production and international trade.

INDIRECT USE VALUE: ECOSYSTEM SERVICES AND TEMPLATES FOR INDUSTRIAL PRODUCTS

The so-called 'ecosystem services' consist of the regulatory functions of ecosystems, such as nutrient recycling, sedimentation processes, waste treatment, water regulation, gas regulation and other functions which obviously support or protect ecological activity. The services depend on the proper functioning of the ecosystem, which, in turn, depends on an ecologically healthy level of biological diversity. In other words, the resilience of the ecosystem – its capacity to recover from external stresses and to maintain the option of change – is defined by the level of its biodiversity (Barbier, 1997, p. 116; Costanza *et al.*, 1997). Thus, biodiversity provides goods and services that are necessary for maintaining the planet's ecological balance, and provide the basis for economic activities and – ultimately – the survival of humanity.

Another aspect of indirect use value is to consider biological resources as a template for the production of other resources. The biochemical compounds of plants may be used as medicines themselves or as a template for the synthetic production of medicines. Ideas for the design of new materials may be derived from natural resources (Norton, 1988, p. 201; Kunin and Lawton, 1996, p. 294).

LONG-TERM OR OPTION VALUE

The long-term, optional value of biological diversity lies in the amount of information contained in life forms, which has accumulated in the process of evolution. A community of animals and plants, which has co-evolved over millions of years, contains an encapsulated history of information, and this cannot be exactly reproduced (Swanson, 1994, p. 246). This information secures the continuity of the natural evolutionary process and the adaptation of species to a changing world. In conserving biological diversity, the option remains intact that the future exploration and use of currently unknown resources may generate discoveries of new and useful qualities (Glowka *et al.*, 1994, p. 9; Swanson, 1995, pp. 161–162). The loss of any species, and of even one gene, could thus limit our options for the future (Serageldin, 1997).

Given the present stage of scientific knowledge, it is impossible to assess the future value of any specific species: what is needed is more information to adapt to a changing world. Accordingly – pursuant to the precautionary principle – as many options as possible ought to be maintained, which can only come through the effective management and conservation of biological diversity.

CONCLUSIONS

As mentioned above, the attempt to give a monetary value to biodiversity is both difficult and controversial.[11] Some scientists and economists agree that it is not possible to determine a value in monetary terms of even one species, let alone attempt to assign an economic value to biodiversity in the aggregate (Ehrenfeld, 1988, p. 214; Iltis, 1988, p. 99; Norton, 1988, p. 202; OECD, 1999, p. 27). It is also suggested that a purely economic analysis of the value of biodiversity is insufficient, for it is not sophisticated enough to take into account the enormously complex matrix of interrelationships between the numerous living organisms in any given ecosystem, and the holders and users of those resources. In particular, the assessment of potential future value is difficult, and there is no prediction possible as to the exact, or even minimal,

[11] See the example of the calculation of the value of the Peruvian Amazonian Forest in Kunin and Lawton (1996, pp. 285–286), which also gives an example of the problem of the predominance of the short-term value over the long-term value in the decision-making process.

amount of diversity needed for the ecological health of our planet, to maintain the ability to provide for all of our needs.

It can at least be said that biological diversity of domesticated and wild animals and plants is important for humanity and its long-term survival. Conservation is therefore not only a direct, private concern, but also of global interest.

The loss of biological diversity

WILD BIODIVERSITY

According to the results of scientific research (see Primack, 1995, p. 91), biodiversity *in total* has gradually increased since the beginning of life on Earth. This evolution has not been at a steady rate: periods of fast evolution have alternated with periods of little change or periods of mass extinction. Science describes five periods of extinction. Interestingly, each time the perished species have been replaced by an even larger number of descendants of the surviving species. However, this process has required tens to hundreds of millions of years (Wilson, 1992, p. 29; Primack, 1995, p. 92), and in any case it is thought that global biodiversity reached its absolute peak about 30,000 years ago.

What is clearly known is that biological diversity is currently being lost at an alarming rate, this process having significantly accelerated within the past 150 years (Primack, 1995, p. 96). Biologists warn that the present biodiversity loss is part of the sixth, man-made extinction period, the rate of extinction being 100 to 1000 times higher than the estimated rates for the natural extinction process (Primack, 1995, pp. 95–96).

LOSS OF THE DIVERSITY OF PLANT GENETIC RESOURCES FOR FOOD AND AGRICULTURE

The diversity of PGRFA is in danger of extinction not only in *in situ* conditions, but also in *ex situ* collections, which must be continually upgraded to take account of evolutionary change.

The loss of *ex situ* material. Seeds or other reproductive parts of plant material stored in *ex situ* facilities must be regenerated in order to maintain their viability. According to the FAO's report on the state of the world's PGRFA (1998), two factors have a major impact on regeneration needs and capacity: the storage conditions and the capability to handle the regeneration of cross-pollinated species. The frequency of the regeneration depends, among other things, on the storage conditions: poor storage conditions result in the rapid loss of seed viability and the need for more frequent regeneration. If the finances for the maintenance of the viability of the samples are not available, the sample gets lost. On the other hand, frequent multiplication of the accession can also lead to the loss of the original genetic information of the sample. An entire collection might become lost if a storage institution is closed or restructured (FAO, 1998, p. 113).

The *in situ* loss of biodiversity of plant genetic resources for food and agriculture. The main loss of diversity in crops is the replacement of local varieties by improved or exotic varieties and species. Genetic erosion occurs because the genes and gene complexes found in the diverse farmers' varieties are not (entirely) contained in the modern variety. In addition, the number of varieties used may be reduced when commercial varieties are introduced into traditional farming systems.

Genetic erosion is reported in many countries, both in the developing and the developed world. For example, only 20% of the local maize varieties reportedly known in Mexico in 1930 are now known. In China, the wheat varieties used decreased by a factor of 10 between 1949 and 1970, and in the USA, between 86 and 95% of the vegetable and cereal varieties grown by farmers in the 19th century have since been lost (FAO, 1998, p. 35).

REASONS FOR THE LOSS OF BIOLOGICAL DIVERSITY

Natural scientists generally agree that the present loss of natural biodiversity is predominantly man-made. The main direct causes are the alteration and loss of natural habitat, overharvesting, the introduction of alien species and disease, and pollution and climate change.

These phenomena are driven by a complex array of underlying causes. Therefore it is broadly acknowledged that policies and instruments to conserve biodiversity and traditional knowledge have to address the fundamental pressures that cause their loss. It is important to understand these underlying sociocultural and economic mechanisms.[12]

Consequently, the following provides an overview of the direct causes of biodiversity loss, together with an analysis of some of the underlying causes. For the purpose of our study, the focus will be on the economic line of argument.

Direct or proximate causes for biodiversity loss. The predominant direct cause of biodiversity loss is clearly habitat alteration (see Primack, 1995, p. 140; Wood *et al.*, 2000, p. 5). The best-known example is the destruction of tropical rainforests through logging or clearing for agricultural use. However, other types of ecosystems are equally threatened, such as wetlands, tropical dry forests, mangroves and grasslands. In arid and semi-arid zones, human pressure leads to the damage of ecosystems and finally to desertification.

Loss of biodiversity is not only driven by the complete destruction of habitats, but also by their degradation. Degradation may occur through fragmentation by, for example, the destruction of parts of the original habitat, or by reducing it into smaller areas by building roads, railway tracks, canals, bridges and so on; secondly, by pollution of soil, water and air, by, for instance, pesticides, industrial and private emissions of pollutants; and thirdly, by climate change.

Another direct cause which might be of importance in our context is the loss of animal and plant species caused by their overexploitation through unsustainable fishing, hunting and collecting practices, in order to profit from the demands of international markets.

The loss of agrobiodiversity partly follows similar patterns. As mentioned earlier, the intensification of agriculture and the expansion of its use lead to both the reduction of biological diversity of agricultural ecosystems and of crop diversity. Hence, the dominance of blueprint paradigms and policies, such as industrial agriculture and the Green Revolution, which focus on high yield through the use of monoculture systems and uniform technologies, high-yielding seeds and animal breeds, agrochemicals, irrigation, mechanized equipment and large infrastructure developments, are all identified as fundamental causes of agricultural biodiversity loss (FAO, 1999, p. 20).

Underlying causes. The direct causes clearly indicate that there are also underlying reasons driving people to the unsustainable use of resources. If strategies and instruments for the conservation and sustainable use of biodiversity are to be effective, they must address the underlying causes as well (see Wood *et al.*, 2000).

According to the economic line of argument, the so-called economic failure of the market to adequately recognize the full and true value of biological resources plays a pre-eminent role. Two main reasons are given for this phenomenon. First, the value of biological diversity is not fully reflected in economic decision-making systems. This means that the cost for engaging in activities that lead to the loss of diversity is low. This leads to an unsustainable utilization of the resources, that is to say, the exploitation of biological resources is, relatively speaking, too cheap, and incentives arise for their over-exploitation. Secondly, no market exists for many products of biological diversity, so the value of these resources cannot be appropriated by their owners, and is thus not integrated in the decision-making process about the utilization of those resources. Therefore, the markets do not secure the economically

---

[12] UNEP/CBD/SBSTTA/ 7/11/Add.1, pp. 4–5. See also, for example, OECD (1996, pp. 43–65) and OECD (1999, pp. 20–21).

correct balance between the utilization and conservation and maintenance of biodiversity (OECD, 1996, p. 51).

Economic theory holds that in order for any resource to be properly managed by market mechanisms, the price of that resource needs to reflect all the values that society places upon it. This means internalizing the external benefits and costs associated with using a resource. In environmental law, this is, as a rule, realized by the 'polluter pays' principle: the environmental costs caused by the utilization of a resource (such as the cost of the treatment of sewage) are internalized into its price, thus creating an incentive to use the resource efficiently. According to economic theories, if the resources are not properly valued in this sense, misleading information is created about their scarcity, thus providing inadequate incentives for their utilization and management.

None the less, in the case of biodiversity the concept of cost-internalization does not make sense. Utilization of biodiversity does not create damage to the environment in the same way as, for example, is the case with the utilization of clean air or water for waste disposal. The process of biodiversity loss is difficult to assess economically in a market context. With regard to the use of biodiversity or biogenetic resources, the reverse is true: biological diversity creates benefits accruing to society as a whole, and potential benefits to its owners. Consequently, it is argued that in order to create incentives for its conservation, the *benefits* generated by its use must be internalized. The creation of property rights is proposed as an instrument to achieve this end.

The internalization of the benefits created by biodiversity proves to be difficult because of the problem of 'open access'. The nature of biological diversity is such that once the information contained within it has been acquired, it is impossible for the original owner to prove that it was exclusively his or hers. The information becomes independent from its original source and – legally speaking – falls within the public domain, accessible for everybody.

Due to this 'open-access' situation, scientists within industrial innovation systems (although, in fact, anyone) could use these resources up to now without the consent of their holders and without offering compensation or sharing of any profits made, which have been substantial in several cases. Concerned people and communities speak of 'piracy of traditional knowledge' analogous to the piracy of industrial intellectual property by countries without adequate protection of intellectual property themselves.

According to economic theories, the informational character of biological diversity provides an important reason for its loss. Because its main value cannot be appropriated, the value of biodiversity is ignored by landowner decisions, and diversity of plant species is sacrificed in favour of more lucrative uses of land resources, such as industrialized monocultures (Swanson, 1995, pp. 146–151). Accordingly, the solution of the global biodiversity problem requires the creation of a mechanism to appropriate these information values supplied by evolution. For instance, Swanson proposes the creation of an 'informational resource right' as a specific property right, tailored in analogy with intellectual property rights, to protect the informational value of biogenetic information (Swanson, 1995, p. 173).[13]

Not all values of genetic resources lend themselves to be appropriated by private property rights, as is the case with the direct-use values, which can be directly consumed, traded or used as an input into commercial activities. Biological diversity also creates benefits that accrue to society as a whole: for example, through ecosystem services or long-term option value. In the majority of cases these values do not accrue to an individual economic actor but to society at large, at the

[13] This proposition to create an intellectual property type of right for genetic information is a controversial subject in the literature. However, the discussion in this chapter focuses on protecting informational values generated by, or contained in, human know-how and skill.

local, national, regional or even global level (OECD, 1999, p. 32; Biber-Klemm, 2001).

Agrobiodiversity is maintained by local subsistence farming systems as a side-effect of striving to prevent food shortages. Furthermore, local and indigenous people, in conserving wild resources and using them in a sustainable way, by, for instance, actively nurturing wild plants or giving up the opportunity to make higher profits that would result from a more intensive use of the resources, thereby contribute to the 'global biodiversity services', performing a service to humankind. These biodiversity services correspond to the characteristics of a 'public good'.[14]

The problem of a public good is that although the entire society benefits from its use, free accessibility to it entails that no mechanisms for its protection, such as market mechanisms, exist. Private owners, as a rule, only pay attention to those values that can be privately appropriated. The realization of public good values, such as the commitment to sustainable logging or to maintain landraces, creates costs in terms of private losses. The OECD uses the term 'profitability gap' to describe this situation, which is typical for a public good.

If the public good is to be maintained, specific measures are needed for its provision (Swanson, 1994; Kaul *et al.*, 1999). The public policy implication is that both states and international regimes must play some role in the provision of such goods, otherwise they will be undersupplied (Stiglitz, 1999). Stiglitz identifies two strategies to provide for public good 'knowledge' in the sense of informational values. First, to increase the degree of appropriability of the returns by issuing intellectual property rights; and, second, to grant direct government support.

The 'root causes' study. Wood *et al.* (2000), mandated by the World Wide Fund for Nature (WWF), undertook research into the analysis of the 'root causes' of biodiversity loss. Root causes refer to the set of factors driving biodiversity loss, the distance of which from the actual incidence of the loss, either in space or time, makes them a challenge to identify and remedy (p. 3).

The starting point is the observation that, in spite of many initiatives and developments, the approaches up to now have failed to stop biodiversity loss in any meaningful way (p. 2). One of the assumptions is that biodiversity loss will continue until the indirect causes for it are properly understood and addressed. The premise is that the failure to address biodiversity loss 'stems from an inability to understand and articulate the basic conflict that exists between the promotion of growth and consumption on one side, and activities and incentives promoting sustainable development and conservation of biological diversity on the other'. It is put forward that 'new conservation approaches must begin with an understanding of that conflict and be based on the expectation that the roots of the conflict must be addressed if they are to succeed in the long run' (p. 2). Therefore, the strategies for biodiversity conservation and sustainable use must be based on an analysis of the complexity of factors that drive biodiversity loss (p. 9). One must understand the socio-economic forces that create the incentives for the activities that put pressure on biodiversity, and disincentives for more sustainable behaviour (p. 12).

One of the main challenges is that the causes driving biodiversity loss span different scales: local and global; short term and long term; and different political and economic levels (p. 20). 'Biodiversity loss occurs at the local level as the result of many individual decisions about resource use. The local actor who contributes to biodiversity loss ... is acting within a particular set of social, cultural, political, economic and environmental constraints. To understand biodiversity loss, we must

[14] The notion of 'public good' is defined by two main qualities: (i) its benefits are not subject to rivalry in consumption, i.e. their utilization by one person does not exclude the utilization by others; and (ii) its benefits are non-exclusive, i.e. it is extremely difficult and costly to exclude others from the utilization of the good (Kaul *et al.*, 1999).

understand those limits and possibilities and how they affect resources use' (p. 19).

Accordingly, the case studies carried out in the framework of the project have adopted an interdisciplinary, comprehensive approach. A complex array of socio-economic factors was found to be at work in all of them. Changing domestic and global pressures lead to the increasing use of resources, yet policies and institutions are frequently guided by an inappropriate model for biodiversity protection (p. 63).

The main findings show the interdependence of the causes of biodiversity loss at local, regional and international levels, and are difficult to describe in their entirety because of the complexity involved. A short overview will be given here.

The causes of biodiversity loss are identified as follows:

- On the domestic level, demographic pressures due to population growth and migration, poverty and inequality, and sometimes the isolation of a site, may all lead to degradation of ecosystems due to the increased use of resources.
- On the international level, pressures caused by macroeconomic policies and international trade may indirectly shape the use of resources, in many cases leading to the unsustainable use of those resources.
- These factors lead to government and market responses which predominantly address socio-economic and political pressures, generally paying little attention to environmental problems, and which promote the expansion of the use of natural resources. This frequently occurs coupled with the failure of government efforts designed to protect the environment.
- The result may be greater use of biological resources without corresponding measures essential to protect the environment. The study concludes (pp. 77–78):

> ... the number of domestic and international pressures shaping the local behaviour that leads to biodiversity loss is sometimes overwhelming ... Governments and individuals have to respond to both these domestic and external pressures, not as environmental pressures, but as social, economic and political problems. Solutions to all of these problems, from population growth to balance-of-payments problems, are sought in natural resources use. For individuals and communities, resource use often provides at least short-term alleviation of these pressures. Similarly for governments, the permitting or promoting of resource use provides a temporary solution to economic, social and political pressures.

In view of the factors originating at the international level, the study recommends, 'the international community has to ensure that the context in which countries are being asked to make tough choices about their development is properly supportive of the objectives of sustainable development. International institutions and regimes need to create incentives and remove disincentives for sustainable development, and the conservation of biodiversity in particular'.

Conclusions

Despite measures intended to halt the loss of biological diversity, in fact it appears that the loss is increasing. This is true for the biodiversity of both wild and domesticated plant and animal (genetic) resources, and their corresponding ecosystems.

Human activity is largely responsible for the loss of biological diversity, although natural events also play an important role. A subtle change in climate or introduction of an alien species may have just as devastating consequences for a particular plant or animal as destroying the habitat in which it lives. The various causes of the loss of biodiversity are complex in themselves, and even more complex when considered as a whole. Indeed, the true nature of this process is difficult to perceive as the effects of a local, present-day pressure may become discernible only over a long period of time or in a greater geographical context. Conflicting interests may make it difficult to adopt measures for conservation and sustainable use. Pressing short-term interests for the survival or economic welfare of local people may prevail over the long-term interest of the survival of biological diversity for humanity at the global level.

The 'root causes' discussion shows that biodiversity conservation and sustainable use is a cross-cutting issue, involving different spheres of time, geographical location, societal groupings and politics. A 'biodiversity mainstreaming' would be needed, integrating criteria of sustainable biodiversity use at all social, political and economic levels.

Legally speaking, it is clear that no simple and uniform solution will answer the problem of protecting biological diversity. Any solution must combine a variety of measures that encompass the enunciation of the appropriate rights, together with corresponding instruments and institutions for their implementation and other supporting measures, on all levels and in all areas that have some bearing on the use of natural resources.

In stating this, we fully agree with the final thoughts of Wood *et al.* (2000, p. 93), who state, 'the conflicts and contradictions that drive biodiversity loss represent some of the most intractable problems our societies face ... We fully recognise that taking action ... will be a long and difficult process'.

1.1.3 Traditional knowledge

Introduction

In the past decades there has been growing awareness of the close interrelationship between the level of biological diversity found in a specific region and the cultural distinctiveness of its inhabitants. In studying this overlap, a high correlation has been observed to exist between species richness and language richness as an indicator for cultural richness. Oviedo (2000, p. 6) gives the following explanation for this correlation:

> an historical process of co-evolution of small-scale human groups with their local ecosystems ... Over time, as human communities interact closely with the local environment, modifying it as they adapt to life in specific ecological niches, they acquired intimate and specialised knowledge of the environment and how to use and manage it for individual and group survival. Thus it is generally agreed that small-scale societies with a history of continued and unchallenged occupation of given territories will, over time, tend to develop and maintain detailed and accurate knowledge about their ecological niches, as well as about sustainable ways of extracting and managing natural resources.[15]

Thus, even if the detrimental influence of humankind on the environment is at the forefront of the discussion, it must be pointed out that in many cases the human impact on biodiversity has been beneficial in terms of conserving, fostering and even creating biodiversity (Oviedo, 2000, pp. 7–8).[16]

This new understanding of the specific knowledge of communities living in close relationship to their environment has, under the heading of 'traditional knowledge', been integrated in the Rio Process, and in the process to revise the International Undertaking on PGRFA, and it has recently been confirmed and enhanced by the outcome of the World Summit on Sustainable Development.[17]

So, the interrelationship between traditional knowledge and the maintenance and sustainable use of biological diversity has found its expression in a variety of legal instruments and documents. For instance, principle 22 of the Rio Declaration underlines the vital role indigenous people and

[15] For a critique of the 'ecologically noble savage' hypothesis, including references, see Dutfield (2000, p. 2).

[16] This, by the way, is not only true for indigenous people living in remote areas of tropical rainforests, but also, for example, true for traditional mixed farming systems found in central Europe.

[17] See the Report of the United Nations' World Summit on Sustainable Development and Annexed Plan of Implementation of the World Summit on Sustainable Development (A/CONF.199/20).

other local communities have in environmental management and development, because of their knowledge and traditional practices. The preamble of the CBD (in para. 12) acknowledges the existence of traditional knowledge, innovation and practices which are relevant for the conservation of biological diversity and the sustainable use of its components, and specifically recognizes the vital role that women play in its conservation and sustainable use (para. 13). The value of traditional knowledge for conservation and sustainable use is then specified in Article 8(j). In the field of agriculture, the close interdependence between the maintenance of agroecosystems, the diversity and further evolution of PGRFA, and human intervention is pointed out in the Global Plan of Action.[18] In its recital and in Article 9.1 on Farmers' Rights, the International Treaty on PGRFA explicitly recognizes the 'enormous contribution that the local and indigenous communities and farmers ... have made and will continue to make for the conservation and development of plant genetic resources'.

Terminology and characteristics

THE NOTION OF TRADITIONAL KNOWLEDGE

Neither the CBD, the Bonn Guidelines, nor the ITPGRFA give a definition of the term 'traditional knowledge'. The CBD speaks of 'traditional knowledge, innovations and practices of indigenous and local communities embodying traditional lifestyles relevant for the conservation and sustainable use of biological diversity' (Article 8(j)); the ITPGRFA refers to 'traditional knowledge relevant to plant genetic resources for food and agriculture' (Article 9.2(a)) and of the 'enormous contribution that the local and indigenous communities and farmers of all regions in the world (have made), particularly those in the centres of origin and crop diversity ... for the conservation and development of plant genetic resources' (Article 9.1).

In the relevant literature, different terms are used. According to the given context, it is alternatively referred to as traditional ecological knowledge (TEK) indigenous or aboriginal knowledge, traditional biodiversity-related knowledge (TBRK) or others.[19]

This different terminology alone indicates that there is no simple, commonly agreed upon definition of traditional knowledge. In our study the term traditional knowledge, innovation and practices is understood to include the relevant knowledge of both indigenous and non-indigenous communities such as farming communities, relying on traditional production systems.[20]

Hence, the term 'communities embodying traditional lifestyles' is understood in a broad sense, including both farming communities and indigenous peoples. The latter, who, according to the Study of the Problem of Discrimination Against Indigenous Populations, are understood as communities

> ... which, having an historical continuity with 'pre-invasion' and pre-colonial societies that developed on their territories, consider themselves distinct from other sectors of the societies now prevailing in those countries, or parts of them. They form at present non-dominant sectors of society and are determined to preserve,

[18] Global Plan of Action for the Conservation and Sustainable Utilization of Plant Genetic Resources for Food and Agriculture. Adopted by the International Technical Conference on Plant Genetic Resources, Leipzig, Germany, 17–23 June 1996.

[19] Compare the analysis in WIPO/GRTKF/IC/3/9 p. 6 and the non-exhaustive enumeration of 20 different terms in Annex I of the same document!

[20] The term 'traditional systems' is interpreted in a wider sense than in the definition proposed in the Note of the Executive Secretary of the CBD (UNEP/CBD/WG8J/2/5, 27 November 2001): 'A community that has not adopted practices of biological resource ... use based on modern scientific, technological and commercial methods of natural resource management, but relies primarily on its age-old traditional methods and practices'. We also understand 'traditional' to refer to the way knowledge is acquired and integrated in the system.

> develop and transmit to future generations their ancestral territories, and their ethnic identities, as the basis of their continued existence as peoples, in accordance with their won cultural pattern, social institutions and legal systems (cited from WIPO, 2001, p. 23).[21]

The CBD clearly refers to traditional knowledge in the context of conservation and sustainable use of biological resources.[22] This means that, in particular, traditional knowledge embedded in artistic creations or 'expressions of folklore' is not explicitly considered (WIPO, 2001, p. 25). In turn, it is acknowledged that the maintenance of biological diversity goes hand in hand with the preservation and maintenance of cultural diversity. That is to say, traditional knowledge 'relevant for the conservation and sustainable use of biological diversity' is to be understood in this broad sense.[23]

CHARACTERISTICS OF TRADITIONAL KNOWLEDGE AND ITS MEANING FOR BIODIVERSITY CONSERVATION AND SUSTAINABLE USE

In general. Humankind has used the informal knowledge systems for thousands of years, this information being transmitted and developed by one generation to the next. The tradition of learning and teaching is itself very old, and the knowledge derived from this process may have originated a long time ago, but it is a misconception to think that traditional knowledge must itself be outdated in some way. Rather, it refers to the human tradition of passing down knowledge throughout the ages and improving it with experience over time (WIPO, 2001, p. 212). Posey (2001, p. 382) characterizes traditional ecological knowledge as 'a cumulative body of knowledge and beliefs handed down through generations by cultural transmission about the relationship of living beings (including humans) with one another and with their environment'.

Traditional knowledge is the basis for local-level decision making in areas of contemporary life, such as natural resource management, nutrition, food preparation and health. It is inherently dynamic and constantly evolving through experimentation and innovation, fresh insight and external stimuli (Posey, 2001, p. 382). So, it might not even always be the particular knowledge that persists, but rather the traditional methods of research and application.

Traditional knowledge as the result of an eco-cultural learning process. Since the beginning of the discourse on the interrelation between traditional knowledge and biodiversity conservation, organizations of indigenous peoples have pointed out the 'inextricable link' between biological and cultural diversity.

In the framework of the Rio process, the concept of so-called biocultural diversity has become more widely known and accepted. It is implicitly recognized in the CBD, which acknowledges the important role that indigenous people and local communities continue to play in the conservation and sustainable use of biological diversity.

The connection between traditional knowledge and biological and cultural diversity has recently become apparent with the studies of linguistic diversity. Linguistic diversity, as a central part of cultural diversity, plays an important role in the

[21] Albeit discussed controversially, see WIPO (2001, p. 26). The position of indigenous people has been further explored in international fora such as the International Labour Organization (ILO) and the United Nations Commission on Human Rights, resulting in the ILO Convention No. 169 on Indigenous and Tribal Peoples and the Draft Declaration on the Rights of Indigenous Peoples.

[22] See also Article 10 CBD, which obliges the contracting parties to 'protect and encourage customary use of biological resources in accordance with traditional cultural practices that are compatible with conservation or sustainable use requirements'.

[23] See the 1996 document 'Knowledge, Innovations and Practices of Indigenous and Local Communities: Implementation of Article 8j', UNEP/CBD/ COP/3/19, p. 14, No. 60.

conservation and sustainable use of biological diversity (see Maffi, 2001a, pp. 8–12).

Traditional knowledge is the result of learning processes of specific people found in a specific environment over a long period of time. Mülhäusler (2001, p. 143) points out that an important aspect to any type of management is that one can manage only what one knows, and as a corollary, one knows that for which one has a linguistic expression. Language mirrors the evolution of traditional knowledge and the importance that specific biological resources have for survival: languages are the repositories of past experience. Hunn (2001, p. 125), for instance, describes the elaborated classification of local plant varieties in the Mixtepec Zapotec language of indigenous inhabitants of Oaxaca, Mexico, which is considerably more refined and closer to the biological facts, containing far more nuances of meaning, than the local Spanish vernacular. He concludes that traditional languages are valuable inventories of traditional environmental knowledge, which simply cannot always be translated into another, more commonly understood language, and, as such, complement what Western science has learned of global biodiversity (Hunn, 2001, p. 130).

Holders. Traditional knowledge (TK) is frequently assumed to be communally shared, the communities being characterized by a specific conservation ethic developed from living in a particular ecosystem. This conservation ethic is based on the (forced or voluntary, see Chapter 8) local-scale self-sufficiency, the reliance on locally available natural resources, and the ethical principle of family bonding and cross-generational communication, including links with ancestors and concern for the well-being of future generations. This leads to cooperation in resource usage: rights to lands, territories and resources tend to be considered as collective and inalienable rather than individual and alienable, and resource exploitation is socially controlled and guided by respect for nature and the needs of survival of the living and of future generations (see also Posey, 2001, p. 382).

However, TK must not forcedly be commonly owned, or held in custody for a community by individuals. There are models of individual ownership of traditionally generated knowledge, such as the 'grassroot-innovations' described by Gupta (1997, 2003), or the handling of their knowledge and wisdom by African shamans, as analysed by Nwokeabia (2001).

As to the communities being holders of TK, the (political) debate up to now had a focus on the knowledge of indigenous people and was strongly linked to questions of the rights of minorities over their resources. The knowledge of 'local (farming) communities' was made a subject of (bottom-up political) discussion at a far lesser degree, and rather in the context of the participatory rights as enshrined in the Farmers' Rights of the ITPGRFA, which in any case are rather general and declaratory in nature (see Chapters 2 and 6).

One farmers' document, the People's Plan of Action – the statement of non-governmental organizations (NGOs) at the occasion of the FAO's Fourth International Technical Conference on Plant Genetic Resources[24] – states that farmers' rights and rights of indigenous peoples are of a different nature and should be complementary and mutually supportive. It points out that the central objective of farmers' rights ought to be to ensure control of, and access to, agricultural biodiversity by local communities, so that they can continue to develop their farming systems sustainably.[25] The statement also points out that the farmers' innovations are closely linked to the ecosystem to which they are related.[26] The

[24] Leipzig, 17–23 June 1996, circulated by Third World Network.

[25] http://users.ox.ac.uk/~wgtrr/safehands.htm (accessed 5 November 2002). See also Posey (1996, p. 207).

[26] 'All agricultural biodiversity from time immemorial has been cultivated, developed, maintained and improved by farmers familiar with local soils, water cycles, climate, and other fundamental aspects of each particular ecosystem. The knowledge of farmers and indigenous peoples is human knowledge at its best, and forms an important aspect of the intellectual and biological wealth of the South.'

intellectual contribution of farmers to the diversity of crop varieties and animal breeds is emphasized in a great variety of documents: for instance, in the 'Farmers' Rights Charter', a document drafted by Indian Farmers' Unions.[27] It is stated that 'farmers are the original breeders *and through their breeding science and technology* have produced the rich diversity of crop varieties and animal breeds as a gift to the world'. Therefore, farmers ought to have the right to 'participate fully in any benefits derived from the improved use of these genetic resources' and, of course, in the ITPGRFA (Preamble para. 7 and Article 9.1). Some documents stress that farmers' innovations take place collectively and cumulatively, and that therefore farmers' rights, arising from their role as conservators and breeders, are community rights.

The difference between indigenous people and local communities, in any case, is rather fluid. Local communities might be understood as farming communities in subsistence farming systems, which do not correspond to the definition of 'indigenous' as cited above, or do not wish to use this line of argument to their end. Therefore, the difference between the knowledge held by indigenous people and the knowledge held by farming communities, and their convergence, are not easy to identify.

Therefore, in view of the characterization of TK, in spite of certain differences, common features of both types of knowledge can be identified: the information is frequently not perceived as the creation of individuals, but is understood as the achievement of a specific community, having evolved – and continuing to evolve – in cumulative steps over many generations. It is managed and exchanged according to the customs or customary laws of the community. A close interaction exists between TK of any one kind and the surrounding ecosystem. TK plays a key role in the preservation and sustainable use of the diversity of wild and domesticated plant varieties and animal species. In turn, it depends on the surrounding environment in which it has been created. It is a cross-cutting issue that is embedded in the culture of a people. Thus its existence is dependent on, and determined by, the maintenance of this culture from one generation to the next.

TYPES OF TRADITIONAL KNOWLEDGE AND THE QUESTIONS OF PRIVATE AND PUBLIC DOMAIN

The preceding discussion indicates that different types of TK exist. Differences can exist in view of, first, the types of knowledge, and second, the degree of publicity it has within a community or society. Third, it can be differentiated taking account of the different holders or its geographical distribution. Finally, a distinction can be made in view of the method of keeping it as a tradition (for details see Chapter 4).

It is frequently assumed that TK is communally shared, such communities being characterized by a strong ethos of the need to share valuable information. Ownership and property rights are considered to be alien concepts in indigenous and traditional societies. This may lead to the conclusion by 'outsiders' that TK is the property of nobody, in analogy with the theory of 'no-man's land' of the territories of native peoples (see Dutfield, 2000). Dutfield concludes that 'it seems that territories, ecosystems, plant varieties (whether domesticated or not) and traditional knowledge are treated as if they are *res nullius* before their "discovery" by explorers, scientists, governments, corporations and conservation organisations'.

This perception is based on the fact that intellectual property law based on industrial invention deems any information which does not fall within the scope of existing intellectual property law, including information which is no longer protected by it as the period of protection has expired, to belong to the *public domain*, meaning that it is able to be freely accessed and exploited by all.

However, intellectual property law is

[27] Draft Charter by Indian Farmers Unions, circulated by Third World Network, February 1996. Citation from Posey (1996).

not the only basis for determining property rights: customary law may recognize other forms of ownership (Fishman, 2000, pp. 1–3; Correa, 2001, p. 3, citing Dutfield, 2000, p. 285). According to the Four Directions Council, 'indigenous peoples possess their own locally-specific systems of jurisprudence with respect to the classification of different types of knowledge, proper procedures for acquiring and sharing knowledge, and the rights and responsibilities which attach to possessing knowledge' (cited in Dutfield, 2000, p. 4). And as to the exchange of crop varieties in traditional farming societies, Castillo specifies that this exchange is by no means a simple give and take process, but takes place in a reciprocal relationship. He speaks of a 'culture of reciprocity', the basis of which is 'not money but trust, mutuality of benefit and social equivalence in the value of the item or service being exchanged' (Castillo, 1997, p. 26).

The WIPO Fact-finding Mission on Intellectual Property and Traditional Knowledge (WIPO, 2001, p. 57) found that major differences exist between the property law of modern and traditional communities, especially whether the rights are inalienable or not. Being locally specific, the traditional systems display a far greater diversity than those that are available to protect intellectual property generated in the context of industrial R&D (Dutfield, 2000, p. 4; for examples see Tirso in Brush, 2000; Dutfield, 2000, p. 4; Nwokeabia, 2001). Thus, the point is not that the holders of TK do not recognize concepts of intellectual property, but rather, that the formal intellectual property (IP) system is a type of IP system with which they are not familiar, and which does not answer the protective needs of their knowledge (see also WIPO, 2001, p. 221).

It can be concluded that even if certain TK is not subject to the protection of formal intellectual property rights, it is not necessarily public knowledge, for it may be protected by other means, such as customary law, which recognizes other forms of ownership.

In this context it should be pointed out that TK might be possessed by individuals, or by some members of a group, or be available to everyone within that group. However, 'while modern arts and sciences often place individual accomplishment over community development, traditional knowledge systems celebrate the community's cooperative effort'. This means 'an isolated invention, such as the use of a plant to heal wounds, is an element which (generally) vests in the com- munity or collective' (WIPO, 2001, p. 212).

THE LINK BETWEEN TRADITIONAL KNOWLEDGE AND BIODIVERSITY CONSERVATION AND SUSTAINABLE USE

As mentioned above, TK is recognized as being of great value for the conservation and sustainable use of biological diversity (Sheldon and Balick, 1995, p. 45; Woodliffe, 1996, pp. 257–259). This is seen in the context of the special characteristics of TK. The combination of accumulative knowledge and potential for innovation and adaptation of traditional systems – in combination with the equivalent knowledge base and innovative capacity of modern or scientific systems – is capable of offering, albeit unquantifiable, but probably substantial, opportunities for identifying improved techniques for conservation and sustainable use of biological diversity (UNEP/CBD/ TKBD/1/2 1997, RNr 86/88).[28]

TK is especially important in farming systems because of the strong causal connection between domesticated species and human activity. This kind of TK encompasses a profound knowledge of the ecosys-

[28] Compare in this context the enlightening example found in *New Scientist* of 17 October 1998, 18/19. Researchers, after having gathered information on whale populations in the eastern Canadian Arctic by interviewing the Inuit population, came to the conclusion that there are as many as 350 bowhead whales instead of the few tens that scientists had believed. It is pointed out that 'scientists constrained by tight budgets and conventional techniques gain only a fragmentary understanding of wildlife and its environment. However the Inuit, whose ancestors have been living in close contact with the land for centuries, have accumulated a wealth of detailed knowledge based on year-round observation over a wide area' (Edwards, 1988). See also McDonald *et al.* (1997).

tem conditions and breeding techniques adapted to local conditions, including selection for tolerance or resistance to biotic and abiotic stresses. It further comprises information on the useful properties contained in the genotypes that have been selected and conserved. Unlike most natural biodiversity, PGRFA require continuous active human management. Human intervention is needed for the maintenance of agroecosystems, the maintenance of the diversity of PGRFA and their further evolution and adaptation to novel circumstances. TK plays an important role in all of these processes.[29]

Value of traditional knowledge

TK related to plant biological resources is valuable for many reasons. It forms an integral part of the conservation and further development of the world's biological resources, but it is equally important for the fulfilment of other needs. This includes the provision of a livelihood to indigenous and local communities. TK has both a short-term, or direct, use value, for the present survival of a great number of people living in subsistence economies who depend on biodiversity for their day-to-day survival, and a long-term value for the survival of humanity as a whole. TK is also a valuable source of information for the commercial use of the components of biodiversity.

Not all TK has direct commercial value. It may be that other values are held to be more important, resulting in the benefits derived from the knowledge being kept within the community. A plant may be used in spiritual rituals, promoting spiritual well-being of the members of the community who use it, so that its commercialization for this or other purposes may be out of the question.

In particular, the indigenous people insist on a holistic approach to biological diversity and TK. They maintain that issues about biological diversity cannot be separated from their lands and territories, spiritual, social and cultural life, or from their identities. Accordingly, they reiterate that their lands and territories and the diversity contained within them are imbued by a variety of values: social, cultural, spiritual and economic.[30] Given this background, indigenous people have expressed their concern about the economic orientation of the CBD.[31] They point out that their

> ... knowledge is not merely a commodity to be traded like any other in the market place. Our knowledge of biodiversity is indivisible from our identities and our laws, institutions, value systems and cosmovisions as Indigenous Peoples. For generations, our peoples have been and continue to be custodians of nature upon which we all depend. We are therefore fully committed to the first two objectives of the Convention, that is, the conservation and sustainable use of biodiversity. However, any discussion of the third objective, that of access and benefit sharing, must recognise our fundamental rights to control our own knowledge, our right to free, prior informed consent as peoples, and our collective land and territorial security.[32]

This viewpoint is to be kept in mind in the following discussion of the economic aspects of TK.

THE SHORT-TERM OR DIRECT USE VALUE OF TRADITIONAL KNOWLEDGE

Medicine and pharmaceuticals. TK is indispensable in satisfying the daily needs of most of the world's population. The World Health Organization has estimated that 80% of the people in developing countries rely on traditional medicine for their primary health-care needs, and that about 85% of traditional medicine involves the use of

[29] For more information and the discussion on the importance of the influence of farmers' breeding skills in the evolution of landraces, see Chapter 4.

[30] See, for example, the Statement of International Indigenous Forum on Biodiversity at the Ad Hoc Open-Ended Working Group on Access and Benefit Sharing, CBD, 22–26 October 2001, Bonn, Germany, reprinted in TEBTEBBA Briefing Paper No. 8.

[31] This had been negotiated without the participation of indigenous peoples.

[32] Statement of the International Indigenous Forum on Biodiversity, see Note 31 above, RN 6.

plant extracts (Farnsworth, 1992, p. 91; ten Kate and Laird, 2000; WHO, 2002a,b). More than 4 billion people rely on plants as an indispensable source of medicine, and traditional medicine is often the only treatment available to people living in remote areas, or who are too poor to afford the more expensive remedies offered by modern medical and pharmaceutical companies (Correa, 2001, p. 2, citing Balasubramanian, 1997, p. iii; WHO, 2002a,b). About one-third of the world's population lacks regular access to affordable essential medicines, meaning that modern medicine is never likely to be a realistic treatment option for them. The growing demand for herbal medicines in the developed world[33] is a further indication of the great value of traditional medicine. The European market alone was calculated to be worth US$11.9 billion in 1999, and the world market about US$43 billion and rising (Correa, 2001, p. 3, citing WHO, 2000, p. vi; and Pranato, 2001, p. 2). Such figures give an indication of the value of traditional ecological knowledge, quite apart from the many non-economic benefits to society as a whole.

Food and agriculture. TK plays an important role in assuring food security for traditionally living peoples or farmers living in subsistence systems. According to Posey (2001, p. 383), most traditional people who inhabit forests or areas close to forests rely extensively upon hunted, collected or gathered foods and resources, a significant portion of which are influenced by humans to meet their needs. Farmers in subsistence systems maintain a high diversity of crop species in order to assure their annual harvests and thus to guarantee a minimal level of production and to prevent food shortage, even at the cost of a higher average productivity. Sophisticated crop rotation systems lead to sustainable and durable production and resource use.

The use and improvement by farmers of landraces in the production and development of food and agriculture remains essential for many people. In many instances seed production relies on the informal sector, being based on the knowledge about seed selection and storage (often held by women) and the exchange of seeds between farmers and farms.

Whereas the value of semi-domesticated or non-domesticated resources 'has systematically been undervalued and overlooked by science' (Posey, 2001, p. 383), the value of the diversity of landraces or farmers' varieties is receiving increasing acknowledgement. Modern agricultural practices depend on crop species that promote productivity and resistance to disease, which can only be maintained with the continuous input of new germplasm. The diversity of landraces and the associated information of their specific qualities contribute invaluable information to formal breeding processes.

THE LONG-TERM VALUE OF TRADITIONAL KNOWLEDGE

The preceding discussion shows that TK is an important asset in the maintenance of biological diversity. Biological diversity of domesticated and wild animals and plants is important for humanity and its long-term survival as a whole. Its conservation is therefore of global interest. The maintenance and development of knowledge of the qualities of biological (plant and animal) resources secures the option of new uses in the future. Local and indigenous people, in using and conserving wild resources in a sustainable way, by, for instance, actively nurturing wild plants, thus contribute to 'global biodiversity services'. The same is true for agrobiodiversity that is maintained by local subsistence farming systems (see above).

[33] The widespread use of traditional medicine and the tremendous growth in the market for herbal products has also led to the threatened extinction of certain biological resources – even when dealing with 'natural products', the goals of conservation and sustainable use must never be forgotten.

The loss of traditional knowledge

In addition to the threat to biological diversity by extinction, the loss of TK is also increasing.[34] Given the close interrelationship between biodiversity and its associated knowledge, this phenomenon is not surprising: if a plant variety becomes extinct, then the entire body of knowledge about its properties becomes irrelevant. In turn, if the knowledge about the properties of a specific plant becomes lost, its usefulness to humankind is severely reduced. Short-term financial benefits may then prevail over the long-term value of the sustainable use of the plant, and thus lead to its loss.[35]

This process is well documented regarding the loss of knowledge integrated in landraces, and is being increasingly investigated in connection with the loss of the diversity of languages. Maffi (2001a) investigates the loss of TK in the context of the loss of cultural and linguistic diversity. She points out the close interdependence of linguistic, cultural and biological diversity, and submits that all these 'diversities' serve the same function of ensuring the perpetuation of life on Earth. However, analogous with biological diversity, cultural diversity and the diversity of languages are going through an extinction crisis of unprecedented scale and pace.

As mentioned above, it is broadly acknowledged that policies and instruments to conserve biodiversity and TK have to address the fundamental pressures that cause their loss.[36] In the debate on biodiversity loss, it is generally accepted that a close interrelationship exists between the maintenance of biodiversity and TK. However, the mechanisms leading to the loss of TK are not further analysed. As the focus in the present study is on instruments to support conservation and sustainable use of TK in the context of international trade, the questions to be asked are whether the loss of TK follows the same pattern as the loss of biological diversity, and if the problem of the maintenance of TK can be treated in a similar fashion.

Swanson's (1994) analysis of the decline of farmers' varieties and associated TK provides an example of this process and, at the same time, illustrates the relationship between the existence of the resource and of the related TK. Swanson treats the issue under the heading of the so-called 'insurance value'. Landraces characteristically have a relatively lower, but secure, output. This quality provides the subsistence farmers with insurance both against food scarcity and for survival. The abandonment of the old varieties can occur due to the following situations:

- The substitution of traditional varieties with new ones: in becoming accustomed to the new varieties and new production methods, farmers lose their knowledge and experience of traditional varieties. Their new knowledge serves as their insurance for food security and survival. The knowledge of the specific qualities of the old varieties becomes lost over a period of about two generations. With the loss of this knowledge the old variety loses its usefulness, and is thus abandoned.
- The substitution of other assets for biological ones: improved roads and better means of transportation from rural areas to cities facilitate the change from producing simple 'food crops' to high-yielding 'cash crops'. In addition, the centralization of populations in cities provides alternative insurance against food scarcity, through the opportunity to make use of surplus human labour as another source of income.

---

[34] Gupta (2003) estimates that this erosion of knowledge is sometimes a greater threat than erosion of the resource itself. He argues that information of potential value might easily be lost forever, because, as a rule, traditional knowledge is handed down orally.

[35] See Note 11 above.

[36] See, for example, OECD (1996, pp. 43–65); OECD (1999, pp. 20–21); and UNEP/CBD/ SBSTTA/7/11/Add. 1 ss. 4 and 5.

- The development of industry in urban centres further severs the dependency of farmers on income derived from farming. This means that, even in times of crop failure, markets exist for other products produced by farmers, which can also be translated into income.

The consequence is that individual motivation to maintain biodiversity as a form of insurance against hunger is, in fact, reduced. The ability to pursue the highest average yield provides another kind of insurance against hunger, and from the perspective of the individual farmer, this concept makes sense. The problem is that while the individual farmer continued to maintain biodiversity for his own food security needs, he or she, at the same time, performed a valuable task in the interest of the entire society: the conservation and maintenance of crop diversity as a 'global public good'. The abandonment of crop diversity as an individual form of insurance against food scarcity has, not surprisingly, led to global disinterest in a high level of crop diversity.

The consequence of these processes is described by, for example, Gupta (2003) and WIPO (2001, p. 214). They observe the lack of respect for traditional knowledge due to too little appreciation of its value, including its financial worth, and the failure to acquire a true understanding of TK, which is often overlooked within the modern approach to science. This causes the younger generation to be reluctant to learn the 'old ways', in the erroneous belief that only modern science can provide them with economic and social well-being, which WIPO calls the 'absence of willing heirs' (Alcorn, 1995; WIPO, 2001, p. 214).

There are two lines of argument to explain this process: the economic approach and the sociocultural approach.

THE ECONOMIC LINE OF ARGUMENT

The economic line of argument has two facets: the problem of open access and the theory of the global public good.

As with genetic resources, in dealing with TK of any kind one is essentially dealing with information: this is either information contained in the TK, or the genetic information contained in the seeds of plant varieties or in domesticated animals. The information found in both TK and plant genetic resources for food and agriculture (PGRFA) has the following attributes:

- TK, once it has been revealed, is accessible for everybody. Although it is true that TK is frequently protected by customary laws within the cultural framework in which it is found, these laws are not sufficient to regulate conditions of exchange and trade in the broader context of the growing international interdependence of cultures and the globalization of markets (see Girsberger, 1999; Dutfield, 2000).
- Genetic resources are self-propagating in nature and can be produced and traded as goods for consumption. The pertinent information, which results from the skill of generations of breeders, is contained in each seed. By the use of modern biotechnology, access to this information can be gained on the basis of only a small quantity of material: one seed is sufficient to reproduce the information.

Due to this 'open-access' situation of plant genetic resources and related TK, scientists within industrial innovation systems could use and patent the information for industrial utilization, often without further improvement or 'inventive step'. This has frequently happened without the consent of the holders of the resources and without compensation or the sharing of profits. This phenomenon has been dubbed the 'piracy' or 'biopiracy' of TK and resources.[37]

Intellectual property rights have been created to protect the informational value of new plant varieties and other inventions arising from the formal research and development (R&D) process. However, as a rule,

[37] Pat Mooney of the civil society organization, Rural Advancement Foundation International (RAFI; now known as the ETC Group), is said to have first coined the phrase 'biopiracy'.

these instruments presently cannot protect landraces and traditional knowledge, as they do not correspond to the criteria requested by, for instance, patents or plant breeders' rights.[38] From the legal point of view, once genetic diversity is distributed or traditional knowledge is revealed, it falls within the public domain and becomes accessible to everybody.

Since the direct-use value of the information contained in TK and PGRFA cannot be appropriated, the holders of TK may lack interest to further its use and so it is accordingly abandoned.

Further, TK clearly has the characteristic of a global public good. The intellectual input of generations of farmers and traditionally living communities has maintained and enhanced a good – being wild and domesticated biodiversity and the knowledge about it – which serves humankind as a whole. The value generated has no market value but is freely accessible, corresponding to a 'public good' similar to biodiversity. The entire society, both present and future, benefits from its use, but as it is openly accessible, no economic mechanisms for its provision exist.

THE SOCIOCULTURAL LINE OF ARGUMENT

Generally speaking, TK is neither written nor documented in any enduring way. Its existence depends on the cultural context in which it is embedded, and on the language by which it is transmitted.[39] Rather alarmingly, around 90% of the more than 6000 currently spoken languages, and the cultural expression found in them, face extinction within the next 100 years (Oviedo *et al.*, 2000, p. 6). One reason for this so-called 'language-shift' is the enormous pressure upon indigenous and local communities to assimilate into 'mainstream society', whereby they are required to abandon their native languages in favour of majority languages (Oviedo, 2000, p. 5).

Maffi argues that as a consequence of the loss of languages, much of the knowledge and wisdom of their speakers will also disappear, as much of it cannot be adequately translated into the newer, majority languages. This erosion of language and knowledge has been called the 'extinction of experience' (Nabhan and St Antoine, cited in Maffi, 2001b, p. 30). The extinction of language and the knowledge encompassed by it is just as permanent as the extinction of biological resources.

Maffi emphasizes that traditional ecological knowledge[40] is especially at risk of disappearing, due to its dependence on an intact ecosystem. Specifically, if local or indigenous people are removed from their land, absorbed by a market economy with little room for traditional subsistence practices, or forced to subsist in severely degraded ecosystems, valuable ecological knowledge may be lost. (Maffi, 2001a, p. 7; also compare WIPO, 2001, p. 214). She therefore strongly argues that 'the preservation of the world's linguistic diversity, and of the distinct forms of local knowledge that indigenous and minority languages encode, must be incorporated as an essential goal in

[38] In the case of patents, the innovation has to be new, involve an inventive step or, synonymously, to be non-obvious and be useful (Article 27.1 TRIPS). The protective criteria for plant breeders' rights require that a plant variety be new, distinct, homogeneous (uniform) and stable. TK frequently lacks the criteria of novelty, the necessary inventive step, or the level of creativity seen as prerequisites for protection by industrial intellectual property rights (IPR). Landraces, as developed by informal breeding systems of farming communities, fail to fulfil the criteria of stability and uniformity, and thus fail to meet the protective criteria of the plant breeders' rights. See Chapter 3 for more details.

[39] For the role and importance that language has in the conceptualization of the world, and for interpreting, understanding and changing it, the intrinsic and defining role of language in human biocultural diversity, see Maffi (1999, p. 21).

[40] Maffi defines traditional ecological knowledge as local peoples' classification, knowledge and use of the natural world, their ecological concepts and their resource management institutions and practices. She points out that its value for sustainability has been documented by social scientists over several decades (see sources in Maffi, 2001a, p. 6).

bioculturally-oriented diversity conservation programmes' (Maffi, 1999, p. 21).

Conclusions

The discussion of the pattern of loss between biodiversity of traditional plant genetic resources for food and agriculture (PGRFA) and of TK reveals both similarities and some important differences:

1. A close relationship exists between the loss of biodiversity and the loss of the traditional knowledge related to it. The loss of biodiversity itself results in a downward spiral of decreasing economic opportunities and declining economic growth as biological resources become more and more degraded (IUCN, 2001).
2. Both the loss of knowledge and the loss of biodiversity are irreversible: once the information is lost, it cannot be retrieved. Interestingly enough, it seems that the loss of knowledge occurs at an even faster rate than the loss of diversity – the AIDS pandemic in Africa, which is eradicating a great part of an entire generation of knowledge holders, and thus interrupts the transfer chain, may be a case in point.
3. Both the loss of biodiversity and the loss of TK are due to a vast array of complex, intertwined causes.
4. Biodiversity in a broad sense, TK and traditional PGRFA all have direct-use value (such as medicinal knowledge and knowledge of the qualities of a specific grain variety), and optional value as pools from which future generations can draw for adaptation to changing circumstances. Nevertheless, the specific information contained in TK and in traditional PGRFA presently does not have a market value. Further, the direct-use value, as a rule, does not accrue to individuals but to a more or less specified society.
5. The resulting 'profitability gap' for the individual user of biological resources can be defined as the difference between the profits made by its sustainable and unsustainable utilization. In respect of TK, this 'profitability gap' has to be defined in a different way. It is true that the maintenance or abandonment of TK sometimes depends on a conscious decision being made by its holder – sophisticated industrially produced medicines may be preferred to traditional ethnobotanical preparations; high-yielding varieties may be planted instead of traditional landraces; and so on. But, in many cases, the loss of TK is due to indirect processes and influences.
6. The reduction of negative externalities causing the loss of biodiversity and traditional knowledge by economic instruments proves to be complicated. Here there is a complexity and mix of influences: involved are many stakeholders who are responsible for a certain level of damage; the damaging process spans a long period of time; the transaction costs are high (see OECD, 1996, p. 52). The integration of the positive externalities seems to be more promising.

Accordingly, the creation of measures or instruments to support the maintenance of TK in the context of international trade is only likely to be of value where a market exists, or can be created, for this type of knowledge or for the resulting product. In particular, traditional ecological knowledge, in being closely linked to ecosystems, is not marketable in the context of international trade.

What is required is commercial interest for the creation of a market. This could be envisaged for the trade of traditional medicinal knowledge or of traditional PGRFA. However, it must be kept in mind that this forms only a small part of existing traditional knowledge. Indirect support for TK and agrobiodiversity, or biodiversity generally, can be created by economic or market incentives which promote the sustainable utilization and the traditional processing of natural resources (such as labelling and certification devices).

None the less, indirect measures typically do not provide sufficient incentives for conservation, particularly if the commercial value of biodiversity and TK might not be as high as supposed (Simpson, cited in UNEP/CBD/SBSTTA/7/11).

Given this, and the realization that the public good value of TK and traditional

PGRFA cannot adequately be covered by market mechanisms, it is necessary to create other means for their support.

The loss of TK follows an even more complicated pattern than the loss of biological diversity. Consequently, Posey and others insist upon a holistic approach to support traditional knowledge. He maintains that the role of external support has primarily become that of creating more space for *indigenous control*, based on certain rights, including self-determination, territorial and land rights, the right to development, collective rights, community empowerment, prior informed consent and privacy, control of access, religious rights and religious freedom, and the right to a unique language (Posey, 2001, p. 387).

The consideration of ways to maintain TK cannot be limited to the economic aspects alone, but must also promote respect, equity and mutual trust. Neither is it appropriate to insist that the holders of TK stick to their present way of life, aiming at creating 'museums of diversities'. This would imply a denial of the right of those people to development and to the furthering of their well-being. The measures to promote TK should be designed in a participatory way, leaving the basic decision as to its use to the holders of the TK themselves.

1.1.4 Conservation and sustainable use of biodiversity and traditional knowledge: the interface with trade[41]

Introduction

The goal of the research papers in this book is to examine measures to support maintenance and sustainable use of biodiversity and related TK in the context of international trade. The core question is which mechanisms, instruments and institutions could be created to make international trade (more) supportive of biodiversity including crop biodiversity and TK?

It has been pointed out above that maintenance and evolution of TK is closely linked to the characteristics of the ecosystems it is related to. In looking at the interface between TK and trade, this close interaction between the generation and maintenance of knowledge and the intact condition of the natural environment is to be taken into account. Since biological resources are primarily traded as goods for consumption, the system of trade in raw materials has an important influence on biological diversity, and thus indirectly also on the maintenance of TK.

These insights link the goal of maintaining biodiversity for present and future uses closely to the livelihood of people living in marginalized subsistence systems in the poor countries of the South, which cannot afford to put the interests of conserving biodiversity over their interests in daily survival and well-being.

Accordingly, a World Conservation Union (IUCN) study concludes that economic factors lie at the heart of biodiversity loss. Almost all forms of economic activities have potential to cause biodiversity loss. They impact on biological resources, ecosystems and their diversity, e.g. through using up non-renewable resources, by converting resources and habitats to other uses and by adding waste and effluent to the environment (IUCN, 2001).

The liberal economic theory holds that free trade is generally desirable in that it theoretically maximizes economic efficiency, as competitive markets allow the most efficient allocation for privately appropriable goods. This is true if world market prices reflect the true values of biological resources, i.e. when biodiversity costs and benefits are fully internalized and the prices reflect the scarcity of the resources.

However, if this is not the case, trade liberalization and trade regimes may have negative impacts on biodiversity. In this case trade can magnify the inefficient allocation of resources and put more pressure on valuable resources. So, trade can be both detrimental and advantageous to biodiversity (Janssen, 1997; FAO, 1999).

Biodiversity differs in important fea-

[41] Author: Susette Biber-Klemm.

tures from other, more conventional and simpler resource management problems, such as environmental pollution for example, where the relation between impacts and causes is relatively easy to understand, and thus measures are, at least theoretically, easy to design. In the context of trade, the following features of biodiversity are of specific significance and concern (Young and Gunningham, 1997; OECD, 1999, pp. 25–27).

First, biodiversity is a concept that describes the number and variety of living organisms on the planet, defined in terms of genes, species and ecosystems, and refers to the intrinsic indivisibility of ecological phenomena. Thus, viewed from a trade point of view, 'biodiversity' as such is not a tradable good.

Secondly, there is a great amount of uncertainty as to the amount of existing diversity and its value and function within the biosphere. Furthermore, adverse impacts on biodiversity accumulate and can create losses in the long run. This means that political decisions as to its conservation are 'decisions under uncertainty', encompassing all the problems that are linked to the implementation of a precautionary approach to resolve environmental problems (OECD, 1999, p. 21; Barbier in OECD, 1997, pp. 128–131).

Thirdly, the complexity of biodiversity brings about a mix of pressures, arising from different economic sectors, being the result of different underlying causes. The OECD identifies as the three main (and interlinked) underlying driving forces: (i) patterns of consumption and production; (ii) population growth; and (iii) economic failure (1996, p. 48). It is rarely possible to identify a single cause for the decline of biodiversity – the underlying causes may not even be clearly identifiable. The loss of biodiversity is a cross-cutting issue, driven by complex and diffuse causes that involve many different sectors and forms of economic activity.

Fourthly, from an economic point of view, the elements of biodiversity can be both of private and public value. An additional problem in this context consists of the fact that often not all the private use benefits can be captured due to a situation of open access.

Regarding TK, Maffi (2001a) underlines that, within the general phenomenon of erosion of the linguistic diversity, traditional ecological knowledge is at an especially high risk of disappearing, due to its place-specific and subsistence-related nature. As reasons for this 'extinction of experience' Maffi enumerates the following: (i) people being removed from their land; (ii) subsisting in severely degraded ecosystems; or (iii) being absorbed into the market economy with little room for traditional subsistence practices (Maffi, 2001a, p. 7). Generally speaking, it is submitted that current global socio-economic processes also bring about a culturally homogenized trend that contributes to the loss of TK.

From the *economic* point of view, different mechanisms lead to the loss of biodiversity and the consequent loss of TK, or vice versa: local and global market failure; poorly defined or missing property rights; uncertainty and information failure; intervention or institutional failure; government integration failure; and international trade are all expressed as causes (OECD, 1996, p. 51). The conclusion drawn is that, as the problem is a cross-cutting issue, a mix of instruments is necessary for its solution (see OECD, 1996, p. 83 and criteria p. 92, 1999, p. 36). In the trade context, so-called economic failure as an underlying cause of biodiversity loss, and the dual public and private character of biodiversity, play a pre-eminent role.

Trade liberalization favours the highest possible yield at minimal cost. This promotes an intensification of the utilization of the resources, as is the case, for instance, in timber production or in so-called blue-print farming.[42] In leading to this intensification, the expansion of global markets and recent

[42] Industrial agriculture focuses on maximizing commercially important yields and productivity through the use of monoculture systems and uniform technologies, including high-yielding seeds, agrochemicals, irrigation, mechanized equipment and large infrastructure developments (see FAO, 1999, p. 20).

patterns of trade liberalization have a double effect on traditionally sustained ecosystems and their biological diversity: on the one hand, agricultural (crop) biodiversity is homogenized by standardizing food production and consumption (FAO, 1999). On the other hand, the conversion and degradation of habitats is promoted by intensifying the production (e.g. forests) or through abandonment of their utilization.

The CBD follow-up process has undertaken to analyse the impact of trade liberalization on the conservation and sustainable use of agricultural biological diversity. In the preliminary findings it is concluded that, given the high complexity and dynamics of the relationship between international agricultural trade, trade liberalization and biological diversity, it is difficult to put forward solid data, but that an analysis is at present forced to rely largely on stylized observations[43] (UNEP/CBD/COP/6/INF/2, No. 89). Further, it is pointed out that the international agricultural markets have undergone a profound restructuring that is independent of trade policy reform effects. The documents name as examples: first, changes in product demands, i.e. increasing demand for meat and processed foods caused by income growth in developing countries; secondly, changes in production methods, i.e. a change to larger-scale, modern or industrial-type agricultural and livestock production which is typically associated with high rates of specialization; and thirdly, the decrease in transportation costs, which leads to extension of transportation infrastructure and the establishment of new production locations (Nos 14–17). It is clear that, even if not being their primary cause, trade liberalization may have a reinforcing impact on these processes.

Given this, it is concluded that, particularly, changes in the location of agricultural production, between regions in general, and between developed (contracting) and developing (expanding) countries, will have an important impact on biological diversity, as they are bound to lead to changes in land use and to intensification of production (Nos 94–102).

With regard to trade and market mechanisms, the creation of incentives stays in the foreground as a means of supporting sustainable use and the maintenance of biological diversity. As incentives 'tend to use market mechanisms to influence decision making', they offer opportunities to integrate the respective goals into the pertinent law-making processes (OECD, 1996, p. 70, 1999, p. 73ff.).

Some background information on incentives and financial mechanisms will be elaborated upon here as a basis for the discussions in the following chapters.

Economic failure as reason for the loss of biodiversity and traditional knowledge

Economists argue that economic failure is the reason behind the loss of the diversity of biological resources and TK. Economic failure means that the interplay of market forces does not secure the economically correct balance of utilization and conservation and maintenance of a resource (OECD, 1996, p. 51). Two main reasons are given for this phenomenon. First, the value of biodiversity is not fully reflected in economic decision-making systems. This means that the cost for engaging in activities that lead to the loss of diversity is low. This leads to an unsustainable utilization of the resource. Secondly, for many outputs and values of diversity, no market exists.

The OECD (1996) examines different elements that could lead to economic failure and thus to the loss of biodiversity. The following are of interest in our context: (i) local and global market failure; (ii) missing or poorly defined property rights; and (iii) international trade[44] (also see OECD, 1999, p. 74).

[43] See the synthesis of the frameworks to assess the impact of trade liberalization on agricultural biological diversity: UNEP/CBD/COP/7/INF/15.

[44] Additionally: temporal failure; uncertainty and information failure; intervention, or institutional, failure; government integration failure (OECD, 1996, p. 51 and pp. 51–65).

LOCAL AND GLOBAL MARKET FAILURE

According to economic reasoning, *missing or incomplete local markets* are one reason for the loss of biodiversity. This means, on the one hand, that the biodiversity costs, i.e. the negative externalities, which arise through the direct or indirect utilization of the biological resource, are not accounted for by the users of the resource.

On the other hand, neither are the positive externalities produced by biodiversity integrated into the market price. Prices for natural resources only reflect the cost of labour and capital, but do not compensate the additional goods and services that are rendered to society as a whole through this maintenance of diversity.

This is even more true for TK, for which generally no market exists, or for which the returns are minimal, such as traditional medicinal knowledge; and for traditional PGRFA, which are traded or bartered as food for consumption in the absence of a formal seed market.

MISSING PROPERTY RIGHTS

One reason for the market failure of biodiversity and TK lies in its dual character, which is discussed under the heading of 'public' and 'private' goods in economic theory. This means that biodiversity and TK create benefits that can be appropriated by private owners and single economic actors, such as individuals or companies, as well as benefits that accrue to society as a whole. The private benefits, as a rule, result from goods provided by biodiversity which can be traded in a market. They correspond to the direct-use values described above.

However, almost all elements of biodiversity, at the same time, contribute to public existence values, being the functioning of the whole of biological diversity and life-supporting systems, and to the public goods and services of biodiversity, such as its optional value or ecosystem resilience. In the majority of cases, these values do not accrue to an individual economic actor but to society at large, at the local, national, regional or even global level (OECD, 1999, p. 32; UNEP/CBD/SBSTTA/7/11). As a rule, private owners will pay attention to those values that can be privately appropriated. The realization of public good values, for instance the decision for sustainable logging or for maintaining landraces, creates costs in term of private losses. The OECD uses the term 'profitability gap' to describe this situation. It is the typical situation of a public good.

With regard to TK and traditional PGRFA, at present, hardly any market value exists for the specific information they contain. This leads to a 'profitability gap' for the individual user of the biological resource, which can be defined as the difference in the profit between its sustainable and unsustainable utilization.

In combining these conclusions with the outcome that the public value of biodiversity TK and traditional PGRFA cannot be covered by market mechanisms, it becomes clear that other mechanisms must be created. Economic measures for the compensation of this profitability gap are needed, and incentives for the maintenance of biological diversity and the integration of the use of traditional knowledge and traditional PGRFA in the development process must be created.

Supporting conservation and sustainable use of biodiversity and traditional knowledge: financial aspects

Sustainably using biological resources frequently implies renouncing a higher short-term profit and even long-term economic development for a future global benefit. Justice and equity demand that this service, which is mostly rendered by the biodiversity-rich, but economically poor, parts of the world, should be rewarded and compensated.

The specific characteristics of biodiversity and TK entail that the (financial) resources available for their maintenance ought to answer specific criteria. The vast timelines of the processes, in particular, make a steady, long-term and predictable flow of means necessary.[45]

[45] See also doc. UNEP/CBD/COP/3/7.

The CBD and the ITPGRFA have both created instruments and mechanisms to compensate both the direct use of biological resources and associated traditional knowledge, and the long-term public interest in their maintenance. In the framework of the CBD, the access and benefit-sharing regime (Article 15, see Chapters 2 and 6) is meant to compensate the direct-use value. In addition, the developed country Contracting Parties are obliged to provide 'new and additional financial resources to meet the agreed full incremental costs' caused by the fulfilment of the obligations of the CBD (Article 20). The ITPGRFA has created the Multilateral System of Access and Benefit-Sharing (ABS) (Articles 10–13; see Chapters 2 and 6) and provides for an additional funding strategy (Article 18).

The idea underlying the ABS concept is to finance sustainable use of genetic resources through trade mechanisms. The basic idea is that goods and services that flow from biological diversity can provide incentives for its conservation and sustainable use (UNEP/CBD/SBSTTA/5/13 No. 5).

However, in evaluating the ABS system in regard to traditional PGRFA and TK, the following points must be considered:

- Incentives in the context of international trade reach only a small segment of the resources, namely where a market for this type of information exists. In particular, traditional ecological knowledge as such, being closely linked to the ecosystems, is not marketable in an (international) trade context.
- Direct influences are possible where a commercial interest and a market exist or can be created. This could be envisaged for trade in medicinal traditional knowledge, or in traditional PGRFA. However, it must be kept in mind that this only forms a small part of the existing knowledge.
- An indirect support for TK and agrobiodiversity, or biodiversity in general, can be created by economic or market incentives which promote, for instance, the sustainable utilization and/or the traditional processing of natural resources (e.g. labelling and certification devices).
- The commercial value of biodiversity and TK might not be as high as envisaged (Simpson, cited in UNEP/CBD/SBSTTA/7/11 p. 25).
- In the CBD system on ABS, the contracting parties are responsible for determining access to genetic resources and for defining the basic conditions in national legislation (Article 15). Access to TK and sharing of benefits arising from its utilization is subject to national legislation too. Therefore, there is a risk that benefits may not accrue to the holders of TK but rather be attributed to the State, the holders of TK frequently belonging to marginalized groups. Thus the benefits are not generated at the level where the information is generated (see also Chapter 6).

In sum, it may be concluded that direct influences are possible where a commercial interest and a market exist or can be created. This could be envisaged for trade in traditional medicinal knowledge or in traditional PGRFA. However, it must be kept in mind that this forms only a small part of the existing knowledge. Therefore it can be concluded that direct incentives typically do not provide sufficient incentives for conservation, especially since the commercial value of biodiversity and TK might not be as high as hoped for (Simpson, cited in UNEP/CBD/SBSTTA/7/11 p. 25).

With regard to public funding, the latest decisions of the CBD COP on financial resources speak clearly about this. In Decision VI/16 (2002) the Conference of the Parties takes note of the importance of additional financial resources and emphasizes the need for a substantial increase of international financial support towards the implementation of the Convention.

Additional measures that aim at increasing the returns on sustainable use of the components are necessary and of specific importance.

The creation of incentives: the Convention on Biodiversity discussion

INTRODUCTION

Article 11 of the CBD provides for the creation of 'economically and socially' sound measures that act as incentives for the conservation and sustainable use of biological diversity. In the CBD follow-up process the implementation of this article was intensely discussed. In its Decision III/18, the Conference of the Parties underlined the importance of incentive measures for the realization of the objectives of the Convention. In Decision V/15, the Conference of the Parties decided to establish a programme of work intended to promote the development and implementation of incentive measures.

The deliberations were supported by research within the OECD. The following thoughts are based on the report of the Workshop on Incentive Measures for the Conservation and Sustainable use of Biological Diversity, 17–19 September 2001, Montreal, Canada (UNEP/CBD/SBSTTA/7/11/Add.1) which integrates the former documents and results of research.

In analysing this CBD process, it must be considered that the elements elaborated refer to incentives for biodiversity protection in a general, practical sense. Their relevance in the broader context of world trade has to be evaluated carefully.

DEFINITION AND GENERAL CONCEPTS

The CBD Secretariat defined incentives as 'a specific inducement designed and implemented to influence government bodies, business, non-governmental organizations, or local people to conserve biological diversity or to use its components in a sustainable manner' (UNEP/CBD COP/3/24). Their purpose is 'to change institutional and individual behaviour in order to achieve the objectives of the Convention on Biological Diversity', in our case the conservation and sustainable use of PGRFA and the maintenance of the associated knowledge.

It differentiates between four types of incentives: (i) positive incentives, such as incentive payments, (beneficial) taxation and fiscal measures; (ii) disincentives, for instance environmental liability, user fees; (iii) indirect incentives, such as biodiversity prospecting deals and eco-labelling; and (iv) removal of perverse incentives, such as reduction of agricultural support harmful to biodiversity.

The Conference of the Parties sets out from a broad notion of 'incentive measures'. It explicitly includes not only economic but also social and legal measures (Decision V/15). Hence, according to the CBD documents, a programme of incentive measures typically ought to include three components: (i) formal policy instruments, i.e. economic and legal instruments, regulations and public investment; (ii) social and institutional measures, i.e. information provision, capacity-building and stakeholder participation; and (iii) compliance mechanisms, i.e. measures to encourage both socially enforced compliance and legal enforcement (UNEP/CBD/COP/5/15, pp. 2–3 No. 6–9). The concept also encompasses a carefully designed framework and institutional environment, the so-called 'institutional incentives'.

According to doc. UNEP/CBD/COP/3/24, pp. 4–7, this institutional framework comprises the following interactive components: (i) the formal constraints, which are the written instruments that provide a legally enforceable framework for the economic and social activities of a society, i.e. laws, government policies and property rights (No. 19); (ii) the social conventions, i.e. the unwritten rules that govern everyday human behaviour in economic and social exchange (No. 22); and (iii) compliance, i.e. the degree to which individuals and organizations respect and adhere to the existing constraints, both formal and social (No. 25). The institutional incentives are the product of a complex interaction between the full range of these elements.

Hence, within the CBD process a holistic concept of incentives has been developed. It significantly broadens the conventional terminology which defines an incentive measure as 'an economic or legal instrument designed to encourage beneficial activities (a positive incentive) or to discourage harmful activities (a negative incentive)' (UNEP/CBD/COP/5/15 p. 2 No. 5).

The following conceptual guidelines are considered to be relevant for the design of programmes on economically, socially and ecologically sound incentive measures: (i) internalization, i.e. to incorporate external costs and benefits into the decisions of producers and consumers; (ii) efficiency, i.e. expected social benefits are greater than or equal to the cost of implementation, administration and enforcement; and (iii) the precautionary principle, i.e. incentive measures 'err on the side of caution' when scientific knowledge is uncertain or where ecological consequences might be irreversible (UNEP/CBD/COP/5/15, pp. 3–4).

As shown above, TK is even more dependent on the sociocultural environment than biodiversity, for which this concept has been developed. Therefore, the institutional approach gains even more importance in this context. In sum, it may be concluded that incentives for the maintenance of TK and agrobiodiversity involve more than economic measures alone. However, in a trade context, possible instruments are limited to economic measures in a market context.

INCENTIVES TO MAINTAIN TRADITIONAL KNOWLEDGE IN A TRADE CONTEXT: TYPES AND CRITERIA

The CBD process proposes different types of incentives. In our context the positive, indirect and negative incentives could be relevant.[46]

Positive incentives are economic, legal or institutional measures designed to encourage beneficial activities, as for instance direct payments for organic farming, taxation and fiscal measures. Indirect incentives include trading mechanisms that create or improve upon markets, including *inter alia* property right mechanisms, biodiversity prospecting, or schemes for ecolabelling initiatives. Perverse incentives, finally, are often unanticipated side-effects of policies designed to attain other objectives, such as, for instance, the subsidizing of modern plant varieties or the prohibition of the use of landraces by governments in order to enhance food production (see Girsberger, 1999, pp. 44–45).

The following elements for the design of incentive measures[47] are considered to be of importance and applicable *mutatis mutandis* in our context (UNEP/CBD/SBSTTA/7/11 pp. 35–42).

- Mix of measures: it is assumed that no single measure is likely to cover all the different protective needs given by the different types and characteristics of information implied.
- The efficiency principle: (programmes on) incentive measures should be designed to ensure that expected benefits are greater than, or equal to, the cost of implementation, administration and enforcement.
- Underlying causes: incentives should be designed to address the underlying causes of biodiversity loss.
- Equity and distributional impacts: it is important to ensure that the definition of beneficiaries is equitable and inclusive. Incentive measures should aim to take into account both those who benefit and those who assume the cost of the measure.
- Capturing value for local and indigenous communities: the value of biodiversity/TK for subsistence, cultural or commercial purposes should be recognized and incentive measures designed so that they support the social and economic development needs of indigenous and local communities.
- Raising awareness of biodiversity (and TK) values: the identification and assessment of the value of biodiversity (and TK) can be an incentive in itself and supports the design of other incentive measures. Raising awareness among all stakeholders improves the chances for incentive measures of being successful.

[46] The fourth category, i.e. the disincentives or mechanisms that internalize the costs of use and/or damage of biological diversity, is less relevant in our context.

[47] See a more concise list of elements to be considered for the design of incentive measures for biodiversity conservation and sustainable use in UNEP/CBD/SBSTTA/7/11/Add.1, pp. 10–25.

Conclusions

Given this, it may be concluded that the task to create incentives is to be considered within a broad legal, economic and social context. Given the complexity of the causes for loss of biodiversity and TK, a broad range of instruments is necessary. The creation of legal instruments by itself is not enough. Information and capacity building are necessary prerequisites; the necessary institutions must be created; perhaps a market must be developed; and the cost-effectiveness of the endeavour evaluated.

Subsequent research therefore ought to answer the following questions:

- Can instruments and institutions be created in the context of international trade that form a basis for the creation of a mix of incentives at the national or even local level? What would be these instruments and institutions?
- What are the important elements for the design and implementation of incentives as identified by the CBD process that ought to be taken into consideration?

1.2 Goals of Instruments and Measures to Protect Traditional Knowledge and Plant Genetic Resources: Basic Principles and Ethical Reflections[48]

1.2.1 Introduction

In general

In the preceding section an economic and sociocultural line of argument was followed to explain the value and loss of biodiversity and traditional knowledge. It led to the conclusion that biological diversity is to be preserved in order to secure the survival of mankind.

The international agreements, in dealing with the loss of biodiversity, stipulate general goals indicating the rationale and framework for the proposed measures. Instruments or measures designed to contribute to the overreaching goal of the maintenance of biodiversity ought to be assessed in these terms.

Goals as laid down in international legal instruments concerning genetic resources and traditional knowledge

The relevant international instruments, the Convention on Biological Diversity (CBD), the FAO's International Treaty on Plant Genetic Resources for Food and Agriculture (International Treaty or ITPGRFA), and related documents specify the following goals:

- The *conservation* of biological diversity in general (Article 1 CBD) and of plant genetic resources for food and agriculture (PGRFA) in particular (Article 1.1 ITPGRFA), and the *maintenance*, *preservation* and *protection* of traditional knowledge (Article 8(j) CBD, Article 9.2(a) ITPGRFA) as the basic concern.
- The *sustainable use* of the components of biological diversity, i.e. of the diversity within species (genetic diversity), between species and of ecosystems (Articles 1 and 2 CBD), and of plant genetic resources for food and agriculture (Article 1.1 ITPGRFA).
- The *fair and equitable* sharing of the benefits resulting from the application of the traditional knowledge (Article 8(j) CBD), and arising from the use of PGRFA and of the genetic resources in general, respectively (Articles 1.1 and 9.2(b) ITPGRFA; Article 15.7 CBD), as well as of the results of research and development (Article 15.7 CBD).
- *Security* of PGRFA in order to assure *basic needs* for sufficient food, shelter, clothing, fuel, ornamental plants and medicinal products (Resolution 3 of the Nairobi Conference for the Adoption of the Agreed Text of the Conven-

48 Author: Susette Biber-Klemm. I thank Ch. aus der Au and G. Dutfield for their input into an earlier version of this section.

tion on Biological Diversity)[49]; and *food security* in particular (Article 1.1 ITPGRFA).[50]

- To *respect* traditional knowledge (Article 8(j) CBD) and to *recognize* the contribution of farmers to the preservation and improvement of PGRFA (ITPGRFA, recital para. 7, Article 9.1); to recognize the right of farmers to *participate* in making decisions on matters related to the conservation and sustainable use of plant genetic resources (Article 9.2(c) ITPGRFA), and to *involve* the holders of TK in its wider application (Article 8(j) CBD).

These goals encompass, on the one hand, principles that are to govern the reciprocal relationships generated by the implementation of the Agreements: equity, fairness and mutual acknowledgement are the corresponding keywords. On the other hand, in view of the use of natural resources, the basic principle is stated to be sustainability. One of the principal goals in this context is the promotion of food security.

These issues are interrelated and highly abstract. In order to be practicable for the evaluation of proposed instruments and mechanisms, they must be differentiated in the given context. In the following section the basic principles of equity, fairness and sustainability will be considered. The implications of the goals of food security, or rather of the fulfilment of basic needs – as an application of the above more abstract principles – will then be spelled out.

1.2.2 Equity and fairness

Introduction

The concept of equity has only recently appeared in the discourse on sustainable development and environment, in particular in connection with North–South issues. It seems to signal one of the leitmotifs, or visions, of a new global partnership. The term, as a rule, is not spelled out more precisely. It is widely used in different contexts and disciplines, and therefore has a multifaceted character.

Equity plays a key role in the context of the CBD and the ITPGRFA. It appears in the formula of 'fair and equitable' benefit sharing. The following analysis will take account of this wording, considering first the notions of equity and fairness as such, and then analysing the particulars of the combined formula.

Legal sources in the biodiversity context: conventions and soft law

In the CBD, equity is used in the formula of 'fair and equitable sharing of benefits'. Article 8(j) speaks of the benefits arising from the utilization of knowledge, innovations and practices of indigenous and local communities, and Article 15.7 regulates the sharing of both the results of research and development and the benefits arising from the commercial and other use of genetic resources. In Article 19.2 'fair and equitable' is used in the context of access to results and benefits of biotechnologies achieved on the basis of resources provided by the party.

In the definition of the objectives of the International Treaty, the formula of the fair and equitable sharing of benefits (arising from the use of PGRFA) is linked to the goals of sustainable agriculture and food security. In Article 9.2 the notion of equity alone is used to specify the rights of farmers to *participate* in sharing benefits, and Articles 10 and 13 integrate and differentiate the formula of 'fair and equitable benefit sharing' in view of the Multilateral System of Access and Benefit Sharing.[51]

The Rio Declaration[52] is one of the basic soft-law documents spelling out the elements of 'sustainable development'. It

[49] Reprinted in *Secretariat of the Convention on Biological Diversity* (2001) pp. 281–289.

[50] For the context of international trade in the framework of the WTO compare para. 6 of the Preamble of the Agreement on Agriculture. Commitments under the programme to reform agricultural trade are to be made having regard to food security as one of the so-called non-trade concerns (see Chapter 8).

[51] For details on the Multilateral System see Chapter 2.

[52] Declaration on Environment and Development, proclaimed at the occasion of the World Summit on Environment and Development. United Nations General Assembly doc. A/CONF.151/26 (Vol. I).

therefore sets out from a broader, more general understanding of the term. It evokes in its preamble the goal of 'establishing a new and *equitable* global partnership'. This partnership is to be achieved 'through the creation of new levels of cooperation among states, key sectors of societies and people'. Principle 3 declares, 'the right to development must be fulfilled so as to *equitably* meet developmental and environmental needs of present and future generations'.

If we compare and analyse the above instruments, we can state the following. In all instruments the term 'equity' is used to specify the desired relationship between the relevant players: in the CBD and the ITPGRFA the relationship between providers and purchasers of genetic resources and TK, and in the Rio Declaration the relationship between different stakeholders on a global level, including the relationship between present and future generations.

The CBD/ITPGRFA context departs from the equity–fairness combination. The formula is not defined in the definitions part of the CBD. According to Byström *et al.* (1999, p. 20) up to now neither has it been widely used in international law, and there are still few attempts being made to define or interpret the concept. With regard to interpreting the formula, the question is whether the words are synonymous with one another or rather are meant to indicate different sides of the same coin, the nuances of their meaning being of significance. In agreeing with Byström *et al.* (1999, p. 20), we part from the working hypothesis that the frequent use of both terms in the formula of 'fair and equitable' implies that a broad meaning is intended, encompassing the nuances of both terms.

Equity

DEFINITIONS AND GENERAL USAGE IN (SCIENTIFIC) DEBATES ON RESOURCE USE

Byström *et al.* (1999) understand the term 'equity' as an ethical concept. They analyse it in the context of the 'fair and equitable benefit sharing' formula.

For its definition they refer to *Webster's New Collegiate Dictionary*, where equity/equitable is spelled out as '[implying] a less rigorous standard than "just" and usually a fair and equal treatment of all concerned. ("Just" implies a precise following of a standard of what is right and proper. "Fair" implies an elimination of personal feelings, interests or prejudices so as to achieve a proper balance of conflicting needs, rights or demands.)' This definition of equity encompasses two elements: the element of equality (in the sense of equal treatment), which is further specified by the element of fairness. So, in this definition, fairness is included in the notion of equity: the fair and equitable benefit-sharing formula therefore would be a tautology.

In connection with conservation and sustainable use of PGRFA, the terms 'equity' and 'equitable' are used to describe a system that assures the balance between the compensation of small farmers for their accomplishment in conserving the diversity of crop varieties parallel to the benefits generated by formal breeders. Swaminathan (1997, p. 18) insists that the fact that 'primary conservers suffer from poverty, while those who utilise their knowledge and the products of their in situ on farm conservation traditions are prosperous' is strikingly inequitable. He maintains that a balance is to be found between the poverty of the rural conservers, on the one hand, and the breeders and genetic engineers working in commercial enterprises on the other (Swaminathan, 1997, p. 10).[53]

The Brundtland Report[54] (1987, p. 48) puts equity in the context of the sustainability of the use of natural resources. The argument is that many problems with sustainable use of natural resources are due to the unequal provision of access. In the context of food security, this would mean that

[53] Also see Brush (1994, pp. 10–11). For equity in connection with bioprospecting: Zerner and Kennedy, cited in Byström *et al.* (1999); Straughan in IPGRI (1997, p. 178).

[54] Report of the World Commission on Environment and Development, entitled *Our Common Future* (see Brundtland, G., 1987).

food security is not only a question of raising average productivity and incomes. What is also needed is equity in food production and distribution (p. 141).

Brown Weiss (1989) elaborates on the aspect of equity between generations. She points out that intergenerational equity is based on the inherent relationship that each generation has to other generations past and future, in using the common patrimony of natural and cultural resources of the planet. Accordingly, resources are understood as a global, common good (i.e. not a good belonging to specific entities such as states), from which common, global (planetary) rights and common, global (planetary) obligations are derived. In this relationship, each generation is both a custodian and a beneficiary of the global resources (pp. 21–27). Present generations have rights and duties as an 'intermediary' between past and future generations. The goal is to achieve and protect the welfare and well-being of every generation. This includes the maintenance of both the ecological processes sustaining life and the cultural resources important for the survival and well-being of the human species (p. 23). Brown Weiss points out three types of possibly inequitable situations: (i) the depletion of resources for future generations; (ii) the degradation of the quality of resources for future generations; and (iii) the access to use and benefits of the resources received from past generations. She concludes that as to the equitable use of resources: first, equity *between* generations implies equity *within* generations. Otherwise the intergenerational obligations arising from the principle of equity could be assigned to one portion of the present generation, and the rights to a different segment (p. 21). And secondly, that all members of the present generation are entitled to equitable access to the legacy received from former generations (p. 28). From this line of argument she deduces three basic principles of intergenerational equity: the conservation of options (of the natural and cultural resources base); the conservation of quality; and the conservation of access, providing equitable rights of access to the legacy of past generations and to conserve this access for future generations (p. 38).

The following can be concluded: 'equity' is not always used in a distinct and specific way. Sometimes it is employed synonymously with equality, or the inherence of fairness is implied. The term often seems to signal the necessity to balance the distribution of resources, income and wealth between the 'rich and poor'. This characterization is frequently, but not only, meant in a North–South context. However, equity is not only perceived as an issue in view of the distribution of (existing) wealth or income, but also as a question of access to opportunities to generate income and wealth.

THE TRADITIONAL CONCEPTS OF EQUITY

The argument of (global) equity is not a completely novel concept. It takes up elements of traditional theories of equity as existing, for example, in the context of law, economics and social psychology.

In *economics*, equity is often used as a corrective or complementary element to efficiency. According to economic theory, efficiency of the market is achieved by an ideal market situation with perfect competition. It leads to so-called pareto-optimal situations. Pareto-efficiency means that, given the correct conditions, a competitive market will guarantee the efficient allocation of resources among consumers and the efficient usage of factors of production. A situation is pareto-optimal if nobody's situation can be improved without worsening somebody else's situation (Byström *et al.*, 1999, p. 21). However, this optimum does not relate to questions of distribution. A situation can be pareto-efficient both if one person gets all and the other gets nothing, or if there is an equal distribution between the negotiating parties. Therefore an additional criterion is needed to indicate the morally better situation (Sen, 1985, cited in Byström *et al.*, 1999).[55] Equity thus introduces an element of distributive justice as a corrective measure of pure market efficiency.

---

[55] For further economic arguments and conclusions see Byström *et al.* (1999, pp. 21–25).

Equity theory in *social psychology* is a method for investigating the concepts of distributive justice held by a specific society. It identifies the principles of distributive justice evolving under different social conditions and explores its evaluation by the members of the societies. At the foundations of equity theory lies the insight that the balance between giving and receiving is the precondition for judging a relationship as 'equitable'. So, equity theory is about the determination of just distributions or of just allocations of positive and negative outcomes, and also about the appropriate mechanisms or procedures for achieving just distribution. It identifies the principles of distributive justice that evolve under differing social conditions and specifies when such principles are perceived as fair or just by the members of the social system (Cook and Messick, 1983, p. 1).

In law, the notion of equity has a long tradition. The *traditional* understanding of equity has its roots in the Aristotelian concept of *aequitas*, which defines equity (*aequitas*) as the corrective of law *in casu*, which allows the judge to except certain cases from the strict application of the law in order to achieve a just result.

In the *Anglo-Saxon legal system*, equity is a system of rules that historically developed from the principle of *aequitas*. The principles of equity are applied in cases where the inflexibility of the common law offers no relief to new or more complex situations. Thus the principle was created or acknowledged to support change and progress. According to T. Cottier (Equity, unpublished 1986, on file with author) it is this historically dynamic quality (compared to the German notion of *Billigkeit* and the French concept of *équitié*) which may account for much of the prominence it has attracted on the international stage in the present time of great social change.

The principle has then made its way into *international public law*. There, it has various meanings. On the one hand it is used in the sense of corrective justice, corresponding to the original Aristotelian argumentation. In this context, Schachter (cited by Girsberger, 1999, p. 188) distinguishes the following uses of the term: (i) equity as a basis for 'individualised justice tempering the rigours of strict law'; (ii) as a consideration of fairness, reasonableness and good faith; or (iii) as a basis for certain specific principles of legal reasoning associated with fairness and reasonableness (to wit, estoppel, unjust enrichment and abuse of rights).

According to Cottier (1986, p. 30), in the aftermath of decolonialization and the coming to independence of an increasing number of states in the Third World, the meaning of equity has expanded. In the context of the Third World countries' concern for economic self-determination, the dynamic and antithetical function of equity to common law, combined with its meaning as being corrective of ideal market theories, has resulted in the principle increasingly becoming a symbol for global justice. The 'Declaration on the Establishment of a New International Economic Order', the corresponding Action Programme and the Charter of Economic Rights and Duties of States (1974) have moved equity to centre stage. Equity now appears in the context of the sharing of benefits resulting from the utilization of common goods, and as a necessary qualification of international economic relations and cooperation. Equity became a prominent feature in the United Nations Convention on the Law of the Sea (1982): marine resources are to be used in an 'equitable and efficient manner'. The achievement of such goals is meant to 'contribute to the realisation of a just and equitable international economic order' (Preamble).

CONCLUSION

Equity is an important keyword used in the discussion on genetic resources and TK. However, it is frequently used in a non-specific, intuitive way. As a rule, it signals the necessity of balancing the use of limited resources between social groups. In the 'new' meaning found in the framework of international public law, equity indicates issues of economic and political distributive justice. In this context it is frequently used in connection with the distribution of resources and wealth in the North–South context.

The described disciplinary interpretations indicate the necessity of defining equity in the context of a given situation. Thus, as with the traditional meaning of 'equity' in the sense of justice in the given case, an inductive approach is to be sought.

However, in order to be operational, a more differentiated interpretation is needed. So, we agree with Byström *et al.* (1999) that further and deeper reflection on a philosophical, economic and legal level is both necessary and inevitable. Such fundamental analysis would, however, exceed the scope of this study. Nevertheless, reflections on the meaning of the combination of equity with the principle of fairness might bring understanding of the issue one step nearer.

Fairness

As elaborated above, equity is closely linked to fairness. In the literature the two notions are frequently used concurrently. For example, Brown Weiss (1989) uses the terms as synonyms in stating that 'intergenerational equity arises in the context of fairness among all generations' (p. 21). Franck (1993, p. 97), in his analysis of fairness in the international legal system, considers equity as a legal recourse to introduce elements of flexibility and fairness into the legal system. The two notions seemingly express complementary concerns and requests. The following section attempts to differentiate equity from fairness.

JUSTICE AS FAIRNESS: THE THEORY OF RAWLS

In his famous work *A Theory of Justice*, Rawls (1971) sought to identify the principles that govern 'the basic structure' or the fundamental political and economic arrangements of a just society. He sought to formulate a general set of principles or laws that we all could mutually agree to. His ultimate political ideal was the realization of fairness, encompassing a right to basic welfare and a right to equal opportunity (Hellsten, 1998, p. 821). So, ultimately, his theory is about the just distribution of basic goods and rights.

According to Rawls's theory, justice as fairness is 'a conception of justice that would be agreed upon in an initial situation of equality, between persons as moral equals, and that regulates all further agreements about how institutions are to be set up or reformed' (Loizou, 1998, p. 347). To this end, Rawls adapted the theory of the social contract in proposing a thought experiment designed to model an initial position of equality between prospective citizens who, as 'free and rational persons', are to engage in a reasoned debate over principles of justice in order to arrive at a final and binding consensus. The starting point for the debate is an initial position of equality, the so-called 'original position', where participants are imagined behind a 'veil of ignorance' that occludes information about the actual and future social position (see Rawls, 1971, p. 11).

The specific conditions of Rawls's approach are that all parties are – with equal rights and duties – involved in the process of finding a consensus (p. 19). 'Fairness' in Rawls's theory thus encompasses both a 'method' and a process of finding justice, on the one hand, and the result achieved by this method, on the other hand. Fairness in procedure means that all parties participate in an equal position in the process of finding consensus. The resulting theory says that 'social and economic inequalities are to be arranged so that they are both: (i) to the greatest benefit of the least advantaged, consistent with the just savings principle; and (ii) attached to offices and positions open to all under conditions of fair equality of opportunity' (Loizou, 1998, p. 352).

The applicability of Rawls's theory to justice as fairness on the international level, in particular in connection with a 'world community', is discussed controversially. In our context, however, the question is to be asked in a different manner: of interest here is the meaning of the term fairness that can be deduced from Rawls's theory, and not its applicability to international relationships. In this context it is relevant that Rawls's theory not only encompasses a result, but also a method or procedure to achieve this result. His principles of justice are the result of a debate between individuals who are in a position of equality. Accordingly, it seems legitimate to conclude that 'fairness', in the sense of Rawls, (also) signals a procedural aspect of justice, which is based on the unbiased participation of all involved parties on an equal footing. The coming together with the prerequisite of the 'equal footing' in the given context is not easy to achieve. It requires, for instance, the creation or the acknowledgement of rights, possibly language skills, financial means and an administrative and communication infrastructure, and capacity building of all involved stakeholders.

EQUITY AS FAIRNESS? THE THEORY OF FRANCK

Franck (1993), in his analysis of fairness in the international legal and institutional system, also closely connects fairness with equity. He considers equity in its traditional sense, as a legal recourse to introduce elements of flexibility and fairness into the legal system (p. 97). He sees equity as either a correction or exception to the non-equitable rules of law, or in the broader sense of being the dominant rule of law in itself, in governing resource allocation (p. 81).

He perceives the element of fairness as a necessary attribute of rules governing the allocation of scarce resources within a given community. Franck considers 'fairness' to indicate a specific quality of a process, incorporating an element of the equalization of life chances and of access to goods (p. 99). Fairness leads to a democratic entitlement (p. 123) linked to the questions of self-determination of peoples (p. 125). It is thus correlated to questions of empowerment of individual and collective people and their participation in the relevant processes.

'FAIR AND EQUITABLE'

In common law, 'fair and equitable' is a formula frequently used in various areas of law (although nowhere spelled out). In contract law, the formula appears in connection with agreements where one party needs specific protection, such as in consumer law or employment law, or it indicates a fair and equitable balance of right and privileges. In the area of public law, the functioning and outcome of (state) systems has to be fair and equitable for the subjects, if possible including fair and equitable opportunities. This last element includes the aspect of access to opportunities and of opportunities to participate. Finally, the social and distributive aspect of the formula appears in the statement of fair and equitable distribution of wealth.

It is conceivable that it is from these roots that the formula has made its way into international public law, comparable to the assimilation of the concept of equity. In this way it could thus signal the need for specific consideration for the weaker party, allowing for equal opportunities of participation or access.

Conclusions

Given this, the following may be concluded. The fair and equitable formula seems to indicate a specific quality of both process and outcome in questions of distributive justice. In this we agree with Byström *et al.* (1999) who conclude that the nuances in meaning may be taken to suggest that 'fair' refers mainly to process, and 'equitable' mainly to outcome. In other words, equity indicates the goal of distribution, and fairness an attitude towards reaching this goal. From the fact that the CBD text almost exclusively uses both words, Byström *et al.* (1999, p. 24) conclude that such a distinction was intended. Similarly Franck (1993, p. 32) describes 'fairness' as 'a human, subjective, contingent quality which merely captures in one word a *process of discourse, reasoning and negotiation* leading – if successful – to an agreed formula that locates itself at a conceptual intersection between various plausible formulae for allocation'.

Equity rather indicates the quality of the result of this process. It indicates a necessity to balance a situation, frequently in relation to the distribution of (basic) goods. In view of future generations, it is closely interrelated with sustainable development. In view of North–South equity, 'fair and equitable' is closely interrelated with empowerment, participation and capacity building.

1.2.3 Sustainability

General: term and interpretation

The significance of the term 'sustainability' in the international discourse on environment and development can be traced back to its use in the so-called 'Brundtland Report',[56] the report of the UN General Assembly Commission on Environment and Development. The Commission was mandated to 'propose long-term environmental strategies for achieving *sustainable development*', including the recommendation of 'ways in which concern for the environment may be translated into greater cooperation among developing countries and between countries at different stages of economic and social development'.[57]

In this report, 'sustainable development' found its famous and much-cited definition as 'meeting the needs of the present without compromising the ability of the future generations to meet their own needs' (p. 43).

Historically, the term has its roots in forestry and fisheries, meaning the maintenance of the long-term productivity of the respective resources.[58] Forerunners of the term of sustainable development were the concepts of 'wise use' found in the Ramsar Convention (1971) and of 'sound strategies' coined in the follow-up process of the Stockholm Conference in 1972. The interpretation of the 'wise use' concept by the Ramsar follow-up process for the first time linked the issues of conservation and human use of resources.[59]

The term 'sustainability' was taken up by the 1992 World Conference on Environment and Development and found its way – then and since – into a vast array of international instruments dealing with resource use and development. The Rio Declaration places human beings at the centre of concerns for sustainable development and asserts their entitlement to a healthy and productive life in harmony with nature (Principle 1). The recent Johannesburg Declaration on Sustainable Development confirmed the commitment to sustainable development, that is, to 'build a humane, equitable and caring global society cognisant of the need for human dignity for all' (paras 1 and 2).

However, the two international treaties of interest in our context, the CBD and the Seed Treaty, speak of sustainable *use* of the components of biodiversity (Article 1 CBD), and of plant genetic resources for food and agriculture (Article 1.1 ITPGRFA), respectively. The CBD takes up the Brundtland pattern in defining sustainable use as 'the use of components of biological diversity in a way and at a rate that does not lead to the long-term decline of biological diversity, thereby maintaining its potential to meet the needs and aspirations of present and future generations' (Article 2, para. 16).

At this point it might be asked what is the difference and the connection between the two variations of the term sustainability – sustainable *use* and sustainable *development* – and what can be concluded for our research question. Putting the question in a more general way, as the terminology around 'sustainability' is used in such varied ways, it is important to work out the important points of the concept for the given context.

Thus, first the concept of 'sustainability' will be analysed. The specific meaning

---

[56] Also World Commission on Environment and Development (WCED) Report, Brundtland (1987).

[57] United Nations, General Assembly, Process of Preparation of the Environmental Perspective to the Year 2000 and Beyond, A/RES/38/161, 19 December 1983.

[58] In German the equivalent term is *Nachhaltigkeit*. Also compare Becker (1997, p. 4).

[59] Biber-Klemm (1992, p. 112).

of the term 'sustainable *use*' in the relevant legal texts will then be investigated and put in relation to the aspect of 'sustainable *development*'.

SUSTAINABILITY

In a general sense, sustainability can be defined as 'the capacity that something has to be sustained into the future' (Dower, 1998b, p. 767). Originally, the concept was used in the context of productivity, to describe either specific ecosystem features, such as the 'ability of a system to maintain productivity in spite of a major disturbance' (Becker, 1997, p. 3, citing Conway, 1983) or as 'sustainable yield' of agricultural crops (Becker, 1997, p. 3, citing Plucknett and Smith, 1986).

In the present political discourse, however, the term is used in varied ways encompassing the following elements:

- Sustainability is understood as integrating the different components that define the interrelationship between humanity and nature; that is, the economic, social and ecological factors.
- Sustainability comprises different spatial emphases. In particular, the ecological aspect also encompasses what could be called the 'global thinking' or the awareness of: (i) the interdependence of all natural processes; (ii) the fact that all human activities and development make use of natural resources and (can) have global effects; and, therefore, (iii) the effects on humanity as a whole.
- The link between the need for the preservation of natural resources and their utilization.
- The different time frames, in particular the inclusion of future generations into ecological thinking and decision-making, which opens up the time horizon. The consequence is the development of a 'planetary ethos which recognises that we share the planet with all communities throughout time' (Brown Weiss, 1989, p. 291).

Two conclusions can be drawn from this. First, it becomes apparent that the term is, in most instances, understood to have a normative character. And secondly, this normativity points to the common responsibility and thus to the necessity of *intragenerational* and *intergenerational equity*, that is, the necessity of reconciling sustainability with concerns for equity and justice.

SUSTAINABLE USE

The 'sustainable use' formula in the CBD is – for the time being – the result of the paradigm change in the politics of nature conservation. Whereas up until the middle of the 20th century conservation of nature was seen in a segregated way, implying the necessity of geographically separating conservation from the use of 'nature', later on the awareness of the cross-cutting character of conservation of nature developed.[60]

Accordingly, the CBD takes up this issue in postulating the integration of conservation and sustainable use of biological diversity into relevant sectored or cross-sectored plans, programmes and policies (Article 6(b)). Conservation and sustainable use of biodiversity are to be integrated into decision making, measures are to be developed to avoid or minimize adverse impacts, and customary use of biological resources compatible with conservation and sustainable use are to be protected and encouraged (Article 10). However, the relevant provision on *in situ* conservation remains relatively vague: Article 8(i) asks Contracting Parties to 'endeavour to provide the conditions needed for compatibility between present uses and the conservation of biological diversity and the sustainable use of its components'. Further, Article 8(j) asks for the recognition of the contributions of indigenous and local communities to this end.

Another important aspect is the combination of conservation and sustainable use of the components of biological diversity

[60] In international law this evolution is apparent, for example, in the relevant consecutive Resolutions and Declarations (Stockholm 1972, World Charter of Nature 1982, Rio 1992, Johannesburg 2002).

with economic issues, as in Article 11 on the creation of incentives and Article 15 on access and benefit sharing.

Article 6 of the Seed Treaty spells out the elements of sustainable use of PGR in a more differentiated way. Sustainable use of PGRFA may include such measures as the promotion of diverse farming systems, of plant breeding efforts adapted to local conditions with the participation of farmers, and of the use of local and locally adapted crops.

The case of agrobiodiversity illustrates the insight that 'sustainability' does not necessarily mean abstention from the use of natural resources, but frequently implies the necessity of human intervention.

SUSTAINABLE DEVELOPMENT

The combination of the terms sustainability and development was coined in the IUCN World Conservation Strategy (1980), but gained paradigmatic appeal only by its use and interpretation in the Brundtland report (Becker, 1997, p. 6).

This report identified five components necessary for sustainable development: (i) economic growth; (ii) equity, in the sense of the fair allocation of resources; (iii) increasingly democratic, in the sense of participatory, systems; (iv) adoption of lifestyles within the planet's ecological means to sustain human life; and (v) population levels corresponding to the productive potential of ecosystems (Carpenter, 1998, p. 279).

Regarding intragenerational equity, Carpenter (1998, p. 279) assesses *Our Common Future* as a 'commendable and constructive compromise between conflicting agendas of the developed and less-developed countries'. However, the policy emphasis, which is upon economic growth, has been amply criticized.

At the centre of the discussion is the meaning of the term 'development', which in itself is a topic of intense debate.[61] Engel and Kothari criticize the purely economic approach and defend a broader understanding, including sociocultural facets of human life. Engel expands the meaning of development to 'evolution, unfolding, growth and fulfilment of any and all aspects of life', and proposes a comprehensive social paradigm, which consists of five operational principles: (i) integration of conservation and development; (ii) satisfaction of basic human needs; (iii) achievement of equity and social justice; (iv) provision of social self-determination and cultural diversity; and (v) maintenance of ecological integrity. Kothari proposes that sustainable development be conceived as an ethical ideal consisting of 'equity, based on autonomy and self-reliance of diverse entities ... an emphasis on participation; and an accent on the importance of local conditions and the value of diversity' (cited from Carpenter, 1998, p. 279).

ANALYSIS

In the discourse on sustainability, the terms 'sustainability', 'sustainable use' and 'sustainable development' are often used interchangeably. However, it is appropriate to distinguish them clearly. The term 'sustainability' encompasses 'sustainable use' and 'sustainable development'. Sustainable *development* is concerned with the evolution of human populations. It involves the persistence through time of the diversity of human communities and ethical ideals of human flourishing, the dynamically balanced development of economic enterprise, and the preservation and regeneration of ecological systems and resources that sustain that development (Carpenter, 1998, p. 276). Sustainable *use* of natural resources is (one of) the foundations of sustainable development.

One of the core rationales of the present research is to find ways to support and maintain the diversity of PGR in the context of general economic development, in generating resources to support the development of local and indigenous communities. Therefore it is proposed to take the broad term of 'sustainable development' as a starting point.

[61] See for instance the overview given by Dower (1998a, pp. 757–766).

In the present political discussions, the notion of 'sustainable development' is strongly connected to economical issues. But, as has been pointed out, it also encompasses aspects of *intergenerational* and *intragenerational* equity and justice. Thus, it not only has an economic, ecological and sociocultural nature, but also implies a moral valuation. In this context, the meaning of the term development ought particularly to be further investigated.

At this stage of the research, the term is understood as a signal for the integration of development with the maintenance of mechanisms supporting the existence and further evolution of TK and PGRFA. It strongly indicates the necessity of reconciling traditional and modern knowledge and technologies, in order to maintain biological and cultural diversity and TK. Development in this sense naturally includes economic and technological aspects. But it has to take place with the participation of the involved people, societies and communities, taking account of the local sociocultural conditions.

The particulars of biological diversity and traditional knowledge with a view to sustainable use and development

SUSTAINABLE USE OF BIODIVERSITY

In the context of sustainable use of biodiversity, it is important to be aware of its specific characteristics:

- Biological diversity is both a *renewable* and an *absolutely non-renewable* resource: if used sustainably it can (theoretically) be maintained (subject to the natural extinction rate). However, once it has been lost it cannot be retrieved and – unlike almost all natural resources that have substitutes – biodiversity cannot be replaced.
- Biodiversity is ultimately a *concept*. It can be understood as the existing variety of animals, plants and ecosystems, and the interrelationships between these elements. However, at the same time biodiversity also means the 'variability of the living', with its evolutionary capacity through space and time.
- Biodiversity loss is driven by manifold, complex and diffuse causes. Often the effects of a local present-day pressure only become apparent on a long-term scale or in a higher geographical context.
- However, biological diversity is also dependent on sustainable human intervention, either in the field of the domestication and breeding of animals and plants, or by the transformation of originally uniform ecosystems and habitats, such as in sustainable farming systems.[62]

For the sustainable use of biological diversity, these features imply that:

- Biodiversity is not only to be conserved in its present state: it is important to maintain its evolutionary potential and the option for evolutionary processes to take place. This implies the *in situ* conservation of the components of biological diversity.
- Sustainable human intervention and management is essential in many cases. Examples are: (i) traditional plant breeding (intercropping, cross-breeding with wild species and *in situ* conservation); (ii) sustainable land management in diverse agricultural systems; and (iii) continuation of the management of the so-called partner species in forest ecosystems and domestic gardens. Thus, sustainable use of biodiversity is closely linked to the maintenance of cultural diversity.
- As the connection between cause and effect is frequently difficult to discern, the precautionary principle is of major importance in the sustainable use of biodiversity.

TRADITIONAL KNOWLEDGE AND SUSTAINABLE DEVELOPMENT

The value of TK for the sustainable use of plant genetic resources has been discussed earlier in this chapter. However, at this point the question is to be asked the other

[62] Other examples include forest systems and pastoralism.

way around – that is to say, *what does sustainable development mean for TK*?

According to Brown Weiss (1989, p. 24), intergenerational equity includes the notion of conserving the improvements made in the natural and cultural resource base made by preceding generations and to pass them on to future generations. These improvements encompass the knowledge, information and technologies relevant for sound resource use in harmony with human well-being (Rawls, 1971, p. 289).

One method to save traditional knowledge from being lost is its documentation. However, as has been explained above, TK is not static, but in a permanent process of evolution and adaptation to new circumstances. Therefore, as with crop varieties, the static *ex situ* conservation of TK by methods of registration and so on is not apt to cover its entire scope. In the context of sustainable development it is therefore crucial to maintain the evolutionary power of TK and to promote its further development in the dialogue, with arising problems and new technologies.

Accordingly, in a recent document of the CBD's Working Group on Article 8(j)[63] it is pointed out that

> one way to ensure the maintenance of such knowledge at the local level is to make sure *that it retains its relevance in today's world* – the principle of 'use it or lose it' [emphasis added] ... It is ... important that indigenous and local communities build on and capitalise on their traditional knowledge by using it, for example, to create new products and processes ... In order for this to happen, however, indigenous and local communities will need to build their capacity for research and development.

1.2.4 Food security and basic health care

In general: theoretical background

Food security and basic health care are considered to be the most basic of the 'basic needs' of human beings. The concept of 'basic needs' is employed in social theories as one of the central ideas involved in a full account of justice. Basic needs are the things that are necessary for the most fundamental projects involved in living a human life, and which are essential to living or to functioning normally (Wolf, 1998, p. 336). The objective of the 'basic needs' concept is to provide all human beings with the opportunity for a full life.

The definition of these basic needs or basic goods is controversial. A minimal list of basic needs contains – according to Hicks and Streeten (cited in Wolf, 1998, p. 338) – health, education, food and clothing, water supply, sanitation, drinking water, social security, housing and participation in government activities.

This list corresponds to Article 25 of the Declaration of Human Rights, which enumerates these goods systematically in the context of social and cultural rights, the rights to which everyone 'as a member of society' is entitled. Article 22 characterizes these rights as being indispensable for dignity and the free development of personality.

Rawls (1971, p. 62) speaks of 'primary goods' which include rights, liberties, powers and opportunities, income and wealth, and the social basis of self-respect as (social) primary goods. Possession of these goods enables a person to pursue his or her individual plans (p. 93). Rawls developed his political concept of social justice based on the efficiency and equity of the distribution of these goods (Sen, 1987, p. 5).

Sen points out two different approaches to the assessment of the value of 'basic goods'. Are they intrinsically valuable in themselves, or are they valuable as instruments serving other more basic goals, such as the enhancement of the happiness or welfare of individuals? He criticizes the commodity-based growth theory that 'values material goods and commodities as intrinsically good ... Thus the mere means are transformed to ends' (in Hellsten, 1998, p. 826). He puts forward the so-called capability approach, which emphasizes that material goods and commodities are not good in themselves but are good only by

[63] UNEP/CBD/WG8(j)/2/4, 27 November 2001, p. 15, Nos 76 and 77.

virtue of what they can do for human beings. He argues that instead of mechanically focusing on the equal distribution of individual rights, or other primary goods and commodities, we should rather look for the final goal of this distribution, that is, human well-being. That is to say, we should ask the question 'What can goods do for people?', or even 'What can people do with these goods?'[64]

The differences in these lines of argument are similar to those in the sustainability discussion, in which the broad interpretation of sustainability also includes social aspects and basic needs. As is the case with the sustainability argument, the argument of 'basic needs' is closely related to the goals of justice and equity.

Food security[65]

Food security is defined by the FAO as the physical and economic access for all people at all times to sufficient, safe and nutritious food that meets their dietary needs and food preferences for an active, healthy life with no risk of losing such access. As a policy issue it encompasses responsibilities not only on a global, regional and national level, but also on household and individual levels.[66]

Food security encompasses two dimensions in the context of the above-discussed principle of sustainable development. First, the dimension of nutritional security for the present world population, i.e. feeding the poor in the present and in the immediate future.[67] Secondly, the aspect of feeding the future world population which is: (i) steadily increasing; and (ii) in all probability will live under climatic conditions which differ from those of the present.

FACTUAL BACKGROUND

More than 800 million people throughout the world do not have enough food to meet their basic needs. Despite the fact that food supplies have increased substantially during the past decades – in 1995, world per capita supplies of food for direct human consumption were 18% greater than in 1965 – during the same period the world population increased by 70%.[68] Projections indicate that although the annual world agricultural growth rate has decreased from 3% in the 1960s to 2% in the past decade, world food supply will continue to outpace world population growth, at least until 2020. Worldwide, per capita availability of food is projected to increase by around 7% between 1995 and 2020, or 9% for developing countries (Pinstrup-Andersen, Pnadya-Lorch and Rosegrant, 1999, cited from Serageldin and Persley, 2000).

According to projections of the FAO, food insecurity is expected to diminish in some parts of the developing world such as East Asia, South Asia and Latin America; however, to accelerate in sub-Saharan Africa, West Asia and North Africa. Sub-Saharan Africa and South Asia are expected to remain the main focus of hunger in the developing world (cited from Nuffield Council on Bioethics, 1999, p. 59).

It becomes clear that food security depends not only on the production of enough food.[69] Food security presents itself

---

64 See also FAO (1992) *International Conference on Nutrition, World Declaration on Nutrition*: (1) 'We recognise that access to nutritionally adequate and safe food is a right of each individual.' (11) 'We recognise that the nutritional well being of all people is a pre-condition for the development of societies.'

65 For basic background information, see The Crucible II Group (2001) *Seeding Solutions*, Vol. 1, pp. 6–7; FAO (1992) *International Conference on Nutrition: World Declaration on Nutrition and Plan of Action of Nutrition;* FAO, World Food Summit (1996) *Rome Declaration on World Food Security and World Food Summit Plan of Action* (Rome Declaration and Plan of Action).

66 FAO Rome Declaration (1996) p. 3.

67 See FAO Rome Declaration (1996) para. 2, which schedules pledges to reduce the number of undernourished people by half by the year 2015.

68 FAO, Sustainable Development Department, Population Programme Service FAO Women and Population Division, *SD Dimensions*, October 1999. http://www.fao.org/sd/wpdirect/wpan0040.htm

69 The World Declaration on Nutrition in 1992 even recognized that 'globally there is enough food for all and that inequitable access is the main problem'.

as a complex issue, involving access to food, its processing, and distribution on different levels, including between rural and urban surroundings. Food security is not only a question of the amount of food available, but also of the content or quality of the food (Serageldin and Persley, 2000, p. 2). The Rome Declaration points out that constraints on access to food, the continuing inadequacy of household and national incomes to purchase food, and the instability of supply and demand (as well as natural and man-made disasters) all prevent basic food needs from being fulfilled.

According to a FAO (1999) analysis, the possible global increase in production matters only up to a certain point because of the heterogeneity of situations between industrialized countries with slow or no population growth, and which have the most productive and progressive agricultural systems, and developing countries, with a rapidly growing population and in which agricultural production is lagging. Another crucial point is the inland inequality to the access to food. The population segment which is threatened the most by food insecurity is the poor, whether resource-poor farmers found in marginal areas, or the growing urban population – which is a result of rural–urban migration for reasons of food insecurity. The main challenge is to enable these people to satisfy their food needs through a stable system of production and exchange.

FOOD SECURITY AND TRADE

Food security is one of the so-called non-trade concerns that have to be taken into account in the continuation of the process to reform agricultural trade (Article 20, Agreement on Agriculture). In the current negotiations, non-trade concerns are raised by several groups of participants.[70]

They point out the specific characteristics of agriculture in developing countries, where a majority of the population depends on agriculture for their livelihood. There physical access to food can be ensured only through a certain minimum level of self-sufficiency. Therefore it is critically important for them that agriculture remains a viable source.

They argue that, according to the analysis of the FAO,[71] the promotion of economic growth, reduction of poverty and the enhancement of food security are all closely linked to the necessity of developing the potential capacity of agriculture. As the subsistence farmers participate only marginally in international trade, it is argued by India, for example, that WTO arrangements should be more flexible and allow for the support and protection of agricultural and rural development in the framework of the Green Box. According to the line of argument of the Indian proposals, the high vulnerability of agriculture in developing countries justifies the extension of special provisions to the developing country members for ensuring their food and livelihood security concerns. The proposal includes provisions for the general development of agriculture, including its diversification in

[70] Committee on Agriculture, Special Session, 9 November 2000. Note on Non-Trade Concerns. Submission to the Special Session of the WTO Committee on Agriculture by Barbados, Burundi, Cyprus, Czech Republic, Dominica, Estonia, the European Communities, Fiji, Iceland, Israel, Japan, Korea, Latvia, Liechtenstein, Madagascar, Malta, Mauritania, Mauritius, Mongolia, Norway, Poland, Romania, Saint Lucia, Slovak Republic, Slovenia, Switzerland, and Trinidad and Tobago. G/AG/NG/W/36/Rev.1

Committee on Agriculture, Special Session, 15 January 2001. Negotiations on WTO Agreement on Agriculture: Proposals by India in the areas of: (i) Food Security, (ii) Market Access, (iii) Domestic Support, and (iv) Export Competition. G/AG/NG/W/102.

Committee on Agriculture, Special Session, 14 February 2001. Negotiations on WTO Agreement on Agriculture: Proposal by Nigeria. G/AG/NG/W/130.

[71] FAO (1999) Commodity Policy and Projections Service, Commodities and Trade Division, Issues at stake relating to Agricultural Development, Trade and Food Security. FAO Symposium on Agriculture, Trade and Food Security: Issues and Options in the forthcoming WTO Negotiations from the Perspective of Developing Countries, Geneva, 23–24 September 1999, Paper No. 4.

developing countries, in the Green Box provisions, and/or to exempt product-specific support given to low-income and resource-poor farmers from the Aggregate Measures of Support (AMS)[72] calculations.

FOOD SECURITY AND CLIMATE CHANGE[73]

The report of the Intergovernmental Panel on Climate Change (IPCC Report, 2001, p. 6)[74] describes two factors influencing the effects of climate change on natural systems: (i) the *sensitivity* of a system, i.e. the degree to which a system is affected either adversely or beneficially by climate-related stimuli; and (ii) the *adaptive capacity*, i.e. the ability of a system to adjust to climate change, to moderate potential damage, to take advantage of opportunities, or to cope with the consequences. Vulnerability is the degree to which a system is 'susceptible to or unable to cope with adverse effects of climate change'.

According to the latest report of the Intergovernmental Panel on Climate Change, the natural systems that are especially vulnerable to climate change, because of their limited adaptive capacity, include boreal and tropical forests, polar and alpine ecosystems, prairie wetlands and remnant native grasslands. Human systems that are sensitive to climate change include water resources, agriculture (especially food security) and forestry. The vulnerability of these systems varies with geographic location and temporal, social, economic and environmental conditions.

In respect of food security, the ICC projects the following adverse impacts: (i) a general reduction in potential crop yields in most tropical and sub-tropical regions, assuming that the envisaged increases in temperature take place; and (ii) decreased water availability for populations in many water-scarce regions, particularly in the sub-tropics and tropics. Of benefit would be the predicted increase in crop yields in some regions at mid-latitude (p. 6).

The climate-induced yield losses and adaptation of livestock production systems both involve costs. Autonomous agronomic adaptation can mitigate – but not entirely neutralize – the impact (p. 11). The panel concludes that the impacts of future changes in climate extremes are expected to fall disproportionately on the poor (pp. 6 and 11).

Adaptation of human systems to the effects of climate change depends on factors such as wealth, technology, education, information, skills, infrastructure, access to resources and management capabilities. The developing countries are generally poorest in this regard. 'Policies that lessen pressures on resources, improve management of environmental risks, and increase the welfare of the poorest members of society can ... reduce vulnerability to climate and other stresses' (IPCC, 2001, p. 8; The Crucible II Group, 2001, Vol. 1, p. 11).

VALUE OF DIVERSITY FOR FOOD SECURITY

PGRFA are the source for tailoring food production in order to meet future challenges. The Leipzig Declaration[75] summarizes the reasons for the importance of their conservation as follows: PGRFA have an essential importance for the food security of present and future generations. They are 'the basis of natural and directed evolution in the plant species most critical to the survival and well being of human beings. All countries require plant genetic resources if they are to increase food supplies and agricultural production sustainably and meet the related challenges of changes in the envi-

[72] The AMS is the basis for the calculation of the authorized domestic support of agricultural production. In principle, all support measures are to be included; exemptions exist for the so-called Green Box and some development measures. For details see Chapter 8.

[73] For basic background information, see The Crucible II Group (2001) *Seeding Solutions*, Vol. 1, pp. 11–12.

[74] Working Group II of the Intergovernmental Panel on Climate Change (2001). http://www.grida.no/climate/ipcc_tar/vol4/english/pdf/wg2spm.pdf (accessed 19 September 2005).

[75] Leipzig Declaration on conservation and sustainable utilization of plant genetic resources for food and agriculture. Adopted by the International Technical Conference on Plant Genetic Resources. Leipzig, Germany, 17–23 June 1996.

ronment, including climate change' (para. 3; also see Hawkes, 1983; Brush, 2000, pp. 3–5; Thrupp, 2000).

With a view to the climate change scenarios, adaptation is a necessary strategy to complement the efforts to mitigate climate change (IPCC, 2001, p. 8).

Biological diversity, in particular the diversity of PGRFA, is the key to adapting PGRFA to climate change.

PROPOSED STRATEGIES

As has been demonstrated above, food security and food insecurity is a complex issue, due to multifaceted underlying causes. With a view to the conservation of plant genetic resources and TK, food security encompasses the following policy issues as specified in the FAO World Food Summit Plan of Action (1996):

- Poverty eradication in order to improve *access*: either with a view to the production of food, given the means of production of land, water, improved seeds and plants, appropriate technologies and farm credit and so on; or to the means enabling the purchase of food.
- Promotion of creative initiatives of food producers; support of their full involvement and enablement as crucial elements for success in achieving food security (p. 4).
- Increase of food production through sustainable management of natural resources. 'Increased production, including traditional crops and their products, in efficient combination with food imports, reserves, and international trade can strengthen food security and address regional disparities.'
- Long-term investment in research and in cataloguing and conserving genetic resources.
- To minimize the vulnerability to, and impact of, climate fluctuations, pests and diseases (p. 4).

In general, the solution to food insecurity is discussed very controversially. Food insecurity is either seen as the consequence of the quantity of food available, or rather as the problem of access and distribution. Therefore, one line of argument sees the solution as increasing the production of food, in creating varieties with specific qualities that increase yield, tolerance for higher temperatures and/or drought, and pest resistance, mainly making use of the possibilities created by the techniques of genetic engineering. The background argument is that from the point of view of sustainability, the surface of arable land ought not be expanded.[76]

The other line of argument supports the proposition that emphasis ought to be put on improving local farming systems. It is argued that poor farmers cannot afford genetically modified crops, which are anyway not designed to serve their specific needs. Traditional multiple cropping systems provide as much as 20% of the world food supply. It is put forward that family farming systems with multiple cropping have potential for increasing production if given the necessary technical, financial and educational support. This has the advantage of generating income, preventing migration, and mitigating problems of accessibility and distribution.

At this point no decision on this issue is necessary. In both cases, conservation of the diversity of plant genetic resources and support of subsistence farming systems are a necessary precondition, either as a means of supporting food security in itself or as a means of conserving genetic resources which could prove to be crucial for creating varieties adapted to new needs, whether by methods of traditional breeding or genetic engineering.

Health care, medicinal care, traditional medicine

WHO describes 'traditional medicine' as a way of protecting and restoring health, which existed before the arrival of modern medicine. Systems of traditional medicine

[76] See Nuffield Council of Bioethics (1999) (and 2003 follow-up paper: http://www.nuffieldbioethics.org/go/textonly/ourwork/gmcrops/introduction); De Greef (2000).

have had to meet the needs of the local communities for many centuries.

They belong to the traditions of each country and have been handed down from generation to generation. Herbal medicine is considered as one part of traditional medicine, besides acupuncture, traditional birth attendants and mental healers. WHO points out that medicinal plants are the oldest known health-care products. They are still very important as a large proportion of the population in a number of developing countries relies on traditional medicine, and herbal medicines in particular are assuming greater importance in the primary health care of many individuals and communities. Traditional medicine has maintained its popularity in a number of Asian countries, such as in China, where traditional herbal preparations account for 30–50% of the total medicinal consumption. During the past decade there has also been a growing interest in traditional systems of medicine in many developed countries (WHO, 1996, 2002a,b).

According to WHO, the reasons for the inclusion of traditional healers in primary health care include: (i) their knowledge of the sociocultural background of their peoples; (ii) the fact that they are highly respected and experienced in their work; (iii) economic considerations; (iv) the vast distances to be covered in some countries; and (v) the shortage of health professionals, particularly in rural areas.

From this it can be concluded that, in analogy with food security, physical and economic accessibility and availability are core elements of basic health care. This entails: (i) the conservation or preservation of the sources of herbal medicine in a sufficient quantity; (ii) the support of local, 'decentralized' and low-cost health systems; and (iii) the support of the continual passing down of knowledge, which entails technology transfer, appropriate financing and other incentives to do so.[77]

1.2.5 Conclusion

In short, as a common denominator for the discussed goals, their linkage to and dependence on participatory processes is proposed. Such a participatory approach would have to provide mechanisms that allow the involved communities to take control over the developments that affect them.

Food security and basic health care are prerequisites to the above-mentioned ends. Participation and creative bottom-up processes are closely intertwined with the fulfilment of these basic needs. A decentralized system of food production is more conducive to facilitating access to food. Similarly, a decentralized system of health care is conducive to facilitating basic health care, particularly for local and indigenous communities.

Food security and basic health care are related to the eradication of poverty, which would improve access to the means of production – land, water, improved seeds, technology and finance – or the means to purchase food. In this context, the discussed goals gain yet another meaning. With a view to the climate change predictions, which indicate a higher vulnerability

[77] Article from Zimbabwe News Online, Edition #18, 15 April 1998: Traditional healers in Zimbabwe plan to set up a college. The Zimbabwe National Traditional Healers' Association (Zinatha) plans to build a college to educate traditional healers on plant medicines, following a marked increase in the demand for traditional medicine. According to the chairman of Zinatha, Professor Gordon Chavhunduka, it is expected that the college will open in 2003, at a time when traditional medicines are fast gaining popularity, especially in the face of the AIDS pandemic. At least US$10 million is needed to establish the college, which will initially enrol 50 students. Professor Chavhunduka said there were many people aspiring to be traditional healers, hence the need to have such a college. If established, the college will be the first one of its kind in southern Africa. The professor said the focus during training would be on two branches, namely plant medicine and spiritual healing. The college has already been approved by the Ministry of Higher Education and Technology. He said traditional medicine was lagging behind so-called modern medicine because before independence indigenous Africans were 'brainwashed' into thinking that the use of traditional medicine was taboo.

of the regions which are already poor, they can be interpreted to indicate a culture of sharing or access to (technological) achievements.

So, the formula for the equitable and fair sharing of the benefits indicates both a process and an outcome. Equity indicates justice, or rather a balance in the distribution of 'benefits' or resources and goods in the broader sense, on a case-by-case basis adapted to a given situation. In the context of sustainability, this justice is not only to be considered for the present, but also between generations. Equity in this context means that not only the goods, but also the obligations, are to be borne in an equitable way. In our case, this would mean that the obligation to maintain the biological and intellectual resources is to be adapted to the different capacities (and in principle rather by developed than by developing countries, taking account of the stage of economic development and the resources used to this end).

According to the 'equitable and fair' formula, equity is to be seen in the context with fairness. Fairness by itself indicates a specific participative, communicative procedure which takes account of the needs, but also of the contributions of all participants. It is a human, subjective, contingent quality, which captures in one word a process of discourse, reasoning and negotiation, leading to an agreed formula. Accordingly, the term implies a discursive process, on a basis of equal footing, a mutual attitude of respect and consideration. In the combination 'fair and equitable' it indicates a specific protection of the weaker parties, and/or the grant of equal opportunities. To this end, special treatment for the less privileged may be a prerequisite. So, participation includes (mutual) capacity building, and empowerment of the weaker party.

Considered together with the sustainable development formula, which combines ecological, economical and social elements, the sociocultural aspects gain momentum.

Combined with the precautionary principle, the necessity to take account of the global interdependence of the processes becomes evident. The principle points towards the necessity of finding ways and means to maintain as many options as possible, including knowledge, technologies and the gene pool for the generations to come. Therefore, the evolutionary and creative powers of TK, and the specific capacities and capabilities of its holders, are to be taken account of in the light of changing circumstances. An open discourse between 'traditional' and 'formal' problem solving and scientific methods is to be sought and furthered. TK and traditional PGRFA need to retain their relevance in the world today – here the adage 'use it or lose it' applies.

Bibliography

Alcorn, J.B. (1995) Ethnobotanical knowledge systems – a resource for meeting rural development goals. In: Warren, D.M., Slikkerveer, L.J. and Brokensha, D. (eds) *The Cultural Dimension of Development. Indigenous Knowledge Systems.* Intermediate Technology Publications, London, pp. 1–12.

Barbier, E.B. (1997) Ecological economics, uncertainty and implications for policy setting priorities for biodiversity conservation. In: Organization for Economic Co-operation and Development (ed.) *Investing in Biological Diversity: The Cairns Conference.* OECD Proceedings, Paris.

Becker, B. (1997) Sustainability assessment: a review of values, concepts, and methodological approaches. *Issues in Agriculture* 10. Consultative Group on International Agricultural Research, Washington, DC.

Biber-Klemm, S. (1992) Rechtsinstrumente zum Schutze der natürlichen Lebensräume von Tieren und Pflanzen. Dissertation, University of Basel, Switzerland.

Biber-Klemm, S. (2001) The community-based conservation and management of genetic diversity in agroecosystems: the role and function of law. In: Engels, J.M.M, Ramanatha Rao, V., Brown, A.H.D. and Jackson, M.T. (eds) *Managing Plant Genetic Diversity.* IPGRI, Rome.

Brown Weiss, E. (1989) *In Fairness to Future Generations: International Law, Common Patrimony, and Intergnerational Equity.* The United Nations University, Tokyo; Transnational Publishers, New York.

Brundtland, G. (1987) *Our Common Future: The World Commission on Environment and Development.* Oxford University Press, Oxford, UK.

Brush, S. (1994) *Providing Farmers' rights through* in-situ *conservation of crop genetic resources.* FAO, Commission on Plant Genetic Resources, Background Study Paper No 3. FAO, Rome.

Brush, S. (ed.) (2000) *Genes in the Field: On-Farm Conservation of Crop Diversity.* IPGRI, Rome; IDRC, Ottawa; and Lewis Publishers, London.

Byström, M., Einarsson, P. and Nycander, G.A. (1999) *Fair and Equitable. Sharing the Benefits from Use of Genetic Resources and Traditional Knowledge.* Swedish Scientific Council on Biological Diversity, Stockholm. http://www.internat.environ.se/documents/nature/biodiver/final.pdf (accessed 20 September 2005).

Carpenter, S.R. (1998) Sustainability. In: Chadwik, R. *et al.* (eds) *Encyclopedia of Applied Ethics,* Vol. 4. Academic Press, San Diego, California, pp. 275–293.

Castillo, G.T. (1997) Whose ethics and which equity? Issues in the conservation and use of genetic resources for sustainable food security. In: International Plant Genetic Resources Institute (ed.) *Ethics and Equity in Conservation and Use of Genetic Resources for Sustainable Food Security.* Proceedings of a workshop to develop guidelines for the CGIAR, Foz do Iguaçu, Brazil, pp. 19–31.

Cook, D.M. and Messick, K. (eds) (1983) *Equity Theory: Psychological and Sociological Perspectives.* Praeger, New York.

Correa, C. (2001) *Traditional Knowledge and Intellectual Property – Issues and Options Surrounding the Protection of Traditional Knowledge – A Discussion Paper.* Quaker United Nations Office, Geneva.

Costanza, R., d'Arge, R., de Groot, R., Farber, S., Grasso, M., Hannon, B., Limburg, K., Naeem, S., O'Neill, R.V., Paruelo, J., Raskin, R.G., Sutton, P. and van den Belt, M. (1997) The value of the world's ecosystem services and natural capital. *Nature* 387, 253–260.

Daes, E.I. (1997) *Protection of the Heritage of Indigenous People.* United Nations, New York.

De Greef, W. (2000) My turn: agricultural biotechnology and moral imperatives. *In vitro* cellular and developmental biology. *Plant* 36, 309–311.

Dower, N. (1998a) Development ethics. In: Chadwik, R. (ed.) *Encyclopedia of Applied Ethics,* Vol. 1. Academic Press, San Diego, California, pp. 755–766.

Dower, N. (1998b) Development issues environmental. In: Chadwik, R. (ed.) *Encyclopedia of Applied Ethics,* Vol. 1. Academic Press, San Diego, California, pp. 767–779.

Dutfield, G. (2000) The public and private domains: intellectual property rights in traditional ecological knowledge. *Science Communication* 21(3), 274–295. Available at: http://scx.sagepub.com/cgi/reprint/21/3/274

Edwards, R. (1988) The price of arrogance. *New Scientist,* 17 October, pp. 18–19.

Ehrenfeld, D. (1988) Why put a value on biodiversity? In: Wilson, E.O. (ed.) *Biodiversity.* Washington, DC National Academy Press, pp. 212–216.

Engelmann, F. and Engels, J.M.M. (2002) Technologies and strategies for *ex situ* conservation. In: Engels, J.M.M, Ramanatha Rao, V., Brown, A.H.D. and Jackson, M.T. (eds) *Managing Plant Genetic Diversity.* IPGRI, Rome.

Farnsworth, N.R. (1988) Screening plants for new medicines. In: Wilson, E.O. (ed.) *Biodiversity.* Washington, DC National Academy Press, pp. 83–97.

FAO (1992) World Declaration on Nutrition and Plan of Action of Nutrition. International Conference on Nutrition, Rome, 5–11 December, 1992. http://www.fao.org/docrep/u9920t/u9920t0a.htm and http://www.fao.org/docrep/u9920t/u9920t0b.htm (accessed 19 September 2005).

FAO (1996) Rome Declaration on World Food Security and World Food Summit Plan of Action. World Food Summit, Rome, 13–17 November, 1996. http://www.fao.org/documents/shav_cdr.asp?url_file=/docrep/003/w3613e/w3613e00.htm (accessed 19 September 2005).

FAO (1998) *The State of the World's Plant Genetic Resources for Food and Agriculture.* FAO, Rome.

FAO (1999) Food and Agriculture Organization/Netherlands Conference on the Multifunctional Character of Agriculture and Land, 12–17 September, Maastricht, Netherlands. Background Paper 1: Agricultural Biodiversity.

FAO (2001) *Report of the Panel of Eminent Experts on Ethics in Food and Agriculture.* FAO, Rome.

Fishman, S. (2000) *The Public Domain. How to Find and Use Copyright-Free Writings, Music, Art and More.* NOLO, Berkeley, California, pp. 1–3.
Franck, T.M. (1993) *Fairness in the International Legal and Institutional System.* Hague Academy of International Law, Collected Courses, III.
Freese, C.H. (1998) *Wild Species as Commodities. Managing Markets and Ecosystems for Sustainability.* Island Press, Washington, DC.
Gaston, K.J. (1996) What is biodiversity? In: Gaston, K.J. (ed.) *Biodiversity. A Biology of Numbers and Difference.* Blackwell Science, Oxford, UK.
Girsberger, M.A. (1999) *Biodiversity and the Concept of Farmers' Rights in International Law. Factual Background and Legal Analysis.* Peter Lang, Berne, Switzerland.
Glowka, L., Burhenne-Guilmin, F. and Synge, H. (1994) *A Guide to the Convention on Biological Diversity.* Environmental Policy and Law Paper No. 30. IUCN, Gland, Switzerland.
Gupta, A.K. (1997) *Roots of Creativity and Innovation in Indian Society: A Honey Bee Perspective.* International Conference on Creativity and Innovation at Grassroots, Ahmedabad, India. Indian Institute of Management, Centre of Management in Agriculture.
Gupta, A.K. (2003) Securing traditional knowledge and contemporary innovations: can global trade links help grassroots innovations? In: Cottier, C. and Mavrodis, P. (eds) *Intellectual Property: Trade, Competition and Sustainable Development.* Proceedings of the World Trade Forum, Berne, 27–28 August 1999. University of Michigan Press, Ann Arbor, Michigan.
Hawkes, J.G. (1983) *The Diversity of Crop Plants.* Harvard University Press, Cambridge, Massachusetts.
Hellsten, S. (1998) Theories of distributive justice. In: Chadwik, R. (ed.) *Encyclopedia of Applied Ethics*, Vol. 1. Academic Press, San Diego, California, pp. 815–827.
Henne, G. (1998) *Genetische Vielfalt als Ressource: Die Regelung ihrer Nutzung.* Nomos verlagsgesellschaft, Baden-Baden, Germany.
Hunn, E.S. (2001) Prospects for the persistence of 'endemic' cultural systems of traditional environmental knowledge: a Zapotec example. In: Maffi, L. (ed.) *On Biocultural Diversity: Linking Language, Knowledge, and the Environment.* Smithsonian Institution Press, Washington, DC.
Iltis, H. (1988) Serendipity in the exploration of biodiversity: what good are weedy tomatoes? In: Wilson, E.O. and Frances, P.M. (eds) *Biodiversity.* National Academy Press, Washington, DC, pp. 98–105.
IPCC (Intergovernmental Panel on Climate Change), Working Group II (2001) *Climate Change 2001: Impacts, Adaptation, and Vulnerability.* Summary for Policymakers. Available at: http://www.ipcc.ch
IUCN, The Economics Unit (2001) *Incentive Measures for Biodiversity. A Training Module for Natural Resources Managers.* Draft 26 February 2001. Available at: http://www.biodiversityeconomics.org/applications/library_documents/lib_document.rm?document_id=560 (accessed 20 September 2005).
IUCN, UNEP, WWF (1980) *Stratégie Mondiale de la Conservation: La Conservation des Ressources Vivantes au Service du Développement Durable.* IUCN, Gland, Switzerland.
Janssen, J. (1997) Property rights on genetic resources: economic issues. In: Biber-Klemm, S. (ed.) *Legal Claims to Biogenetic Resources.* NFP 42 Working Paper No. 8. Swiss National Science Foundation, Berne, Switzerland.
Kaul, I., Grunberg, I. and Stern, M.A. (1999) Defining Global Public Goods. In: Kaul, I., Grunberg, I. and Stern, M.A. (eds) *Global Public Goods.* Oxford University Press, New York, pp. 2–19.
Krebs, A. (1999) *Ethics of Nature: A Map.* Walter de Gruyter, Berlin.
Kunin, W.E. and Lawton, J.H. (1996) Does biodiversity matter? Evaluating the case for conserving species. In: Gaston, K.J. (ed.) *Biodiversity. A Biology of Numbers and Difference.* Blackwell Science, Oxford, UK.
Loizou, A. (1998) Theories of Justice: Rawls. In: Chadwik, R. (ed.) *Encyclopedia of Applied Ethics*, Vol. 4. Academic Press, San Diego, California, pp. 347–363.
Mendelssohn, R. and Balick, M.J. (1995) The value of undiscovered pharmaceuticals in tropical forests. *Economic Botany* 49.
Maffi, L. (1999) Linguistic diversity. In: Posey, D.A. *et al.* (eds) *Cultural and Spiritual Values of Biodiversity.* Intermediate Technology, London, pp. 21–35.
Maffi, L. (2001a) On the interdependence of biological and cultural diversity. In: Maffi, L. (ed.) *On Biocultural Diversity: Linking Language, Knowledge, and the Environment.* Smithsonian Institution Press, Washington, DC, pp. 1–50.

Maffi, L. (ed.) (2001b) *On Biocultural Diversity: Linking Language, Knowledge, and the Environment.* Smithsonian Institution Press, Washington, DC.

McDonald, M. Arraguitainaq, L. and Novalinga, Z. (eds) (1997) *Voices from the Bay, Traditional Ecological Knowledge of Inuit and Cree in the Hudson Bay Bioregion.* Canadian Arctic Resources Committee, Ottawa; Environmental Committee of Municipality of Sanikiluaq, N.W.T.

McNeely, J.A., Miller, K.R., Reid, W.V., Mittermeier, R.A. and Werner, T.B. (1990) *Conserving the World's Biological Diversity.* World Bank Publications, Washington, DC.

Mülhäusler, P. (2001) Ecolinguistics, linguistic diversity, ecological diversity. In: Maffi, L. (ed.) *On Biocultural Diversity; Linking Language, Knowledge, and the Environment.* Smithsonian Institution Press, Washington, DC, pp. 133–144.

Norton, B. (1988) Commodity, amenity and morality – the limits of quantification in valuing biodiversity. In: Wilson, E.O. and Frances, P.M. (eds) *Biodiversity.* National Academy Press, Washington, DC, pp. 200–205.

Nuffield Council on Bioethics (1999) *Genetically Modified Crops: The Ethical and Social Issues.* Nuffield Council of Bioethics, London. Available at: www.nuffieldfoundation.org (accessed 20 September 2005).

Nwokeabia, H. (2001) *Why Industrial Revolution Missed Africa: A 'Traditional Knowledge' Perspective.* Economic Commission for Africa (ECA/ESPD/WPS/01/02).

OECD (Organization for Economic Cooperation and Development) (1996) *Saving Biological Diversity. Economic Incentives.* OECD, Paris.

OECD (1997) *Investing in Biological Diversity: The Cairns Conference.* OECD Proceedings. OECD, Paris.

OECD (1999) *Handbook of Incentive Measures for Biodiversity. Design and Implementation.* OECD, Paris.

Oviedo, G. (2000) *Indigenous and Traditional Peoples of the World and Ecoregion Conservation.* WWF, Gland, Switzerland, p. 116.

Oviedo, G., Maffi, L. and Larsen, P.B. (2000) *Indigenous and Traditional Peoples of the World and Ecoregion Conservation: an Integrated Approach to Diversity.* W.W.F. International and Terralingua, Gland, Switzerland.

Posey, D.A. (1996) *Traditional Resource Rights.* IUCN, Gland, Switzerland.

Posey, D.A. (2001) Biological and cultural diversity. In: Maffi, L. (ed.) *On Biocultural Diversity: Linking Language, Knowledge, and the Environment.* Smithsonian Institution Press, Washington, DC.

Primack, R.B. (1995) *Naturschutzbiologie.* Spektrum Akademischer Verlag, Heidelberg, Germany.

Rawls, J. (1971) *A Theory of Justice.* The Belknap Press of Harvard University Press, Cambridge, Massachusetts.

Rural Advancement Foundation International (RAFI) (1994) *Conserving Indigenous Knowledge: Integrating Two Systems of Innovation.* United Nations Development Programme, Canada.

Secretariat of the Convention on Biological Diversity (2001) *Handbook of the Convention on Biological Diversity.* CBD, UNEP, Earthscan Publications Ltd, London and Sterling, Virginia.

Sen, A. (1987) Food and Freedom. Sir John Crawford Memorial Lecture. 29 October, Washington, DC. http://www.worldbank.org/html/cgiar/publications/craw.html (accessed 20 September 2005).

Serageldin, I. (1997) Equity and ethics: twin challenges, twin opportunities. In: International Plant Genetic Resources Institute (IPGRI) (ed.) *Ethics and Equity in Conservation and Use of Genetic Resources for Sustainable Food Security.* IPGRI, Rome.

Serageldin, I. and Persley, G.J. (2000) *Promethean Science. Agricultural Biotechnology, the Environment, and the Poor.* Consultative Group on International Agricultural Research, Washington, DC.

Sheldon, J.W. and Balick, M.J. (1995) Ethnobotany and the search for balance between use and conservation. In: Swanson, T. (ed.) *Intellectual Property Rights and Biodiversity Conservation: An Interdisciplinary Analysis of the Values of Medicinal Plants.* Cambridge University Press, Cambridge, UK, pp. 45–64.

Stettler, N. (2002) *Natur erforschen.* Chronos Verlag, Zurich, Switzerland.

Stiglitz, J.E. (1999) Knowledge as a global public good. In: Kaul, I., Grunberg, I. and Stern, M.A. (eds) *Global Public Goods.* Oxford University Press, New York, pp. 308–325.

Straughan, R. (1997) Analysis of ethical principles arising from the Workshop. In: International Plant Genetic Resources Institute (ed.) *Ethics and Equity in Conservation and Use of Genetic Resources for Sustainable Food Security.* Proceedings of a workshop to develop guidelines for the CGIAR, 21–25 April, Foz do Iguaçu, Brazil. IPGRI, Rome.

Swaminathan, M.S. (1997) Implementing the benefit sharing provisions of the convention on biological diversity: challenges and opportunities. *Plant Genetic Resources Newsletter* 112, 19–27.

Swanson, T.M. (1994) *The International Regulation of Extinction.* New York University Press, New York.

Swanson, T.M. (1995) The appropriation of evolution's values: an institutional analysis of intellectual property regimes and biodiversity conservation. In: Swanson, T.M. (ed.) *Intellectual Property Rights and Biodiversity Conservation: An Interdisciplinary Analysis of the Values of Medicinal Plants.* Cambridge University Press, Cambridge, UK, pp. 141–175.

Ten Kate, K. and Laird, S.A. (2000) *The Commercial Use of Biodiversity: Access to Genetic Resources and Benefit Sharing.* Earthscan Publications, London.

The Crucible II Group (2001) *Seeding Solutions.* International Development Research Centre, Ottawa; International Plant Genetic Resources Institute, Rome; Dag Hammarskjöld Foundation, Uppsala, Sweden.

Thrupp, L.A. (2000) Linking agricultural biodiversity and food security: the valuable role of agrobiodiversity for sustainable agriculture. *International Affairs* 76(2), 265–281.

UNFPA (United Nations Population Fund) (2001) *Footprints and Milestones: Population and Environmental Change – The State of the World Population.* Thoraya Ahmed Obiad, Executive Director.

United Nations (2002) Report of the World Summit on Sustainable Development. Johannesburg, South Africa. 26 August–4 September 2002.

WHO (World Health Organization) (1996) *Traditional Medicine.* Fact Sheet N 134, September. Available at: http://www.who.int/inf-fs/en/fact134.html (accessed 20 September 2005).

WHO (2002a) *WHO Traditional Medicine Strategy 2002–2005.* WHO, Geneva.

WHO (2002b) *Traditional Medicine – Growing Needs and Potential.* WHO Policy Perspectives on Medicines No. 2, May. WHO, Geneva.

Wilson, E.O. (1992) *The Diversity of Life.* Penguin Books Ltd, London.

WIPO (World Intellectual Property Organization) (2001) *Intellectual Property Needs and Expectations of Traditional Knowledge Holders.* WIPO Report on Fact-finding Missions on Intellectual Property and Traditional Knowledge (1998–1999). WIPO, Geneva.

Wolf, C. (1998) Theories of justice: human needs. In: Chadwik, R. (ed.) *Encyclopedia of Applied Ethics*, Vol. 4. Academic Press, San Diego, California, pp. 335–345.

Wood, A., Stedman-Edwards, P. and Mang, J. (eds) (2000) *The Root Causes of Biodiversity Loss.* Earthscan Publications, London.

Woodliffe, J. (1996) Biodiversity and indigenous peoples. In: Bowmann, M. and Redgwell, C. (eds) *International Law and the Conservation of Biological Diversity.* Kluwer Law International, The Hague, pp. 255–269.

World Commission on Environment and Development (1987) *Our Common Future.* Oxford University Press, Oxford, UK.

Young, M.D. and Gunningham, N. (1997) Mixing instruments and institutional arrangements for optimal biodiversity conservation. In: OECD (ed.) *Investing in Biological Diversity: The Cairns Conference.* Proceedings of the OECD International Conference on Incentive Measures for the Conservation and the Sustainable Use of Biological Diversity in Cairns, Australia, 25–28 March 1996. OECD, Paris, pp. 141–165.

2 The Current Law of Plant Genetic Resources and Traditional Knowledge

Susette Biber-Klemm,[a] Thomas Cottier,[b] Philippe Cullet[c] and Danuta Szymura Berglas[a]

[a]World Trade Institute, University of Berne, and Faculty of Law, University of Basel, Maiengasse 51, CH 4056 Basel, Switzerland; [b]World Trade Institute, University of Berne, Hallerstrasse 6/8, CH 3012 Berne, Switzerland; [c]School of Law, School of Oriental and African Studies, University of London, Thornhaugh Street, London WC1 0XG, UK

2.1 Introduction

This chapter seeks to lay out the main legal foundations in existing international law for the protection of plant genetic resources (PGR) and traditional knowledge (TK). It is, primarily, an exercise of stocktaking, of providing a survey and of identifying issues to be dealt with in successive chapters.

The chapter first introduces the general principles of international law, which, today, are enshrined by the doctrine of permanent sovereignty over natural resources. While there was a trend to address the heritage of biological diversity in terms of a common public good and ascribe it to the limited domain of common heritage, it is well established, since the 1992 Rio Conference and subsequent instruments, that the matter does pertain to national sovereignty and therefore national regulation within the bounds of international law. A survey of international agreements, in particular the Convention on Biodiversity (CBD), affirms this view. Equally, a survey of agreements which seek to provide an open, multilateral international system within the International Undertaking on Plant Genetic Resources (IUPGR) (International Undertaking; IU) and the new International Treaty on Plant Genetic Resources for Food and Agriculture (ITPGRFA), administered by the Food and Agriculture Organization (FAO), confirm the point by way of introducing mutually granted concessions of unimpaired facilitated access to plant genetic resources. Similarly, the efforts within the Consultative Group on International Agricultural Research (CGIAR) in supporting an open system confirm the overall principle. This chapter then provides a short survey of World Trade Organization (WTO) law and seeks to identify its main implications for PGR and TK. Given its comprehensive and complex structure, the Agreement on Trade-related Aspects of Intellectual Property Rights (TRIPS Agreement) will be examined in detail in subsequent chapters of this book. The same is true for other instruments addressing property rights, in particular the *Union internationale pour la protection des obtentions vegetales* (UPOV)

Convention and the ITPGRFA. Finally the survey addresses pertinent national legislation by way of example. A short description of the European and US intellectual property rights (IPR) system provides the necessary background, as a brief introduction to access legislation currently existing or partly enacted in developing countries, under the doctrine of permanent sovereignty over natural resources.

At this stage, it is already clear that the problem of PGR and TK is formed by a host of different principles, rules and agreements, most of them administered by different international organizations and fora, and thus also by different national authorities and constituencies: from departments of the environment, to agriculture, to foreign trade, human rights, human health, plant and animal health, to intellectual property, competition, and perhaps additional ones. They all face the problem of policy coordination at home, and even more so at the international level. The system is still highly fragmented, organized in different chapters, often without much coordination or interaction between these. It will be one of the major goals to bring about structures in the field that allow meaningful interaction, cooperation and joint decision-making.

2.2 Permanent Sovereignty over Plant Genetic Resources[1]

State sovereignty remains the fundamental principle around which all inter-state relations are organized.[2] In effect, international law is based on the principle that all states are juridically equal and that there is no authority superior to states. One of the specific elements of state sovereignty concerns the control of the natural, biological and genetic resources found in areas under their jurisdiction.

The question of sovereignty over natural resources was the object of intense debate in the aftermath of decolonization. Newly independent states, after gaining political independence, sought to gain control over resources found under their jurisdiction, with a view to gaining control over the exploitation of economically valuable natural resources. The recognition of States' Permanent Sovereignty over Natural Resources was first formalized in a UN General Assembly Resolution.[3] This principle refers to the right of each state to freely exploit and develop its natural resources. It constitutes the basic principle for allocating rights and responsibilities in international law in this field. The principle of permanent sovereignty has since been reiterated in numerous treaties and other international documents. Its binding status in international law is unquestioned (Schrijver, 1997).

One of the attributes of sovereignty is that states can freely choose to restrict it by way of agreement. In the field of the environment, the scope of sovereign rights has slowly been circumscribed. In fact, one of the most generally recognized principles of customary international law in the field of the environment is Principle 21 of the Stockholm Declaration, which provides that states have

> the sovereign right to exploit their own resources pursuant to their own environmental policies, and the responsibility to ensure that activities within their jurisdiction or control do not cause damage to the environment of other States or of areas beyond the limits of national jurisdiction.[4]

[1] Author: Philippe Cullet.

[2] UN Charter, San Francisco, 26 June 1945 and UN General Assembly Res. 2625 (XXV), Declaration on Principles of International Law Concerning Friendly Relations and Co-operation among States in Accordance with the Charter of the United Nations, 24 October 1970, reprinted in 9 ILM 1292 (1970).

[3] UN General Assembly Resolution 1803 (XVII), 14 December 1962, Permanent Sovereignty over Natural Resources, reprinted in 2 ILM 223 (1963).

[4] Declaration of the United Nations Conference on the Human Environment, Stockholm, 16 June 1972, reprinted in 11 ILM 1416 (1972).

This basic limitation on states' rights to use their territories and resources as they see fit was confirmed in the Rio Declaration.[5]

Further refinements have been added over time. Specific issues have given rise to special principles. In the case of biodiversity, states negotiating the CBD found a specific compromise. The convention 'reaffirms' states' sovereign rights over their biological resources but increases their responsibility to conserve and sustainably use the resources they control.[6] The responsibility that states have towards the international community is encapsulated in the principle of 'common concern'. Common concern implies that the sovereign rights that are recognized as belonging to states are tempered by the international community's interest in the conservation and sustainable use of a global good (Boyle, 1994). In other words, states retain control over their biological resources but have a duty to cooperate with other states in resolving issues of interest to the international community.

Apart from the special principle of 'common concern' devised for biological resources, distinct rules have also evolved in the case of genetic resources. The development of rules for the control of genetic resources has followed a different trend from biological and natural resources found under the jurisdiction of states. Plant genetic resources were traditionally seen as a common heritage of humankind.[7] The main implication of the common heritage status in the case of genetic resources was that there should be no restriction on availability and, consequently, appropriation under private right schemes was deemed inappropriate. This constituted the core principle of the 1983 International Undertaking. The notion of common heritage has progressively given way to appropriation under the guise of both private rights and sovereign rights. First, disagreements over common heritage status in the context of the International Undertaking have led states to adopt interpretations of the International Undertaking that could be accepted by all FAO member states. The 1989 resolutions maintained the principle of common heritage but intimated that both plant breeders' rights and farmers' rights were compatible with the principle of common heritage.[8] Secondly, the CBD recognized states' sovereign rights over both biological and genetic resources that fall under their jurisdiction. Since the principle of common heritage had by 1993 become redundant, it became necessary to renegotiate the International Undertaking to take into account the new trends towards appropriation and benefit sharing. The resulting ITPGRFA makes no mention of the principle of common heritage.

The regulation of genetic resources presents further difficulties. Given their nature, a state's control over its genetic resources is even more difficult to assert than control over biological resources. The rapid development of genetic engineering has led to calls for new rules to regulate access to genetic resources. The CBD specifically addresses this issue. Article 15 recognizes states' sovereign rights to regulate access to their genetic resources. However, it goes on to impose an obligation on states not to unduly restrict access by other parties. In other words, the CBD gives states the authority to ultimately determine con-

[5] Principle 2, Rio Declaration on Environment and Development, 14 June 1992, Rio de Janeiro, reprinted in 31 ILM 874 (1992).

[6] Preamble, Convention on Biological Diversity, Rio de Janeiro, 5 June 1992, reprinted in 31 ILM 818 (1992).

[7] The concept of common heritage was first developed in the context of the law of the sea concerning deep seabed resources. See, for example, UN General Assembly Res. 2749 (XXV), Declaration of Principles Governing the Seabed and the Ocean floor, and the Subsoil Thereof, Beyond the Limits of National Jurisdiction, 17 December 1970, GAOR 25th Session, Supp. 28 (A/8028).

[8] Res. 4/89, Agreed Interpretation of the International Undertaking, 29 November 1989, Report of the Conference of FAO, 25th Session, Rome 11–29 November 1989, Doc. C89/REP and Res. 5/89, Farmers' Rights, 29 November 1989, Report of the Conference of FAO, 25th Session, Rome 11–29 November 1989, Doc. C89/REP.

ditions for access to their resources but at the same time strongly encourages them not to simply deny access to other contracting parties. While states are asked to provide regulated access to their genetic resources, the CBD makes provision for two important factors. First, access must be subject to prior informed consent. Secondly, the benefits directly or indirectly derived from accessing genetic resources must be fairly shared. This ensures that while states are strongly encouraged to provide access to resources that are under their sovereignty, they also participate in the benefits arising from this limitation of their sovereign rights.

Finally, the special situation of genetic resources held in the gene banks of the Consultative Group on International Agricultural Research (CGIAR) should be mentioned. These accessions have, in most cases, been provided voluntarily by states party to these *ex situ* gene banks, on the understanding that all designated germplasm held by the various International Agricultural Research Centres (IARCs) would be freely accessible. The CGIAR has been under pressure to revise its intellectual property guidelines in recent years. The basic principle remains that the IARCs do not apply intellectual property protection to their designated germplasm and require recipients to observe the same conditions. They also refrain from asserting IPRs over the products of their research. An exception to this rule is made in case the assertion of IPRs facilitates technology transfer or otherwise protects developing countries' interests. The CGIAR also imposes the condition that any IPRs on the IARCs' output will be assigned to the Centre and not an individual. While the guiding principles on intellectual property generally seek to contain to an extent the monopoly elements of IPRs such as patents, plant breeders' rights are specifically welcomed (Consultative Group on International Agricultural Research, 1999). Recipients of germplasm can apply for plant breeders' rights as long as this does not prevent others from using the original materials in their own breeding programmes.

This brief review of the evolution of the concept of sovereignty in the specific field of biological and genetic resources can be summarized as follows. First, at least since the 1960s, the principle that states have sovereign rights over their natural and biological resources has been reiterated on numerous occasions. Secondly, genetic resources that used to be considered a common heritage of humankind are now firmly included in the same category as biological resources. Thirdly, the relative importance of the notion of sovereignty should not be overestimated, since there has been a parallel trend towards much stronger private property rights over biological and genetic resources, in particular IPRs. Overall, the main trend over the past few decades has been towards the appropriation of economically valuable resources either through sovereign or private property rights.

2.3 International Agreements and Institutions

A number of international agreements constitute the law concerning PGR and TK at the international level. This section highlights some of the most significant treaties and institutions dealing with the management of biological and genetic resources, IPRs and issues surrounding the conservation and use of TK.

2.3.1 The Convention on Biological Diversity[9]

The CBD, even more than most environmental treaties, is the result of the progressive development of international environmental law since the early 1970s. Since the 1972 Stockholm Conference, states have signed various regional and international agreements designed to deal with various environmental issues, from

[9] Author: Philippe Cullet, with the help of Danuta Szymura Berglas.

the protection of specific species, habitats or ecosystems, to treaties dealing with the use and exploitation of environmental resources, as well as treaties addressing some of the consequences of industrial activity, such as air pollution or hazardous wastes. However, before 1992, states had not managed to adopt a comprehensive legal framework for the conservation and management of biological resources. The CBD fills this gap and provides the first umbrella agreement addressing both the conservation and use of all biological resources.

The CBD was adopted in 1992 in the form of a binding framework treaty. Its three main goals are the conservation of biological diversity, the sustainable use of its components,[10] and the fair and equitable sharing of the benefits derived from the use of genetic resources (Article 1 CBD). The Convention reaffirms the cornerstone principle of state sovereignty over resources that grants states sovereign rights to exploit their own resources pursuant to their own environmental policies, together with the responsibility to ensure that activities within their own jurisdiction or control do not cause damage to the environment of other states or areas beyond the limits of national jurisdiction. The Convention, however, innovates in bringing a new qualification to the principle of sovereignty. It introduces the notion that the conservation of biological diversity is a 'common concern of humankind', whereby states have a duty to cooperate in the sustainable management of resources found under their jurisdiction.

The CBD provides a number of general obligations for member states. In particular, these include a commitment to develop national strategies, plans or programmes for the conservation and sustainable use of biological diversity. Member states must also integrate, as far as possible, and as appropriate, the conservation and sustainable use of biological diversity into relevant sectoral or cross-sectoral plans, programmes and policies. Generally, member states are required to promote the sustainable use of biological resources by: integrating consideration of the conservation and sustainable use of biological resources into national decision-making; adopting measures relating to the use of biological resources to avoid or minimize adverse impacts on biological diversity; protecting and encouraging customary use of biological resources in accordance with traditional cultural practices that are compatible with conservation or sustainable use requirements; supporting local populations to develop and implement remedial action in degraded areas where biological diversity has been reduced; and encouraging cooperation between its governmental authorities and its private sector in developing methods for sustainable use of biological resources (Article 10 CBD).

Conservation under the CBD is to be achieved in two main ways. First, the Convention emphasizes *in situ* conservation which proposes the conservation of genes, species and ecosystems in the surroundings where they have developed their distinctive properties by establishing protected areas, rehabilitating degraded ecosystems and adopting legislation to protect endangered species.[11] *In situ* conservation implies, among other things: the development of guidelines for protected areas; the regulation of biological resources; the promotion of the protection of ecosystems and natural habitats and the maintenance of viable populations of species in natural surroundings; the promotion of environmentally sound and sustainable development in adjacent

[10] Sustainable use is defined under Article 2 of the CBD as the use of components of biological diversity in a way and at a rate that does not lead to the long-term decline of biological diversity, thereby maintaining its potential to meet the needs and aspirations of present and future generations.

[11] Article 2 of the CBD defines *in situ* conditions as 'conditions where genetic resources exist within ecosystems and natural habitats, and, in the case of domesticated or cultivated species, in the surroundings where they have developed their distinctive properties'.

areas; the rehabilitation and restoration of degraded ecosystems and the promotion of the recovery of threatened species; controlling the risks associated with the use of living modified organisms; controlling alien species; seeking compatibility between present and future use; developing necessary legislation to protect threatened species or populations; regulating any processes or activities found to have an adverse impact; and providing financial support for *in situ* conservation, especially to developing countries (Article 8 CBD).

Secondly, supplementary *ex situ* conservation outside the natural habitats of the protected biodiversity components is also proposed. *Ex situ* conservation requires the use of gene banks and zoological and botanical gardens to conserve species, which can contribute to saving endangered species. *Ex situ* measures are preferably undertaken in the country of origin. They include a duty to maintain facilities for the conservation of, and research on, plants, animals and microorganisms; to seek the rehabilitation of threatened species and their reintroduction into their natural habitats; to regulate the collection of biological resources from natural habitats for *ex situ* conservation so as not to unnecessarily threaten ecosystems and *in situ* populations of species; and to provide financial support for *ex situ* conservation, especially to developing countries (Article 9 CBD).

The CBD addresses a number of other issues of specific interest in the context of plant genetic resources. It imposes a duty on all member states to: respect, preserve and maintain knowledge, innovations and practices of indigenous and local communities embodying traditional lifestyles relevant for the conservation and sustainable use of biological diversity; promote their wider application with the approval and involvement of the holders of such knowledge, innovations and practices; and encourage the equitable sharing of the benefits arising from the utilization of such knowledge, innovations and practices (Article 8(j) CBD). The Conference of the Parties (COP) has addressed the implementation of Article 8(j) in different ways. An *ad hoc* open-ended inter-sessional working group was established at COP 4 to give more importance to issues related to TK.[12] The Working Group is mandated with the task of giving advice on legal and other means of protection of the knowledge, innovations and practices of indigenous and local communities embodying traditional lifestyles relevant to the conservation and sustainable use of biological diversity. Further, COP 5 adopted a specific programme of work which aims at fostering the participation of local and indigenous communities in all aspects of the implementation of Article 8(j).[13] Member states are also requested to protect and encourage customary use of biological resources in accordance with traditional cultural practices that are compatible with conservation or sustainable use requirements (Article 10 CBD).

The CBD also provides a general legal framework regulating access to biological resources and the sharing of benefits arising from their use. The question of access is closely related to the issue of property rights and the legal status of biological resources under international law. Access has progressively become a contentious issue as the development of genetic engineering provides new ways to acquire IPRs over inventions derived from biological resources. As a result, since much of the world's biodiversity is found in developing countries,[14] the question of access became one of central importance in the CBD negotiations.

---

[12] Decision IV/9. Implementation of Article 8(j) and related provisions, in Decisions Adopted by the Conference of the Parties to the Convention on Biological Diversity at its Fourth Meeting, Bratislava, 4–15 May 1998, UN Doc. UNEP/CBD/COP/4/27 (1998).

[13] Programme of Work on the Implementation of Article 8(j) and Related Provisions of the Convention on Biological Diversity, in Decisions Adopted by the Conference of the Parties to the Convention on Biological Diversity at its Fifth Meeting, Nairobi, 15–26 May 2000, UN Doc. UNEP/CBD/COP/5/23 (2000).

[14] An important exception is Australia, which is both developed and 'megadiverse'.

The Convention attempts to provide a framework which both respects donor countries' sovereign rights over their biological and genetic resources and facilitates access by users. Access must therefore be provided on 'mutually agreed terms' and is subject to the 'prior informed consent' of the country of origin (Article 15 CBD). Further, the CBD provides that donor countries of microorganisms, plants or animals used commercially have the right to a fair share of the benefits derived from their use. Benefit-sharing, as conceived under the Convention and the Bonn Guidelines adopted in 2002, can take the form of monetary benefits, such as access fees; up-front payments; payment of royalties; licence fees in case of commercialization; research funding; and joint ventures. Benefit-sharing can also take the form of non-monetary benefits such as the sharing of research and development results; collaboration, cooperation and contribution in scientific research and development programmes; participation in product development; admittance to *ex situ* facilities of genetic resources and to databases; training related to genetic resources; and access to scientific information relevant to conservation and sustainable use of biological diversity (Bonn Guidelines, 2002).

The Convention goes further than the general regime provided by Article 15 and specifically provides for technology transfer as an invaluable instrument for the effective implementation of the Convention. In fact, Article 16 specifically recognizes the need to facilitate the transfer of technologies that are relevant to the conservation and sustainable use of biological diversity, or make use of genetic resources and do not cause significant damage to the environment.

The CBD is noteworthy in the context of international environmental agreements for offering a Statement on the relationship between the management of biological and genetic resources and IPRs. Article 16 clearly indicates that IPRs are not to undermine the working of the Convention. The actual relationship of the CBD with the TRIPS Agreement is an issue that has not yet found a specific answer. This is partly due to the fact that a clear-cut answer to this question would remove some of the grey areas that currently allow different states to take different views on this matter.

The agreement reached on the substantive provisions of the Convention was partly dependent on the willingness of developed countries to provide financial resources to subsidize developing countries' compliance with the proposed regime. As a result, the Convention requests the allocation of 'new and additional financial resources' to enable developing countries to meet the 'agreed full incremental costs' of implementing measures which fulfil the obligations of this Convention (Article 20 CBD). The importance of financial commitments for developing countries is illustrated by the fact that developing countries have the possibility, under the Convention, of making the implementation of their commitments dependent on the effective implementation by developed countries of their commitments related to financial resources and transfer of technology. In practice, developed countries discharge their financial commitments through the Global Environment Facility (GEF), which has operated the financial mechanism since the entry into force of the CBD.[15]

The institutional structure of the CBD includes a number of bodies. These include the Conference of the Parties, a Subsidiary Body on Scientific, Technical and Technological Advice (SBSTTA), and the Secretariat. The Conference of the Parties, which brings together all member states, is generally mandated with keeping the implementation of the Convention under review. More specifically, it reviews progress under the Convention, identifies new priorities to be pursued, sets work plans for members, amends the Convention, creates expert advisory bodies, reviews progress reports by member nations and collaborates with other international organizations and agree-

[15] See Chapter 6 for further developments on the GEF.

ments. Periodic state reports to the Conference of the Parties constitute one of the main monitoring instruments instituted under the Convention. State parties must report on the means they have adopted to implement the objectives of the Convention and the level of success of such measures. The Conference of the Parties has launched a number of thematic programmes covering, for instance, the biodiversity of inland waters, forests, marine and coastal areas, dry lands and agricultural lands, agricultural biodiversity, and cross-cutting issues such as the control of alien invasive species, strengthening the capacity of member countries in taxonomy, and the development of indicators of biodiversity loss.

The SBSTTA has been established to provide expert advice to other organs of the Convention. It is a multidisciplinary expert body which has the mandate to: provide scientific and technical assessments of the status of biological diversity and of the effects of types of measures taken in accordance with the Convention; identify innovative, efficient and state-of-the-art technologies and know-how relating to the conservation and sustainable use of biological diversity, and advise on the ways and means of promoting development and/or transferring such technologies; and provide advice on scientific programmes and international cooperation in research and development related to conservation and sustainable use of biological diversity. The CBD functions on a daily basis through a Secretariat situated in Montreal, which organizes meetings, drafts documents, assists member governments in the implementation of programmes, and coordinates with other international organizations to collect and disseminate information.

The CBD provides for the adoption of protocols. To date, only one instrument, the protocol on biosafety, has been adopted.[16] The Cartagena Protocol has been adopted in response to concerns over the negative environmental and health effects of some genetic engineering developments. The Protocol seeks to regulate some aspects of the transboundary movement of living modified organisms, in particular bio-engineered agricultural goods. Its aim is to ensure that a sufficient level of protection is achieved so that the transfer of living modified organisms does not entail adverse effects on the conservation and use of biodiversity. The Protocol also recognizes that the risks to human health posed by modified organisms are closely related to the risk they pose to biodiversity. The Protocol is put in practice through a procedure for 'advanced informed agreement'.[17] This imposes a duty on the exporter of a living modified organism falling under the scope of the Protocol to provide at least the information listed in Annex I to the Protocol. In the case of seeds or live animals, importing countries have the right to restrict imports in order to minimize possible adverse effects on the conservation and sustainable use of biodiversity.[18] One of the most significant elements of the Protocol is that a decision to refuse an import can be taken even in the absence of full scientific certainty regarding the extent of the potential adverse effects. In the case of commodities to be used as food or feed, the procedure is generally less onerous for the exporter and does not include an obligation to notify the importing country of each shipment, even though the importing country can generally restrict importations of a specific commodity.

2.3.2 The International Treaty on Plant Genetic Resources for Food and Agriculture[19]

The international legal regime for the conservation and use of agricultural plant

[16] Cartagena Protocol on Biosafety to the Convention on Biological Diversity, Montreal, 20 January 2000, reprinted in 39 ILM 1027 (2000).
[17] Article 7 ff. of the Cartagena Protocol, Note 16 above.
[18] Article 11 of the Cartagena Protocol, Note 16 above.
[19] Author: Philippe Cullet, with the help of Danuta Szymura Berglas.

genetic resources has been marked by significant changes over the past few decades. Traditionally, plant genetic resources for food and agriculture (PGRFA) were freely exchanged on the basis that PGRFA constituted a common heritage of humankind. As a result, PGRFA could not be appropriated. These principles were reflected in the practice of the CGIAR and were embodied in the International Undertaking adopted by the FAO Conference in 1983 (FAO Resolution 8/83).

It affirmed the principle that plant genetic resources are a heritage of humankind that should be made available without restriction to anyone. This covers not only traditional cultivars and wild species but also varieties developed by scientists in laboratories. The International Undertaking was adopted as a non-binding conference resolution. However, the emphasis on the free availability of PGRFA proved to be unacceptable to some developed countries that already had interests in genetic engineering. Broader acceptance of the International Undertaking was only achieved after the FAO Conference passed interpretative resolutions in 1989 and 1991.[20] These resolutions affirm the need to balance the rights of formal innovators as breeders of commercial varieties and breeders' lines on the one hand, with the rights of informal innovators of farmers' varieties on the other. Resolution 4/89 recognizes that plant breeders' rights, as provided for in UPOV Convention, are not inconsistent with the Undertaking, and simultaneously recognizes farmers' rights as defined in Resolution 5/89. Resolution 3/91 recognizes the sovereign rights of nations over their own genetic resources.

Further revision of the International Undertaking has been prompted by the growing importance of biological resources at the international level. In 1992, Agenda 21 called for the strengthening of the FAO Global System on Plant Genetic Resources, and its adjustment in accordance with the outcome of negotiations on the CBD. Resolution 3 of the Final Act to the CBD recognized that certain matters which the Convention had not addressed, such as the issue of access to *ex situ* collections not acquired in accordance with the Convention, and the realization of farmers' rights, were to be dealt with by the FAO's Global System on Plant Genetic Resources, of which the International Undertaking was the cornerstone. The FAO Conference provided the framework for the revision of the International Undertaking in its Resolution 7/93, which called on member states to generally harmonize the Undertaking with the provisions of the CBD, to address the issue of access to plant genetic resources, including the question of *ex situ* collections not addressed in the CBD, and specifically to ensure that the new instrument would foster the realization of farmers' rights. The negotiations for the revision of the Undertaking in harmony with the Convention began with the First Extraordinary Session of the Commission on Plant Genetic Resources in November 1994 and continued until the adoption of the Undertaking as a Treaty on 3 November 2001.[21]

The new Undertaking is now known as the International Treaty on Plant Genetic Resources for Food and Agriculture (ITPGRFA).[22] The Treaty was the object of arduous negotiations that led to a final consensus text that was acceptable to all the states present, apart from the United States and Japan, who abstained from voting upon the text.

The overall objectives of the ITPGRFA are significantly different from those of the 1983 Undertaking. The Treaty, reflecting the new orientation given by the CBD, emphasizes the conservation of PGRFA, their sustainable use and benefit sharing. The guiding principles for these three

[20] Res. 4/89, Note 8 above and Res. 5/89, Note 8 above.

[21] Negotiating documents can be found on the web site of the Commission on Genetic Resources for Food and Agriculture at http://www.fao.org/ag/cgrfa/default.htm

[22] International Treaty on Plant Genetic Resources for Food and Agriculture, Rome, 3 November 2001 (hereafter ITPGRFA).

objectives are the promotion of sustainable agriculture and food security.

The ITPGRFA addresses a number of issues pertaining to the conservation and use of plant genetic resources. In other words, the Treaty focuses on issues not addressed in other international treaties, such as farmers' rights, but it does not directly address patents or plant breeders' rights as covered in the TRIPS Agreement and the UPOV Convention, respectively.[23] The ITPGRFA is the first treaty providing a legal framework that not only recognizes the need for conservation and sustainable use of plant genetic resources for food and agriculture, but also delineates a regime for access and benefit sharing (ABS), and in this process provides direct and indirect links to intellectual property right instruments. Secondly, it directly links plant genetic resource conservation, IPRs, sustainable agriculture and food security. Thirdly, the ITPGRFA provides the first multilateral system of ABS for PGRFA. Fourthly, it includes a section on farmers' rights. The ITPGRFA gives recognition to farmers' contribution to conserving and enhancing plant genetic resources for food and agriculture. It further gives broad guidelines to states concerning the scope of the rights to be protected under this heading, but overall devolves the responsibility for realizing farmers' rights to member states. This includes the protection of TK, farmers' entitlement to a part of benefit-sharing arrangements and the right to participate in decision-making regarding the management of plant genetic resources. However, the treaty is silent with regard to farmers' rights over their landraces. In fact, the 'recognition' of farmers' contribution to plant genetic resource conservation and enhancement does not include any property rights. In this context, the only rights that are recognized are the residual rights to save, use, exchange and sell farm-saved seeds.

An important consequence of the Treaty is that guidance concerning the management of CGIAR collections will, in the future, come from the Treaty's Governing Body.[24] This is significant because the CGIAR has historically largely worked on the basis of the sharing of resources and knowledge. In recent years, following the adoption of the TRIPS Agreement in particular, the CGIAR has progressively modified its position with regard to the grant of IPRs.[25] New guiding principles on intellectual property were adopted to harmonize CGIAR's core principle that designated germplasm is held in trust for the world community with the recognition of various forms of property rights, including sovereign rights, farmers' rights and private rights.[26] In principle, the IARCs do not apply intellectual property protection to their designated germplasm and require recipients to observe the same conditions. They also refrain from asserting IPRs over the products of their research. An exception to this rule is made in cases where the assertion of IPRs facilitates technology transfer or otherwise protects developing countries' interests. The CGIAR also requires that any IPRs on the IARCs' output should be assigned to the Centre and not an individual. While the guiding principles generally seek to contain the monopoly elements of IPRs such as patents, plant breeders' rights are specifically welcomed. Recipients of germplasm can apply for plant breeders' rights as long as this does not prevent others from using the original materials in their own breeding programmes.

In practice, the ITPGRFA is of greater

[23] On the relationship between the Treaty and intellectual property right instruments, see, for example, Article 12.3.f of the ITPGRFA, Note 22 above.

[24] See Article 15 of the ITPGRFA, Note 22 above.

[25] See, for example, CGIAR, 'Progress Report on IPR Matters and Proposal for Review of Plant Breeding', Mid-Term Meeting, 1999, Beijing, CGIAR Doc. MTM/99/20.

[26] See CGIAR (1999) *CGIAR Center Statements on Genetic Resources, Intellectual Property Rights, and Biotechnology*. CGIAR, Washington, DC. Individual IARCs have also adopted intellectual property right policies; see, for example, International Maize and Wheat Improvement Center, Policy on Intellectual Property (2000).

importance for PGRFA covered under the Multilateral System. Concerning the resources covered, member states must adopt an integrated approach to the exploration, conservation and sustainable use of PGRFA. Their specific obligations include the need to inventory PGRFA, promote their collection, promote farmers' and local communities' efforts to manage and conserve their PGRFA on-farm, promote *in situ* conservation of wild-crop relatives and wild plants for food production, and co-operate to promote the development of an efficient and sustainable system of *ex situ* conservation. Further, member states are to take measures to eliminate threats to PGRFA. Alongside conservation obligations, member states must also promote the sustainable use of PGRFA. Measures include the promotion of agricultural policies that foster the development and maintenance of diverse farming systems; an emphasis on research which enhances and conserves biological diversity by maximizing intra- and inter-specific variation for the benefit of farmers; the promotion of plant breeding efforts which strengthen the capacity to develop varieties particularly adapted to social, economic and ecological conditions; the promotion of the use of local and locally adapted crops, varieties and underutilized species; and the support of the wider use of diverse varieties and species in on-farm management, conservation and sustainable use of crops.

The main institutional innovation of the ITPGRFA is found in the novel scheme devised to regulate ABS of PGRFA covered under the Treaty. The Multilateral System is a consequence of the policy reversal, which has seen PGRFA pass from the domain of a shared resource to one that is under the sovereignty of states. This is a direct consequence of the link between the Treaty and the CBD that provides for national sovereignty over biological resources. The underlying reason for the inclusion of a system of facilitated access is that the sovereign rights of states over their PGRFA are qualified by the recognition that these resources are a common concern of humankind and that all countries depend largely on PGRFA that originated in other countries. As a result, donor countries have full control over their PGRFA but there are strict limitations on their ability to restrict access to other states.

Under the Multilateral System, a series of crops (listed in Annex I) which account for most of – but not all – human nutrition are covered by a provision under which member states agree to provide facilitated access. As per the ITPGRFA, access is to be provided only for the purpose of utilization and conservation for research, breeding and training for food and agriculture. As a result of the recognition of PGRFA as a common concern, access has to be accorded expeditiously. Member states must also make available all passport data and, subject to applicable law, any other associated available non-confidential descriptive information. Concerning material that is under development by farmers or breeders at the time when access is requested, the Treaty gives the country of origin the right to delay access during the period of development. This was one of the most difficult parts of the Treaty negotiations related to the treatment of IPRs. The compromise solution is that recipients of PGRFA cannot claim IPRs that limit the facilitated access to the PGRFA, or their genetic parts or components, in the form received from the Multilateral System. Further, PGRFA accessed under the Multilateral System must also be made available to other interested parties by the recipient under the conditions laid out by the Treaty. This provision, which stops the appropriation of isolated components from material accessed under the Multilateral System, was strongly opposed by some countries that had determined that this would stifle innovation. On the other hand, when intellectual property or other property rights already protect the PGRFA in question, access can only take place in conformity with the treaties regulating those particular kinds of property rights. As is the case with some other treaties, like the Biosafety Protocol, the ITPGRFA refuses to establish a hierarchy between itself and other related treaties, such as IPRs treaties. This leaves the door open for conflicting interpretation at the time of implementation.

The question of access is closely related to that of benefit sharing. In fact, the benefit-sharing regime constitutes another part of the bargaining process that seeks to make PGRFA a common concern of humankind. Access is to be facilitated and, as a result, donors are granted the right to receive some form of benefits in return. In fact, the Treaty goes one step further in asserting that the premise for benefit sharing is, in the first place, the fact that access to PGRFA constitutes in itself an important benefit for recipient countries. Different types of benefit-sharing mechanisms are provided for under the Treaty: these include the exchange of information, access to and transfer of technology, capacity building, and the sharing of the benefits arising from commercialization. With regard to the sharing of information, the Treaty envisages that member states will, for instance, provide catalogues and inventories, information on technologies, and the results of technical, scientific and socio-economic research. The Treaty provides a specific information system, the Global Information System, which will generally serve to provide all member states with relevant information concerning crops in the Multilateral System. Concerning technology transfer, the Treaty only provides a general obligation to facilitate access to technologies for the conservation, characterization, evaluation and use of PGRFA, which is further qualified by the fact that access to such technologies is subject to applicable property rights. In the case of developing countries, specific mention is made of the fact that even technologies protected by IPRs should be transferred under 'fair and most favourable terms', in particular in the case of technologies for use in conservation as well as technologies for the benefit of farmers in developing countries. Beyond information and technology transfers, benefit sharing can take the form of capacity-building measures. These include the establishment of programmes for scientific and technical education and training in conservation and sustainable use of PGRFA, and the development of facilities for conservation and sustainable use of PGRFA. Finally, the Treaty provides for the sharing of monetary benefits. These include, for instance, the involvement of the private sector in developing countries in research and technology development. Further, the standard Material Transfer Agreement, through which facilitated access will be implemented, will include a requirement that an equitable share of the benefits arising from the commercialization of a product that incorporates material accessed through the Multilateral System will have to be paid to the Trust Account set up under the Treaty. The benefits that arise under the benefit-sharing arrangements must be primarily directed to farmers who conserve and sustainably use PGRFA.

The ITPGRFA deals with a number of other issues of relevance in the context of the management of plant genetic resources. Importantly, the Treaty specifically addresses the issue of *ex situ* collections held in trust by the International Agricultural Research Centres (IARCs) of the CGIAR. The Centres, which have signed agreements with the FAO concerning their collections, are now invited to sign new agreements with the Treaty's Governing Body. These agreements will determine that the access provisions of the ITPGRFA will govern the collections of the Centres that are part of the Annex I list. This will, however, only cover materials collected after the entry into force of the Treaty and that fall within its scope. The Centres are also obliged to provide preferential treatment to countries that provide material to their gene banks and are not to request any Material Transfer Agreement if a country of origin wants to access its own material. Generally, the Centres recognize the authority of the Governing Body to provide policy guidance relating to their *ex situ* collections.

The Governing Body, a body comprising all member states, will oversee the implementation of the ITPGRFA. Its functions will generally be to promote the full implementation of the Treaty. This will include the adoption of plans and programmes for the implementation of this Treaty, the establishment of subsidiary bodies as may be necessary and the consideration and adoption of amendments – by consensus – to the Treaty.

Overall, the Treaty, which constitutes the outcome of many years of negotiations, is noteworthy for providing the first international recognition of farmers' rights in a binding instrument. The provisions concerning ABS, which will constitute the most important part of the Treaty in practice, are largely lacking in specificity. This reflects the difficult balancing that the negotiators had to provide between the interests of developed and developing countries, big private seed companies and small farmers and a number of other actors in between.

2.3.3 The law and policy in the World Trade Organization[27]

From GATT to the WTO

With the advent of the World Trade Organization (WTO), which succeeded the General Agreement on Tariffs and Trade (GATT) 1947, the legal framework bearing on PGR, and indirectly on TK, has significantly changed by its actual or potential implications (Cottier, 1992). The stronger rules governing trade in goods (GATT), the General Agreement on Trade in Services (GATS), and in particular the TRIPS agreement set a new context for transactions involving PGR. A relatively well-developed dispute settlement mechanism accords this framework greater efficacy. However, while trade in PGR falls under current WTO law, the protection of TK so far has not been addressed in treaty law. It was placed on the agenda for further discussions under the TRIPS Agreement at the 2001 Doha Ministerial Conference.[28]

The WTO system today rests on three major pillars: the rules affecting trade in goods, trade in services and the protection of IPRs.[29] The WTO system is an important part of the law on PGR and the protection of TK. Most of the WTO law potentially applies to goods and services relating to these subjects. This is true for the GATT 1994 and the sectoral agreements, in particular the Agreement on Agriculture and the Agreement on Sanitary and Phytosanitary Measures.[30] Of paramount importance is the TRIPS Agreement, as it directly addresses intellectual property protection of PGR.

International trade in PGR has always been subject to the multilateral disciplines of the GATT 1947. In principle, this agreement also applied to trade in agricultural commodities derived from PGR, and the GATT 1994 continues to do so. The most-favoured nation treatment (Articles I, II) and national treatment (Article III) principles have to be respected. The prohibition of quantitative import restrictions also applies, with narrowly defined exemptions (Article XI), as well as general exceptions (Articles XX, XI). The rules on import licensing (Article XIII) and other pertinent provisions, in particular the disciplines on tariffs (Articles II, VII and XXVIII), transit (Article V), fees and formalities, (Article VIII), anti-dumping (Article VI), marks of origin (Article IX), transparency and judicial review (Article X), state trading and monopolies (Article XVII), and balance of payment exemptions (Articles XII, XVIII) all apply to agricultural products. The same was true of important sectoral agreements, such as the old Agreement on Technical Barriers to Trade. However, under GATT 1947 there were many exemptions, and disciplines remained weak. Key provisions on

[27] Author: Thomas Cottier.

[28] WTO, Ministerial Declaration, para. 19, WT/MIN(01)/DEC/1 (20 November 2001).

[29] There are 12 side agreements to the GATT 1994 and no fewer than six understandings, all of which elaborate further on its provisions. The rules on services trade have been further developed after the Uruguay Round to cover the regulation of financial services and telecommunications. There are four plurilateral agreements, of which the Agreement on Government Procurement is particularly important.

[30] Other agreements not dealt with in this survey may also be of importance, such as the Agreement on Import Licensing Procedures and the Agreement on Safeguards. The GATS is relevant for consulting and other professional services in connection with PGR and TK.

subsidies to agricultural products were weak.[31]

The system has witnessed substantial structural changes since the 1970s, which accelerated with the Uruguay Round and will further evolve with forthcoming rounds of trade liberalization. These are of importance in the present context. First, international trade law has progressively developed beyond market access and begun to increasingly entail harmonization of domestic regulations. It has moved towards a global law of integration. Secondly, WTO law now has a relatively effective dispute settlement mechanism which no longer leaves non-compliance without costs and disadvantages for governments and nations. Both of these aspects place WTO law in the forefront of international legal developments. It is worth recalling these developments before turning to the contents of WTO law pertinent to PGR and TK.

From tariffs to farm support levels and intellectual property protection

Trade policy in the early post-Second World War era focused on the gradual reduction of tariffs and the elimination of preferential systems. The GATT 1947 provided, in essence, a framework for a gradual process of tariff reduction. The provisions of the GATT 1947 were originally designed to accompany this process and to avoid frustration and circumvention by other means: in particular quantitative restrictions, subsidies, or discriminatory taxes on imports. Indeed, the gradual reduction, and even elimination, of tariffs within customs unions and free trade areas increasingly shifted the emphasis on non-tariff measures. Quantitative restrictions, export subsidies, anti-dumping duties, technical standards, balance of payment measures, labelling requirements, import licensing, rules on government procurement, and 'voluntary' export restraints (VERs) all became additional and widely used instruments. With their beginnings in the Kennedy Round, such non-tariff barriers became, besides the classical process of tariff-reduction, the main objects of trade negotiations in the Tokyo Round.

A third generation of trade barriers was brought in focus during the 1980s. It encompasses a number of issues, ranging from domestic farm support to restrictive regulation of service industries, of investments and, finally, the protection of intellectual property. These subjects stem from different fields of law, public and private, but have in common that they are all mainly part of the general domestic legal system. They are not directly geared to classical international trade relations but they have been affecting such relations more and more in terms of limiting market access. There is a link between the above factors and some of the great political difficulties in reaching final agreement in key issues of the Uruguay Round, such as reduction of agricultural support systems, and conditions for investment and liberalization of trade in services. The main difficulties to be settled in the area of TRIPS have similar roots. The issue of patentability of pharmaceuticals, foodstuffs, or of living matter – namely, the problem of how far exceptions to patentability should be allowed to go, or, in copyright, how the relationships of authors, producers, performers and users should be arranged – are perhaps the most prominent examples in showing how far international negotiations and regulations have penetrated socially, ethically or culturally sensitive issues of the domestic political process.

More than ever before, international trade regulation today seeks to provide fair competition in a globalizing market econ-

[31] The USA, for example, benefited from a general waiver. The European Community, in defence of its Common Agricultural Policy (CAP), often retaliated by using its veto power on the establishment of dispute settlement panels or on the adoption of reports, leaving them without much consequence in practice. Switzerland enjoyed major benefits in its protocol of accession, making it one of the most protectionist Members in agriculture, despite a relatively high level of food imports.

omy. Foreign and domestic economic affairs can no longer be separated; and the increasing importance and attention paid to foreign policy in general, formerly of real interest often merely to a few, is a natural effect of such developments. This is also true in the context of trade in PGR and the protection of TK, which entail an inextricable linkage of domestic and foreign policy issues.

International dispute settlement and enforcement

The dispute settlement mechanism of the WTO is a unique and powerful instrument in international economic relations. The Dispute Settlement Body (DSB) of the WTO has now developed a substantial jurisprudence that is unique in a highly fragmented and decentralized system of international law.[32] In a nutshell, the WTO dispute settlement mechanism consists of three phases. Upon the filing of a complaint, consultations are to be held among the Parties. Failing a successful conclusion by negotiations (the result in approximately two-thirds of all disputes), the complaining Party is entitled to ask the DSB for the establishment of a panel.[33]

A panel, once established, regularly consists of trade officials and/or academics knowledgeable in the field of law at issue, most commonly economists and lawyers. It is assisted by the WTO Secretariat (specifically the Legal Affairs Division and the operational divisions concerned). Parties submit legal briefs to the panel and two or three oral hearings are held at which interested Third Parties may also take part and express their legal arguments. Panel proceedings are conducted in a judicial manner and on the basis of a set of customary and written procedural rules developed over time. The panel may also conduct hearings with independent experts in the field. The panel drafts a report which is circulated to the Parties for comments before a final version is submitted to the DSB. If the report is adopted in the absence of a reverse consensus, it proceeds to implementation. However, the Parties, in particular the losing Party, may object and file an appeal to the Appellate Body, a standing panel of seven senior international trade lawyers. Appeals are generally limited to specific points, which may also be raised by the winning Party. The jurisdiction of the Appellate Body is limited to questions of law. It does not extend to factual assessment and evidence, apart from certain exceptions. The report of the Appellate Body is again submitted to the DSB. The losing Party no longer has the right to object, and the report is adopted unless there is consensus (including the winning Party) to reject the report. It goes without saying that this procedure in practical terms entails automatic adoption of the report, and the losing Party is obliged to comply. The entire proceedings from the filing of the complaint are subject to very tight time limits. Proceedings are usually completed, including the appellate phase, within 18 months. Longer periods have been found to be necessary in more complex cases, but even those are far shorter than most domestic, let alone international, judicial proceedings.

The decision forms the basis for implementation. The losing Party, in principle, is obliged to implement the recommendation. However, it has the option to offer compensation which is rarely accepted and often not of interest to the winning Party. Often the amount of compensation is set by a panel of arbitration – which is normally composed of the original panel – in terms of surcharge tariffs equivalent to the financial losses incurred since the unlawful measure

[32] From 1995 to the beginning of 2001, 103 disputes were resolved. Of these, 48 or almost half were settled without recourse to adjudication (see Duk Park and Umbricht, 2001).

[33] Importantly, complaints on nullification and impairment of benefits do not entail only violations of treaty provisions, but may also comprise so-called non-violation complaints. This course of action, albeit rare, envisages the protection of legitimate expectations as to conditions of competition, and bars Members from taking measures in bad faith that nullify trade concessions granted in previous negotiations.

was put in place. Failing implementation and compensation, the winning Party may obtain the right to suspend concessions (retaliation). Countries are free to target areas not directly linked to the subject matter in dispute. Retaliation for failure to comply with the GATT 1994 and the related sectoral agreements may thus eventually result in measures in the field of services or intellectual property, and vice versa. Targeted goods may also be alternated in order to increase pressure to comply. The USA, for example, operates special *carousel* legislation to this effect.

The enforcement system benefits large markets with retaliatory powers. It is one of the weaknesses and inequities of the system that it leaves small markets and weaker countries with little power to retaliate. The system has not reached a level of multilateral implementation that would encourage all Members to participate in such action. It is therefore not surprising that large countries rank among the main users. At the same time, it is important to emphasize that the WTO dispute settlement mechanism has also been used successfully by developing and smaller countries alike, as compliance with adopted reports has also proved to be in the long-term interest of large Members. Except for three major cases, compliance so far has shown an impressive and successful record. In difficult cases, the new system has shown the efficacy of its novel features, unheard of so far in international law: it eliminated veto or exit powers, and made rulings subject to monitoring of implementation and effective enforcement. It is not a coincidence that two of the cases that have proved to be major stumbling blocks, the banana dispute[34] and the dispute on hormone-treated meat,[35] essentially addressed problems of production and trade in agriculture which – besides services – is one of the most highly regulated and protected sectors. Tensions with national or regional law abound. The mandatory nature of the system, and the imposition of substantial surcharge tariffs following denial of implementation by the European Communities, almost brought about crises and major trade wars with the USA. For the first time, problems could no longer be evaded by veto, and the Parties had to go through a long and protracted process of negotiating a political settlement. This exit no longer exists, and the tensions show, more than anything else, the deep and fundamental changes of the rules of the game brought about by the advent of the WTO.

The system applies to all areas of WTO law alike. While panels are barred from creating additional obligations or rights, the process of interpretation inherently entails a refinement of the law. Indeed, in between trade rounds, the main emphasis of WTO legal activities is on dispute settlement and the development of case law. In the field of PGR, no cases have so far been decided. Yet, related areas, such as sanitary and phytosanitary measures, have been repeatedly scrutinized and a substantial body of law has evolved. This is also true for intellectual property protection, which until 1995 had never been the subject of mandatory dispute settlement. The DSU fully applies to the TRIPS Agreement.[36] So far, nine requests for consultations in the field have been filed, many of them relating either to rules on patent protection for pharmaceuticals and agricultural chemical products or enforcement of IPRs.[37] In some of these cases, mutually agreed solutions between the Parties involved were notified (Baracol, 2001, pp. 37–41). The relative success of consultations may be because the TRIPS provisions are detailed and provide good guidance in assessing rights and obligations.

[34] European Communities – Regime for the Importation, Sale and Distribution of Bananas, WT/DS27/AB/R (9 September 1997).

[35] European Communities – Measures Concerning Meat and Meat Products, WT/DS26 and 48/AB/R (16 January 1998).

[36] The only exception concerned so-called non-violation complaints, which were excluded until the year 2000.

[37] Cf. see above, Annex 1, giving an overview of WTO disputes in the field of intellectual property rights.

The panel stage was reached by two disputes relating to obligations of countries to provide a so-called mailbox for the filing of patent applications even before the general transitional period of 10 years elapsed. In India–Patent Protection for Pharmaceutical and Chemical Products,[38] the Panel and the Appellate Body ruled upon a complaint by the USA (and also one by the European Communities) that administrative practices, short of a legal basis, did not amount to an adequate compliance with these obligations. On substance, important cases addressed the scope of fair use exceptions in patent law and in copyright law. A further case addressed the protection of trademarks and trade names under US law in the context of economic sanctions against Cuba.

The Agreement on Agriculture

The advent of the WTO was accompanied by a comprehensive Agreement on Agriculture, the main feature of which was the regulation of import restrictions (Horber, 1995, p. 51). Member states were obliged, by way of their tariff schedules, to reduce tariffs in agriculture by 36% on average (at least 15% per tariff line) by 2001. This leaves most countries with still very high bound tariffs, the reduction of which will be the subject of further rounds of negotiations.[39] Independently of tariff protection, a minimum market access right beginning with 3% of domestic consumption and increasing to 5% had to be realized by 2001. Domestic support by product-related subsidies – formerly typically a domestic issue – had to be reduced by 20%. Further decreases are to be expected in the Doha Round. On the other hand, Members obtained full autonomy in the use of non-product-related payments (the Green Box), which has been mainly used to finance direct assistance, and has become the basis of a new agricultural policy in both the European Communities and Switzerland. The shift from product-related support (often causing excess production) to direct payments has paved the way for ecological reforms. Overall, they bear the potential for sustainable agriculture, which, short of pressures from the global trading system, would have been difficult to bring about domestically. Finally, the agreement obliges countries to reduce export subsidies by 36% (on the basis of 1986/1988 rates) on agricultural products and transformed products (such as chocolates or biscuits) and a quantitative reduction by 21% of subsidized products. The Doha Ministerial Declaration of November 2001 pledged to work towards further reduction, with a view to eliminating all export subsidies in agriculture, given their often detrimental impact on developing countries.[40]

The Agreement on the Application of Sanitary and Phytosanitary Measures

A new Agreement on Sanitary and Phytosanitary Measures (SPS Agreement) regulating foodstuffs was adopted, replacing the application of the Technical Barriers to Trade (TBT) Agreement in this field. It was elaborated during the Uruguay Round and addresses the problem of excessive foods standards that may unreasonably impair trade in agricultural products. Such standards also exist in the field of genetic resources, in particular seeds. The Agreement essentially seeks to encourage the use of international foods standards developed by international bodies, such as the joint Codex Alimentarius Committee of the FAO and the World Health Organization (WHO). However, it allows Members to apply their own and stricter standards, provided that they are in a position to scientifically demonstrate that such higher standards are required in order to respond to the level of risk that has been defined. The complex

[38] WT/DS50/AB/R (19 December 1997).

[39] For example, bound agricultural tariffs in the European Communities amount to an average of 17.3% and in Switzerland to 34%. WTO, Trade Policy Review: European Union 2000, xix (2000) and WTO, Trade Policy Review: Switzerland and Liechtenstein, xxi (2000).

[40] WTO, Ministerial Declaration, para. 13, WT/MIN(01)/DEC/1 (20 November 2001).

agreement allows Members to invoke the precautionary principle. Since 1995, it has given rise to a number of complex disputes which sought to define the proper levels of protection and of judicial review of such measures, the *Hormones* case being the most prominent and controversial. It is important to note that it has remained unclear whether the SPS Agreement also applies to genetically modified organisms (GMOs). Moreover, it may partly overlap and conflict with the Cartagena Protocol on Biosafety which allows trade restrictions mainly, but not only, for reasons of environmental protection.

The Agreement on Trade-related Aspects of Intellectual Property Rights

CHANGING PARADIGMS AND THE ADVENT OF THE AGREEMENT

While the law of PGR outside trade-related matters has been defined primarily by the instruments discussed above, in particular the CBD, the ITPGRFA and the CGIAR System and its rules and policies, the inclusion of intellectual property protection in the WTO system with the adoption of the TRIPS Agreement has fundamentally altered previous equations. New conditions and incentives for investment have been set in global law, exclusive rights granted, and a new relationship between protected products and those pertaining to the public domain has emerged.

It is important to note that international efforts to bring about adequate IPR protection did not start with the TRIPS Agreement. The international dimension of intellectual property is not new. In fact, as early as the 1880s, it ranged among the first efforts in international economic law to bring about shared perceptions. The many agreements and unions, in particular those administered by the World Intellectual Property Organization (WIPO), provide ample evidence of this tradition. Other organizations and forums, in particular the OECD and the UN General Assembly, also show a considerable record of work and recommendations to the extent that IPRs are related to the transfer of technology or to competition.[41] Yet, due largely to differing perceptions as to the role of intellectual property in the process of economic development, as well as diverging traditions among industrialized countries, substantive global standards, in particular concerning industrial property, remained largely of an open and permissive nature. Throughout the 1970s and the early 1980s attempts to introduce more specific norms largely failed due to such differences; the unsuccessful effort to revise the Paris Convention is a case in point. Moreover, an effective, swift and simple system of dispute settlement is lacking within the WIPO, and no standards existed in the treaty system on the effective enforcement of IPRs. The thrust of worldwide cooperation therefore has focused mainly on achieving national treatment for foreign nationals and, with the exception of international copyright law, on the important field of acquisition and administration of rights, namely by using effective ways to simplify the international registration of national trademarks, designs and patents.

The difficulties mentioned in achieving global progress in traditional international forums have led to the increased linkage of IPR protection with trade policies, and the marriage has attracted considerable scholarly attention ever since.[42] The negotiations

[41] The negotiations on the law of the sea in the 1970s (UNCLOS III) provide a classic example of extensive efforts to negotiate on intellectual property rights in order to bring about such transfers in the field of deep seabed mining.

[42] Outside the regional context of the European Communities, such linkages were first established by the 1984 US Trade Act, and further reinforced in 1988 with controversial provisions and procedures allowing for retaliation in market access in goods and services against countries having unfair and distorting regimes on intellectual property, whether or not these were in conflict with current international obligations. The IPR protection–trade linkage in domestic US law, and bilateral and regional efforts, such as the Caribbean Basin Initiative, were certainly a source of inspiration in bringing the subject into the GATT system upon the initiative of the US when the Punta del Este Mandate for the Uruguay Round was drafted and agreed upon in 1986.

that led to the adoption of the TRIPS Agreement were initiated by the USA.[43] The initiative was particularly based upon the estimates made by the USA, whose industries in 1986 suffered losses of some 24 billion dollars due to inadequate protection of intellectual property and related investments. In the beginning, the European Community was reluctant to commit itself on the issue, but in the course of the negotiations, it gradually became one of the major *demandeurs*.

The 7-year negotiations went through three phases. A long process of fact-finding, which also worked as a tool for mutual education, led to comprehensive proposals by various Contracting Parties. After the mid-term review of Montreal in December 1989 and the April 1990 ministerial decision, intensive negotiations took place. Work was successfully concluded in December 1991, when the draft TRIPS Agreement was wrapped up as part of the 'Dunkel Text'. During the last phase, efforts were aimed at keeping the results achieved and at avoiding their being jeopardized by counter-productive new proposals in the field of intellectual property, or by a deadlocking of negotiations regarding other issues such as agriculture or audio-visual services. Unlike during the first two phases, the last phase was marked by conflicting interests between the USA and the European Communities on copyright issues (cultural aspects, blank tapes levies). The results achieved were made possible due to various factors, many of which are interesting for the purposes of a general assessment of the importance of intellectual property. To a great extent they exceeded the initial expectations.

THE SUBJECT MATTER OF THE AGREEMENT

Signed on 15 April 1994, the TRIPS Agreement lays down the essential elements for the minimum level of protection that countries must accord to intellectual achievements. It covers all fields of intellectual property: copyright, topographies of integrated circuits, trademarks and service marks, geographical indications, industrial designs, patents for inventions and trade secrets. It defines rights and obligations under these different forms of intellectual property. Generally speaking, such rights include the exclusive right to make, use, sell, offer to sell, and import products protected by IPRs. In addition, member states undertake to provide detailed rules guaranteeing the procedures and remedies that must be available at the national level for IPRs to be effectively enforced. Members had to implement the TRIPS Agreement by the end of the transitional period, which ended for developing countries in 2000. The transition period will end in 2005 for rules on product patents, except for least developed countries. With the adoption of the TRIPS Agreement, intellectual property is now an integral part of the multilateral trading system.

GENERAL PROVISIONS

The Agreement strengthens the national treatment principle as contained in the relevant existing international conventions (Article 3). However, it includes a concept of trade law that is new in the field of intellectual property law: the most-favoured nation (MFN) principle (Article 4). A WTO Member must accord automatically to other WTO Members the same favourable treatment it has granted to a third country on a bilateral basis. The MFN particularly strengthens the position of smaller countries *vis-à-vis* the big trading powers. As a result of the TRIPS Agreement and its MFN clause, a bilateral agreement concluded, for example, between the USA and Japan on the improvement of registration procedures will also be extended *erga omnes* so as to avoid any incompatibility with the TRIPS obligations.

[43] At the end of the Tokyo Round, the USA and the European Communities unsuccessfully launched a draft Anti-Counterfeit Code. Subsequent work did not materialize due to opposition by developing countries. Eventually, the protection of intellectual property rights was included in the 1986 negotiating mandate of Punta del Este, and – as a political compromise – limited to its 'trade-related aspects'.

The TRIPS Agreement contains two general provisions that are of potential importance for PGR and TK. First, Article 7 states:

> The protection and enforcement of intellectual property rights should contribute to the promotion of technological innovation and to the transfer and dissemination of technology, to the mutual advantage of producers and users of technological knowledge and in a manner conducive to social and economic welfare, and to a balance of rights and obligations.

This general provision was proposed by developing countries and intended to safeguard the proper balance between private rights and public goods against one-sided and detrimental effects that IPRs may have, and in order to promote their use as incentives for foreign investment and transfer of technology. The provision is of a horizontal nature and applies to all provisions of the agreement. Given its nature, it is not independently operational but forms the interpretation of other provisions, in particular when defining the proper relationship between exclusive rights and public goods. We shall return to the impact of the provision on PGR and TK.

Secondly, the TRIPS Agreement also recognizes in Article 8 that intellectual property regulations do not prevent Members from adopting measures necessary to protect public health and nutrition, and to promote the public interest in sectors of vital importance to their socio-economic development. Members are equally entitled to take measures against the abuse of IPRs or against practices that unreasonably restrain trade or adversely affect the international transfer of technology. The effects of intellectual property protection may entail some potential for conflicts with antitrust law. However, Article 8 is constrained by the proviso (introduced upon the motion of industrialized countries) that such measures are required to be consistent with the other provisions of the TRIPS Agreement. Article 8.1 therefore does not allow deviation from the operational provisions of the TRIPS Agreement. As to Article 8.2, addressing the abuse of IPRs and competition, it is important to note that the authority of Members to take measures is further detailed in Article 40 of the TRIPS Agreement, which preserves the rights of Members to take action under their own national or regional competition rules. The issue of restrictive business practices is addressed in the TRIPS Agreement only in terms of the matter being left to national jurisdiction. It is expected that it will be brought into focus again in future negotiations, when a fine balance will have to be struck between protection of investments, exclusive rights resulting from those investments, and the efficient functioning of global competition.[44]

Article 8, like Article 7, forms interpretation and has to be taken into account in defining the scope and impact of rights and obligations under the TRIPS Agreement. The provisions of Article 8.1 played an important role in recent disputes and discussions relating to compulsory licensing of patents over new essential medicines (for HIV/AIDS). The Ministerial Declaration adopted at Doha confirmed its importance, while leaving operational rules, in particular on compulsory licensing of patents (Article 31), unchanged.[45]

COPYRIGHT AND RELATED (NEIGHBOURING) RIGHTS

Copyright entails the protection of expressions of mental activity not only in the arts and sciences, but also in modern technology. Its scope is limited and it does not pro-

[44] The Doha Ministerial Conference adopted the following work programme on competition in para. 25, without opening negotiations on the subject. 'In the period until the Fifth Session, further work in the Working Group on the Interaction between Trade and Competition Policy will focus on the clarification of: core principles, including transparency, non-discrimination and procedural fairness, and provisions on hardcore cartels; modalities for voluntary cooperation; and support for progressive reinforcement of competition institutions in developing countries through capacity building. Full account shall be taken of the needs of developing and least-developed country participants and appropriate flexibility provided to address them'.
[45] WTO, Ministerial Declaration, para. 17, WT/MIN(01)/DEC/1 (20 November 2001).

tect substance, ideas, procedures and methods (including mathematical concepts). It inherently does not require a high level of originality. The same is true for neighbouring rights that address the rights of performing artists, film broadcasters and phonogram producers.

Copyright and neighbouring rights protection is important for the protection of TK to the extent that it is expressed. Traditional music and performances are at the forefront. Moreover, it is important for electronic data collections on PGR and for programming specifically designed for this field. It covers books and publications on the subject. Otherwise, copyright is not suitable for addressing PGR and knowledge *per se*. It is rather by protecting the cultural background and its diversity that copyright indirectly makes a contribution to their valuation and reward.

The provisions of the TRIPS Agreement on copyright amend the traditional rules of the Berne Convention and essentially focus on new technologies: the protection of computer programs, including exclusive rental rights, and of electronic databases (Articles 10 and 11); the protection of rights of interpreters and broadcasting organizations against unauthorized reproduction, the worldwide protection of phonograms against unauthorized copying and reproduction, and the recognition of exclusive rental rights over phonograms (with the possibility of preventing the rental in particular of new copies, subject to existing rules which are limited to an obligation of equitable remuneration) (Article 14); and the protection of cinematographic works to avoid widespread copying materially impairing the exclusive right of reproduction (Article 11). The human rights dimension of copyright was explicitly excluded (due to pressure from the US film industry), as the TRIPS Agreement (and dispute settlement) does not encompass moral rights under Article 6bis of the Berne Convention (Article 9.1).

The agreement prescribes a minimum duration of protection of 50 years (Article 12), except for broadcasters' rights, for which the period is 20 years (Article 14.5) from the end of the calendar year of authorized publication, making, or broadcasting. In other words, unlike in many national laws, there is no *post mortem auctoris* term of protection.[46] The protection of performers' rights, which was traditionally limited to European civil law countries, is for the first time dealt with in a multilateral treaty of a wider geographical coverage.

TRADEMARKS

Trademark protection is of key importance in identifying and individualizing products originating from PGR and/or based on TK. Seeds and varieties can be sold under trademarks, and TK can be protected indirectly by linking it to trade names and supplementary qualifications, using terms not in the public domain.

The function of trademarks consists in identifying the origin of the product and allowing the consumer to distinguish different products. For the first time in international economic law, the TRIPS Agreement provides a globally uniform definition of trademarks (for goods and services), including protection of combinations of colours (but not requiring protection for sound marks) (Article 15). Protection of internationally well-known trademarks (i.e. without registration) applies not only to products but also to services. Trademarks cannot be used to bar parallel imports unless there is a risk of confusion for the consumer (Article 16.1). The term of protection is no less than 7 years, with indefinite possibility of renewal (Article 18). A possibility of cancellation of trademarks exists only after a minimal time period of non-use of 3 years (Article 19). There is a prohibition of other requirements for the use of trademarks (e.g. requirement for use to be only in combination with another trademark), and (unlike in patents) a complete prohibition of compulsory licensing (Articles 20, 21). The TRIPS Agreement entails a significant

[46] The provisions are of particular significance for producers of computer programs and phonograms, against counterfeiting and piracy.

improvement of trademark and service mark protection in countries other than in Western Europe and North America. The protection of internationally well-known trademarks is thereby enhanced.

PROTECTION OF GEOGRAPHICAL INDICATIONS

Geographical indications are a prime instrument for protecting products derived from PGR and TK, as this protection (other than trademarks) requires the quality, reputation or other characteristic of a product to be essentially attributable to its geographical origin. This allows the protection not merely of the uniqueness of the product but also the uniqueness of its origin and TK at that place.[47]

As a general rule, protection is ensured by an obligation to introduce laws protecting against deception of the consumer as to the true geographical origin and quality of the product and against other forms of unfair competition (Article 22). The same rule also applies to the use of names in the course of trade to the extent that they are used in a manner to mislead the public (Article 24.8). Members are barred from registration of trademarks which mislead the public as to the origin of the product (Article 22.3). Protection thus takes place on a case-by-case basis without prescribing registration in international law.

A higher (absolute) level of protection was introduced for wines and spirits (Article 23). These products enjoy privileged protection. The use of geographical names for products produced elsewhere is barred, even where the true origin of the product is indicated or where it is accompanied by qualifications such as 'kind, type, style, imitation' or the like. However, such protection does not extend to denominations in customary use. Homonymous indications, namely similar names for different origins, are protected, subject to measures against consumer deceptions. In particular, denominations of wines and spirits continuously using foreign locations, but in use before 1984 and in good faith, are grandfathered and cannot be challenged (Article 24.4). Generally, no protection exists against use of denominations if they are in generic use in the common language of a Member. The same is true for products of the vine if they coincide with the customary name of a grape variety (Article 24.6).

The TRIPS Agreement contains obligations to further develop protection of geographical indications. An international system for notification and registration for geographical indications of wines should be negotiated (Article 23.4). Member states pledged to enter into negotiations in order to further extend the coverage of protection, in particular to foodstuffs. In this respect, the Council for TRIPS initiated preparatory work by the end of 1998, but no results have been achieved.[48] The Doha Ministerial Conference reiterated the agreement to negotiate on the subject of an international registration of wines, but a similarly strong commitment exists for the further development of protection for other geographical indications. The Conference merely noted that the TRIPS Council would deal with these issues pursuant to procedures adopted for implementation issues.[49]

INDUSTRIAL DESIGNS

The Agreement adopts, for the first time in international economic law, a common def-

[47] For example: Flocons de Quinoa, 'Quinoa Real', a grain cultivated in the Andes (Bolivia) 3000 m above sea level, which has specific qualities due to cultivation at this altitude.

[48] See the press release on: http://www.wto.org/wto/new/pu101298.htm

[49] Doha Ministerial Declaration, para. 18: 'With a view to completing the work started in the Council for Trade-Related Aspects of Intellectual Property Rights (Council for TRIPS) on the implementation of Article 23.4, we agree to negotiate the establishment of a multilateral system of notification and registration of geographical indications for wines and spirits by the Fifth Session of the Ministerial Conference. We note that issues related to the extension of the protection of geographical indications provided for in Article 23 to products other than wines and spirits will be addressed in the Council for TRIPS pursuant to paragraph 12 of this Declaration'.

inition of industrial designs. However, countries remain free to require either novelty or originality (Article 25). There is an obligation for Members to provide for protection of creations in the field of textiles so as to take into account the short life cycle (fashion) of such design creations. The term of protection must be no less than 10 years (Article 26.3). The provision on textile designs is of particular importance to the design industry. Registration may become important for traditional crafts and the preservation of cultural identity and tradition, to the extent that they bring about innovative patterns that qualify, in terms of originality, for design protection. Moreover, copyright protection can also be used in many countries as a possible way to protect designs without any registration requirements.

TOPOGRAPHIES OF INTEGRATED CIRCUITS

The protection of integrated circuits is not of any relevance here except for the fact that it is an example of *sui generis* protection, derived for a specific technology. The TRIPS Agreement, building upon the Treaty on Intellectual Property in Respect of Integrated Circuits (Washington Treaty), grants exclusive rights, including the right to prevent the importation of goods containing the topographies. Limited exceptions (only stock in hand or ordered before notice of an unlawful act) exist in favour of the *bona fide* buyer of counterfeited goods (Articles 36, 37). The term of protection extends to at least 10 years (Article 38). It entails a prohibition of reciprocity treatment in the national laws of industrialized countries through the MFN clause.

PATENTS FOR INVENTIONS

The provisions on patents for inventions of the TRIPS Agreement define, to a large extent, the legal conditions applicable to genetic resources and TK. The provisions of the TRIPS Agreement considerably enhance the protection for breeding and genetic engineering, and have changed the legal relationship between genetic resources and TK on the one hand, and newly appropriated technological advances on the other. They are of considerable significance not only for the chemical and pharmaceutical industries, but also for all other sectors with active patent-oriented strategies. In the field of biotechnology, the limitation of patentability to microorganisms reflects the minimal consensus in a worldwide controversial discussion, which will continue as the present state of the law neither satisfies the long-term needs of the industry nor of developing countries.

The TRIPS Agreement establishes the principle of non-discrimination as to the fields of technology, the place of invention, and the country of production (Article 27.1), thus barring differential treatment of different fields of technology. Members are not allowed to apply different rules on patent protection depending on different fields of technology: no special rules for genetic engineering can be adopted. The minimum term of protection is 20 years (Article 33). The Agreement entails an obligation to provide product protection in the field of pharmaceuticals and foodstuffs. Developing countries are obliged to introduce such protection by 2005. In the meantime, inventions are to be protected by so-called 'pipeline' protection, which amounts to quasi-exclusive rights by means of administrative measures and marketing approval until full patent protection takes effect (Articles 70.8, 70.9).[50]

The TRIPS Agreement recognizes general exceptions from patenting in order to protect the *ordre public* and morality, including the protection of human, plant and animal life and health or to avoid serious prejudice to the environment. Importantly, exceptions can only operate if at the same time the Member prohibits commercial exploitation of the product. On the other hand, it is not sufficient to preclude patenting only because the law prohibits the exploitation of the invention. Beyond these general exceptions, the TRIPS Agree-

[50] The scope of these rules was expounded in India-Patent Protection. WT/DS50/AB/R, 19 December 1997.

ment recognizes special exemptions for which Members may qualify. Negotiations on these points were extremely difficult, and they resulted in what was considered a provisional solution. Plants and animals other than microorganisms, and essentially biological processes for the production of plants and animals, may be excluded from patentability. But there is an obligation to grant patent protection of inventions relating to microorganisms and essentially biological processes. It entails an obligation to provide an effective *sui generis* protection for plant varieties. Members may choose to operate under the UPOV Convention. They may choose to design new and innovative schemes of protection, taking into account considerations of benefit sharing and access regulation under the CBD. Finally, they may, in addition, opt for special or general patent protection, either exclusively or in accumulation. We shall return to these distinctions in a close examination of patenting of life forms in different countries (see Section 2.4).

Article 27.3(b) was meant to be of a provisional nature and the provision was subject to a review 4 years after the entry into force of the TRIPS Agreement in 1995. The review has not produced any substantive results. The Doha Ministerial Conference reiterated the mandate to pursue the review. Importantly, this mandate includes an examination of the relationship of the TRIPS Agreement, the CBD, the protection of TK and folklore, and other new developments raised by Members:

> 19. We instruct the Council for TRIPS, in pursuing its work programme including under the review of Article 27.3(b), the review of the implementation of the TRIPS Agreement under Article 71.1 and the work foreseen pursuant to paragraph 12 of this declaration, to examine, *inter alia*, the relationship between the TRIPS Agreement and the Convention on Biological Diversity, the protection of traditional knowledge and folklore, and other relevant new developments raised by Members pursuant to Article 71.1. In undertaking this work, the TRIPs Council shall be guided by the objectives and principles set out in Articles 7 and 8 of the TRIPS Agreement and shall take fully into account the development dimension.[51]

This statement implies recognition of the fact that the scope of patentability cannot be settled independently of other relevant instruments and the interests at stake, and of the importance to obtain an overall balance in accordance with the goals set forth in Article 7 of the Agreement.

PROTECTION OF TRADE SECRETS (UNDISCLOSED INFORMATION)

Protection of undisclosed information is of importance to TK to the extent that the circle of knowledgeable people is restricted. This may be of importance to healers and shamans in the use of genetic resources.

Protection of undisclosed information was introduced by the TRIPS Agreement and recognized as an intellectual property right (Article 2) going beyond mere protection against unfair competition. The notion of undisclosed information is defined. A secret is protected if it is not generally known or readily accessible to persons within circles that normally deal with the information. It must have commercial value and is subject to reasonable measures of precaution in order to keep the secret (Article 39). The TRIPS Agreement establishes the right to take action against infringement of trade secrets, including damages to be paid by enterprises which knew, or should have known, that the information given included trade secrets (Article 39.2, note 10; Article 45). It entails an obligation to protect test data concerning pharmaceuticals or agricultural chemical products that utilize new chemical entities in the course of an approval procedure against unfair commercial use (Article 39.3).

The adoption of provisions on the protection of trade secrets – for the first time in a multilateral treaty – is of particular importance in the light of growing use of trade secrets by enterprises in various fields. Pro-

[51] WTO, Ministerial Declaration, para. 19, WT/MIN(01)/DEC/1 (20 November 2001).

tection of test data against unfair commercial use will need to be developed in clearer terms (as for example in European Community law through a 5-year limitation of protection). The provision is likely to result in licensing of test data and therefore in sharing the burden *ex post* of research efforts. The protection against further disclosure of test data by the competent authorities is of particular interest to the agrochemical and pharmaceutical industries, whose products are subject to marketing approval procedures.

ENFORCEMENT AND TRANSPARENCY

The TRIPS Agreement entails a code of detailed rules on enforcement. The right to have access to review by judicial authorities of final administrative decisions is established (Articles 42, 41.4, 62.5). The TRIPS Agreement stipulates the right to obtain injunctions (Article 44), damages (Article 45) and other remedies, including destruction of products and means of production (Articles 46, 59). Protection against misuse of procedures by the right holder (harassment) is established (Article 48).

The right to obtain provisional measures (Article 50) and suspension of the release of *prima facie* counterfeit or pirated goods for a period of ten working days, and if need be, against payment of security, is established (Articles 51–57). An obligation to make laws and agreements publicly available provides for transparency. In particular, the obligation to publish or make publicly available final judicial and administrative decisions of general application (precedents) expresses fundamental precepts of the rule of law.

The enforcement provisions require legislative amendments, particularly in developing countries and in Central and Eastern Europe. The adoption and implementation of such provisions are of key importance to an effective protection and enforcement of IPRs. From a long-term perspective, they improve the means by which to tackle intellectual property problems through judicial channels. Before the adoption of the TRIPS Agreement, only diplomatic interventions were possible. The requirement to publish precedents will improve transparency in many countries, including Western Europe.

WTO law and genetic resources

In conclusion, the law of the WTO is of paramount importance for genetic resources and TK. The regulatory level and the effective dispute settlement mechanism place it at the heart of future developments. The brief survey above indicates that regulation of the two areas is affected by a number of agreements, including the GATT 1994, the Agreement on Agriculture, the Agreement on Sanitary and Phytosanitary Measures and the TRIPS Agreement. We realize that the impact of WTO rules is not limited to the particular problem of intellectual property and patenting. It also affects other forms of IPRs. Moreover, other regulatory areas, in particular tariffs and other trade regulations, such as phytosanitary measures, affect it too. There is an overall potential for promoting the use of, and the protection of, genetic resources and TK in the different policy instruments of the WTO. But there is a potential to render such protection and promotion more difficult if it is not sufficiently taken into account in future law-making. It will be a matter of examining these problems in the full context of other agreements and domestic regulations. The ministerial mandate of Doha to examine the issue of patenting life forms in a wider context, including the protection of TK, is encouraging. It is a necessary first step towards a larger coherence and has to be expanded to all pertinent issues alike.

2.3.4 The International Union for the Protection of New Varieties of Plants (UPOV)[52]

The UPOV Convention was first signed in 1961 to provide a form of legal protection of

[52] Author: Philippe Cullet with the assistance of Danuta Szymura Berglas.

plant varieties for Western European countries.[53] It seeks to protect new varieties of plants both in the interest of agricultural development and of plant breeders. Although it did not introduce patents, UPOV sought from the outset to provide incentives to the private sector to engage in commercial plant breeding, by introducing so-called plant breeders' rights. Despite the distinction between patents and plant breeders' rights, the two share several basic characteristics: they provide exclusive commercial rights to holders, reward an inventive process, and are granted for a limited period of time, after which they pass into the public domain.

More specifically, UPOV recognizes the exclusive rights of individual plant breeders to produce or reproduce protected varieties, to condition them for the purpose of propagation, to offer them for sale, to commercialize them, including exporting and importing them, and to stock them with a view to production or commercialization (Article 14.1 UPOV). Protection under UPOV is granted for developed or discovered plant varieties that are new, distinct, uniform and stable (Article 5 UPOV). The concept of novelty under UPOV is noteworthy because it differs from the approach under patent law. Under UPOV, a variety is novel if it has not been sold or otherwise disposed of for purposes of exploitation of the variety. Novelty is thus defined in relation to commercialization and not by the fact that the variety did not previously exist. UPOV gives a specific time frame for the application of novelty. To be novel, a variety must not have been commercialized in the country where the application is filed for more than a year before the application, and in other member countries for more than 4 years.[54] The criterion of distinctness requires that the protected variety should be clearly distinguishable from any other variety whose existence is a matter of common knowledge at the time of the filing of the application. Stability is obtained if the variety remains true to its description after repeated reproduction or propagation. Finally, uniformity implies that the variety remains true to the original in its relevant characteristics when propagated.

The UPOV Convention has been revised several times since 1961: in 1972, 1978 and 1991. At present some states are parties to the 1978 Act and some states are parties to the 1991 Act. Non-member states that wish to join the UPOV regime at present must join under the 1991 Act, but there is no obligation for existing member states to ratify the latest version of the convention if they do not wish to.

The revisions to the Convention have generally served to progressively strengthen plant breeders' rights. The 1991 version, for instance, extends breeders' rights to all production and reproduction of their varieties and to species as well as general and specific plant varieties. This now also includes so-called 'essentially derived varieties' (Article 14.5 of UPOV). Protection of an essentially derived variety is obtained if the variety is predominantly derived from the initial variety and retains its essential characteristics. It must also be clearly distinguishable from the initial variety while conforming to the initial variety in the expression of the essential characteristics.

One of the main distinguishing features of the original UPOV regime is that the recognition of plant breeders' rights is circumscribed by two main exceptions. First, under the 1978 version of the Convention, the so-called 'farmer's privilege' allows farmers to re-use propagating material from the previous year's harvest and to freely exchange seeds of protected varieties with other farmers. Secondly, plant breeders' rights do not extend to acts done privately and for non-commercial purposes or for experimental purposes, and do not extend to the use of the protected variety for the purpose of breeding other varieties and the

53 Note that this section introduces the UPOV Convention in general. Further developments concerning plant breeders' rights can be found in Chapter 3.

54 Article 6 of the UPOV Convention. In the case of other member countries, the relevant timeline is 6 years for trees and vines.

right to commercialize such other varieties. The 1991 version of the Convention, by strengthening plant breeders' rights, has conversely limited existing exceptions. The remaining exceptions include acts done privately and for non-commercial purposes, experiments, and for the breeding and exploitation of other varieties. Breeders are now granted exclusive rights to harvested materials and the distinction between discovery and development of varieties has been eliminated.[55] Further, the right to save seed is no longer guaranteed as the farmer's privilege has been made optional.

As noted, plant breeders' rights were first conceived as an alternative to patent rights. As a result, UPOV originally provided that the two kinds of IPRs should be kept separate. Under UPOV-1978, member states can, for instance, only offer protection through one form of IPRs. The grant of a PBR on a given variety implies that no other intellectual property right can be granted to the same variety. This restriction has been eliminated under UPOV-1991 and double protection is now allowed.

The UPOV Convention was first negotiated and ratified mostly by developed countries. It is only since the adoption of the TRIPS Agreement that more developing countries have progressively joined the Convention.[56] Even though developing countries did not participate in the development of this legal regime which is tailored for mechanized and large-scale agriculture, the rationale for joining it is that the UPOV regime is generally held to fulfil the conditions of a *sui generis* system as required under Article 27.3(b) of the TRIPS Agreement.[57]

2.4 Selected Regional and National Legal Frameworks[58]

2.4.1 Developing countries

Developing countries have adopted different strategies to respond to the need to introduce IPR frameworks in the field of plant genetic resources as required by the TRIPS Agreement. This section reviews a few of the initiatives that have been taken at the broad level of a whole continent (the African Model Legislation), at the regional level (Andean Community) and at the national level (Peru and India). This serves to illustrate different ways in which developing countries have faced the challenge of introducing IPR in the field of plant genetic resources.

The African continent: the OAU Model Legislation[59]

In the African continent, states have tried to respond individually and collectively to the challenge that the implementation of the TRIPS Agreement presents for most of them. Before the adoption of the TRIPS Agreement, most sub-Saharan African states, in particular the least developed ones among them, had not invested significant time and effort in developing IPR regimes in the field of plant genetic resources. The sudden need to implement the TRIPS Agreement has led African states to pursue different strategies. Some states have tried to develop national frameworks while others have tried to cooperate in the framework of existing regional IPR organizations.[60] Most states realized that the difficulties involved in devising a new framework for access and control over plant

[55] See, for example, Nijar and Ling (1994, p. 277).

[56] Overall, as of July 2002, out of 51 member states there were 16 developing country members, an overwhelming majority of them being Latin American countries, and with only four developing country members altogether from Africa and Asia.

[57] See, for example, WTO, Review of Article 27.3(b): The View of Switzerland, WTO Doc. IP/C/W/284 (2001).

[58] Authors: Philippe Cullet, Susette Biber-Klemm and Danuta Szymura Berglas.

[59] Author: Philippe Cullet.

[60] See, in particular, Annex 10 of the Agreement to revise the Bangui Agreement on the Creation of an African Intellectual Property Organization of 2 March 1977 and 24 February 1999, Bangui.

genetic resources and related knowledge necessitated an effort at a broader level. As a result, the Organization of African Unity (OAU) took on the task of negotiating among its member states a model law adapted to the African region that could be relied upon when introducing national legal frameworks concerning plant genetic resources.

The African Model Legislation for the Protection of Rights of Local Communities, Farmers, and Breeders and for the Regulation of Access to Biological Resources (Model Legislation) was finally adopted in 2000 after a period of consultations and negotiations. The Model Legislation pursues a number of inter-related goals. It generally seeks to ensure the sustainable management of biological resources. Within this broad objective, the Model Legislation focuses on a number of different issues. These include the question of access to biological resources, community rights, farmers' rights and plant breeders' rights.

The Model Legislation first seeks to reassert countries' sovereign rights over their biological resources and proposes a system whereby access can only be granted with prior and informed consent of the state of origin of the biological resource, as well as of concerned communities.[61] Among the conditions laid down in the Model Legislation is one specifying that patents over life forms are not recognized on accessed biological resources. As a result, individuals or legal entities that collect biological resources are barred from applying for patents over biological resources accessed in accordance with the Model Legislation.[62]

After dealing with the issue of access and the conditions on which access can be granted, the Model Legislation defines three types of related property rights. First, it recognizes the collective rights of communities to their biological resources and the right to collectively benefit from their use, rights to their innovations, practices, knowledge and technology as well as the right to benefit collectively from their utilization.[63] In practice, these rights allow communities to prohibit access to their resources and knowledge, but only in cases where access would be detrimental to the integrity of their natural or cultural heritage.[64] The right to control access is strengthened with a right to receive at least 50% of the benefits derived from the commercial use of their resources or knowledge. The duty to channel the benefits back to the communities is put on the state.[65] The Model Legislation also provides for the recognition of community IPRs. These rights include the rights of communities to community innovation, practice, knowledge or technology.[66] Apart from the specific indication that registration is not a condition for the protection of community IPRs, the Model Legislation does not provide a complete framework for community IPRs, a task that must be undertaken by individual countries adopting the Model Legislation.

The Model Legislation goes on to define farmers' rights. Farmers' rights include the protection of TK relevant to plant and animal genetic resources; the right to an equitable share of benefits arising from the use of plant and animal genetic resources; the right to participate in making decisions on matters related to the conservation and sustainable use of plant and animal genetic resources; the right to save, use, exchange and sell farm-saved seed or propagating material; and the right to use a commercial breeder's variety to develop other varieties.[67]

Besides the recognition of community

[61] See Articles 3–8 of the African Model Legislation for the Protection of Rights of Local Communities, Farmers and Breeders and for the Regulation of Access to Biological Resources, 2000.

[62] Article 9 of the African Model Legislation, note 61 above.

[63] Articles 17–19 of the African Model Legislation, note 61 above.

[64] Article 20 of the African Model Legislation, note 61 above.

[65] Article 23 of the African Model Legislation, note 61 above.

[66] Article 24 of the African Model Legislation, note 61 above.

[67] Articles 25–27 of the African Model Legislation, note 61 above.

rights and farmers' rights, the Model Legislation also defines plant breeders' rights along the lines of the model provided by the UPOV Convention. The Model Legislation seeks to provide a comprehensive regime that includes not only the rights of farmers and communities but also the rights of commercial breeders. The underlying reason for including plant breeders' rights in these provisions is that the Model Legislation proposes a plant breeders' rights regime that is much more balanced than the UPOV version. The Model Legislation thus provides broad exemptions to the rights of breeders and also gives significant latitude to member states to restrict the rights for reasons of public interest.[68]

On the whole, the Model Legislation seeks to provide a framework that takes into account the requirements of Article 27.3(b) of the TRIPS Agreement. At the same time, it tries to go further and to recognize the specific needs of African countries by emphasizing the need for a strict regime on ABS, as well as the need for the recognition of other rights, such as the collective rights of communities and farmers' rights. While the Model Legislation constitutes a source of inspiration for all states that are developing legal frameworks in this field, it has not been widely adopted in the form it received at the level of the OAU. This is probably due, in part, to the fact that while the plant breeders' rights regime proposed under the Model Legislation goes some way towards meeting the criteria of an 'effective' *sui generis* system under the TRIPS Agreement, the rather stringent restrictions imposed on the rights of commercial breeders, and the introduction of farmers' rights as well as community rights, make it a rather controversial instrument at the international level. This is the case although it probably constitutes an appropriate starting point for most African countries, and is much more adapted to local circumstances and needs than a regime solely based on the UPOV Convention.

Andean region[69]

THE ANDEAN COMMUNITY'S REGIME ON GENETIC RESOURCES AND TRADITIONAL KNOWLEDGE

The Andean region has been at the forefront of efforts to introduce legal frameworks to implement the access and benefit-sharing provisions of the CBD. This section surveys the region-wide regime on access to genetic resources and traditional knowledge and gives an overview of its implementation in the Peruvian legislation.

Two Decisions of the Andean Community are relevant for its regime on access to genetic resources and TK: the 1996 Decision on a Common Regime on Access to Genetic Resources,[70] and Decision 486 on the Common Intellectual Property Regime.[71]

The 1996 Decision 391 of the Cartagena Agreement,[72] established a common regime on access to genetic resources. Its purposes are: (i) to establish the conditions for a just and equitable participation in the benefits of access; (ii) to lay the foundations for the recognition and valuation of the genetic resources and their by-products, and of their associated intangible components; (iii) to promote conservation and sustainable use of biological diversity; (iv) to promote the development of scientific, technological and technical capacities at all levels; and (v) to strengthen the negotiating

[68] Articles 43 and 45 of the African Model Legislation, note 61 above.

[69] Author: Susette Biber-Klemm.

[70] Andean Community Commission, Decision 391: Common Regime on Access to Genetic Resources, Caracas, Venezuela, 2 June 1996 http://www.comunidadandina.org/normativa/dec/D391.htm (English and Spanish).

[71] Decision 486: Régimen Común sobre Propiedad Industrial, Lima, Perú, a los catorce días del mes de setiembre del año dos mil.

[72] The Cartagena Agreement of 1969 (the Andean Subregional Integration Agreement), as revised by the Protocol of Trujillo of 1996, created the Andean Community. The Andean Community aims at the economic and social integration of its members in view of the creation of a Latin American Common Market. The members of the Andean Community are Bolivia, Colombia, Ecuador, Peru and Venezuela.

capacity of the member countries (Article 2).

The 'strategic value' of the know-how, innovations and practices of the native Afro-American and local communities, their historic contribution to the conservation and development of biological diversity, and the sustained use of its components are explicitly recognized in the preamble (paras 5 and 6). The close interdependence between the native populations and communities and their biological resources, and the need for its reinforcement, are acknowledged (para. 7).

The Decision's subject matter encompasses the genetic resources, their by-products and so-called 'intangible components'. By-products (*Producto derivado*) are defined as 'a molecule, a combination or mixture of natural molecules, including crude extracts of live or dead organisms of biological origin that come from the metabolism of living beings' (Article 1.7). The 'intangible components' are defined as all know-how, innovation or individual or collective practice with a real or potential value that is associated with the genetic resource, its by-products or the biological resource that contains them, whether or not protected by intellectual property regimes (Article 1.18). Thus the Decision explicitly includes not only genetic, but also biochemical information[73] and TK.

The Decision is based on the concept that genetic resources originating in a state, and their by-products, are goods belonging to the heritage of the state. As such, they are 'inalienable, not subject to prescription and not subject to seizure or similar measures'. This exclusive right of the state is independent of the property regimes applicable to the biological resources that contain the genetic or biochemical resources, or to the land on which they are located (Article 6). The state is the exclusive holder of the property rights over the genetic resources. This means that the genetic information contained in domesticated plants and crops bred by farmers and indigenous communities falls under the authority of the state, the individual holders not being allowed to decide about the use made of the genetic information.[74] Farmers and local communities have no rights either to the genetic information contained in their varieties, or to the biochemical information contained in, for instance, medicinal plants.

Neither does the Andean Community's regulation of Plant Breeders' Rights recognize any rights of farmers to their varieties.[75] The 'Common Provisions on the Protection of the Rights of Breeders of New Plant Varieties'[76] closely follows the UPOV regulations. Breeders' certificates are to be granted to persons who have created new, uniform, distinct and stable plant varieties: the term 'creation' meaning 'the production of a new variety by the application of scientific skills' (Article 4).

TK, however, remains under the authority of its authors. Member countries are obliged to 'recognize and value the rights and the authority of the native, Afro-American and local communities to decide about their know-how, innovations and traditional practices associated with genetic resources and their by-products' (Article 7).

This differentiation of the rights to genetic and biological resources and associated TK leads to a complex system and various 'layers' of access contracts in the Decisions ABS regime.[77]

Access contracts are to be concluded between the applicant and the state (Article

[73] The question being whether medicinal plants as such (i.e. biological resources) fall under the regulation or not (Ruiz Muller, 2003).

[74] As to the consequences of this regime on the rights of grassroot stakeholders in the ABS process, see Biber-Klemm, Chapter 7; similarly, Ruiz Muller (2003) and Rosell (1997).

[75] A narrow farmers' privilege for own (non-commercial) use is granted (Article 26).

[76] Decision 345: Régimen Común de Protección a los derechos de los Obtentores de Variedades Vegetales, Bogotá, Colombia, 21 October 1993.

[77] As to the problems generated by this system regarding the marketing of the information see Biber-Klemm, Chapter 7; similarly, Ruiz Muller (2003).

32), the latter defining the conditions for access (Article 17). In addition, so-called ancillary contracts defining the conditions for access have to be concluded between the applicant and the owner of the biological resource (land owner, possessor, manager; *ex-situ* facility; owner of the biological resource [Articles 41–44]). These contracts do not confer an independent right of access but depend on the main access contract with the National Authority (Article 42). If 'intangible components' are included in the research, the conditions for access have to be negotiated with their respective holder and to be included in an annex to the access contract (Article 35). So, in order to be allowed to access one resource, various layers of contracts with various stakeholders in different locations might have to be negotiated.[78]

In the case of failure to comply with the stipulations of the annex, the entire access contract can be nullified (Article 33).

The national intellectual property offices have to control compliance with the access regulations. Sanctions such as fines, confiscation, closing-down of establishments or disqualification from applying for new accesses may be imposed upon unauthorized access or transactions that are not protected by corresponding contracts (Articles 46 and 47). IPRs that have been obtained or developed on the basis of illicit access are not to be acknowledged by the Member Countries.

This control is reinforced by Decision 486 on the Common Intellectual Property Regime,[79] which, revising an earlier version, was adopted in 2000.[80] The Decision is specifically meant to implement the TRIPS Agreement and the CBD. With respect to access to genetic resources and traditional knowledge, the Decision stipulates the following: as in decision 319, the member states recognize the rights and faculties of local, indigenous and Afro-American communities to decide over their collective knowledge (Article 3). It narrows the possible scope of patents in excluding not only life forms, in whole or in part, as they are found in nature, natural biological processes and biological material which exist in nature, but also material that can be isolated from any life form, including genomes or germplasm (Article 15).

It provides a certain control of the legitimacy of access, as any application for a patent on an invention, obtained or developed from genetic resources, their derived products, or from TK shall include a copy of the access contract, or the copy of the document that accredits a licence or authorization of use from the community, respectively (Article 26). Patents granted on inventions obtained or developed from genetic resources or traditional knowledge, without presentation of a copy of the proper access contract or licence from the community, are to be nullified (Article 75), and any mark referring to elements of the cultures of indigenous, Afro-American or local communities shall not be registered without the community's express consent (Article 136).[81]

THE PERUVIAN REGIME ON ACCESS TO GENETIC RESOURCES AND TRADITIONAL KNOWLEDGE

In implementing the Andean framework, Peru has issued regulatory instruments on access to both genetic resources, and to TK. These are, with regard to the genetic resources, the 'Law on the Conservation and Sustainable Use of Biological Diversity'[82] and the 'Draft Regulation on Access to Genetic Resources'; with regard to TK,

---

78 See the enumeration in Ruiz Muller (2003), Box 1.

79 Decision 486: Régimen Común sobre Propiedad Industrial, Lima, Perú, a los catorce días del mes de setiembre del año dos mil.

80 The following is based on information from GRAIN: Andean Community Adopts New IPR Law, 5 October 2000. http://www.mtnforum.org/resources/library/grain00a.htm (23 May 2004).

81 See the critical assessment of this decision in the interview with Margarita Florez in the GRAIN information above, note 80.

82 Ley sobre la conservación y aprovechamiento sostenible de la diversidad biológica, Ley N° 26839. Accessible in Spanish on: http://www.grain.org/brl/brl-ley-peru-es.cfm

the 'Law introducing a Protection Regime for the Collective Knowledge of Indigenous Peoples derived from Biological Resources'.[83]

The Law on Conservation and Sustainable Use of Biodiversity is the implementation of the Convention on Biological Diversity. It is based on the designation in the constitution (Artículo 66°) of all natural resources (renewable and non-renewable) as the heritage of the state. It was enacted in 1997. The law adopts the Andean Pact regime of the state being owner of the genetic resources, independently of the ownership of the biological resources. Accordingly, the state is party to the access procedures (Artículo 27, 28). Access can be restricted or denied for reasons including conservation of resources and their ecosystems, and/or potential negative effects on essential elements of the cultural identity of the peoples involved, and for 'strategic' genetic resources[84] ('Recursos genéticos ... calificados como estratégicos') (Article 29).

The 'Proposal of a Regulation on Access to Genetic Resources' establishes the procedure for gaining access.[85] It fixes the content of the request, determines the competent authorities and establishes the period of time granted to the authorities for the individual steps. It further spells out the minimum conditions the agreements must contain. As parties to the negotiations, it determines a state agency, and for the negotiations of the ancillary contracts the provider of the genetic resource – which can be the state, but also, for instance, an *ex situ* gene-bank or an indigenous people, including landowners or owners of the biological resources – and the purchaser/applicant. The conditions of compensation are to be negotiated with the provider[86] of the resource (Article 34). However, the main contract has to be negotiated with the competent state agency. The ancillary contracts 'can' (*podrán*) be concluded with the other involved stakeholders, the wording leaving open whether there is an obligation to involve them or not (Article 9). The financial benefits resulting from the access contract are to be set aside for a fund for the conservation, investigation and development of genetic resources. A directive council administers the fund, encompassing representatives from the involved ministries, universities teaching genetics, NGOs and indigenous peoples (Article 51). The fund is meant to support projects for the conservation and use of genetic resources, as well as to improve the scientific capacities of the universities conducting research

[83] Ley 27811: Régimen de protección de los conocimientos colectivos de los pueblos indígenas vinculados a los recursos biológicos (10 August 2002); http://www.concytec.gob.pe/infocyt/ley27811.html; http://www.grain.org/brl/peru-tk-2002–en.cfm. See also the overview in WIPO (without document number), Descriptions of National and Regional Experiences with Existing *sui generis* Measures and Laws for the Protection of TK; advance copy, WIPO Intergovernmental Committee on Intellectual Property and Genetic Resources, Traditional Knowledge and Folklore, Fifth Session, July 7–15, 2003. For medicinal plants there is specific legislation (Ley 27300 de aprovechamiento sostenible de las plantas medicinales, 8 July 2000).

[84] The draft regulation on Access on Genetic Resources defines 'Strategic Genetic Resources' as: 'Todo recurso que es nativo o no del Perú o que el Perú es su centro ancestral de diversificaciôn, que serve paragraphsatisfacer al menos una necesidad humana básica, que se sosepcha que tiene un potencial ceonico inusitado, que es naturalmente escaso a nivel mundial o endémico y natrualmente insustituible. La calida de recurso gehético stratégico es declarado expresamente por el Estado Peruano' (Article 2.39). However, the interdiction of the exportation of viable or primarily processed parts of the – commercially interesting – Maca plant (*Lepidium meyenil)* is based on the argument of the conservation of the national genetic heritage allowing exceptions of free trade (Decreto Supremo N° 039–2003–AG, 8 December 2003).

[85] Propuesta de Reglamento sobre Acceso a los Recursos Geneticos. Version Abril-Julio 2001. Accessible under http://www.inrena.gob.pe/divbiol/reglamento-final-06–de-agosto.pdf (23 May 2004). The proposal is presently (summer 2004) in the parliamentary process; the final version not yet being known (personal information Begoña Venero, INDECOPI).

[86] The draft regulations define as provider the person, institution, indigenous people or other facility in the framework of Decision 391 and the present regulations ... including the state.

in genetic resources, but also institutions that enable the negotiating capacity of the indigenous people (Article 49).

So, the proposed Peruvian regime on access to genetic resources hardly strengthens the position of local farmers and other holders of genetic resources and is even a rather weak implementation of the participatory farmers' rights as stipulated in the ITPGRFA.

In turn, the right of indigenous peoples to their collective knowledge is explicitly recognized and regulated in the 'Law introducing a Protection Regime for the Collective Knowledge of Indigenous Peoples derived from Biological Resources'.[87] This law is based on a provision in the Peruvian Law on Intellectual Property that gives the Ministry of Industry, Tourism, Integration and International Trade the competence to 'establish a regime to protect and possibly register the knowledge of native and farming communities'.[88] The term 'indigenous peoples' is broadly defined as 'aboriginal peoples holding rights that existed prior to the formation of the Peruvian state, maintaining a culture of their own, occupying a specific territorial area and recognising themselves as such. These include ... rural and native communities' (Article 2(a)).

The law creates a *sui generis* right to the *collective knowledge* that belongs to the *communities* of the indigenous peoples (and explicitly not to its individual members) (Article 10). This collective knowledge is considered a part of the cultural heritage of the indigenous peoples (Article 11). Therefore the rights of the indigenous peoples to their collective knowledge are inalienable and not extinguishable (Article 12). The present generations preserve, develop and administer their collective knowledge for their own benefit and for that of future generations (Article 9). Consequently, the knowledge can only be licensed. The licence does not confer exclusive rights, nor is it able to prevent others from using the knowledge or other licences from being awarded on the same knowledge. Nor do the licences affect the right of the peoples to use and further develop their collective knowledge (Article 32).

The regime creates three instruments: a system to control *access* to the collective knowledge, a *system of registers* of collective knowledge, and an additional *funding system*, which is to contribute to the integral development of the indigenous peoples.

Access to TK is dependent upon obtaining the prior informed consent of the representative organizations of the indigenous peoples possessing the collective knowledge. Other indigenous peoples possessing the same knowledge are to be informed by the organization. Their interests and concerns are to be taken account of in the negotiations (Article 6). The law defines the obligatory contents of the licence contracts, such as information on the intended research, its risks and implications, the statement of the compensation and the obligation of the licensee to inform (Article 27). The contracts are to be registered by the National Intellectual Property Institute (INDECOPI, Instituto Nacional de Defensa de la Competencia y de la Protección de la Propiedad Intelectual), which verifies the legitimacy of the licence contracts. The conditions for gaining access are different for access for scientific and industrial use respectively. In respect of access for *scientific* use, only the prior informed *consent* of the people concerned is needed. In the case of access for *commercial* or *industrial* application, in addition, a licence contract has to be concluded. This contract has to fix the conditions for the payment for access and an equitable distribution of the benefits (Article 7). Further, the contract of licence must contain the obligation to periodically inform the licensor in general terms about the advances in the investigation, industrialization and marketing of the products

[87] Note 83 above.

[88] For the description of the legislation process, in particular in view of the participation of indigenous communities, see Tobin and Swiderska (2001).

developed from the collective knowledge (Article 21(e)).

The *Fund for the Development of Indigenous Peoples* is meant to support the comprehensive development of the indigenous peoples through the financing of projects and other activities (Article 37). The fund is autonomous and administered by an Administrative Committee which comprises five members of representative organizations of indigenous peoples and two members of the National Commission for the native peoples. The law states explicitly that the Committee shall, to the extent possible, proceed according to the traditional mechanisms and customs developed by the indigenous peoples for allocating and distributing collectively generated benefits.

The fund is made up of a percentage of the economic benefits resulting from the marketing of goods that have been developed on the basis of TK; further, of contributions from the state budget, international technical cooperation and donations (Article 41).

An interesting detail is that in cases where the collective knowledge has passed into the public domain within the previous 20 years, a percentage of the benefits resulting from the marketing of the goods developed on the basis of this knowledge has to be set aside for the fund. Collective knowledge is considered to be in the public domain either when it has been made accessible to persons other than the indigenous peoples by mass communication media, or where it has become extensively known outside the confines of the indigenous peoples and communities (Article 13). The Public Domain TK is registered in the Public National Register.

The registration system consists of three types of registers: the Public National Register, the Confidential National Register and the Local Registers (Article 15) (for details see Biber-Klemm, Chapter 7). The registers have a twofold purpose: on the one hand they are meant as instruments to preserve the collective knowledge of indigenous peoples; on the other hand they are intended to serve as a tool to defend the interests of indigenous peoples regarding their TK (Article 16).

The registration does not constitute rights over TK. All knowledge that is collective in nature, developed by indigenous peoples and which is not in the public domain, irrespective of whether it is registered or not, is protected against disclosure, acquisition or use without the consent of the indigenous peoples who possess it, and against unauthorized disclosure involving a breach of a duty of reserve. So there is a certain degree of protection but no exclusive rights are conferred (Article 42).

The Andean Region's Experience in Perspective

The Andean Pact legislation on access to genetic resources and the Peruvian law on *sui generis* rights to TK have been hailed as landmarks in the process of implementing the CBD. In particular, the Peruvian legislation has been at the forefront for the creation of *sui generis* rights to TK associated with genetic resources. The Law 27811 provides a regulatory framework through which indigenous peoples can assert their rights over their collective knowledge (Ruiz Muller and Lapeña, undated, p. 15).

Yet, up to now, no access contracts have been concluded on the basis of Peru's TK legislation, and according to Ruiz, the Andean Pact Decision 391 has had little practical impact on the actual conclusion of access contracts (Ruiz Muller, 2003, para. No. 10).

Even if it may be too early to evaluate the Peruvian legislation, it might be interesting at this point to have a closer look at the possible reasons for this occurrence.

Ruiz Muller (2003) analyses the issue in view of the Andean Pact Decision 391. He identifies the creation of incentives as one of the most important factors to promote research into biodiversity. He considers it as an absolute necessity to create a legal framework that is clear, simple, practical, without inherent contradictions, and clearly promoting bioprospection.

He perceives the following as obstacles. First, the system of different layers of contracts, given by the regime of differentiation in ownership described above. Secondly, the fact that the state is the principal negotiator, and, as a rule, has the lead in the negotiations in matters concerning scientific topics and technical and legal issues regarding genetic resources; in fact matters over which specialized institutes have greater capacity and expertise. Thirdly, the differences in access procedures for the different resources such as microorganisms, PGRFA and genetic resources for medicinal use; and finally, the high transaction costs (time, money, effort) the Decision generates, which might be higher than the expected benefits.

These observations, which in our opinion are also at least partly true for other countries and regions,[89] have led to a process aiming at evaluating and improving the regime in the near future (Ruiz, personal information). Ruiz advocates a proactive rather than a defensive approach to legislation in matters of ABS and pleads for a system that is simple, practical and flexible enough to be adapted to the variability of real access situations.

India[90]

India has made significant progress in recent years towards setting up a legal regime for the management of plant genetic resources. The proposed regime is laid out in three separate legislative instruments. These are: the Protection of Plant Varieties and Farmers' Rights Act, the Biodiversity Bill and the Patents (Amendment) Act of 2002. The separation of various elements into three is partly due to India's international legal obligations. The Plant Varieties Protection Act constitutes a response to Article 27.3(b) of the TRIPS Agreement,[91] the Biodiversity Act constitutes India's attempt to implement the CBD,[92] and the Patents (Amendment) Act seeks to put India in conformity with its TRIPS obligations in the field of patent protection.[93] The proposed regime reflects its fragmented parentage insofar as there are a number of inconsistencies or overlaps among the three bills. Since each instrument was drafted separately, they are examined in turn.

THE PLANT VARIETIES PROTECTION AND FARMERS' RIGHTS ACT

The Protection of Plant Varieties and Farmers' Rights Act constitutes the government's response to its obligations under Article 27.3(b) of the TRIPS Agreement. The Indian government has chosen not to introduce patents over plant varieties but rather to devise its own system of IPRs. Further, it was agreed that the UPOV Convention should not be ratified but that a law suited to the specific conditions of the country should be drafted. In its present form, the Act focuses on the establishment of plant breeders' rights and farmers' rights. The regime for plant breeders' rights largely follows the model provided by UPOV and the criteria for registration are the same as those found in UPOV, namely novelty, distinctness, uniformity and stability. The Act incorporates elements from the 1978 version of UPOV and includes some elements from the more stringent 1991 version, such as the possibility of registering essentially derived varieties.

The second main aim of the Act is to introduce farmers' rights. At this level, the Joint Parliamentary Committee, to which the bill was referred after its introduction in Parliament, has proposed substantial changes.[94] While the version of the bill originally introduced in Parliament only contained a short provision on farmers'

[89] See Biber-Klemm, ABS, Chapter 6 and Registration, Chapter 7.
[90] Author: Philippe Cullet.
[91] Protection of Plant Varieties and Farmers' Rights Act, 2001.
[92] Biological Diversity Act, 2002.
[93] Patents (Amendment) Act, 2002.
[94] See Joint Committee on the Protection of Plant Varieties and Farmers' Rights Bill, 1999, *Report of the Joint Committee* (August 2000).

rights, the Committee decided to add a whole new chapter on farmers' rights. As adopted, the Act seeks to put farmers' rights on a par with breeders' rights. It provides, for instance, that farmers, like commercial breeders, can apply to have a variety registered.[95] Generally, the Act envisages that farmers should be treated like commercial breeders and should receive the same kind of protection for the varieties they develop.

The Act also provides two avenues for benefit sharing.[96] The first scheme allows individuals or organizations to submit claims concerning the contribution they have made to the development of a protected variety. The final decision is taken by the Authority established under the Act, which determines the amount, taking into account the importance of the contribution in the overall development of the variety and its commercial potential. The second benefit-sharing avenue allows an individual or organization to file a claim on behalf of a village or local community. The claim relates to the contribution that the village or community has made to the evolution of a variety.

THE BIODIVERSITY ACT

The Biodiversity Act was drafted in response to the CBD. The Act does not aim at providing a general regime for biodiversity management but focuses on some specific elements that are of concern at present. The Act thus reflects the government's strong reaction to biopiracy.[97] The Act is also partly premised on the desire to avoid a direct confrontation with WTO obligations in the field of plant genetic resources.

Generally, the Act focuses on the question of access to resources.[98] Its response to the current challenges is to assert the country's sovereign rights over natural resources. It therefore proposes to put stringent limits on access to biological resources or related knowledge for all foreigners. The Act's insistence on sovereign rights reflects current attempts by various countries to assert control over national resources or knowledge. While the Act focuses on preserving India's interests *vis-à-vis* other states in rather strong terms, its main impact within the country will be to concentrate power in the hands of the government. Indeed, Indian citizens and legal persons must give prior intimation to the state biodiversity boards of their intention to obtain biological resources.[99] The Act is even more stringent in terms of IPRs since it requires that all inventors obtain the consent of the National Biodiversity Authority before applying for such rights.[100] The impacts of this clause are, however, likely to be limited since patent applications are covered by a separate clause.[101] Further, the Authority has no extra-territorial authority and cannot monitor applications for IPRs outside India.

Overall, the Biodiversity Act implicitly takes the position that India cannot do more than regulate access by foreigners to its knowledge base. It does, however, attempt to discipline the IPRs system in some respects. As noted, it requires inventors who want to apply for IPRs to seek the Authority's permission. It also authorizes

[95] Section 16.1(d) of the Plant Varieties Act, note 94 above.

[96] See Sections 26 and 41 of the Plant Varieties Act, note 94 above.

[97] See, for example, US Patent No. 5,401,504, *Use of Turmeric in Wound Healing*, issued 28 March 1995 and European Patent No. EP 0436257, *Method for Controlling Fungi on Plants by the Aid of a Hydrophobic Extracted Neem Oil*.

[98] The only substantive chapter of the Biological Diversity Act – Chapter II – is entitled Regulation of Biological Diversity.

[99] Section 7 of the Biological Diversity Act, note 95 above.

[100] Section 6 of the Biological Diversity Act, note 95 above.

[101] Permission of the National Biodiversity Authority must be obtained before the sealing of the patent but can be obtained after the acceptance of the patent by the patent authority. See Section 6.1 of the Biological Diversity Act, note 95 above.

the Authority to allocate a monopoly right to more than one actor. Further, the Authority is also entitled to oppose the granting of IPRs outside India.[102] The Act also seeks to address the question of the rights of holders of local knowledge by setting up a system of benefit sharing. The benefit-sharing scheme is innovative insofar as it provides that the Authority can decide to grant joint ownership of a monopoly intellectual right to the inventor and the Authority or the actual contributors if they can be identified.[103] However, the sharing of IPRs is only one of the avenues that the Authority can choose to discharge its obligation to determine benefit sharing. Further, it is in the Authority's power to allocate rights to itself or to a contributor, such as a farmer contributor, and the latter has no right to demand the allocation of property rights. Other forms of benefit sharing include technology transfers, association of benefit claimers with research and development, or the location of production, research and development units in areas where this will facilitate better living standards for the benefit claimants.

PATENTS (AMENDMENT) ACT

The Patents Act adopted in 1970 dealt with patents in general and was not specifically related to biological resources. However, it addressed a number of issues that are of relevance in the context of PGR management. It rejected, for instance, the patentability of all methods of agriculture and was generally much more restrictive than similar laws in Western countries. TRIPS has imposed significant alterations to this Act. Thus, where formerly only a process patent could be obtained for no more than 7 years for food- or medicine-related inventions,[104] TRIPS now requires the availability of product and process patents for 20 years.

The Patents (Amendment) Act, 2002 generally seeks to modify the Act to allow compliance with TRIPS. There was not much scope for diverging from the rather precise TRIPS obligations if India wanted to avoid further confrontation with the WTO.[105] The Amendment Act thus brings the duration of the rights to a uniform 20-year period and also substantially modifies the sections concerning the working of the patents by, for instance, doing away with licences of rights. The provision that seeks to oblige patentees to manufacture their inventions in India was also struck out because of the TRIPS requirement that imports should not be treated differently from products produced locally.[106] With regard to environmental protection, the Amendment Act includes some of the TRIPS exceptions related to environment and health. It also addresses the question of biopiracy by imposing the disclosure of the source and geographical origin of biological material used in a patented invention. Further, non-disclosure of the geographical origin or the anticipation of the invention in local or indigenous knowledge constitutes grounds for opposing or revoking a patent.

---

[102] Section 18.4 of the Biological Diversity Act, note 95 above.

[103] Section 21.2(a) of the Biological Diversity Act, note 95 above.

[104] Section 53 of the Indian Patents Act, 1970.

[105] India was taken to the WTO dispute settlement mechanism by both the USA and the EU for its failure to implement in due time its obligations relating to Exclusive Marketing Rights. For the US complaint, see, for example, India – Patent Protection for Pharmaceutical and Agricultural Chemical Products (US complaint), Report of the Panel, 5 September 1997, WTO Doc. WT/DS50/R, and India – Patent Protection for Pharmaceutical and Agricultural Chemical Products (US complaint), Report of the Appellate Body, 19 December 1997, WTO Doc. WT/DS50/AB/R.

[106] Article 27 of the Agreement on Trade-Related Aspects of Intellectual Property Rights, Marrakesh, 15 April 1994, reprinted in 33 ILM 1125 (1994).

2.4.2 Developed countries

Europe[107]

European states have sought harmonization of their intellectual property laws for a long time, both in the context of the European Union (EU) and in broader forums. The European Patent Convention (EPC) of 1973 currently brings together 27 states and provides a general framework for the harmonization of patent law throughout Europe.[108] It has become one of the focal points of interest in the development of patent law in the field of biotechnology. In the context of the EU, the most significant instrument with regard to recent developments in genetic engineering is undoubtedly the Biotechnology Directive of 1998.[109]

EUROPEAN PATENT CONVENTION

The EPC generally provides a uniform patent granting procedure that applies throughout the member states. In the context of the law of plant genetic resources, the EPC is particularly noteworthy with regard to the scope of patentability. The general conditions for patentability are substantially similar to those of the TRIPS Agreement and include the criteria of novelty, inventiveness and industrial applicability.[110] Some specific exceptions to patentability are provided in Article 53. This generally includes inventions whose exploitation would be contrary to public morality. The notion of public morality has been interpreted as covering public peace or social order or serious prejudice to the environment.[111]

The EPC also excludes the patentability of plant or animal varieties (but not plants and animals in general, as in the TRIPS agreement), or essentially biological processes for the production of plants or animals, with the exception of microbiological processes or their products that are patentable.[112] The exclusion from patentability of plant varieties was first interpreted by the European Patent Office (EPO) as being a consequence of the existence of national and international regulations for the protection of plant breeders' rights.[113] The consequence was that only the patenting of plants or their propagating material in the genetically fixed form of the plant variety was prohibited, but not the patentability of the plant or the animal itself. Since 1995, the interpretation of Article 53.b is that claims on plants are not acceptable because transgenic plants encompass plant varieties. However, plant cells have now been determined as being patentable.[114]

EUROPEAN UNION BIOTECHNOLOGY DIRECTIVE

The Directive on the Legal Protection of Biotechnological Inventions is now the central instrument in the EU context concerning the patentability of GMOs.[115] It has been one of the most contentious pieces of legislation to go through the European Parliament. After lengthy debates it was adopted in June 1998, and subsequently unsuccessfully challenged.[116]

107 Author: Philippe Cullet.

108 Convention on the Grant of European Patents, Munich, 5 October 1973 (hereafter EPC).

109 Directive 98/44/EC of the European Parliament and the Council of the European Union on the Legal Protection of Biotechnological Inventions, 6 July 1998, 1998 OJ L 213 (hereafter Biotechnology Directive).

110 Article 52.1 of the EPC, note 111 above.

111 *Plant Genetic Systems vs Greenpeace*, EPO Technical Board of Appeal, 21 February 1995 (T 356/93).

112 Article 53.b of the EPC, note 111 above.

113 Ciba-Geigy/Propagating Material, EPO Technical Board of Appeal, 26 July 1983 (T49/83), and Lubrizol/Hybrid Plants, 10 November 1988 (T 320/87).

114 See *Plant Genetic Systems vs Greenpeace*, EPO Technical Board of Appeal, 21 February 1995 (T 356/93).

115 Biotechnology Directive, note 112 above.

116 See *Netherlands vs Council of the European Union*, Judgment of 9 October 2001 (Case C-377/98).

The Directive is premised on the recognition of the importance of biotechnology and the necessity of providing legal protection to biotechnological inventions in the European Community. It complies with other EU legal instruments and the EPC. The principle is that biotechnological inventions are to be protected by patent rights. It specifically excludes the patentability of plant and animal varieties, and essentially biological processes for the production of plants or animals. However, patentability is provided for inventions concerning plants or animals if the technical feasibility of the invention is not confined to a particular plant or animal variety. This significantly narrows down the exception concerning plant and animal varieties.

The Directive follows the EPC and TRIPS in prohibiting patentability in situations where the commercial exploitation of an invention would be contrary to *ordre public* or morality. It provides specific examples of inventions deemed to breach these norms. These include processes for cloning human beings and processes for modifying the genetic identity of animals which are likely to cause them suffering without any substantial medical benefit to human or animal, and also animals resulting from such processes.

The scope of protection provided under the Directive is extremely broad. A patent on biological material possessing specific characteristics gives the patent holder rights over any biological material derived from that biological material through propagation or multiplication in an identical or divergent form, and possessing those same characteristics. However, the Directive includes a limited exception in the case of agricultural use. It provides a form of farmers' privilege modelled after the plant breeders' rights model, authorizing farmers to use the product of their harvest for propagation or multiplication on their own farm.[117] This exception extends to protected livestock as well.

Finally, the question of the Directive's relationship with other international treaties in the field of patents and biodiversity must be highlighted. Article 1 specifically indicates that the provisions of the Directive are without prejudice to the obligations of the member states pursuant to international agreements, and in particular the TRIPS Agreement and the CBD. In its ruling on the application for annulment of the Directive brought by The Netherlands, the European Court of Justice had to give an opinion on the compatibility of the Directive with the CBD. In rejecting The Netherlands' contentions, the Court made an important statement concerning the theoretical impact of the Directive on the protection of TK. It stated that:

> [i]t cannot be assumed, in the absence of evidence, which is lacking in this case, that the mere protection of biotechnological inventions by patent would result, as is argued, in depriving developing countries of the ability to monitor their biological resources and to make use of their traditional knowledge, any more than it would result in promoting single-crop farming or in discouraging national and international efforts to preserve biodiversity.[118]

The Court further dismissed a claim concerning the absence of benefit-sharing provisions in the Directive. It noted that the CBD does not specifically impose on member states the obligation to take into account the interests of the country from which the genetic resources used in the patented invention originate. However, the Court also acknowledged that Article 1.2 imposes on member states the obligation to implement the Directive in such a way that they do not breach their obligations under the CBD.

---

[117] See also Council Regulation (EC) No 2100/94 on Community Plant Variety Rights, 27 July 1994, OJ 1994 L227/1.

[118] C-377/98, note 119 above at § 65.

USA[119]

The USA has been at the forefront of the development of patent law, in particular in the fields of plant varieties and genetic engineering. The USA was the first country to enact a specific Plant Patent Act, in 1930, that provides protection for certain asexually reproduced plants.[120] Protection is granted to whoever invents or discovers and asexually reproduces distinct and new plant varieties. The scope of protection has been the subject of recent litigation. In *Imazio vs Dania*, the Court of Appeals determined that the term variety should be understood as providing protection only to a specific plant and not to a range of plants.[121]

The Plant Patent Act has been supplemented by the 1970 Plant Variety Protection Act, which grants protection to novel varieties of sexually reproduced plants.[122] The concept of variety is here informed by the definition found in UPOV-1991 and differs from the concept of variety found in the Plant Patent Act.[123] The rights granted under this Act are plant breeders' rights and not patents. One of the main differences between the two is that farmers can make use of seeds generated from protected stock, an exception that is not available under the patent regime. The scope of the exceptions in favour of farmers has been the object of significant debate. In *Asgrow vs Winterboer*, a seed company sued farmers for selling the crop produced from protected stock to other farmers for use as seed.[124] The main issue in this case concerned the scope of the farmer's privilege under 7 U.S.C. 2543. This section clearly protects farmers who use part of the seeds produced from protected stocks on their farm or who sell them. However, the court held that farmers are not protected in cases where the saved seed is grown for the purpose of sale for replanting because this constitutes a violation of 7 U.S.C. 2541(3).[125] The privilege under section 2543 is to be understood exclusively as authorizing a farmer to sell the seed s/he has set aside for the purpose of replanting his own acreage.

The lead of the USA in the context of plant variety protection has been extended in the past two decades to the field of genetic engineering. The USA was the first country to accept the patentability of artificially created life forms, thereby paving the way for the rapid development of the new biotechnology industry. The decision of the US Supreme Court in *Diamond vs Chakrabarty* triggered one of the most significant changes in the patent regime in recent times.[126] The Court was presented with the novel case of the 'invention' of an artificially created life form. It analysed the case from the point of view of the distinction between an invention and a discovery. In the balancing act, the Court put more weight on the fact that the bacterium was made by human beings (manufacture) than on the principle that products of nature did not constitute patentable subject matter under US law. The Court found support for its position in the existence of a plant patent Act that makes a clear distinction

[119] Author: Philippe Cullet.

[120] Plant Patent Act of 1930, 35 U.S.C. 161 ff.

[121] *Imazio Nursery vs Dania Greenhouses*, 3 November 1995, Court of Appeals, Federal Circuit, 69 F.3d 1560.

[122] Plant Variety Protection Act of 1970, 7 U.S.C. 2321 ff.

[123] *Imazio Nursery vs Dania Greenhouses*, 3 November 1995, Court of Appeals, Federal Circuit, 69 F.3d 1560, 1568.

[124] *Asgrow Seed Company vs Denny Winterboer and Becky Winterboer*, dba Deebees, 18 January 1995, Supreme Court, 513 U.S. 179.

[125] 7 U.S.C. 2541: 'Except as otherwise provided in this subchapter, it shall be an infringement of the rights of the owner of a protected variety to perform without authority, any of the following acts (...): (3) sexually multiply, or propagate by a tuber or a part of a tuber, the variety as a step in marketing (for growing purposes) the variety'.

[126] *Diamond vs Chakrabarty*, 16 June 1980, Supreme Court, 447 U.S. 303.

between products of nature and human-made inventions rather than between living and inanimate things.

The 1980 decision left open the question of the patentability of plants and living organisms. In the 1985 decision *Ex parte* Hibberd, the patentability of plants was accepted.[127] After significant debates, the Patent and Trademark Office (PTO) granted in 1998 the first patent for a transgenic animal.[128] In the past decade, there has been a flurry of patent applications for transgenic plants and animals. Finally, in *J.E.M. vs Pioneer Hi-Bred*, the Supreme Court confirmed the possibility of inventors of new plant varieties applying for patents. The Court reasoned that the possibility of protecting new plant varieties under the Plant Variety Protection Act did not preclude protection through patents because the application conditions for the latter are more stringent and the rights granted more extensive. These developments must be understood in the context of further developments concerning the patentability of genes. In recent years, patents on genes that claim cover on an isolated gene, or constructs that include the gene, have been increasingly often admitted (Barton and Berger, 2001). This has a direct impact on transformed plants that include such constructs. In other words, the increasing scope of patentability in the field of plant variety must be understood not only with regard to patents on plant varieties themselves but also with regard to patents on genes that make up these varieties.

Australia[129]

FACTUAL BACKGROUND TO THE AUSTRALIAN POSITION

Megabiodiversity. Australia is a country – and continent – of tremendous biodiversity or 'megabiodiversity', holding about 13% of the world's biodiversity (Voumard, 2000, p. 7) and of the estimated 44,000 species of plants growing there, 88% are found only in Australia (Blakeney, 1997, 1999, p. 85). In terms of endemic species, Australia has the most mammals and reptiles, the second most birds, and fifth most higher plants and amphibians.[130]

> The isolation of the Australian continent has resulted in the survival of a number of simplified plant and animal types which have been particularly suitable for genetic modification, or which have provided a valuable reservoir of characteristics for the genetic modification of non-Australian plants and animals. For example, the aridity and salinity of the continent has resulted in the development of organisms with particular resistance to these conditions (Blakeney, 1999, p. 85).

The value of Australian indigenous traditional ecological knowledge and the problem of 'biopiracy'

Duboisia plant. The TK of Australian indigenous people about the properties of Australian plants has often been used in the commercialization of some of those plants,[131] but not always with adequate recognition of their contribution. For instance, the *Duboisia* plant, used by Queensland Aborigines for therapeutic purposes, was developed by a British surgeon at the end of the nineteenth century as a substitute for atropine, and later found to contain hyoscine, used as a sedative in

---

[127] *Ex parte* Hibberd *et al.*, 18 September 1985, Patent and Trademark Office Board of Patent Appeals and Interferences, 227 U.S.P.Q. 443.

[128] U.S. Patent No. 4,736,866, Transgenic Non-Human Mammals, 12 April 1988.

[129] Author: Danuta Szymura Berglas.

[130] Communication from Australia IP/C/W/310 2, October 2001 point 15.

[131] 'A recent examination of plant breeders' certificates issued by the Plant Breeders' Rights Office found that of 36 certificates issued over Australian native plants, 16 had traditional known uses. Two of these certificates issued to Australian Native Produce Industries Pty. Ltd. pertained to traditionally used food species: muntries, *Kunzea pumifera* (AU 96/031), and sea celery, *Apium prostratium* (AU 96/026). Allegedly, no breeding work had been carried out on these species' (Fourmile-Marrie, 1998, p. 2).

treating motion sickness and as a 'truth drug'. According to Blakeney, 'the cultivation of *Duboisia* is currently conducted on a commercial scale in Northern New South Wales and Southern Queensland, but without any recompense to indigenous peoples' (Blakeney, 1999, p. 86).

Native bush food. There is increasing interest in Aboriginal knowledge about bush food as a valuable source of nutrition. The Australian Native Bush Food Industry Committee reported that in 1996 bush food sales were over AU$14 million and rising (Woodley, 1998, p. 329, cited by Fourmile-Marrie, 1998, p. 2). In fact, a non-indigenous Australian, Les Hiddens, known as 'the Bush Tucker Man', has been very successful in bringing the value of Aboriginal knowledge about Australian bush food to the minds of all Australians, through his television programmes, books and CD-ROMs.[132] Unfortunately, this success has also led to him being condemned by some as 'the very public face of (biopiracy)' (Fourmile-Marrie, 1998, p. 2).[133] Apart from the question of whether indigenous Australians have been properly compensated for the commercialization of their TK remains the fundamental issue of whether they agreed to the commercialization of their knowledge by others in the first place.[134]

Acacias. Aboriginal knowledge about the food value of acacias was instrumental in a recent study conducted by the Australian Tree Seed Centre of the Commonwealth Scientific and Industrial Research Organization (CSIRO) Division of Forestry in several African dry-land countries (Senegal, Niger, Burkina Faso, Somalia, Kenya and Zimbabwe) upon the usefulness of planting acacias in those countries. 'With respect to food potential, what is currently known about the food value of acacias has been largely the result of tapping into Aboriginal knowledge' (Devitt, 1991). 'And yet', according to Fourmile-Marrie, 'Aboriginal people are not involved in the subsequent research, development and application processes regarding those overseas projects', contrary to Australia's obligations under the CBD (Fourmile-Marrie, 1998, p. 1).

Smokebush. The endemic plant smokebush (genus *Conospermum*) provides an interesting illustration (Janke, 1999, pp. 24–25). Smokebush grows in the coastal areas between Geraldton and Esperance in Western Australia, and its healing properties have been known and used by indigenous Australians for a long time. During the 1960s the Western Australian government granted the US National Cancer Institute (NCI) a licence to collect plants for screening purposes, and in 1981 specimens were tested for the presence of cancer-fighting properties. None were found, but the specimens were held in storage until the late 1980s when they were again tested, this time in the quest to find a cure for AIDS. Out of 7000 plants screened from around the world, Smokebush was one of only four plants found to contain the active constituent *conocurovone*, which laboratory tests showed could destroy low concentrations of the HIV virus. This 'discovery' was subsequently patented. The US National Cancer Institute has since awarded Amrad, a pharmaceutical company from Victoria,

[132] For instance, *Explore Wild Australia with the Bush Tucker Man* (1999) ABC Books/Viking, Sydney; CD-ROMs *From the Rainforest to Cape York* and *From Arnhem Land to the Kimberley*.

[133] For Aborigines, 'reproduction is unreal, while recreation is real. The fixation on the written word has implications for the practice of cultural heritage' (Janke, 1999, p. 7).

[134] Compare different points of view of indigenous Australians: Les Malzner stated at the Indigenous Reference Group Meeting, September 1997: 'Indigenous ownership is seen more in terms of responsibility for culture rather than excluding others from its use', whereas Francis Kelly Jupurrula, spokesperson for the Central Land Council submitted: 'We should keep it [knowledge of traditional medicines and bush tucker] for ourselves, not give it away. We are just like giving our land away altogether. I think we should all keep it because it's confidential ... We should keep it for our generations' (speaking at Akarnenhe Well, 14–16 October 1997, CLC Submission January 1998) (Janke, 1999, pp. 44 and 46).

an exclusive worldwide licence to develop the patent.

In the early 1990s the Western Australian government also awarded Amrad rights to the smokebush plant in order to develop an anti-AIDS drug, having power to grant exclusive rights to Western Australian flora and forest species for research purposes pursuant to amendments made in 1985 to both the *Conservation and Land Management Act 1984 (Western Australia)* and the *National Parks and Wildlife Act (Western Australia)*. According to Blakeney (1997, p. 196), Amrad paid $1.5 million to the Government of Western Australia to secure access to smokebush and related species, based on projected royalties of $100 million per year by 2002 in the case of successful commercialization.

The concern of the indigenous Australians is that they have not received any acknowledgement, financial or otherwise, for their role in having first discovered the healing properties of smokebush. According to the Centre for Indigenous History and Arts (Western Australia):

> The current legislation disregards the potential intellectual property rights that the Indigenous peoples in Western Australia have in flora on their lands. Furthermore, multinational drug companies could be sold exclusive rights to entire species of flora, preventing anyone from using those species for any other purpose without the consent of the companies. Indigenous peoples in Western Australia now face the possibility of being prevented from using any of the flora which is the subject of an exclusive agreement. It is therefore vital that any reform of the intellectual and cultural property laws include provisions for the recognition of indigenous peoples as the native title owners of all the biological resources of the flora and fauna that are on their lands (Centre of Indigenous History and the Arts, submission to *Our Culture, Our Future*, October 1997 (Janke, 1999)).

In short, it is clear that indigenous Aboriginal ecological knowledge is valuable,[135] but a more just determination of the legal rights to Australia's biological resources and associated TK is still in the process of development.

ABORIGINAL INTELLECTUAL PROPERTY

Aboriginal acknowledgement of their IPRs. The *Julayinbul Statement on Indigenous Intellectual Property Rights* has expressly acknowledged the existence of 'Aboriginal intellectual property' which is found 'within Aboriginal Common Law', and as such 'is an inherent inalienable right which cannot be terminated, extinguished, or taken'.[136]

Clause 5 of the *Julayinbul Declaration*[137] includes 'the right [of indigenous people] to control subsequent use of and access to the genetic make-up within the flora and fauna of the forests'. Further, indigenous Australians would like to see the principle of prior informed consent, which applies to sovereign states under the CBD, to be extended to them (claim 7).[138]

The basis for the Aboriginal definition of their IPRs. Article 2(viii) of the Convention Establishing the World Intellectual Property Organization, 14 July 1967, defines *intellectual property* as the rights relating to literary,

[135] Although the value of indigenous knowledge to the biotechnology industry in Australia has not yet been estimated (Janke, 1999, p. 15).

[136] The Julayinbul Conference on Intellectual and Cultural Property, held at Jingarrba, in the Daintree Forest region of North-Eastern Australia on 27 November, 1993, produced the *Julayinbul Statement on Indigenous Intellectual Property Rights* and a *Declaration Reaffirming the Self-Determination and Intellectual Property Rights of the Indigenous Nations and Peoples of the Wet Tropics Rainforest Area,* cited by Blakeney (1999, pp. 93–95).

[137] *Declaration Reaffirming the Self-Determination and Intellectual Property Rights of the Indigenous Nations and Peoples of the Wet Tropics Rainforest Area,* produced by the Julayinbul Conference on Intellectual and Cultural Property, held at Jingarrba, in the Daintree Forest region of North-Eastern Australia on 27 November, 1993, cited by Blakeney (1999, pp. 93–95).

[138] See comprehensive list of indigenous claims in Janke (1999, pp. XX–XXI).

artistic and scientific works; performances of performing artists, phonograms and broadcasts; inventions in all fields of human endeavour; scientific discoveries; industrial designs; trademarks, service marks and commercial names and designations; protection against unfair competition; and all other rights resulting from *intellectual activity* in the industrial, scientific, literary or artistic fields (emphasis added).

The basis of this definition is 'intellectual activity', whether it stems from an indigenous or non-indigenous person, and TK – of both indigenous and non-indigenous people – is one form of intellectual activity within intellectual property.

The United Nations Special Rapporteur, Erica Irene Daes, stated in a 1993 report 'all elements of heritage should be managed and protected as a single, interrelated and integrated whole'.[139] According to the submissions made to the Australian Government by representatives of Aboriginal communities in their comprehensive report, *Our Culture, Our Future* (Janke, 1999), heritage forms the basis of any discussion on TK and IPRs held by Australian indigenous people. The report states that

> [indigenous] heritage consists of the tangible and intangible aspects of the whole body of cultural practices, resources, and knowledge systems developed, nurtured and refined by Indigenous people and passed on by them as part of expressing their cultural identity. (Janke, 1999, Chapter 1, p. 11)

This includes agricultural, scientific, medicinal and ecological knowledge, which are relevant in our context.

The report uses the term, 'indigenous cultural and intellectual property rights' to refer to indigenous Australians' rights to their heritage, adopting the terminology of Article 29 of the Draft Declaration of the Rights of Indigenous Peoples, 1994 (Janke, 1999, p. 7), which states that:

> Indigenous peoples are entitled to the recognition of the full ownership, control and protection of their cultural and intellectual property. They have the right to special measures to control, develop and protect their sciences, technologies and cultural manifestations, including human and other genetic resources, seeds, medicines, knowledge of the property of fauna and flora, oral traditions, literatures, designs and visual and performing arts.[140]

The explanation given by indigenous Australians for adopting 'indigenous cultural and intellectual property rights' rather than the World Intellectual Property Organization (WIPO) definition is to 'take into account the apparent distinction between "property", which suggests commercialisation and the protection of commercial rights, and "heritage", which implies preservation and maintenance issues' (Janke, 1999, p. 7). Further, the indigenous Australian definition intends to highlight the apparent distinction between the arts and culture on the one hand and science on the other, a split largely attributable to Western thinking.[141] It is not that 'property' and 'heritage' are mutually exclusive, but that heritage is another issue to consider.

> Traditional knowledge is regarded as common heritage and not as a commodity to be patented for commercial exploitation, perhaps to the exclusion of traditional owners. As with many other aspects of indigenous culture, knowledge of different plants and their healing properties is restricted to a particular class of people. Knowledge about the therapeutic

[139] *A Study on the Protection of the Cultural and Intellectual Property of Indigenous Peoples,* July 1993, E/CN.4/sub.2/1993/28,28 July 1993, para. 31, p. 9.

[140] Report of the Sub-Commission on Prevention of Discrimination and Protection of Minorities at its 46th session, *Draft United Nations Declaration on the Rights of Indigenous Peoples,* United Nations Documents E/CN.4/1995/2 and E/CN.4/Sub.2/1994/56.

[141] 'The development of specific legislation in the Australian framework has focused on arts and cultural expression only and has tended to neglect other areas of heritage, such as biodiversity knowledge, resources and scientific and medicinal application of cultural knowledge ... This is because western culture tends to separate arts from science ... in indigenous cultures, the division is less distinct' (Janke, 1999, p. 195).

> properties of plants is passed on by word of mouth. Indigenous people get access to such knowledge when they have attained the appropriate level of initiation. Just as practitioners of western medicine must study medicine before they can practise it, so a certain degree of knowledge is required before a plant can be used safely in indigenous society (Janke, 1999, p. 24).

Indigenous rights, like TK, are not something static, but subject to change. 'How will ... [any reforms] meet the needs of artists who no longer belong to a traditional community, or whose communities can no longer say they own a particular design?' (Stephen Gray, unpublished, 1994). The same may be said for traditional ecological knowledge. 'Any legislation should not attempt to freeze Indigenous culture but should aim at allowing both so-called traditional and contemporary rights to be recognised and protected' (Janke, 1999, p. 186).

The communal nature of Aboriginal knowledge and resources. Australian Courts have recognized that classic intellectual property law is inadequate to deal with the communal nature of indigenous resources and TK.

In *Yumbulul vs Reserve Bank of Australia* (1991) 21 *Intellectual Property Reports* 481 (cited by Blakeney, 1999, pp. 91–92) the trial judge admitted that 'Australia's copyright law does not provide adequate recognition of Aboriginal community claims to regulate the reproduction and use of works which are essentially communal in origin' (p. 490). This case involved the reproduction by the Reserve Bank of the design of a Morning Star Pole on a commemorative banknote. The Galpu Clan of north-eastern Australia sought to prevent this on the basis of the communal obligation of the artist as a clan member to prevent the design of the pole being used in a culturally offensive manner – the authority and knowledge to use the design having been obtained through initiation and revelation – but the trial judge found that he had disposed of his IPRs in a legally binding document.

In *Milpurrurru vs Indofurn Pty. Ltd* (1995) 91–116 *CCH Australian Intellectual Property Cases* 39,051 (cited by Blakeney, 1999, pp. 91–92) the court was prepared to consider cultural harm in its assessment of damages for breach of copyright awarded to a number of Aboriginal artists whose designs were wrongfully reproduced on carpets. None the less, the idea of also compensating the communities of the artists who were no longer alive was rejected, for 'the statutory remedies [of Australia] do not recognise the infringement of ownership rights of the kind which reside under Aboriginal law in the traditional owners of the dreaming stories' (at 39,077). This is another example of Australian courts attempting to determine the extent to which Aboriginal customary law may be recognized under modern Australian law.

The problem is that in seeking to translate indigenous TK into modern intellectual property terms one has to deal with the Aborigines' notion of property, which is said to be 'quite different' (Janke, 1999, Chapter 5 and p. XXII). E.I. Daes has stated

> Indigenous peoples *do not view* their heritage in terms of property at all ... but in terms of community and individual responsibility. *Possessing* a song, story or medical knowledge carries with it certain responsibilities to show respect to and maintain a reciprocal relationship with the human beings, animals, plants and places with which the song, story or medicine is connected. (Daes, 1993, emphasis added)

It is not that proprietary rights are rejected, but that they do not necessarily attach to any given individual, and in any case have certain communal responsibilities attached to them.[142] Aboriginal intellectual property includes the right to control the disclosure, dissemination, reproduction and recording of indigenous knowledge, ideas, and innovations concerning medicinal plants, biodiversity, and environmental management (claim 16)

[142] See list of indigenous claims in Janke (1999, pp. XX–XXI).

– which is similar to Western concepts – and *community* cultural and intellectual property rights (claim 5) – the communal element being foreign to Western notions of intellectual property, and so, not taken properly into account.

A related question is the role of the individual within the context of communal ownership. More correctly said it is 'a great number of generations *[of individuals]* [who] contribute to the development of indigenous cultural heritage'. In the case *Deceased Applicant vs Indofurn (1994) 30 Intellectual Property Reports 209 (the Carpets Case)* it was held that an individual or group is often the custodian or caretaker of a particular item of heritage. Artists may have the authority to depict a traditional, pre-existing design in their artwork by virtue of their birth or by initiation, but they hold this knowledge on trust for the rest of the clan. Further, sufficient evidence was found of individual artistic interpretation even though the artworks in question followed pre-existing traditional designs. The value of this case in our study is the exploration of the interrelationship between the individual and the community of which he or she forms a part, and the relationship between individual and community rights. This is fundamental to the discussion on how TK, which is mostly communal in nature, may be better integrated and protected by the largely individualistic system of Western intellectual property law.

Incidentally, the means of appointing one or more individuals as representatives of their clan is not problematic: *Bulun Bulun and Milpurrurru vs R. and T. Textiles Pty. Ltd.* Federal Court of Australia, unreported, 3 September 1998, van Dousa J. (WIPO, 2001, p. 71). In this way one or more community representatives of a defined indigenous or local community may advocate the rights of that community over its TK in much the same fashion as an individual would seek protection of his or her rights.

Public domain knowledge versus knowledge vested permanently with its holder. A problem is that, according to classical notions of intellectual property law, once ethnobiological information has been published, and patent or copyright protection has expired, it falls within the public domain (Blakeney, 1999, pp. 92–93; see Chapters 1 and 4, this volume), contrary to the customary law of indigenous people, who consider themselves the custodians of it in perpetuity. Indeed, 'the establishment of a public domain collecting society for indigenous works is not favoured because this supports the current legal assumption that indigenous cultural and intellectual property out of copyright is in the public domain and free for all to use and exploit' (Janke, 1999, pp. XXXIX and 208). Rather, an 'indigenous collecting society' is preferred, which recognizes the custodianship of indigenous and local communities over their TK in perpetuity.

For instance, the book *Bush Food* (Isaacs, 2000) provides a neat summary of the most important Australian biological resources known and used by Aboriginal Australians, compiled in collaboration with several indigenous Australians, who are photographed and named and arguably joint inventors for the purpose of the book itself. None the less, it was presumably not intended that this project would bring all the Aboriginal TK contained in the book within the classic public domain, even if the book was first published before the CBD. Indeed, the whole purpose of this type of book is to bring the value of indigenous knowledge to the attention of a wider audience – to educate others about the value of Aboriginal ethnobotanical and ethnobiological knowledge – but without in any way abrogating Aboriginal rights over it.

Henrietta Fourmile-Marrie states that 'a considerable amount of [Aboriginal] TK concerning [their] use of plant and animal species for food and medicine has already been published with the copyright to such information being held by non-indigenous collectors and institutions' (Fourmile-Marrie, 1998, p. 2). Examples include Cribb, A.B. and Cribb, J.W. (1974) *Wild Food in Australia*, Collins, Sydney; Cherikoff V.S. and Isaacs, J. (1989) *The*

Bush Food Handbook, Ti Tree Press, Balmain; Low, T. (1988) *Wild Food Plants of Australia*, Angus and Robertson, Sydney; Bindon, P. (1996) *Useful Bush Plants*, Western Australia Museum, Perth (note: only the last example is post-CBD).[143]

What is sought is that, first, the right as originator, custodian or holder of such knowledge be recognized; and secondly, that this right be used as a basis for claiming a right to compensation, which could take the form of some collecting society or communal fund, given the communal nature of indigenous TK. The use of registers to document important indigenous knowledge, with or without confidentiality, and perhaps despite prior publication, is another proposal (Janke, 1999, pp. 140, 229).[144]

AUSTRALIA'S OBLIGATIONS PURSUANT TO NATIONAL AND INTERNATIONAL LAW

The basis of the Australian legal system. Modern Australia is represented by about 200 years of European settlement, superimposed on an Aboriginal culture more than 60,000 years old. The basis is British and Australian common law, and the development of a modern democracy within a sophisticated social, economic, political and legal system.

The common law notion of *terra nullius*, in which the rights of Aborigines to their land were not recognized upon European settlement, was overturned in an important High Court decision, *Mabo vs Queensland (No. 2) 1992 (Commonwealth) 175 CLR 1*, which led to the enactment of the *Native Title Act 1993 (Commonwealth). Mabo* held that Aboriginal people and Torres Strait Islander ownership of land survived the colonization of Australia, and recognized a form of property called 'native title' which 'has its origin in and is given its context by the traditional laws acknowledged by and the traditional customs observed by the indigenous inhabitants of a territory' (Justice Brennan in *Mabo vs Queensland*, p. 58; WIPO, 2001, p. 78). This has helped to pave the way for indigenous customs to be recognized by modern law, which is an ongoing process.

National and international legal obligations. Australia is party to numerous international conventions, and is at the forefront in developing legislation to take better account of the special needs of its indigenous people. Work is being done on two levels: the further development of modern law, and the coordination of indigenous rights and responsibilities within that modern law. Australia has developed a draft legislative model for regulating access to genetic resources, and considers existing legal and administrative mechanisms as the most appropriate basis for the protection of TK (Communication from Australia IP/C/W/310 2 October 2001 point 1).

Australia ratified the CBD on 18 June 1993, and is party to TRIPS and the International Treaty on Plant Genetic Resources for Food and Agriculture. The *Plant breeders' rights Act 1994 (Commonwealth)*[145] gives effect to Australia's obligations as contracting party to the 1991 text of the International Convention for the Protection of New Varieties of Plants, known as the UPOV Convention, which Australia joined as a Member on 1 March 1989 and ratified on 20 January 2000. Section 17 of the Act recognizes the 'farmers' privilege', allowing seeds to be saved for replanting.

Further, there is an abundance of Com-

[143] Some indigenous Australians believe that indigenous people should receive compensation for aspects of culture that were applied commercially in the past (Janke, 1999, p. 15).

[144] 'Consideration should be given to the establishment of a national register which identifies the owners of ICIP. Any established register should not be a means of evidencing title. The register should only be used to provide contract details for subsequent users of indigenous material to contact the relevant community for prior consent. The register should be designed, managed and controlled by indigenous people' (Janke, p. XLI and Chapter 22).

[145] The *Plant breeders' rights Amendment Bill 2002 (Commonwealth)* seeks to clarify the rights of plant breeders to remuneration regarding public interest restriction and enhance the access of plant breeders to the Plant breeders' rights Scheme (Bills Digest No. 164 2001–02).

monwealth, State and Territory legislation and policy recommendations relating to this topic.[146] Section 51(xviii) of the Australian Constitution gives the Commonwealth, meaning Federal Government, the power to make special laws regarding copyright, patents or inventions and designs, and trade marks, and the common law provides the remedies of breach of confidence and passing off. Under Australian law, State and Territory governments (corresponding with *Canton* level) are primarily responsible for access to genetic resources found in non-Commonwealth land and waters, but it becomes a Commonwealth concern to the extent that federal matters become involved, such as trade and commerce, corporations, external affairs, export control and laws for the 'people of any race'. In any case, the State and Territory governments are similarly bound by Australia's obligations under the CBD pursuant to the external affairs clause in the Federal Constitution (Fourmile-Marrie, 1998, p. 5).

Recommended Action 6.1.7 of the *National Strategy for the Conservation of Australia's Biological Diversity* (an instrument of public administrative law) proposes that within Australia one ought to:

> Recognise the value of the traditional knowledge and practices of Aboriginal people and Torres Strait Islanders and integrate this knowledge and those practices into biological diversity research and conservation programmes by:
>
> Encouraging the recording (with the approval and involvement of the indigenous people concerned) of traditional knowledge and practices;
>
> Assessing their potential value for nutritional and medicinal purposes, wildlife and protected management and other purposes; and
>
> Applying traditional knowledge and practices in ways which ensure the equitable sharing of the benefits arising from their use (National Strategy, 1992, p. 4, cited by Blakeney, 1989, p. 89).

In the Report, *Biodiversity. The Role of Protected Areas* (1993), the House of Representatives Standing Committee on Environment, Recreation and the Arts commented that the identification of traditional practices entails more than the gathering of information, as it raises 'questions of authenticity, knowledge and power'. The Committee recommended that Action 6.1.7 be amended so that indigenous communities are encouraged 'to undertake or otherwise collaborate in research projects which utilise TK and practices in the study of biodiversity and in conservation' (paras. 4.70 and 4.72).[147]

On ABS: The Environment Protection and Biodiversity Conservation (EPBC) Act 1999 (Commonwealth). The EPBC Act 1999 replaces the National Parks and Wildlife Conservation Act 1975 (Commonwealth), but does not affect operation of the Native Title Act 1993 (Commonwealth), which is discussed below.

S301 of the EPBC Act 1999 provides the mechanism to implement Australia's international and domestic obligations for the control of access to biological resources in Commonwealth areas:

1. The regulations may provide for the control and use of access to biological resources in Commonwealth areas.
2. Without limiting subsection 1, the regulations may contain provisions about all or any of the following:
(a) the equitable sharing of the benefits arising from the use of biological resources in Commonwealth areas;
(b) the facilitation of access to such resources;
(c) the right to deny access to such resources; and

[146] *Designs Act 1906; Copyright Act 1968; Circuit Layouts Act 1989; Patents Act 1990; Trade Marks Act 1995; Plant Breeders Rights Act 1994* (and *Plant breeders' rights Amendment Bill 2002).*
[147] Also see the discussion paper entitled *Access to Australia's Biological Resources,* prepared by the Coordination Committee on Science and Technology (CCST), Department of the Prime Minister and Cabinet, Office of the Chief Scientist, AGPS, Canberra, March 1994, pp. 31–32, cited by Blakeney (1999, pp. 89–91).

(d) the granting of access to such resources and the terms and conditions of such access.

In 2000 a Commonwealth Public Inquiry over access to biological resources in Commonwealth areas advised upon a scheme with legal effect that could be implemented through regulations under s. 301 of the EPBC Act 1999 (Voumard, 2000). In essence, the scheme seeks to implement the CBD goals of the conservation and sustainable use of biological resources and the equitable sharing of its benefits, together with the protection of TK, especially that held by indigenous people.

Specifically, the scheme provides for an access permit to native biological resources found in Commonwealth areas, which may be granted or refused by the Government in consultation with the relevant Government Agency or landowner, depending on whether the collection of materials would be ecologically sustainable; and a benefit-sharing contract based on a model contract developed and agreed upon by governments, industry, indigenous organizations and other stakeholders. The benefit-sharing contract between the parties must address prior informed consent, mutually agreed terms, adequate sharing of any benefits derived and the value and protection of indigenous knowledge and environmental benefits in the area from which the resource was obtained.

Applications for Access Permits may be made on-line; the Application fee may only be 'moderate'; and 'as far as possible (subject to concerns about confidentiality), information about access and benefit sharing agreements would be made public' (A Guide to the Draft Amendments to the Environment Protection and Biodiversity Conservation Regulations, 2000 – Access Permits and Benefit Sharing Arrangements, cited in Communication from Australia). The Minister may attach conditions to the Access Permit, such as the obligation to lodge voucher specimens in Australian public collection institutions, or the provision of information about the specimens.

Interestingly, the decision of the traditional owners of biological resources to deny access to their resources is explicitly not reviewable (Voumard, 2000, pp. viii and 21). This is in order to prevent 'legal bullying' which may wear down their decision to deny access, but conceivably could be a problem if the denial of access conflicts with Australia's obligations to provide access for 'environmentally sound uses' pursuant to the CBD.

Where biological resources are in a Commonwealth reserve that is not Commonwealth-owned land, such as Uluru-Kata Tjura (Ayers Rock), Booderee National Parks (Jervis Bay) and parts of Kakadu National Park, the rights of the land owner as owner of biological resources may be regulated by the EPBC Act or regulations made under the Act in relation to Commonwealth reserves (Voumard, 2000, pp. 47, 68 and 71). As Commonwealth 'areas' under s. 525 of the EPBC Act, they are owned by indigenous Australians and administered by Environment Australia under certain leasing arrangements. For example, s. 354 (1) (a) requires that a sample of a native species be taken in accordance with a management plan for the reserve, but this would not affect the exercise of 'traditional' rights to use Aboriginal land in accordance with the *Aboriginal Land Rights (Northern Territory) Act 1976* (s. 71), or native title rights in accordance with the *Native Title Act 1993*. Section 8(2) of the EPBC Act 1999 provides that the EPBC Act does not affect the operation of those Acts.

In short, after determining who owns or who is primarily responsible for the (Commonwealth) land upon which the biological resources may be found, the ABS provisions of the EPBC Act come into play, which themselves do not abrogate the inherent indigenous rights over that land and its resources as recognized by the Native Title Act.

On land rights: The Native Title Act 1993 and Native Title (Amendment) Act 1998 (Commonwealth). The *Native Title Act 1993 (Commonwealth)*, which arose in response to the *Mabo* decision mentioned earlier, preserves native land title and associated rights from confiscation by the government.

The issue of land rights is clearly of central concern to local and indigenous communities as custodians of their biological resources and related TK. The connection between TK and land rights is recognized in Section 223(1) of the Native Title Act, which defines 'native title' as 'the communal, group or individual rights and interests of Aboriginal peoples or Torres Strait Islanders in relation to land or waters' provided that three conditions are fulfilled:

1. The rights and interests must be possessed under the traditional laws acknowledged and customs observed by the Aboriginal and Torres Strait Islanders.
2. The Aboriginal or Torres Strait Islanders must, by those laws and customs, have a connection with the land and waters.
3. The rights and interests must be recognized by the common law of Australia, being the most difficult condition to prove amongst these three (Blakeney, 1999, p. 95).[148]

Section 223(2) refers to 'hunting, gathering or fishing rights and interests' as being included within Native Title, and has been interpreted as meaning that, first, others are not excluded from also exercising these rights, and secondly, that even where traditional medical remedies might have become the subject of a patent or plant variety registration, Aboriginal people would still be permitted to continue to use such traditional medical remedies (Blakeney, 1999, p. 87).

Furthermore, according to current Australian legal opinion, native title rights can include the right to access and use of biological resources, and to control the right of others to access and use of biological resources in relation to particular land, including a corresponding denial of access, but native title would not (yet) encompass a right to control *all uses of a resource wherever it was located* (emphasis added) (Voumard, 2000, p. 72). The Aboriginal custodians of a plant found on *their* land would not be able to extend their rights to the *same plant* found growing on *someone else's* land. Each case of who owns or is primarily responsible for specific land, upon which certain biological resources are found, must be considered individually. This is a practical approach, reflecting the attempt in Australia to take into account competing claims based upon different systems of law, meaning both the rights found under modern Australian law and indigenous customary law.

An important element of modern Australian bioprospecting agreements[149] is that they are made with both the relevant State government and indigenous authorities, and that the benefits are shared with the Aborigines as custodians of the biological resources.[150] For instance, Section 14 of the Kakadu National Park Management Plan[151] reserves the right of the traditional owners, the Bininj/Mungguy, to continue to exercise their traditional rights to gather plants and plant material for food and for making cultural artefacts.

The Uluru-Kata Tjuta National Park Management Plan provides that in respect

[148] *Mabo* was confirmed in the recent High Court decisions of *Western Australia vs Ward; Attorney-General (Northern Territory) vs Ward; Ningarmara vs Northern Territory [2002] HCA 28 (8 August 2002)* found at http://www.austlii.edu.au/au/cases/cth/high_ct/2002/28.html, and *Members of the Yorta Yorta Aboriginal Community vs Victoria and Others* (M128/2001) http://www.austlii.edu.au, which was also mentioned in the *Neue Zürcher Zeitung*, 17 December 2002, p. 52.

[149] For instance, ExGenix Ltd (formerly Amrad Discovery Technologies) has agreed to this (Voumard, 2000, pp. 94–95).

[150] Although the CSIRO submitted to a Canberra Hearing on 30 May 2000 that the intrinsic value of biological resources and associated traditional knowledge can be quite limited. The Queensland government commented that the chance of discovering a new pharmaceutical may be 1 in 1,000,000 samples screened, costing about AU$100 million, taking 8–15 years to develop. The possible monetary benefits from bioprospecting agreements are often overestimated (Voumard, 2000, p. 99).

[151] This has legal effect according to principles of Administrative Law.

of bioprospecting agreements,[152] the agreements need not be exclusive; the indigenous owners may continue to use their resources and knowledge; the indigenous owners may not be forced to divulge their knowledge; and ABS may be determined in accordance with traditional laws and customs (Voumard, 2000, Chapter 6).

The significance of these examples is to show how rights to traditional genetic resources and related TK may be given legal recognition through the existing legal system (Communication from Australia point 29).

The inextricable link between rights over biological resources and TK, including ABS, and rights to land. Perhaps the real difficulty is not to devise an ABS scheme for biological resources in Commonwealth areas, or even to recognize the rights of indigenous Australians to their TK and resources, but to identify the true holder of the resources where competing claims are made over them. 'Commonwealth areas' refers by definition to Commonwealth waters and land owned or leased by the Commonwealth and Commonwealth Agencies, and land in external territories, which may conflict with indigenous claims pursuant to their customary law. The central issues in litigation have become: who has control over the land upon which the biological resources may be found? Are such land rights exclusive? And can they be suspended or extinguished? This conflict reflects the competing claims made by the interaction of modern law with indigenous law. For instance, the *Commonwealth Aboriginal Land Rights (Northern Territory) Act 1976* deals with agreements made between the Northern Territory Government and the Aboriginal people to the effect that the communities own the land, and in some cases lease it back to the government. Occasionally there are disputes between the Aboriginal people and the State or Territory government over who owns the biological resources or minerals found there (WIPO, 2001, p. 78 citing Roundtable, Darwin, Australia, 15 June 1998, Northern Land Council Darwin).

Milirrpum vs Nabalco (1971) 17 FLR 141 recognized Aboriginal customary law as a 'system of law' within the Australian legal system. The fundamental connection which Aborigines have with their land is spiritual (per Blackburn, J. at 167), but neither spiritual connection nor cultural knowledge alone is sufficient to constitute native title to land, which also requires a physical presence on the land {s. 223 (1) (b) of the Native Title Act; *Western Australia vs Ward; Attorney-General (NT) vs Ward; Ningarmara vs Northern Territory (Ward's case)* [2002] HCA (8 August 2002) (Full Court of the High Court of Australia) para. [964.6] and [964.8] per Callinen J.}. Abandonment of traditional lifestyle was found to sever the connection of the Yorta Yorta people to certain land in the case of *Members of the Yorta Yorta Aboriginal Community vs Victoria and Others* (M128/2001), although this decision is not considered final.

Ward's case involved different land tenure including vacant Crown land, pastoral leases, Crown land in or about the town of Kununurra, the Ord River irrigation area, Lake Argyle and the Argyle Diamond Project and the Keep River National Park, and certain waters. Even if native title is found to exist, it may be partially or wholly suspended or permanently extinguished[153] by such things as the granting of pastoral leases or mining leases and the resumption of land for the carrying on of public works: [453], [454], [468.3] and [468.4]. Pastoral leases may provide exclusive possession with a reservation in favour of Aborigines to enter and enjoy, for subsistence purposes. 'To the extent that rights and interests granted by the pastoral lease were not inconsistent with native title rights and interests, the rights and interests under the

[152] Major companies involved in bioprospecting in Australia include AstraZeneca, ExGenix (formerly AMRAD), Australian Institute of Marine Science, CSIRO and BioProspect Limited (Voumard, 2000, p. 93).

[153] Under s 237A 'extinguishment' of native title is permanent, otherwise it is a mere suspension of rights that may be revived.

lease prevailed over, but did not extinguish, native title rights' [464] and [468.10]. Conversely, if exclusive possession is validly granted then there is no need to consider inconsistency, because by definition, exclusive possession is inconsistent with native title: per Callinen J. [964.3]. A lease granted for a national park may protect traditional Aborigine use of the land where native title is not wholly extinguished.

A lease granted in perpetuity may give the lessee a right of exclusive possession that extinguishes all native title: *Wilson vs Anderson* [2002] HCA 29 8/08/02. Aborigines may in some cases simply have the right of occupation to live on the land and seek their subsistence from it – like a lodger in a house owned and exclusively possessed by another (*Ward* [555]).

Ward held that '[t]he evidence established no native title right to or interest in any mineral or petroleum. [So] no question of extinguishment arises' (*Ward* [482] and [468.22]), although this view was not shared by Kirby J.

The common law recognizes the concept of joint possession. 'But possession that is not exclusive is a contradiction in terms, for the right of general control and exclusion is central to the concept of legal possession' [477]. In a similar case, *Wik Peoples vs Queensland* (1996) 187 CLR 1, the pastoral leases were not thought to confer exclusive possession or else did not necessarily extinguish native title, a controversial result which will be limited to its facts [473–481].

The attempt to integrate Australian indigenous law into the modern common law legal system is regarded sceptically by some. In *Ward's* case, His Honour Callinen J. remarked that it would probably be better to 'redress the wrongs of dispossession by a true and unqualified settlement of lands or money than by an ultimately futile or unsatisfactory, in my respectful opinion, attempt to fold native title rights into the common law' ([970] and [1064]). The Commonwealth Regulations would in any event not alter existing property law in Australia.

His Honour's point was that:

> this Court and other legal bodies are founded on a post-dream time legal order. Although some may contend that we should, we do not in fact recognise Aboriginal criminal law, tort law or any aspects of indigenous laws, nor do we pretend to. The question then is why the common law of property, which had been regarded as settled for more than a century, should have been changed to recognise sui generis interests in land that had no counterpart in our legal system. [1064]

This raises the issue of the compatibility – or incompatibility – of maintaining 'two equal and parallel systems of law', an issue raised by indigenous Australians in Janke (1999).

HOW WOULD ABORIGINAL SUBMISSIONS FIT IN WITH THE EXISTING AUSTRALIAN LAW?

To what extent can 'Aboriginal intellectual property' and 'Aboriginal common law' be reconciled with Western notions of intellectual property law and with modern Australian law in particular? Consideration of certain statements and reports reveals a fundamental division of opinion about certain key issues.

The Final Statement of the South Pacific Regional Consultation on Indigenous Peoples' Knowledge and Intellectual Property Rights, held in Suva, Fiji, in April 1995, in which indigenous Australians took part, accuses the modern IPR system and modern science and technology of 'perpetuat[ing] imperialism ... to control and exploit the lands, territories and resources of indigenous peoples'. This is clearly a rejection of the Western form of IPRs, though not a rejection of IPRs *per se*. None the less, such thinking led to a call for the Pacific Region to be 'a life forms patent-free zone' (Article 1), and 'a moratorium on bioprospecting in the Pacific' (Article 2), at least until 'appropriate protection mechanisms are in place' (Blakeney, 1999, p. 94). What is alarming is the result that research and investment in natural biological resources will thereby be discouraged, and the motivation to cultivate such resources may be so reduced as to bring them into danger of extinction (see Chapter 1 on loss of biological resources and related TK). The Final Statement also declared 'the right of indigenous peoples of the Pacific to self-governance and independence of our lands, territories and resources as the basis for the preservation of indigenous peoples' knowledge', the wider ramifications of which fall outside the scope of this study.

On Access and Benefit Sharing. Indigenous Australians consider there to be 'some things that cannot be sold, such as secret/sacred objects and information', and 'indigenous people should be able to stop commodification of certain aspects of their cultures' (Janke, 1999, p. XIX). In short, they demand the right to control access over their knowledge and resources.

Otherwise, indigenous Australians may be interested in benefiting from the commercialization of some of their knowledge and resources.[154] The difficulty is assessing the worth of the contribution of TK to an invention based upon it.

Blakeney notes that 'it will often be questionable whether the contribution of indigenous knowledge from which a pharmaceutical product is developed can be considered to be the sort of contribution which will allow the indigenous persons to be considered a joint inventor under patents law' (1999, p. 93). This is because merely 'being the first to observe a useful property or effect of an invention' (*Consolidated Aluminum Corp. vs Foseco Int'l Ltd* 10 U.S.P.Q. 2d 1143) does not amount to joint inventorship, which requires each joint inventor 'working toward the same end and producing an invention by their aggregate efforts' [269 F. Supp. 818 (DDC1967)].

Alternatively, where ethnobiologists work together with local or indigenous communities in order to discover both the identity of medicinal plants and the treatments for which they are used, the contribution of TK to the development of a plant-based drug is more evident, and accordingly, the demand for a fair share of the profits is greater (Blakeney, 1999, p. 93).

In fact, a distinction exists between the traditional use of a plant, known for its healing purposes, which is or which ought to remain unaffected, and patent protection on new inventions stemming from that plant, such inventions falling outside the scope of TK. Confusion about this is detrimental to the discussion on the apportionment of the benefits derived from the use of the plant and the TK related to it.

At any rate,

> what is becoming evident ... is that Nature is probably not going to produce the next 'blockbuster drug' directly, but that the chemical structures that Mother Nature provides are the structural leads that chemists will then modify to produce 'improved molecules' that no chemist in his or her right mind would have considered making de novo. (Sub. 39 of a CSIRO report cited in Voumard, 2000, p. 237; also consider the Smokebush example mentioned earlier)

That being so, a case can still be made for compensating indigenous people to the extent that they point chemists in the right direction of the benefits that 'Mother Nature' has to offer.

The integration of modern law and indigenous law. Australian legal thinking on the legal rights to plant genetic resources and related TK, especially that which is held by indigenous Australians, is in a process of evolution and compromise.

Some indigenous Australians would like to see the implementation of 'two parallel and equal systems of law' within Australia (Janke, 1999, Part Three of the Report and p. XXVII). Although there is a certain attractiveness to this idea – recognizing indigenous customs to have equal merit – the notion is rejected for both practical and conceptual reasons (Communication from

[154] Claim 9 in the list of indigenous claims (Janke, 1999, pp. XX–XXI). For example, the Aboriginals of Uluru, formerly known as Ayers Rock, now keep part of the proceeds collected from tourists, in recognition of their custodianship.

Australia points 24–32; 31). First and foremost, it is not possible for any state to pursue two separate systems of law and government simultaneously. Either the notions espoused by indigenous law will be compatible with those found in modern law, in which case such law is superfluous, or else they will conflict, in which case an order of priority will need to be determined. This must take into account not only subjective notions of which law is preferable, but also the correct political and legal means to effect changes in the law, which is, after all, the law of the land, in this case a constitutional democracy for all of its citizens whether indigenous or not.

The common law view is that the sources of law are to be found in an increasing order of importance, and that customary law is already included as a source of law, although legislation and case law take precedence over it (see any standard legal text such as *Legal Institutions* by Professor William Morison, Law Book Company, Sydney). 'Specifically, Australia assesses the existing legal framework, and then seeks to apply practical measures that accommodate indigenous concerns in the day to day management and administration of ICIP' (Communication from Australia point 25). 'An entirely new form of statutory protection for ICIP, as envisaged by the Report Our Culture, Our Future is seen by the Australian government to be a "dramatic step", introducing a level of unnecessary complexity and duplication which the government would rather avoid in favour of developing the existing legal system' (Communication from Australia point 31). The Australian government also uses 'non-legislative means ... to accommodate indigenous concerns in the daily management of intellectual and cultural property' (point 32), including collecting societies (point 33).

If 'two equal and parallel systems of law' are strictly not possible, what, then, is possible? A number of decisions[155] and developments have led to the suggestion that 'Aboriginal and Torres Strait Islander IPRs may be recognisable by Australian law either as part of native title within the scope of the reasoning in the Mabo decision, or by analogy with it' (Gray, 1998, p. 78).

Aboriginal Australians have in turn recommended that a *sui generis* legislative framework be established to protect indigenous cultural and intellectual property rights, including ecological knowledge, and that 'any rights granted should ensure that there are no time limits on protection and no fixed form requirement for protection to be given' (Janke, 1999, p. XXXVI).

Further, 'the legislation should not inhibit the further cultural development of materials within their originating communities. That is, customary and traditional use should not be affected' (Janke, 1999, p. XXXVI).

CONCLUSIONS

The development by Australia of a Model Contract on Access and Benefit Sharing is a valuable contribution 'to the global community ... [on] this important issue' (Communication from Australia point 18). 'It is hoped that the measures taken by Australia to protect indigenous intellectual and cultural property within Australia may be of use to other countries' (point 22).

Clearly, in order to incorporate concepts of Aboriginal intellectual property into modern Australian law it is necessary to think of intellectual property as something more than the protection of commercial interests for a given period only. There must be a more fundamental recognition of the worth of all intellectual activity, whether stemming from indigenous or non-indigenous people. In this way, intellectual property may be considered to be part of cultural heritage, rather than something incompatible with it.[156]

[155] *Native Title Act (1993) (Commonwealth)* – which arose due to *Mabo; Wik Peoples vs State of Queensland* (1996) 141 ALR 129; *Bulun Bulun vs Nejlam Pty. Ltd.* (1989) Federal Court of Australia; *Yumbulul vs Reserve Bank of Australia* (1991) 21 IPR 481; and *Milpurrurru vs Infofurn Pty. Ltd.* (1995) 30 IPR 209.

[156] This was suggested by the Office of National Tourism in their submission to *Our Culture, Our Future*, 1997 (Janke, 1999, p. 8).

Furthermore, any effective system for protecting the legal rights over biological resources and related TK must also address the issue of rights over the land in which they are found. There are features of indigenous ecological knowledge that are worthy of special attention, and Australia is at the forefront of attempting to integrate indigenous customary law into the modern legal system. It may be said that Australia is an interesting example of the process of bringing together both North and South interests.

Bibliography

Baracol, D.S. (2001) Consultations Under Article 4 of the WTO Dispute Settlement Understanding: the Role of Non-Litigious Approaches in Settling Disputes. Manuscript, on file with the authors.

Barton, J.H. and Berger, P. (2001) Patenting agriculture. In: *Issues in Science and Technology Online* (Summer 2001).

Blakeney, M. (1997) Bioprospecting and the protection of traditional medical knowledge. In: *European Intellectual Property Reports*, Vol. 6, pp. 296–303.

Blakeney, M. (1999) Ethnobiological knowledge and the intellectual property rights of indigenous peoples in Australia. In: Blakeney, M. (ed.) *Intellectual Property Aspects of Ethnobiology.* Sweet and Maxwell Ltd, London, pp. 83–100.

Bonn Guidelines (2002) *Bonn Guidelines on Access to Genetic Resources and Fair and Equitable Sharing of the Benefits Arising out of their Utilization.* UN Doc. UNEP/CBD/COP/6/20.

Boyle, A.E. (1994) The Convention on Biological Diversity. In: Campiglio L. *et al.* (eds) *The Environment After Rio – International Law and Economics.* Graham & Trotman, London, p. 111.

Consultative Group on International Agricultural Research (1999) *CGIAR Center Statements on Genetic Resources, Intellectual Property Rights, and Biotechnology.* CGIAR, Washington, DC.

Cottier, T. (1992) Prospects in international trade law and policy: the GATT connection. *Aussenwirtschaft* 47, 79.

Daes, E.I. (1993) *Discrimination against Indigenous Peoples*, para. 26; quoted in Puri, K. (1995) Cultural ownership and intellectual property rights post-Mabo: putting ideas into action, 9. *Intellectual Property Journal* 293, 308 (cited by Blakeney, 1999, p. 92).

Devitt, J. (1991) Acacias: a traditional Aboriginal food source in Central Australia. In: House, A.P. and Harwood, C.E. (eds) *Australian Dry-Zone Acacias for Human Food – Proceedings of a Workshop held at Glen Helen, Northern Territory, Australia, 7–10 August 1991.* CSIRO, East Melbourne.

Duk Park, Y. and Umbricht, G.C. (2001) WTO Dispute Settlement 1995–2000: a statistical analysis. *Journal of International Economic Law* 4, 214.

Fourmile-Marrie, H. (1998) Developing a regime to protect indigenous traditional biodiversity-related knowledge. Available at: http://www.law.murdoch.edu.au/balayi/v1n1/fourmile.shtml (accessed December 2003).

Gray, S. (1998) Recognising Aboriginal intellectual property. Paper to WIPO Report, June 1998, p. 78.

Horber, R. (1995) Die Liberalisierung des Agrarhandels: Beweggründe, Auswirkungen und Massnahmen. In: Cottier, T. (ed.) *GATT Uruguay Round: Nine Papers.* Stämpfli, Berne, p. 51.

Isaacs, J. (2000) *Bush food: Nahrung und Pflanzenmedizin der Aborigines.* Könemann Verlagsgesellschaft mbH, Köln (German translation of the original English version (1987) *Bush Food.* Lansdowne Publishing, NSW Australia).

Janke, T. (1999) *Our Culture, Our Future: Report on Australian Aboriginal Indigenous Cultural and Intellectual Property Rights* by Terri Janke as Principal Consultant, Michael Frankel and Company, Solicitors, for the Australian Institute of Aboriginal and Torres Strait Islander Studies (AIATSIS) and the Aboriginal and Torres Strait Islander Commission (ATSIC) Available at: http://www.icip.lawnet.com.au (accessed December 2003).

Morrison, W. (1983) *Legal Institutions.* Law Book Company, Sydney, Australia.

Neue Zürcher Zeitung (2002) 17 December 2002. Zürich, Switzerland, p. 52.

Nijar, G.S. and Ling, C.K. (1994) The implications of the Intellectual Property Rights Regime of the Convention on Biological Diversity and GATT on Biodiversity Conservation: a Third World perspective. In: Krattiger, A.F. *et al.* (eds) *Widening Perspectives on Biodiversity.* International Academy of the Environment, Geneva, p. 277.

Rosell, M. (1997) Access to genetic resources: a critical approach to Decision 391 'Common Regime on Access to Genetic Resources' of the Commission of the Cartagena Agreement. *Review of European Community and International Environmental Law* 6, 274.

Ruiz Muller, M. (2003) Es necesario un nuevo marco jurídico paragraphla bioprospección en la región andina? Breve revision crítica de la Decisión 391. In: *Serie de Politica Y Derecho Ambiental* No. 14, February. Sociedad Peruana de Derecho Ambiental, Lima, Peru, pp. 1–8.

Ruiz Muller, M. and Lapeña I. (undated) New Peruvian Law Protects Indigenous Peoples' Collective Knowledge. Bridges Comment.

Schrijver, N. (1997) *Sovereignty over Natural Resources – Balancing Rights and Duties.* Cambridge University Press, Cambridge, UK.

Tobin, B. and Swiderska, K. (2001) Speaking in tongues: indigenous participation in the development of a *sui generis* regime to protect traditional knowledge in Peru. *Participation in Access and Benefit-Sharing Policy, Case Study no 2.* International Institute for Environment and Development (IIED), London.

Voumard, J. (Inquiry Chair) (2000) *Access to Biological Resources in Commonwealth Areas – Commonwealth Public Inquiry,* July. Australian Government Publishers, Canberra.

Woodley, J. (Chair) (1998) *Commercial Utilisation of Australian Native Wildlife.* Report of the Senate Rural and Regional Affairs and Transport References Committee, Parliament House, Canberra, p. 329 (cited by Fourmile-Marrie, 1998, p. 2).

WIPO (2001) *Report on Fact-Finding Missions (FFM) on Intellectual Property and Traditional Knowledge.* WIPO, Geneva.

3 Intellectual Property Rights, Plant Genetic Resources and Traditional Knowledge[1]

Philippe Cullet,[a] Christophe Germann,[b] Andrea Nascimento Müller[c] and Gloria Pasadilla[d]

[a]*School of Law, School of Oriental and African Studies, University of London, Thornhaugh Street, Russell Square, London WC1 0XG, UK;* [b]*Institute of European and International Economic Law, University of Berne, Hallerstrasse 6, CH 3012 Berne, Switzerland;* [c]*Legal Advisor, Rue de Lyon 48, CH 1203 Geneva, Switzerland;* [d]*Philippine Institute for Development Studies, NEDA Building, Amorsolo Street, Makati City, Philippines*

3.1 Introduction[2]

3.1.1 Overview

This chapter addresses in its first part the economic rationale of forms of protection that are granted under intellectual property laws and policies. In its second part, this chapter will discuss patent protection, plant breeders' rights (PBR) and *sui generis* forms of protection. It shall in particular explore the impact of these forms of protection on traditional knowledge related to plant genetic resources and, more broadly, on biodiversity and equity (benefit sharing). This assessment under existing laws (*de lege lata*) shall be completed by brief case studies to illustrate the legal issues at stake and shall propose solutions. The last part of this chapter outlines the impact of intellectual property rights on competition laws and policies. It describes certain contractual practices (licensing, patent pools, mergers and acquisitions) that can reduce competition. It also includes a case study pertaining to the seed sector to illustrate consequences of concentrations among economic players that are detrimental to public policies such as the promotion of biodiversity and equity. This analysis is relevant for our purposes as favourable impacts of intellectual property rights on biodiversity may be jeopardized by anti-competitive behaviours without appropriate safeguards.

The question of intellectual property rights is relevant both to promote TK related to plant genetic resources (their function as 'positive rights'), and to prevent misappropriation of these intangible values by third parties to the disadvantage of local communities and individuals (their function as 'defensive rights'). As positive rights, IPR may be used as complementary or alternative instruments with respect to subsidies in order to promote the sustain-

[1] The authors would like to thank Martin Girsberger and Marie Wollheim for their comments on this chapter.
[2] Author: Christophe Germann.

able maintenance and further development of TK. As defensive rights, IPRs, as adapted to the needs of TK holders and in line with applicable public policy goals, may serve to prevent abusive use of classical IPR in the context of TK related to plant genetic resources. Both approaches require an understanding of the rationale underlying the grant of IPR and a stocktaking of the scope of existing forms of protection. The purpose of this chapter shall therefore consist of the provision of a basis in order to elaborate ideas and concepts that would allow the use of IPR for the benefit of the sustainable promotion of TK related to plant genetic resources.

3.1.2 Forms of intellectual property protection

The term 'intellectual property' refers to a set of intangible products of human activity that are legally defined. Sources of law addressing intellectual property can be located on the national, regional and global levels of jurisdiction.[3] The legal competence among national, regional and international jurisdictions to define, grant and enforce intellectual property rights constitutes the international component of intellectual property law. One of the main characteristic features of this system lies in the principle of territoriality: the geographical scope of application of intellectual property laws is the territory of the state or community of states that has generated the corresponding rights and obligations. However, because the protection granted under intellectual property laws intrinsically requires a cross-border extension, the international law component of intellectual property has been essential since the beginning. This international law component was originally based upon reciprocity considerations, and more recently has obeyed the rationale underlying international trade rules. It is materialized by way of bilateral as well as multilateral agreements such as the Berne, Paris and Madrid conventions and, since 1995, the Agreement on Trade-Related Aspects of Intellectual Property Rights (TRIPS) that forms Annex 1c of the Agreement Establishing the World Trade Organization (WTO).[4]

Intellectual property laws provide so-called exclusive or *erga omnes* rights, i.e. a bundle of legally enforceable interests vested with their owner, who can oppose them against any third party. Based on law, the holder of intellectual property rights has the capacity to authorize or prevent others from acting in certain ways with respect to these specific rights. As regards the economic aspects, the holder has the right to perform activities enabling him or her to commercially exploit the results of their research and development efforts on an exclusive ('*erga omnes*') basis that provides for a competitive advantage. In addition, the holder may be granted non-economic rights in certain jurisdictions, such as inalienable moral rights to oppose mutilations of the work under author's rights laws (e.g. Article 6bis of the Berne Convention) or, more generally, the right to be named as author or inventor under the copyright or patent laws of most European countries (see for example Article 62 of the European Patent Convention 'EPC').

Intellectual property laws usually contain provisions concerning their scope of protection, the exclusive rights granted, the

[3] Plant variety protection for example is addressed nationally by plant variety legislation such as the Swiss Act on Plant Variety Protection of 20 March 1975, regionally by supra-national rules such as the Council Regulation EC No 2100/94 of 27 July 1994 on Community Plant Variety Rights, and globally by the UPOV Convention and Article 27.3(b) TRIPS. See also infra-sections II. C. and D. on Plant Breeders' Rights and *sui generis* protection systems.

[4] See the TRIPS Agreement under http://www.wto.org/english/docs_e/legal_e/legal_e.htm. For a detailed commentary on the TRIPS Agreement, see Cottier (2004). For the texts of international treaties administered by the World Intellectual Property Organization (WIPO), see http://www.wipo.int/treaties/index.html; national legislations on intellectual property in English may be downloaded from the Collection of Laws for Electronic Access (CLEA) under http://clea.wipo.int/clea/lpext.dll?f=templates&fn=main-h.htm&2.0

limitations to such rights, the duration of protection, the enforcement mechanisms and the sanctions for infringement. Once the period of protection has expired, intellectual property falls into the so-called public domain. Intellectual property is ubiquitous as opposed to property on tangible property (real estate, movables). This means that this form of property cannot be limited to a particular physical body incorporating the right: it is attached to all material or virtual occurrences of the protected work. Major forms of intellectual property protection include:

- Patent and related industrial design.
- Copyright and related neighbouring rights.
- Integrated circuit layout.
- Geographical indications of origin.
- Trademarks.
- Plant variety.
- Trade secrets (confidential information).

Within these main forms there are various sub-forms.[5] The relationship between these forms and sub-forms may vary over time, and different forms of protection may be available for the same intangible value. In this sense, protection can be assured by sequentially using different forms. For example, a pharmaceutical product may first be protected by patent and, after the expiration of the statutory protection term,[6] enjoy trademark protection for an indefinite duration of time, since the terms of trademark protection can be renewed indefinitely.

3.1.3 The functions of intellectual property rights

The various forms of protection are generally tailored to fulfil various legal, economic and social functions. While creativity or innovation requiring qualified individual efforts is to be encouraged for the benefit of society in a market economy, private actors typically need incentives to invest material resources in such efforts. These incentives may take the form of subsidies, or other advantages such as private property titles. In the latter case, the state grants intellectual property rights (IPR) to creators or innovators upon the fulfilment of certain conditions. These competitive privileges are usually limited in time, i.e. they last for the duration of the statutory terms of protection. They may take the form of patents for technical innovation or copyright for creative achievements related to content (such as text, music, films and, more recently, software, etc). The owner of a patent or a copyright has, subject to certain exceptions (e.g. compulsory licensing or fair use), the exclusive right to use, reproduce, disseminate, market, etc., the protected invention and content, respectively, for a given period of time. These terms of protection can vary from one jurisdiction to another, so long as they comply with the minimum duration set forth in international agreements such as the TRIPS agreement at the global or the EU directive at the regional level.

The World Intellectual Property Organization (WIPO) summarizes the rationale underlying the grant of IPR as follows:

> The primary purpose of most branches of the IP system (excluding trademarks and geographical indications) is to promote and protect human intellectual creativity and innovation. IP law and policy does so by striking a careful balance between the rights and interests of innovators and creators, on the one hand, and of the public at large, on the other. Thus, by granting exclusive rights in an invention, for example, the IP system encourages further innovation, rewards creative effort, and protects the (often substantial) investment necessary to make and commercialise the invention. The patent system also encourages people

[5] For an introduction to these various forms of protection, see WIPO, Intellectual Property Reading Material (status: August 2003), on http://www.wipo.org/news/en/index.html?wipo_content_frame=/news/en/documents.html

[6] The minimum duration of patent protection for Members of the WTO is 20 years according to Article 33 TRIPS.

> to disclose inventions, rather than retain them as trade secrets, thus enriching the store of publicly available knowledge and promoting further innovation by other inventors. Thus, public dissemination is an important IP objective. Copyright and other IP brands work in a similar way. The progress and well-being of humanity rests on its capacity for new creations in areas of technology and culture. The promotion and protection of IP can also spur economic growth, create new jobs and industries, and enhance the quality and enjoyment of life. However, the IP system also responds to the needs of the public at large. Most IP rights are of limited duration, after which the protected creations falls into the public domain (only trademarks may be renewed indefinitely, and geographical indications can subsist indefinitely).[7]

Intellectual property is a form of knowledge that societies have decided can be assigned specific property rights. They have some resemblance to ownership rights over physical property and land. Intellectual property creates a legal mean to appropriate knowledge. A characteristic of knowledge is that one person's use does not diminish another's use of it (e.g. reading a text). Moreover, the extra cost of extending use to another person is often very low or nil (e.g. copying an electronic file). From the point of view of society, the more people who use knowledge the better, because each user gains something from it at low or no cost, and society is in some sense better off. Economists therefore say that knowledge has the character of a non-rival public good. The other relevant feature of knowledge, or product embodying knowledge, is the difficulty of preventing others from using or copying it. Most products can be copied at a fraction of the cost it took to invent and market them. Economists refer to this characteristic as contributing to market failure. If a product takes considerable effort, ingenuity and research, but can be copied easily, there is unlikely to be a sufficient financial incentive from society's point of view to devote resources to invention (CIPR, 2002, pp. 11–14).

In short, IPR are thus entitlements granted under law to reward creative or innovative intellectual efforts. IPR protect intangible wealth, which can often be easily appropriated and reproduced. One of the characteristics of intellectual creativity is that it can be used and reproduced without depriving the creator of possession, as opposed to tangible wealth, which is usually subject to finite limitations of ownership and/or use. Furthermore, the products of intellectual creativity can often be reproduced at very limited marginal costs. Innovators and investors in innovation have therefore called for the development of a mechanism to protect the intangible wealth associated with intellectual creativity.

3.2 The Protection of TK Related to Plant Genetic Resources[8]

3.2.1 Definitions[9]

During the second half of the 19th century, the first modern patent laws were adopted in several European countries – in Britain in 1874, Germany in 1877 and Switzerland in 1887, whereas The Netherlands only reintroduced a patent system in 1910, which became effective in 1912; as well as in the USA (the initial Act of 1793 was revised in 1836, 1870 and 1874) and in Japan (1885).[10] At that time patent legislation was essentially designed to protect mechanical and conventional chemical inventions within industrialized countries. With the passage of time a number of new technologies emerged for which patentability was affirmed, most recently for biotechnology and, at least in certain jurisdictions,

[7] WIPO, Intellectual Property Reading Material, see note 5. http://www.wipo.org/news/en/index.html?wipo_content_frame=/news/en/documents.html

[8] Author: Christophe Germann.

[9] See also Chapter 1.

[10] For an historical survey, see Machlup (1999, p. 224 ff.).

for computer programs. Genetic engineering is part of biotechnology, which generally addresses the technical use of biological processes. Modern genetic technology is considered to have started in 1973, when scientists were first able to bind genetic material *in vitro* and to plant it into bacteria as well as to reproduce it for the first time. Biotechnology includes *in vitro* culture, rhizobium technology, fermentation and more advanced techniques that involve genetic engineering.[11]

We shall confine our analysis to the protection of plant genetic resources (PGR) and their related know-how. Article 2 CBD defines 'genetic resources' as 'genetic material of actual or potential value'. The Convention does not further clarify the meaning of value. According to this same article, 'genetic material' includes 'any material of plant, animal, microbial or other origin containing functional units of heredity'. Genetic resources therefore comprise genetic material of actual or potential value of plant, animal, microbial or another origin. One can distinguish between genetic resources of wild species, cultivated or domesticated species and their relatives, and man-made genetic resources.[12]

Girsberger proposes the making of an additional distinction between PGR and PGR used in the production of food and agricultural products (PGRFA), and offers a definition for a 'traditional' PGRFA, i.e. the result of informal plant breeding activities. He further defines 'formal' and 'informal' plant breeding, the former type meaning plant breeding performed by private seed companies and public research institutions, and the latter type meaning plant breeding performed by individual farmers and indigenous and other rural communities, involving their traditional knowledge and know-how. A distinctive characteristic of informal plant breeding is that the breeding process usually proceeds over a long time period, sometimes over several human generations. Furthermore, the results of informal plan breeding are generally freely accessible to others, for example other informal plant breeders, collectors of PGRFA, or the remainder of the community (Girsberger, 1998, p. 1022).

Accordingly, the analysis will encompass traditional knowledge related to PGR in general and to traditional PGRFA, and to traditional PGRFA themselves, subsumed under the term 'traditional knowledge related to PGR'.

3.2.2 Patent protection

Introduction

In this section we shall outline the scope of patent protection in general, and specifically explore the extent to which patent protection is suitable for traditional knowledge related to plant genetic resources. There is little doubt that patents are appropriate for formal plant breeders' innovations. However, the advancement of biotechnology has focused attention on the intellectual property situation of holders of traditional knowledge and traditional PGRFA. As a matter of fact, certain biotechnological innovations draw substantial value from these sources. It is therefore useful to assess the scope of patent protection for innovations related to traditional knowledge and traditional PGRFA in order to better understand the various issues at stake and discuss the corresponding solutions in later chapters. Relevant questions include: What are the problems at the interface between IPR and TK related to plant genetic resources (see Chapter 7)? What solution can patent protection provide to traditional knowledge holders and holders of traditional PGRFA? What are alternative forms of protection that could contribute to reaching the relevant policy goals? How must patent protection be better adapted to

[11] For detailed definitions of 'genetic engineering' and 'biotechnology' and further relevant terms, see Girsberger (1998, p. 1023); and Girsberger (1999, p. 329 ff.) quoting 'Legal Environment'.

[12] Man-made genetic elements are the genetic resources of cultivated species that are characterized by a high degree of human intervention; see Heitz, Buenos Aires, November 1991, pp. 114 and 118.

fit with the interests of traditional knowledge holders?

With the advancement of biotechnology, the question of patenting life forms became a central topic in the intellectual property law and policy agenda of industrialized countries. In turn, this agenda had, and still has, a significant impact on developing countries, where arguably most valuable sources of traditional knowledge related to plant genetic resources reside. This discussion can be characterized by three main issues. First, there is the dichotomy of discovery *versus* invention; secondly, the question of novelty and the assessment of prior art and of the inventive step related thereto; and last, but not least, the debate on ethics and morality of patenting life forms. The track chosen in this area by the USA was quite straightforward. It was driven by strong business interests, and comparably weak opposition from other concerned parties. In contrast, the legislative process in the EU tended to consider non-business concerns, such as morality and equity, with more emphasis, most recently in the context of the elaboration and adoption of the Biotech Directive.[13]

With respect to the qualification of patentable 'invention' as opposed to non-patentable 'discovery', the landmark US Supreme Court case *Diamond vs Chakrabarty* in 1980 stated that the relevant criterion is human involvement in nature and not the distinction between inanimate and living material. In a nutshell, if an achievement can be shown to be human rather than natural, it should be open to patent protection, since human ingenuity should receive encouragement. A comparable approach also eventually prevailed in Europe and other industrialized regions and countries.[14]

The assessment of novelty remains a highly sensitive ongoing issue, both in the North and in the South. In industrialized countries the trial and error process, driven by patent registration-related administrative practices and case law, mainly focused on the balance pertaining to the appropriate level required for the criterion of inventiveness to be fulfilled. If the required inventive step is insufficient, there is concern for efficiency induced by competition. In contrast, if the required inventive step is prohibitive, patent protection no longer works as an incentive for research and development. In new technological fields, patent offices commonly tend to adopt a lax registration practice during the first stage, by requiring a relatively low inventive step. At a second stage, this practice is usually challenged by competitors through judicial review. Eventually, the resulting case law typically leads patent offices to be stricter and, thus, request a higher degree of inventiveness. This trial and error approach means that the initial broad scope of protection is subsequently fine-tuned in order to reduce it to a level where ideally the incentive function that is underlying the grant of the exclusive rights preserves sound competition among market actors.

The main concern of developing countries resides in opposing their prior art achievements against foreign corporations that take undue advantage of them. This is generally the case when companies want to register patents for inventions that draw from traditional knowledge related to plant genetic resources.[15] Concurrent to this concern is the legitimate aspiration of certain traditional knowledge holders to take advantage of the intellectual property system for equity as well as other policy considerations. In this context, the criterion of novelty can preclude protection for such knowledge that is by definition traditional or incremental. The insufficiencies of the current system, which is based upon national and regional legislations, may lead to a situation where certain parties can obtain patent protection for elements of traditional knowledge as inventions in their

[13] See Directive 98/44/EC of the European Parliament and of the Council of 6 July 1998 on the legal protection of biotechnological inventions, OJ L 213, 30 July 1998, pp. 13–21.

[14] For more detailed references, see text accompanying note 30.

[15] See the case studies below, in particular the turmeric case study, p. 135.

jurisdiction because of an insufficient prior art search, although these parties did not generate the knowledge, whereas the traditional holders of such knowledge may not be granted exclusive rights in other jurisdictions because of the lack of novelty that is inherent in this type of intellectual value achieved over generations. Although this is primarily an issue with respect to the US patent law and its definition of prior art, it may inspire legislators in other jurisdictions, and therefore cause a run to the bottom. Additionally, the difficulty to accede to existing TK when carrying out prior art search may be prohibitive if this TK is not documented in writing or in languages the patent authorities are familiar with. Therefore, even though such searches are carried out correctly, access to TK may not be possible in practice along the classical route. As of today, existing laws do not sufficiently address this issue, either on the national level, or on the international level. Some major industrialized countries are reluctant to introduce anti-biopiracy mechanisms to oblige patent applicants to provide relevant information pertaining to the use and sources of traditional knowledge or genetic material for their invention. These mechanisms could take the form of more efficient novelty-destroying international prior art searches and procedures to assess prior informed consent ('defensive rights'). Switzerland, for example, proposed several measures to combat 'biopiracy', e.g. to establish an international gateway (or internet portal) for traditional knowledge that would link local and national databases, improving access to existing traditional knowledge by patent authorities and thus facilitating prior art searches. Switzerland furthermore recommended amending the PCT abbreviation to explicitly enable the national legislator to require patent applicants to disclose the source of genetic resources/traditional knowledge in patent applications.

Finally, the ethical and moral questions respectively divide proponents and opponents of patent protection for living forms, both in the North and in the South. According to Dunleavy and Vinnola (1999, p. C11), the morality-based provisions in Article 6 of the EU Biotech Directive could arguably be considered to represent the most significant difference between future European practice and established US practice. However, these authors perceive that the USA appears to be moving towards an approach that is similar to that of the EU regarding the patentability of these types of inventions. In addition to the 'morality' or ethical questions that specifically refer to intellectual property rights in connection with living things, the very concept of private ownership, as opposed to public domain or collective ownership with respect to intellectual achievements, lacks legitimacy in certain Southern cultures. The intellectual property system itself is largely accepted in industrialized countries, and divergences of opinion mainly focus upon the appropriate standards of protection. In contrast, the ability to grant private rights to individuals or collectives for works of mind remains contested, or even rejected, by many local communities in developing countries for moral or ethical reasons.

To summarize, the debate concerning the patentability of TK related to PGR addresses not merely technical issues, but presents a remarkable conceptual complexity: it is about the legal meaning of 'invention', 'novelty' and 'property', and beyond that, about the legitimacy and acceptance of the entire system.

The following section aims to provide a brief stock take of the patentability of life forms in general, and specifically focus on legal aspects that are relevant to traditional knowledge related to plant genetic resources. It will furthermore outline the issues at stake in order to prepare the ground for reflection throughout later sections of this book concerning the approaches that can be utilized to adapt the current intellectual property system, and in particular patent protection, for the purposes of maintaining and further developing traditional knowledge related to plant genetic resources.

The rationale of patent protection

Patents offer protection for inventions, i.e. for solutions to specific problems in the field of technology. Patents define their owners' exclusive rights pertaining to inventions. They are statutory privileges that take the form of competitive advantages that are granted by governments to inventors for a specified period of time. Patents provide, at least in certain respects, the most extensive exclusive rights amongst the major forms of intellectual property protection. Patents have gained increased prominence with the development of genetic engineering, which is reliant on PGR to some extent. As a result of the increased activity by individuals and the private sector in this field, the scope of patentability has steadily increased over the past couple of decades. Calls for further expansion of the scope of patentability have been countered in recent years by counter-claims that patenting elements which are commonly used by researchers for further innovations, such as those used in gene technology, may end up hindering rather than promoting innovation.

The rationale underlying the grant of patents is based on equity as well as on pragmatic considerations. Equity considerations materialize the concepts of 'justice' or 'natural rights', whereas the pragmatic justifications refer to the promotion of the public interest. The so-called 'natural-law' thesis assumes that humans have a natural property right to their own ideas. Under the 'exchange-for-secrets' or the 'monopoly-profit-incentive' concepts, an inventor discloses the results of his or her efforts to the public. In exchange, the inventor is granted a monopoly-like privilege for a limited period of time, with respect to the commercial use of the invention. In this way, the state provides the inventor with a competitive advantage in consideration of the disclosure of the inventor's intellectual achievement. This disclosure allows third parties to perform further research and development based on the invention. It further enables them to work freely and use the invention after the expiration of its terms of protection. Thus, technological progress is promoted through a legal incentive, the patent, to diffuse the knowledge relevant to the invention that the inventor could otherwise keep secret. By this line of reasoning, the competitive privileges granted to the inventor may also be considered as an award for the investment in qualified inventive efforts that were already performed, and as an incentive to proceed to future inventive efforts. As a matter of fact, it allows the patent owner to derive a material benefit that compensates him or her for the research and development cost the invention has entailed (Machlup, 1999, p. 231 ff.). In the case of violation of this exclusive rights by another person, who exploits the protected invention without authorization, the patent owner can require the courts to order that the infringement ceases as well as awarding civil, administrative and criminal remedies.

The balance that patent laws try to achieve is, on the one hand, to encourage inventive and innovating efforts on an individual basis and, on the other hand, to disseminate knowledge and the resulting technological achievements for the general economic and social welfare of society as a whole. When investing in research and development, companies usually face what economists have labelled the 'incomplete appropriability of knowledge' (Arrow, 1962, p. 609). According to this concept, the investment in the production of knowledge will be difficult, if not impossible, to protect completely, since part of the newly generated knowledge will diffuse to competitors and into the public domain. For firms investing in the production of knowledge, patent protection provides only a partial solution. It allows the investing firm to appropriate a return on its investment in research and development by protecting its invention against unauthorized duplication. However, since the patentee must disclose her/his new knowledge, this enables competing firms to build upon such knowledge to create other inventions that can be similar, but not identical, to the protected one. It may require costly litigation to determine infringement in such circumstances. The cost of substantially imitating an existing product, with or without improvement, is usually lower and less risky than the originator's cost of creating, developing and marketing the new product. Such a competitor can act in a shorter time than was needed by the patentee, and undercut the return to the patentee. Because of the diminished risk-weighted incentive to the originator, some authors have concluded that 'total welfare, but not the welfare of consumers, would be increased by making it more difficult to produce close substitutes for existing products'.[16]

It may be argued that the grant of a patent encourages investment in the production and distribution of an invention, since the patent monopoly assists the investor in penetration and defence of the market. On the other hand, the neutralization of competition during the terms of protection may trigger a rent in form of a high rate of return in favour of the inventor, inhibiting him from further innovation.

In the seed sector, for example, research and development initially were mainly performed by state-owned entities such as government agencies and universities. These efforts were thus funded by public money. For this reason, there was no demand for IPR protection until the private sector started to gain an interest in this business. As described later under Section 3.2.3, this development began with legislative action in the field of plant variety protection in the USA. In contrast to the legal solutions eventually adopted in Europe, innovators in the USA may claim protection both under plant variety and patent law. Subsequently, with the further growth of biotechnology, the private industry required intellectual protection for life forms and microorganisms. Once this sector began to attract private industry, the protection of intellectual property rights in these fields was put onto the legislative agenda of industrialized countries and regions. Private actors argued that effective protection was required in order to secure a return on large private investments performed for research and development into biotechnological innovations. Patents and other forms of protection grant an advantage over competitors who did not make the same investments into innovation. Furthermore, the industry required that this competitive advantage was to be extended *vis-à-vis* the users, because the relevant inventions were typically life forms that could reproduce naturally, and consequently, they did not need repeated purchases by farmers. In the case of genetically modified rape, for example, the results of this law-making process in the USA eventually raised a polemic, when farmers became aware of their increased dependency upon major producers and distributors. Since biotechnology allows the transformation of genes across species boundaries, the genetic modification of organisms raises not only merely economic questions, but also substantial ethical, environmental and social issues that the relevant laws also have to address as part of an ongoing process.

Whereas certain cultures oppose the concept of granting exclusive rights over living forms, other cultures accept this approach under clearly defined conditions in order to fulfil certain policy goals that are in the public interest. The latter cultures argue that intellectual property rights are a cost-efficient alternative to publicly funded innovation because they function as incentives for privately performed research and development.

Even if one accepts that the intellectual property system should be extended to apply to life forms, one can question the appropriate levels of protection. The incentive rationale primarily covers investments into research and development by granting legally secured competitive advantages during the marketing phase. However, too

[16] See *Hilton Davis Chemicals Co. vs Warner-Jenkinson Company, Inc.*, United States Court of Appeals for the Federal Circuit, 62 F.3d 1512; 1995 U.S. App. LEXIS 21069, 8 August 1995, quoting Besen and Raskind (1991).

much protection can cause a situation where major producers and distributors are able to control the market. In this case, undertakings having a dominant position can drive their smaller and, as the case may be, more innovative competitors out of business. Eventually, in the worst-case scenario, the consumers' choice becomes restricted to the products and services that are most powerfully marketed to the exclusion of the most innovative ones. This development undermines the incentive effect of intellectual property rights in contributing to the technological progress for general welfare.[17]

The allocation of IPR is premised on the notion that innovation is driven by profit. From a societal point of view, IPR strive to balance the private interests of creators, by ensuring that they still have an incentive to create, against those of society at large to have the information available for its use. Even though information does not diminish once it is shared, the role of IPR is to ensure that information providers do not lose rights to the information by disclosing it, since such information can be used by an infinite number of persons simultaneously (Landes and Posner, 1989; Baer, 1995). Indeed, one of the perceived philosophic underpinnings of IPR is to ensure disclosure of the information while maintaining exclusive rights for the creator. It is significant, however, that there is still no consensus, even in developed countries, concerning the social or economic utility of granting intellectual property rights. In fact, there have not been any significant empirical studies demonstrating the beneficial impact of the grant of patents on economic or social development (Abbott, 1989).

The concepts of invention and the protection of plant varieties

On the global level, the TRIPS agreement is currently the most detailed international instrument addressing substantive patent protection. This agreement, that is part of the WTO body of rules resulting from the Uruguay round, entered into force in January 1996 for industrialized countries, and, subject to the transitional arrangements in Articles 65 and 66 TRIPS, 4 years afterwards for developing countries and, if certain additional conditions are fulfilled, to previously centrally planned economies. In the area of patent protection, section 5 of the second part of the TRIPS agreement refers to, and further completes, the Paris Convention in the Stockholm version of 14 July 1967. From the perspective of industrialized countries, the Paris Convention needed updating and the addition of extra, as well as more detailed, rules concerning many aspects of patent protection. In particular, the scope of patent protection was highly controversial during the TRIPS negotiations (Cottier, 2004). Article 27 TRIPS provides that patents shall, subject to certain conditions, be available for any invention, whether a product or process, in all fields of technology, provided that they are new, involve an inventive step and are capable of industrial application. The TRIPS Agreement does not define the term 'invention'.

Article 53(b) of the European Patent Convention of 5 October 1973 (EPC)[18] excludes plant and animal varieties or processes that are essentially biological for the production of plants or animals from patent protection, while not extending this exclusion to microbiological processes or the products thereof. In contrast, Article 27(3)(b) TRIPS provides that Members may exclude plants and animals other than microorganisms from patentability, as well as essentially biological processes for the production of plants or animals other than non-biological and microbiological processes. This latter exception serves as criterion that enables a more concrete

[17] Compare Germann (2003), on the detrimental effect of excessive copyright and trade mark protection on cultural diversity in the audiovisual sector.

[18] Version as amended by the Act revising Article 63 EPC of 17 December 1991 and by the decisions of the Administrative Council of the European Patent Organisation of 21 December 1978, 13 December 1994, 20 October 1995, 5 December 1996 and 10 December 1998.

distinction between discoveries and inventions. Members must provide legal tools for the protection of plant varieties, either by patents or by an effective *sui generis* system, or by any combination thereof. In other words, the TRIPS agreement distinguishes between the protection of plants and animals on the one hand, and the protection of plant varieties on the other. Whereas plant varieties must be protected by way of patents or through equivalent means, plants may be excluded from patent protection. At the European level, the EPC excludes plant varieties from patent protection. The TRIPS agreement contains neither a definition of a 'plant variety' nor of a 'plant'. This term is defined in Article 1(vi) of the UPOV Convention 1991:

> 'variety' means a plant grouping within a single botanical taxon of the lowest known rank, which grouping, irrespective of whether the conditions for the grant of a breeder's right are fully met, can be
> - defined by the expression of the characteristics resulting from a given genotype or combination of genotypes;
> - distinguished from any other plant grouping by the expression of at least one of the said characteristics; and
> - considered as a unit with regard to its suitability for being propagated unchanged.

The EPO Board of Appeal discussed the definition of 'plant variety' as set forth under Article 53(b) EPC in the decisions T 49/83,[19] T 320/87,[20] and, with reference to the above-quoted definition contained in the UPOV Convention 1991, in T 356/93,[21] as well as in T 1054/96.[22] According to the approach adopted under the EPC, plant varieties shall be protected under plant breeders' rights, which are specifically designed for this purpose. This form of protection can therefore be considered as a *lex specialis vis-à-vis* patent protection. The purpose and objective of these rights, as defined by the UPOV Convention, are to promote plant-breeding activities, i.e. the creation of new and improved plant varieties. Another rationale behind plant variety protection is the promotion of the seed trade (Girsberger, 1998, p. 1024, with indication of further sources). As Girsberger stresses, the general perception was decisive for concluding the UPOV Convention that protecting plant varieties by patents did not provide effective incentives for investment in plant breeding activities (Girsberger, 1998, p. 1027). This explains the reason why the form of protection of plant varieties is left open in Article 27(3)(b) TRIPS, and why a *sui generis* form of protection for plant varieties was considered to be more appropriate under the EPC.

In the USA, patent protection for asexually reproduced plant varieties was provided as early as 1930 under the Plant Patent Act. In 1970, the US Congress enacted the Plant Variety Protection Act (PVPA), which provided protection to new varieties of 'sexually' reproduced plants in addition to the Act of 1930. Under this legislation, the developer of a novel variety of plant can apply for protection, based on which he or she can exclude others from selling, reproducing, exporting and importing the protected variety for a period of 18 years.

The case law pertaining to the EPC that addresses the concepts of 'invention', 'novelty' and 'morality' in relation to plant genetic resources provides an insight into the current state of patent practice regarding patentability and scope of protection into the area of biotechnology in Europe. According to decision T 49/83 (OJ 1984, 112) that referred to Article 52(1), Article 53(b), R. 28 and R. 28a, no general exclusion of inventions in the sphere of animate nature can be inferred from the EPC. The landmark T 356/93 defined the concepts of '*ordre public*', 'morality', 'plant varieties',

[19] OJ 1984, 112.
[20] OJ 1990, 71.
[21] OJ 1995, 545.
[22] See summary of these cases at http://www.european-patent-office.org/legal/case_law/e/I_B_3.htm (status: August 2003).

'essentially biological processes for the production of plants', 'microorganisms', 'microbiological processes' and 'the products thereof'.[23] The meaning of the term 'plant varieties' was discussed in T 49/83 (OJ 1984, 112) and held to mean a multiplicity of plants, which were largely the same in their characteristics and remained the same within specific tolerances after every propagation cycle. The board added that plant varieties in this sense were all cultivated varieties, clones, lines, strains and hybrids, which could be grown in such a way as to be clearly distinguishable from other varieties, sufficiently homogeneous, and stable in their essential characteristics. The legislator did not wish to afford patent protection under the EPC to plant varieties of this kind, whether in the form of propagating material or of the plant itself. The board further observed that Article 53(b) only prohibited the patenting of plants or their propagating material in the genetically fixed form of the plant variety. Following T 320/87, the board addressed the meaning of a process that, as a whole, was not 'essentially biological' within the meaning of Article 53(b). It highlighted that the transformation step at stake in this case, regardless of whether or not its performance depended on chance, was an 'essential technical step' (i.e. as opposed to an 'essentially biological' one), which had a decisive impact on the desired final result and could not occur without human intervention.[24]

Article 28 TRIPS determines the exclusive rights conferred by a patent. This provision distinguishes between patents for products and patents for processes. Where the subject matter of a patent is a product, the owner is entitled to prevent third parties who do not have their consent from performing the acts of making, using, offering for sale, selling or importing for these purposes, the protected product. The same rights are granted for process patents that cover the process itself and the product obtained directly by such process.[25] This obviously extends the breadth of patent protection, since one may obtain a title for a new process that leads to a product that is itself already protected or in the public domain. Furthermore, patent owners shall also have the right to assign, or transfer by succession, the patent and to conclude licensing contracts.

Pursuant to Article 64 EPC, the European patent confers the same rights as would be conferred by a national patent granted in that state on its proprietor in each Contracting State in respect of which it is granted. This provision therefore refers to the exclusive rights as defined in the substantial patent laws of the Contracting Parties, except that it specifies within its second paragraph that the protection conferred by a patent on a process shall extend to the products directly obtained by such process. In comparison, under breeders' rights such as those provided by Article 14 of the UPOV Convention 1991, the production or reproduction (multiplication), conditioning for the purpose of propagation, offering for sale, selling or other marketing, exporting, importing, and stocking for the purposes of the previously listed acts, of the propagating material of the protected variety can only be performed with the right holder's authorization.[26]

Whereas breeders' rights address plant varieties, rights conferred by patent protection cover inventions. Article 52(2) EPC gives a negative definition of the latter term,

[23] This decision analysed Article 53 by taking into account the historical documentation (*'travaux préparatoires'*) relating to the EPC, and substantially confirmed the findings of T 49/83, T 320/87 and T 19/90.

[24] See note 22. Case law summaries by the EPO may be found under http://www.european-patent-office.org/legal/case_law/e/I_B.htm

[25] Article 28.1(b) TRIPS sets forth: 'A patent shall confer on its owner the following exclusive rights where the subject matter of a patent is a process, to prevent third parties not having the owner's consent from the act of using the process, and from the acts of: using, offering for sale, selling, or importing for these purposes at least the product obtained directly by that process'. Article 34 TRIPS contains a special rule addressing the burden of proof of process patent containing presumptions that are in favour of the process patent owner.

[26] For a more detailed analysis of the UPOV Convention, see Chapter 2, Section 2.3.4.

according to which discoveries, scientific theories and mathematical methods, aesthetic creations, schemes, rules and methods for performing mental acts, playing games or doing business, and programs for computers, as well as presentations of information, shall not be considered as inventions. The items on this list, which is not exhaustive, are commented upon in the EPO Guidelines[27] and are further clarified through case law.[28] The exclusions on this list are either abstract (e.g. discoveries, scientific theories, etc.) or non-technical (e.g. aesthetic creations or presentations of information). Article 52(4) provides that methods for treatment of the human or animal body by surgery or therapy, and diagnostic methods practised on the human or animal body, shall not be regarded as inventions which are susceptible to industrial application. In contrast to the EPC, the TRIPS agreement does not contain a definition of invention. Although Article 27(3)(a) TRIPS allows Members to exclude diagnostic, therapeutic and surgical methods for the treatment of humans or animals from patentability, this provision does not state that these methods are not inventions. Pursuant to *Black's Law Dictionary*, the term 'invention' can be defined as:

> ... the act or operation of finding out something new; the process of contriving and producing something not previously known or existing, by the exercise or independent investigation and experiment. Also the article or contrivance or composition so invented ... Invention is ... not a revelation of something which exists and was unknown, but is creation of something which did not exist before, possessing elements of novelty and utility in kind and measure different from and greater than what the art might expect from skilled workers ... The finding out – the contriving, the creating of something which did not exist, and was not known before, and which can be made useful and advantageous in the pursuits of life, or which can add to the enjoyment of mankind. Not every improvement is invention; but to entitle a thing to protection it must be the product of some exercise of the inventive faculties and it must involve something more than what is obvious to persons skilled in the art to which it relates.[29]

In the light of this definition of an invention and for the purposes of assessing the patentability of PGR, it is interesting to note that under US patent law the relevant distinction is not made between living and inanimate things, but rather it is between natural products, whether living or not, and human-made inventions. As mentioned above, the US Supreme Court decision *Diamond vs Chakrabarty* of 1980 addressed the question of whether an artificially created life form – a new form of bacterium obtained by genetic alteration – is a patentable subject matter. The Court quoted excerpts from the Congress report issued in the course of the enactment of the Plant Patent Act of 1930 as follows:

> There is a clear and logical distinction between the discovery of a new variety of plant and of certain inanimate things, such, for example, as a new and useful mineral. The mineral is created wholly by nature unassisted by man ... On the other hand, a plant discovery resulting from cultivation is unique, isolated, and is not repeated by nature, nor can it be reproduced by nature unaided by man.[30]

According to this approach, the relevant criterion centres around the human involvement in nature: if an achievement is not nature's but human handiwork, it

[27] See Guidelines for Examination in the European Patent Office, Part C, Chapter IV on Patentability, version of October 2001, p. 50 ff.

[28] For an overview of the relevant case law, see http://www.european-patent-office.org/legal/case_law/e/I_A_1.htm.

[29] *Black's Law Dictionary* (1990), p. 824.

[30] S. Rep. No. 315, 71st Cong., 2d Sess., at 6 (1930); H.R. Rep. No. 1129, 71st Cong., 2d Sess., at 7 (1930); quoted in: *Diamond, Commissioner of Patents and Trademark vs Chakrabarty*, Supreme Court of the United States 447 U.S. 303; 1980 U.S. LEXIS 112 (1980) relevant excerpts in: Abbott *et al.* (1999, p. 36).

should be open to patent protection since human ingenuity should receive encouragement. In the case at stake, the Supreme Court considered that the patentee had produced a new bacterium with markedly different characteristics from any found in nature and one having the potential for significant utility.

The distinction between 'discovery' and 'invention' led, in particular, to detailed case law under the EPC with respect to the interpretation of 'essentially biological' within the meaning of Article 53(b). According to decision T 320/87 (OJ 1990, 71), whether or not a (non-microbiological) process was to be considered as 'essentially biological' within the meaning of Article 53(b) had to be judged on the basis of the essence of the invention, taking into account the totality of human intervention and its impact on the result achieved. The necessity for human intervention alone was not a sufficient criterion for its not being 'essentially biological'. Human interference might only mean that the process was not a 'purely biological' process, without contributing anything beyond a trivial level. Furthermore, it was not simply a matter of whether such intervention was of a quantitative or qualitative character. In this particular case, it was concluded that the claimed processes for the preparation of hybrid plants did not constitute an exception to patentability, because they represented an essential modification of known biological and classical breeders' processes, and the efficiency and high yield associated with the product showed important technological character.

Protective criteria

As mentioned, in order to enjoy legal protection, an invention must fulfil certain conditions as set forth by national, regional or international rules. Based on the patent, both the inventor and subsequent assignee or licensee are entitled to produce and market the invention on an exclusive basis for a determined period of time.[31] The owner of the patent thus has the right to prevent third parties from manufacturing and commercially exploiting the invention without his authorization. The invention may relate either to a product or to a process. Patent laws generally rely upon four essential criteria for the grant of exclusive rights.

NOVELTY

According to the first criterion, the invention must be new or novel. This means that the invention has not been disclosed or described before the date of filling in the patent application. For the purpose of assessing the novelty condition, the patent examiner will perform a so-called 'prior art search'. A patent application will generally be defeated by an anticipating disclosure. There may be a controversy between an inventor and a patent examiner over whether a particular prior art disclosure does in fact anticipate an invention, for instance if a prior inventor described a device that is similar, but not identical, to the subsequent inventor's device.

UTILITY

Pursuant to the second criterion, the invention must be useful or capable of industrial application. This condition is particularly relevant within the fields of biotechnology and chemistry, where it is possible for researchers to develop new compounds with relative ease, yet without, at least initially, any immediate practical application in mind. The criterion of utility again became critical in the evaluation of claims for inventions in the area of biotechnology, in order to prevent 'speculative booking' of exclusive rights.

INVENTIVE STEP

As a third condition, the invention must involve an inventive step or, in other words, it should not be obvious. This

[31] See note 6.

means that the invention must not simply be novel, but must result from a qualified intellectual effort that makes it non-obvious. This criterion thus requires a higher standard of novelty through an inventive step. Strong protection leading to a competitive advantage shall only be granted to inventions that would be an apparent improvement to prior art to a person who is reasonably skilled in the art practised by the invention. This requirement is justified by the 'monopoly-profit-incentive' rationale, according to which strong protection shall only be granted to substantial contributions to the technological progress.

ENABLING DISCLOSURE

Finally, the fourth criterion obliges the inventor to disclose in the patent application either a means for enabling the practice of the invention (generally for Europe), or the best known means for practising the invention (for the USA). One of the reasons for this condition is based on the exchange theory of the award of the patent: the patent applicant is awarded exclusive rights in return for the disclosure to society of a new, useful and non-obvious invention. Without a disclosure that enables other persons to benefit from the invention for their own research and development work, this exchange between the inventor and the society would not make sense. This condition also performs the function of filtering out speculative applications, since it constitutes a reliable assessment of the usefulness of the invention for the purpose of its industrial application. The applicant will typically try to disclose to his competitors as little as possible about the secrets of his invention, and obtain as much protection as possible in return. This behaviour, which often leads to tensions between patent examiners and inventors, is obviously not in the public interest since healthy competition promotes innovation to the benefit of society.

PROTECTIVE CRITERIA ACCORDING TO TRIPS AND EPC

As mentioned, pursuant to Article 27(1) of TRIPS, patents shall be available for any inventions, whether products or processes, in all fields of technology, provided that they are new, involve an inventive step and are capable of industrial application. A footnote to this paragraph states that the WTO Members may interpret the terms 'inventive step' and 'capable of industrial application' as being synonymous with the terms 'non-obvious' and 'useful', respectively. However, the TRIPS agreement does not further define any of these protective criteria. For the time being, there is no WTO case law about the meaning of these criteria either. The Members thus have considerable discretion when implementing Article 27(1) TRIPS in their national laws.

In comparison, Article 52(1) EPC addresses patentable inventions, and provides that European patents shall be granted for any inventions which are susceptible to industrial application, which are new and which involve an inventive step. As regards the criteria of novelty, Article 54(1) EPC states that an invention shall be considered to be new if it does not form part of the state of the Article. Thus, the reference to the state of the art is pivotal in the determination of whether the criterion of novelty is fulfilled or not. Paragraph 2 of this provision clarifies that the state of the art shall be held to comprise everything made available to the public by means of a written or oral description, by use, or in any other way, before the filing date of the European patent application. In comparison, for the purposes of Article 15.1 of PCT, relevant prior art includes everything that has been made available to the public anywhere in the world by means of written disclosure, and which can be of assistance in determining that a claimed invention is novel or non-obvious. Under US patent law it is fairly easy to obtain patent protection for inventions that are based on traditional knowledge, because it does not consider that the 'novelty' requirement has been lost if the knowledge is divulged outside the USA by means of public use and sale. The essential requirement of 'novelty' can be destroyed only through publication.[32] Unlike US patent law, in Europe and most countries in the world, novelty is lost by any type of divulgation in a foreign country, whether it is oral or written. In the USA, indigenous communities in developing countries have little opportunity to bring attention to unwritten knowledge, practices and innovations that demonstrate lack of novelty or non-obviousness. To illustrate this issue, one can quote the neem case as an example of a patent that was allegedly based upon misappropriation of non-published traditional knowledge (see p. 136).

PROTECTION OF TRADITIONAL KNOWLEDGE UNDER THE EPC

In the specific context pertaining to the protection of plant genetic resources and traditional knowledge referring thereto, it is interesting to analyse the meaning of paragraph 5 of Article 54 EPC, which states that its previous paragraphs shall not exclude the patentability of any substance or composition comprised in the state of the art, for use in a method referred to in Article 52(4) EPC, provided that its use for any method referred to in that paragraph is not comprised in the state of the Article. Article 52(4) EPC requires that methods for treatment of the human or animal body by surgery or therapy and diagnostic methods practised on the human or animal body, shall *not* be considered as inventions that are susceptible to industrial application within the meaning of paragraph 1, i.e. they are not patentable. However, this latter restriction shall not apply to products, and specifically to substances or compositions, for use in any of these methods. For example, if a traditional healing method uses a substance or composition that contains plant genetic resources the composition of which is part of the state of the art, while the healing method is not, patent protection arguably should be granted to such a combination of method and substance or composition. This interpretation is relevant for patent protection of plant genetic resources in connection with traditional knowledge in the absence of the ability to protect such resources. This may be because of the absence of genetic engineering, or because it is already part of the state of the art, since traditional knowledge alone may not be patentable pursuant to Article 52 paragraph 2, letters c and d, which excludes schemes, rules and methods for performing mental acts, as well as presentations of information, respectively, from patent protection. Thus, the novel aspect would essentially be vested with the combination of traditional knowledge and plant-related genetic resources.

Scope and limitations of patent protection

The scope of protection for patents is determined by the terms of so-called 'claims' in the patent application that are eventually accepted by the patent office. The technical description of the process or product in the patent application usually serves to construe the claims. The scope of protection is limited by statutory provisions such as compulsory licensing and exhaustion rules.[33]

Rules limiting protection aim to ensure that the public interest remains preserved when granting exclusive rights over a given period of time. Public health (e.g. to provide access to essential medicine to poorer parts of the population) or education (fair use limitations) are typical grounds on which to restrain protection.[34]

---

[32] Section 102 of the Patent Act, 35 U.S.C. § 102: 'A person shall be entitled to a patent unless (a) the invention was known or used by others in this country, or patented or described in a printed publication in this or a foreign country, before the invention thereof by the applicant for patent, or (b) the invention was patented or described in a printed publication in this or a foreign country or in public use or on sale in this country, more than one year prior to the date of the application for patent in the United States'.

[33] See, for example, Articles 6, 7, 8 and 40 of TRIPS for general limitations and Articles 30 and 31of TRIPS specifically for patents.

[34] For the discussion on patents for pharmaceuticals and non-trade concerns, see http://www.wto.org/english/tratop_e/TRIPS_e/pharmpatent_e.htm and http://www.cptech.org/

More fundamentally, the scope of protection is conditioned upon the quest of the intellectual to find a balance that addresses both Garrett Hardin's influential 'tragedy of the commons' and also the 'tragedy of the anti-commons'. According to the theory of the tragedy of the commons (or 'the tragedy of open access'), the absence of property rights or access regulations over resources will lead to their depletion. While individuals accrue benefits as they exploit resources, the costs of everybody's exploitation of the resources at unsustainable levels will have to be met by the community as a whole. In economic terms, resource depletion is a negative externality that results from the absence of individual property rights. As an example, Dutfield (2001) quotes the overexploitation of high seas fishing stocks. On the other hand, the tragedy of the anti-commons is the result of excessive intellectual property protection that hinders further innovation and creativity.

As mentioned above, patents are granted in order to provide an incentive for innovations and for the diffusion (disclosure) of these innovations. Without the protection provided by the patent system, inventors would have no incentive to disclose their inventions, and to keep them secret in order to preserve their economic benefits. With disclosure, patents help avoid the wasteful duplication of innovation efforts, and instead channel resources towards unexplored areas of technology. Furthermore, patents are instrumental for the commercialization of innovative products and processes.

However, patents impose a short-run efficiency cost, because of the deadweight loss corresponding to the grant of quasi-monopoly rights. In standard economic parlance, anything that is not perfect competition pays an inefficiency trade-off, in that the use of the innovation is less than optimal. Moreover, just as patents can help to promote innovations, they can also adversely affect the flow of future inventions, by stunting incentives for related innovation or further improvements, especially where the patents have broad scope and the licensing arrangements for the innovation are prohibitive.

Thus, patent policy is not an optimal solution, but rather a second-best solution to a market failure problem in innovation production. The market failure stems from the fact that knowledge is a public good, which, once available, has zero marginal cost if used by others. From the economic perspective, innovations are *non-rival* in consumption, which means that the use of the knowledge by others does not affect the amount that is available to the inventor. Conversely, if knowledge were a public good, then its producer who bore the cost of innovation would lose out from 'free riders', thus providing no incentive to produce the knowledge in the first place. The result would be an underproduction of knowledge or dearth in innovation.

To summarize, patents that provide the legal means of affecting the excludability attributes of an otherwise public good, solve the market failure problem of the non-appropriability of knowledge. Society pays for this by the economic inefficiency associated with monopolies. That is, the 'static inefficiency' (from a monopoly) is the trade-off for the 'dynamic efficiency' resulting from greater innovation.

Preliminary conclusions on patent protection for holders of traditional knowledge

The progress in the biotechnology field caused the meaning of fundamental notions of patent protection to be questioned. The concepts of 'invention', 'novelty', 'inventive step/non-obviousness' and, last but not least, 'morality', are at the heart of the discussion about patenting life forms. Critical observers argue that accommodating the wishes of certain biotechnology sectors has seriously blurred the distinction between 'discovery' and 'invention' in patent law. They focus on many instances where the actual substance resides mainly in nature or traditional practice, to which biotechnology adds only minor changes (Nijar, 1996, p. 4). The question of an inventive step or non-obviousness is closely connected to this issue. Granting property titles for little added intellectual value lacks legitimacy and runs counter to the public policy goals underlying the whole system. Once one accepts that there is an invention, the questions concerning novelty, and more particularly the inventive step, are dealt with in the course of the examination procedure, as well as through subsequent judicial review mechanisms. Current systemic insufficiencies of law and practice pertaining to international prior art searches are causing a shift from prior *ex officio* patent examination to subsequent party-initiated judicial review procedures. From the perspective of holders of traditional knowledge related to plant genetic resources, challenging a patent because it does not fulfil the novelty or non-obviousness requirements may turn out to be a very costly undertaking, which is beyond the means of most individuals or local communities. Traditional knowledge holders are usually at a disadvantage in cases where patent offices are lax during the international prior art check and when, in practice, the effective test only takes place after examination through judicial review on the initiative of a concerned party. As a matter of fact, for the time being, the patent system is mainly tailored towards those corporations that enjoy substantial human and financial resources that enable them to register their own inventions, to administer the rights and, as the case may be, to challenge the registration of their competitors' inventions. Without appropriate institutional infrastructures, such as adapted rights management systems including databases, or collecting societies as flanking measures (see Chapter 7), today's patent system will remain a game for bigger corporations, and one that hardly provides any genuine level playing field for individuals and local communities holding traditional knowledge related to plant genetic resources. These flanking measures cannot be of a merely technical nature, but also need to address complex interactions between different legal cultures.

If traditional knowledge holders are not able to participate in the intellectual property system on a fair basis, this could contribute to the depletion of this knowledge within this system. In this case, holders of traditional knowledge will miss incentives to maintain and develop the intellectual achievements of their communities that may be misappropriated by other parties in stronger positions. When local communities cannot enjoy returns from, or are even precluded from free access to, inventions based on their traditional knowledge because third parties have become right holders, the rationale of equity underlying the grant of intellectual property rights does no longer work. Moreover, it leads to a perverse result of the system when excessive standards of protection start to hinder innovation because they mainly serve to attract capital for marketing efforts that eventually drive competitors out of business, even if they are more innovative.[35] This negative occurrence typically strikes marginal players such as traditional knowledge holders first, and at the end often leads to heavy concentrations among major producers and distributors that are detrimental to welfare. The term 'morality' can be used to express the third main issue to be addressed in this context. In the public debate within industrialized countries it covers the question of patenting life forms.[36] We extend the meaning of 'morality' in order to encompass the complex interface between various legal orders, where one order rejects a body of rules that another tries to bring to application across the formal limits of its own jurisdiction. There are several legal cultures that reject the very idea of private ownership for intellectual works that are based upon philosophical or religious traditions. How are clashes of legal cultures to be settled? On the one hand, the intellectual property system needs to overcome national barriers in order to be effective. On the other hand, there is no legitimacy behind the imposition of this system upon people who do not accept it because it runs contrary to their basic principles. Both 'morality' issues

35 See Germann (2003).

36 See text accompanying note 52.

exhibit at least one common feature: they challenge the intellectual property system for philosophical or religious reasons rather than for practical ones. The qualities and disadvantages of the system can no longer be discussed in a purely rational way. As is the case concerning questions about the death penalty or abortion, one could even contest the democratic decision-making process that leads to the adoption of an intellectual property system in which the opponents to the system are defeated simply because they are in the minority. The task to find a consensus concerning 'morality' considerations on the international level is more difficult by far than the reaching of a cross-border understanding over the more technical concepts of 'invention' and 'inventive step'. Nevertheless, these two latter concepts also pose substantial difficulties in their adaptation for the purposes of holders of traditional knowledge. These tasks include the determination of appropriate forms of protection, specific protective criteria, ownership, transfer of rights modalities, prior informed consent safeguards and efficient international implementation mechanisms. We will explore the corresponding solutions in the chapters below.

3.2.3 Plant breeders' rights[37]

Patents are not the only form of protection to encourage innovation. Several additional forms of IPR are available, such as trade secrets,[38] plant breeders' rights and *sui generis* protection for plant varieties.

Background

Traditionally, patents did not cover natural products. Over time, the scope of the exception relating to nature has been significantly reduced, but international intellectual property rights treaties have never yet gone so far as imposing patent protection for plant varieties. An alternative form of intellectual right protection for plant varieties has developed progressively.

Various factors have contributed to the difficulties of providing IP protection for plant varieties. One of these is that the notion of inventiveness, which characterized patents, would be diluted if plant varieties were brought on board, because a new plant variety was seen more as an improvement of an existing natural product rather than as a 'scientific' invention (Rangnekar, 2000). Furthermore, seeds had always been deemed to be part of the common heritage of humankind and were freely exchanged among farmers and farming communities (Shiva, 1994).[39] However, it was increasingly recognised that a form of intellectual property rights was necessary in the seed sector in order to encourage private investment.

Partly as a result of the progressive commercialization of the agricultural sector, and partly because of the push for the introduction of some form of IPR protection, plant breeders' rights (PBRs) evolved. These gave breeders specific legal rights to the varieties they developed, in a bid to foster the development of varieties that could, otherwise, easily be reproduced by other farmers or competing breeders. Because the impetus behind this measure came from large commercialized farming systems, PBRs, as they exist today, have been developed for the specific contexts and needs of

[37] Author: Philippe Cullet.

[38] The rationale behind trade secrets, in contrast to patents, is based upon property theory or the doctrines of tort, contract and trust, rather than being motivated by providing an incentive for innovation or creativity (Besen and Raskind, 1991). Its subject matter can be any formula, pattern, etc., which provides a business advantage over others who do not possess the information. Trade secret protection is not limited in time.

[39] The different positions expressed with regard to the introduction of plant breeders' rights have taken on added significance in the wake of the adoption of the TRIPS Agreement. Indeed, while a number of OECD countries had progressively adopted a form of PBR before 1994 to foster the development of their private seed industries, most developing countries had not introduced any form of intellectual property right protection in the agricultural sector.

developed countries. The legal regime for PBRs is quite uniform; thanks to the fact that most countries that have introduced PBRs have either joined the International Union for the Protection of New Varieties of Plants (UPOV) regime, or have modelled their legislation on the UPOV regime (on UPOV, see Section 2.3.4). The following paragraphs consequently analyse PBRs as they have developed in the UPOV context.

What are PBRs?

PBRs can generally be described as patent rights with some missing attributes. PBRs share a number of characteristics with patent rights: they provide exclusive commercial rights to the holder, reward an inventive process and are granted for a limited period of time.

To be more specific, PBRs protect plant varieties. Plant varieties can only be protected by PBRs if they fulfil the four basic criteria of: novelty, distinctness, stability and uniformity or homogeneity. Each of these characteristics is given further content by the UPOV Convention. The concept of novelty requires further elaboration because it differs from its acceptance under patent law. Under UPOV, a variety is novel if it has not been sold or otherwise disposed of to others, by or with the consent of the breeder, for the purposes of exploitation of the variety.[40] Novelty is thus entirely defined by the issue of commercialization and not by the fact that the variety did not exist previously. UPOV gives a specific time frame for the application of novelty. In order to fulfil the requirement for novelty, a variety must not have been commercialized in the country where the application is filed for more than a year before the application, or for more than 4 years in other member countries (6 years in the case of trees and vines). The criterion of distinctness requires that the protected variety should be clearly distinguishable from any other variety whose existence is a matter of common knowledge at the time of the filing of the application.[41] The requirement of stability is satisfied if the variety remains true to its description after repeated reproduction or propagation.[42] Finally, uniformity implies that the variety remains true to the original in its relevant characteristics when propagated.[43]

Over time, the definition of protected variety has evolved, insofar as so-called 'essentially derived varieties' were not protected during the early days of plant variety protection. The latest revision of UPOV has introduced protection for such varieties.[44] Protection as an 'essentially derived variety' is obtained if the variety is predominantly derived from the initial variety and retains its essential characteristics.

Content and limitation of PBRs

The rights conferred upon plant breeders differ from patent rights insofar as they provide much more extensive exceptions to the rights they confer than patents. Breeders have exclusive rights to produce or reproduce protected varieties, to condition them for the purposes of propagation, to offer them for sale, to commercialize them, including exporting and importing them, and to stock them in view of production or commercialization.[45] These rights are restricted by a number of exceptions that are compulsory in the UPOV context. The rights of breeders do not extend to acts done privately and for non-commercial purposes; to acts done for experimental purposes; to the use of the protected variety for the purpose of breeding other varieties; or to the right to commercialize such other varieties

[40] Article 6 of the International Convention for the Protection of New Varieties of Plants, Paris, 2 December 1961, as revised in Geneva on 10 November 1972, 23 October 1978 and 19 March 1991 (Geneva, UPOV, Doc. 221(E), 1996) [hereafter UPOV-1991].
[41] Article 7 UPOV-1991, see note 40.
[42] Article 9 UPOV-1991, see note 40.
[43] Article 8 UPOV-1991, see note 40.
[44] Article 14.5 UPOV-1991, see note 40.
[45] Article 14 UPOV-1991, see note 40.

as long as they are not essentially derived from the protected variety. While the previous exceptions are compulsory, a set of further exceptions exists that has been progressively reduced over time. The so-called 'farmer's privilege' falls into this category. Under UPOV-1978 the rights of breeders were circumscribed in such a way that PBR did not interfere with farmers' use of the legally obtained protected variety for propagating purposes on their own holdings. Under UPOV-1991, the rights of breeders have been extended to the harvested material of the protected variety and the farmer's privilege has been made optional.

With regard to the duration of PBR, their first characteristic is that they are limited in time. The period of protection has evolved over time, but always with the idea in mind that the rights conferred expire at the end of a specific period of protection. Under UPOV-1978, the period of protection is for a minimum of 15 years. For vines, forest trees, fruit trees and ornamental trees, the minimum is 18 years.[46] UPOV-1991 extends the minimum period from 15 to 20 years. For trees and vines, the minimum is of 25 years.[47]

At first, PBRs were conceived as an alternative to patent rights, and it was accepted that the two kinds of intellectual property rights should be kept separate. Thus, under UPOV-1978, member states can only offer protection through one form of intellectual property rights.[48] The grant of a PBR pertaining to a given variety implies that no other intellectual property right can be granted to the same variety. Under UPOV-1991 this restriction has been eliminated and *double protection* is now allowed.

3.2.4 *Sui generis* protection systems[49]

Rationale of sui generis *protection*

The question of *sui generis* intellectual property right protection for plant varieties has become a matter of great importance following the adoption of the TRIPS Agreement. As a result of a negotiating compromise, TRIPS requires the introduction of plant variety protection in all member states, but it does not impose the introduction of patents. Article 27.3.b specifically requires all member states to 'provide for the protection of plant varieties either by patents or by an effective *sui generis* system or by any combination thereof'.[50] The introduction of the *sui generis* concept reflects two broad elements. First, a number of countries in the North and the South have rejected the compulsory introduction of plant patents. Second, negotiators did not manage to agree on one specific alternative to patents. As a result, TRIPS gives member states a wide margin of appreciation in determining how to implement their obligation to introduce plant variety protection.

The question of the introduction of plant variety protection is one that mostly concerns developing countries. Indeed, most developed countries had already introduced either plant patents or PBRs before the adoption of TRIPS. Developing countries that are members of the WTO were left with the choice of either adopting the existing regime proposed in UPOV or of devising their own plant variety protection system adapted to their specific situation. A few countries have joined UPOV since 1994, but the majority have decided to adopt their own plant variety protection laws. In a number of cases, these laws draw directly and significantly from the UPOV regime and generally most existing proposals introduce PBRs. In cases where PBRs are

46 Article 8 of the International Convention for the Protection of New Varieties of Plants, Paris, 23 October 1978 [hereafter UPOV-1978].
47 Article 19 UPOV-1991, see note 40.
48 Article 2 UPOV-1978, see note 46.
49 Author: Philippe Cullet.
50 Agreement on Trade-Related Aspects of Intellectual Property Rights, Marrakesh, 15 April 1994, reprinted in 33 I.L.M. 1125 (1994) [hereafter TRIPS Agreement].

adopted only as one part of the regime, the regime is completed by the introduction of a form of farmers' rights. In fact, existing *sui generis* options can be generally defined as regimes introducing PBRs and farmers' rights.

Country examples of sui generis *protection*

The prominence of the UPOV Convention in the debates concerning *sui generis* plant variety protection is partly linked to the fact that the interpretation of the concept of an 'effective' *sui generis* system in Article 27.3.b TRIPS remains problematic. The only generally agreed-upon interpretation is that UPOV is an effective *sui generis* protection regime under TRIPS. This has led to some countries, such as the member states of the African Intellectual Property Organization, simply adopting a regime modelled after UPOV-1991 and at the same time committing themselves to joining the UPOV Convention.[51]

Some countries, such as, India have decided to implement plant variety protection regimes that seek to provide protection to commercial plant breeders and to farmers. Thus, the Indian plant variety protection regime introduces both PBRs and farmers' rights. While a number of countries have attempted to draw up their own *sui generis* plant variety protection regimes, African states have taken a unique initiative in adopting a Model Legislation for the Protection of the Rights of Local Communities, Farmers and Breeders, and for the Regulation of Access to Biological Resources at the level of the Organization of African Unity (OAU).[52]

The African Model Legislation is premised on the rejection of patents on life or the exclusive appropriation of any life form, including derivatives. Its provisions relating to access to biological resources make it clear that the recipients of biological resources or related knowledge cannot apply for any intellectual property right of exclusionary nature. The model legislation mainly focuses on the definition of the rights of communities, farmers and breeders. The community rights that were recognized include rights over their biological resources, and rights to their innovations, practices, knowledge and technology, and the right to benefit collectively from their utilization. In practice, these allow communities the right to prohibit access to their resources and knowledge, but only in cases where access would be detrimental to the integrity of their natural or cultural heritage.[53] Furthermore, the state must ensure that at least 50% of the benefits derived from the utilization of their resources or knowledge are channelled back into the communities.

In this legislation, the rights of farmers are slightly more precisely defined. They include the protection of their traditional knowledge that is relevant to plant and animal genetic resources, the right to an equitable share in the benefits arising from the use of plant and animal genetic resources, the right to participate in making decisions on matters related to the conservation and sustainable use of plant and animal genetic resources, the right to save, use, exchange and sell farm-saved seed or propagating materials, and the right to use a commercial breeder's variety to develop other varieties. The breeders' rights defined under the model legislation generally follow the definition given in the UPOV convention and the duration of the rights is, for instance, modelled after UPOV-1991.

One specific feature of the plant breeders' rights regime under the model legislation is the rather broad scope of the exemptions granted. Exemptions to the rights of breeders include the right to use a protected variety for purposes other than commerce, the right to sell plant or propa-

[51] See Agreement to Revise the Bangui Agreement on the Creation of an African Intellectual Property Organization of 2 March 1977, Bangui, 24 February 1999.

[52] African Model Legislation for the Protection of the Rights of Local Communities, Farmers and Breeders, and for the Regulation of Access to Biological Resources (2000).

[53] Article 20 of the African Model Legislation, see note 52.

gating material as food, the right to sell within the place where the variety is grown and the use of the variety as an initial source of variation for developing another variety.[54]

In summary, the development of *sui generis* plant variety protection is still in its infancy. Until now, efforts have been made by developing countries to balance their obligations under Article 27.3.b of TRIPS with their specific needs and conditions. Since UPOV is the only model that is generally recognized as fulfilling the criteria of an 'effective' *sui generis* plant variety protection regime, a number of states that have not had the time or resources to devise a completely separate *sui generis* protection regime have decided to take this as a basis for a plant variety protection regime. On top of the PBR system, there seems to be a growing trend towards recognizing farmers' rights and providing for different compensation mechanisms (benefit-sharing). Other *sui generis* protection regimes will probably be developed in future years, particularly by least developed countries, who have had until 2005 to implement their plant variety protection regimes. Furthermore, even countries that are classified as developing countries may amend their legislation over time, as further *sui generis* models evolve. *Sui generis* protection is evolving and significant innovations can be expected in years to come.

3.3 Impacts on Existing PGR, Landraces and TK[55]

The progressive strengthening of intellectual property rights has already had significant repercussions in a number of countries on the management of existing PGR and TK. A number of 'problems' have surfaced between developed and developing countries, mostly linked to the different levels of intellectual property protection in these countries. These different levels of protection have led to a number of instances of appropriation of PGRFA and/or knowledge without compensation. These unidirectional flows of plant resources and knowledge are today often referred to as 'biopiracy', a concept that does not have recognition in international law, but conveys the sense of frustration of a number of developing countries with reference to existing legal arrangements (Odek, 1994).

The cases of appropriation without compensation must be understood within the context of the evolving international legal framework in the field of PGR. First, the appropriation of biological resources is governed by the principle of state sovereignty. The case of PGR is slightly more complicated: for several decades, PGR were considered as a common heritage of humankind and were therefore freely shared and distributed to all actors seeking access. At the international level, the CGIAR and its Centres constituted the vehicles for the implementation of this principle, and in particular with a view to fostering food security at all levels within developing countries. Over time, there has been increasing recognition of countries' rights over their PGR, and the ITPGRFA now sanctions states' sovereign rights over the PGRFA (ITPGRFA, preamble, 2001; see Section 2.3.2).

Secondly, the TRIPS Agreement now provides the minimum standards of intellectual property right protection that all WTO member states must introduce. However, TRIPS is implemented in each WTO member country through domestic intellectual property laws, which may differ significantly between countries. Divergences may stem from the different use of flexibility clauses provided in TRIPS. They may also be linked to the fact that countries can decide to provide standards of protection that go beyond the minimum levels required in TRIPS. There is therefore ample scope for divergences in levels of protection between different WTO member countries.

54 Article 43 of the African Model Legislation, see note 52.

55 Author: Philippe Cullet.

Further, there remain a number of countries that are not members of the WTO and are therefore not bound by the TRIPS Agreement. The differences in levels of intellectual property rights protection and other differences between domestic intellectual property laws provide the background to some of the problem cases highlighted in this section.

Thirdly, as noted above, the international legal system has already reacted to the problem of uncompensated appropriation by progressively developing the concept of benefit-sharing. Benefit-sharing is clearly a response to the existence of different levels of intellectual property protection in different countries and the fact that PGR-rich countries lost control over the germplasm they contributed to the CGIAR system.

3.3.1 Case studies

The turmeric patent

In this case, Suman K. Das and Hari Har P. Cohly, two researchers based at the University of Mississippi Medical Center in Jackson, Mississippi, USA, applied for a US patent on the use of turmeric in wound healing (US Patent No. 5,401,504). More specifically, the application related to the use of turmeric to augment the healing process of chronic and acute wounds. The inventors claimed to 'have found that the use of turmeric at the site of an injury by topical application and/or oral intake of turmeric will promote healing of wounds'. This was based on experimental evidence that showed that turmeric causes endothelial cells to proliferate, indicating that this molecule can be used to augment wound healing. The patent application acknowledged that 'turmeric has long been used in India as a traditional medicine for the treatment of various sprains and inflammatory conditions'.

The specific claims of the inventors were:

1. A method of promoting healing of a wound in a patient, which consists essentially of administering a wound-healing agent consisting of an effective amount of turmeric powder to said patient.
2. The method according to claim 1, wherein said turmeric is orally administered to said patient.
3. The method according to claim 1, wherein said turmeric is topically administered to said patient.
4. The method according to claim 1, wherein said turmeric is both orally and topically administered to said patient.
5. The method according to claim 1, wherein said wound is a surgical wound.
6. The method according to claim 1, wherein said wound is a body ulcer.

Turmeric is one of the most basic ingredients found in Indian households and, besides its use in cooking, its antiseptic properties are widely known. The Indian Council of Scientific and Industrial Research (CSIR) therefore challenged the patent on the ground that the alleged invention was actually part of public domain knowledge in India. The patent was re-examined and all the claims cancelled.

The turmeric patent is noteworthy for two reasons. First, the turmeric patent highlighted one specific limitation of US patent law, in an era where inventions patented in the US can originate in any of the five continents. This relates to the interpretation of prior art in US law. 35 USC § 102 makes a distinction between anticipation in the US and in other countries. Thus, traditionally, prior art in foreign countries has only been recognized if it is described in a printed publication (*Hurlbut vs Schillinger*). In this case, printed materials were available but may not have been presented to the patents officer. Secondly, the turmeric patent provides important lessons in the context of the development of access regulations and TK documentation. This case provided a clear example of how difficult it can be to contest a patent abroad even for a large institution like the CSIR. As a result, there has been increased awareness of the need to document knowledge that is in the public domain and make it sufficiently available so that patent offices around the world check claimed inventions against existing sources of information.

The neem patents

The neem tree, which is a widely planted tree in India, has various uses in Indian households and in agriculture. In particular, farmers have long used leaves from the neem tree to make effective pesticides. In recent decades, the properties of the neem tree have been the object of substantial attention and large-scale research has been carried out to turn some of the neem's properties into commercially viable products (National Research Council, 1992).

Attention has focused specifically on the uses of neem as a biopesticide, because of the commercial potential in this area. The challenge has generally been for manufacturers to extract the active properties of the neem and to find a way to increase the shelf life of the product. Indeed, one of the characteristics of the natural formulation is that the preparation lasts only a few days, thus making commercialization of the leaf extract an impossibility.

A number of neem-related patents have been filed in the US and in Europe by Indian and foreign companies and inventors. Their common characteristic is that the patents generally claim novel processes for making a neem-based or neem-derived pesticide and the resulting product (e.g. US Patents 5,827,521 and 5,885,600). Even though a number of patents have been filed, one patent claiming a method for long-term storage of the active pesticidal ingredient (azadirachtin) became the centre of vigorous debates. In the early 1990s, the United States Patent and Trademark Office issued a patent to W.R. Grace, which covers a method of creating a stabilized azadirachtin in solution and the stabilized azadirachtin solution itself (US Patent 5,124,349). Subsequently, the US Environmental Protection Agency registered Grace's stabilized azadirachtin solution for use on food crops under the name of Neemix (Wolfgang, 1995). W.R. Grace also filed a patent for neem for its use as an antifungal product with the EPO (European Patent 0436 257). This patent claimed the invention of a novel insecticide and foliar fungicide derived from a neem seed extract and the processes used to obtain the neem oil. This pesticide was alleged to have the ability to repel insects from plant surfaces, prevent fungal growth, and kill insect and fungal pests at various life stages. This patent was challenged by Indian NGOs and the Indian Government. Eventually, in 2000, the Opposition Division of the EPO revoked the patents after it was shown conclusively that the claims did not fulfil the requirement for novelty in view of their prior public use in India.

A number of interesting lessons can be learnt from the neem patent cases. First, while one specific patent was revoked in Europe, this has not affected the standing of other similar patents, since each and every patent must be opposed separately. Secondly, the neem patents constitute much less direct cases of appropriation of knowledge than do the turmeric patent. This is due to the fact that it is impossible to commercialize the solution that has traditionally been used as a pesticide in India. Drawing a line between appropriation of TK and novelty in the context of existing patent laws becomes a very difficult exercise. Thirdly, some of the neem patents may have practical implications in India, since the patented solutions require neem tree seeds as their primary material. The need for vast quantities of neem seeds may constitute a positive commercial opportunity for some people. Fourthly, it is striking that a number of neem-related patents have been filed by Indian citizens or companies. This clearly shows that the question of the appropriation of TK through patents is not exclusively a North–South issue nor one determined by political boundaries. From the point of view of TK, it is in fact immaterial whether the application is filed by an Indian or a foreign company.

The kava case[56]

Kava (*Piper methysticum*) is an indigenous plant from islands in the South Pacific, where it is commonly used to prepare a traditional drink for both ceremonial and medicinal purposes.

The kava plant is a large shrub that can reach heights of up to 15 feet. It has green, heart-shaped leaves with stems that can be green, red-and-black striped or spotted. Kava is cultivated in the South Pacific, including the Federal States of Micronesia (Lebot and Lévesque, 1989).

People started to use a wild form (*P. wichmanii*) of the kava plant about 3000 years ago, which was later domesticated in the Pacific Islands of Vanuatu. Later on, the plant spread to other islands in the Pacific.

Traditionally kava has been used in two different ways, one of which is related to a ritual that uses kava as a relaxant. Several nations in the Pacific islands follow a ritual in which kava is used as a ceremonial and social drink (WIPO, 1998–99). Kava has also been used traditionally as a medicinal plant in the region (Lebot *et al.*, 1992). It has been used to treat stress, anxiety, insomnia, muscle and back pain, tension headaches, menstrual pain, asthma, the common cold, urinary infections, stomach problems and other maladies. Once planted, the cultivation of kava requires little labour or capital investment. The kava roots continue to grow perennially, and gardens and plantations are usually passed down through the generations.

The protection of kava has not been achieved through patents. Kava does not have the inventive character required by patents. Furthermore, it cannot be considered patentable if it is already well known to the public. Presently, no Pacific island country has plant patent laws, but even if such laws were to become available, no variety of kava could be patented (Clark, 1999).

Despite patents not being sufficient for the protection of the kava plant and related traditional knowledge, patents have been granted to some companies in order to explore kava. An example in the USA is the patent granted to Natrol, Inc., a US-based company that obtained a US patent for 'Kavatrol', a dietary supplement that serves as a general relaxant, composed of kava, chamomile, hops and schizandra. (Downes and Laird, 1999). In Europe, two German companies, William Schwabe and Krewel-Werke, have patented kava as a prescription drug for treating strokes, insomnia, Alzheimer's disease, and so on. In France, L'Oreal has patented the use of kava for hair loss and to stimulate hair growth.

The existence of patents based on kava raises concerns about the conservation and protection of traditional knowledge related to kava. First, the commercialization of kava-based products has had a negative impact on its conservation. In particular, the increasing exploitation of the plant has led to the harvesting of immature kava, thus jeopardizing the quality of the medicinal product and reducing its resource base (Puri, 2002). Secondly, patents granted which exploit medicine that has already been developed and used for generations by local communities constitute a case of appropriation of traditional knowledge.

In the kava case, patents cannot provide protection for traditional knowledge. In this regard, the need remains to examine other potential instruments of intellectual property rights as tools to protect traditional knowledge, such as geographical indications and trademarks.

3.3.2 Proposed solutions[57]

Biodiversity registers

The progressive appropriation of TK through patents has fostered concerns in source countries over their possible loss of control over plant genetic resources and related knowledge. One of the reactions to cases like the turmeric, neem and kava cases has been the development

[56] Author: Andrea Nascimento Müller.
[57] Author: Philippe Cullet.

of biodiversity registers. Such registers generally seek to document existing knowledge in order to prevent unwarranted patent claims from being accepted by patent offices around the world because they do not have access to TK databases. The registers also provide a tool for asserting benefit-sharing claims in situations where a patented invention is directly derived from TK (Rangachari and Subbarav, 1998).

Most biodiversity registers have been conceived as open documents that are meant to foster better access to existing TK. This may encourage the sharing of knowledge amongst different unrelated communities, which may be beneficial from the point of view of the sustainable management of local resources. In cases where knowledge is directly attributable to an individual or community, the registers provide a tool to establish claims of original ownership. The registers also indirectly constitute an attractive source of information for researchers seeking to build on TK. They therefore provide an incentive for commercial research into TK and at the same time a means to ascertain public domain knowledge and a tool for holders of TK wanting to claim benefit-sharing (Gadgil, 1997).

The approach that seeks to make biodiversity registers open to all and to foster outsiders' use of the registered knowledge is not always accepted. In some situations, individuals or communities may either decide not to register their knowledge in a written form, or may register it but not give access to outsiders. The village of Pattuvam in the south-west state of Kerala in India took the latter approach when it decided to document all its biological resources and TK but decided not to make the results available to outsiders. In this case, the promoters of the register wanted to provide documentation of public domain knowledge but did not consider that this knowledge should be freely offered to the outside world for further use. Interestingly, the register is closed so as to avoid appropriation by outsiders, but the possibility of sharing knowledge with other local communities is recognized.

Benefit-sharing

The notion of benefit-sharing has largely been developed as a consequence of the development of genetic engineering and its increasingly frequent use of TK for the development of other products. In fact, there is increasing recognition that the commercial use of TK in a direct or derived form should be compensated in one form or another. In practice, benefit-sharing can give rise to a number of important questions, as illustrated in the case of Jeevani medicine.

Jeevani was developed after the Kani people of southern Kerala were persuaded by biologists from the Tropical Botanic Garden and Research Institute (TBGRI) to share some of their knowledge concerning a plant called *Aarogyappacha* (Anuradha, 1998). The biologists were intrigued by the strong anti-fatigue properties of the plant's leaves. Though widely used by local people, the plant itself seems to have been unknown to outsiders until 1987. The TBGRI carried out research on the plant and, after identifying the active ingredients of the plant, developed a drug with anti-fatigue properties. The rights to manufacture Jeevani were transferred to a private manufacturer for a licence fee of about US$21,000 (at today's exchange rate) for 7 years and a 2% royalty on sales for 10 years (TBGRI – Arya Vaidya, 1995). TBGRI decided to give 50% of the fee and royalty to the Kanis (Ministry of Environment and Forests, 1998). It is significant that while patent applications were made by TBGRI, no patent was granted in India, because the Indian Patents Act did not allow such patents at the time.

This benefit-sharing arrangement is very progressive from a financial point of view. However, it has not been immune to criticism. First, while the section of the Kani tribe that had had significant interactions with the outsiders – including the specific individuals who passed on the information to the scientists and were rewarded with special financial rewards – were generally happy with the benefit-sharing arrangement, other segments opposed it. This is partly due to the fact that some factions of the tribe felt that they were not involved in the negotiating process and were handed down a decision without proper consultation. This raises an issue concerning the prior informed consent of the providers of knowledge. Secondly, benefit-sharing in the form of money may not be the most desirable form of contribution to the local economy. This aspect is highlighted in this case, because the Jeevani medicine can only be produced from leaves of the tree if it is grown in the area where the Kanis live, and not elsewhere. The Kanis could therefore have been given a stake in the production of the raw material for the medicine, a much more stable and substantial form of benefit-sharing (Gupta, 2001a).

The Jeevani case study highlights issues of broader relevance for benefit-sharing. First, benefit-sharing is not limited to transfers of knowledge between different countries. Secondly, the trigger for benefit-sharing is not necessarily the appropriation of knowledge through intellectual property rights, but is generally the commercialization of TK products or products derived from TK. Thirdly, monetary compensation, however attractive it may be as a form of compensation, may neither be the most effective nor the most appropriate way to reward the contribution of specific knowledge to the commercialization of a new product.

Local innovation

The increasing appropriation of knowledge – whether based on TK or not – through intellectual property rights has often led to attempts to preserve existing rights and to avoid or regulate transfers. In some cases, however, there have been initiatives aimed at using the existing intellectual property right system for the benefit of TK holders. The basic idea behind this is to provide a form of recognition for inventions and creativity that may not qualify for patent protection, but which constitute advances in the specific field of activity.

One such experiment is that of the Honey Bee Network in the state of Gujarat, India (Gupta, 2001b). Honey Bee seeks to foster knowledge dissemination among local communities and in the wider world. The specific approach behind this experiment is that rather than putting shared knowledge into the public domain, the information is clearly titled to the individual holder or inventor. In other words, Honey Bee seeks to provide incentives and benefits to innovators to foster information dissemination. Institutionally, this has been achieved through the Society for Research and Initiatives for Sustainable Technologies and Institutions (SRISTI; for general information on SRISTI, see http://www.sristi.org).

Honey Bee's work comprises a variety of different components. An important part of its work is the documentation of innovations, either of contemporary origin or based on traditional knowledge. The databases now include about 10,000 innovations. Since 1997, Honey Bee databases have also formed the basis for the commercial use of documented innovations. The Gujarat Grassroots Innovation Augmentation Network (GIAN) has been set up specifically to provide an interface between innovations, investment and entrepreneurship. The GIAN has helped to file patents on behalf of grassroots innovators and helped with the transformation of innovations into products that can be commercialized.

More recently, the Honey Bee model has been expanded and a National Innovation Foundation (NIF) has been set up by the Department of Science and Technology of the Government of India (for general information on NIF, see http://www.nifindia.org). The main goal of the NIF is to provide institutional support in the scouting, spawning, sustaining and scaling up of grassroots innovations and to assist in their transition to self-supporting activities. Furthermore, the NIF seeks to strengthen R&D linkages between excellence in formal and informal knowledge systems in a bid to help India become a global leader in sustainable technologies.

3.3.3 Problem analysis

The progressive interest in the exploitation of PGR and related changes in intellectual property laws have important practical and policy implications in a number of developing countries. The development of commercial products based on TK can have beneficial impacts for different actors in different countries depending on the specific situation and the legal framework in place. The Jeevani case illustrates a situation where TK is appropriated by a governmental body and then sold on to a private company of the same country. In this case, the commercialization of the medicine has the potential to benefit the national economy without any international implications. Concerns related to the control of TK and benefit-sharing are, however, similar, whether the situation is one where all the transactions happen within a given country or involves an international element. In both cases, there is a need for a legal and policy framework to regulate the appropriation and use of TK outside its area of origin.

The use of TK in commercial applications by outsiders raises even more complicated issues when intellectual property protection is sought for products derived from TK. In this situation, a number of specific problems arise. Even in the TRIPS era, different countries have different levels of intellectual property protection and, for the foreseeable future, it is likely that the level of patent protection in recipient countries such as the USA and Western European countries will remain much higher than in most developing countries. This implies that it will remain 'easier' to apply for some types of patents in the USA than in India, even when the latter introduces product patents on biotechnological inventions in all fields of technology. One of the consequences of this asymmetry is that an opposition to such a patent can be filed only from outside the country of origin, as in the case of the turmeric patent outlined above. Further, in the case of the US patent system, there is the added difficulty concerning the proof of prior Article. This difficulty will slowly reduce over time as the number of registers of TK increases around the world and as an increasing number of registers are uploaded onto the Internet.

The patentability of products derived from TK raises complicated policy issues at the international level. One of the problems is that appropriation does not follow a single path. In fact, as is illustrated in the case of the neem patents, while a majority of patents in Europe and the USA on TK-derived products may be owned by companies from the North, applications from the countries of origin are also prominent. In other words, the problem of 'biopiracy' is not limited to a North–South issue, but is also replicated at the national level or in relations between different developing countries. The international community has started to address two specific problems linked to the appropriation of TK through intellectual property rights. The question of prior informed consent (PIC) comes first in line. Article 15.5 CBD and the guidelines adopted at COP 6 clearly recognize the need for PIC, but this has not yet been implemented, either at the international level or within the national legislation of donor countries. The second issue is that of benefit-sharing, which is addressed in the CBD guidelines and in the ITPGRFA; however, both at the international and national levels, the scope and impact of these provisions is yet to be established in practice.

One of the most common reactions to TK appropriation in developing countries has been the fostering of the development of biodiversity registers comprising records of PGR, PGRFA and knowledge related to these resources. This remains on the whole a reaction to international trends, which seeks to prove the existence of knowledge already in existence, but does not provide an avenue for asserting claims of ownership over the resources or knowledge. Some initiatives, such as the one taken by the Honey Bee Nework, indicate that there can be other ways for developing countries to react to the expanding scope of intellectual property rights at the international level.

3.3.4 Impacts on PGR diversity

The direct impacts of intellectual property rights on PGR diversity are difficult to estimate since in most cases there is no direct impact. At the outset, the distinction between the impacts on PGR diversity and the impacts on TK must be clearly delineated. The latter relates to the impacts of intellectual property rights on the use of PGR and ownership of TK, while the former focuses on PGR conservation and management.

Intellectual property rights do not have direct impacts on PGR conservation and management because they do not directly deal with the issue of biodiversity conservation and do not address the question of the ownership of biological resources themselves. However, they are relevant in PGR management because the introduction of patents in agricultural biotechnology has, for instance, important impacts for agricultural management, and hence agro-biodiversity conservation.

The experience of developed countries that introduced plant breeders' rights and/or plant patents earlier does not provide many useful analogies, insofar as their socio-economic conditions were vastly different from the conditions under which a vast majority of developing countries operate today. However, it is possible to get a general idea of some of the possible impacts of the introduction of intellectual property rights in agricultural biotechnology by looking at the experience of the Green Revolution in Asian countries.

In effect, the practical impacts of hybrid varieties of the past and the genetically modified seeds of the present are largely similar if one excludes the specific biosafety concerns linked to the latter. Both seek to provide yield increases and both often require agricultural management techniques that largely differ from traditional practices by necessitating a number of external inputs, such as irrigation or chemical fertilizers.

A general assessment of the Green Revolution indicates that areas where high-yielding varieties were introduced witnessed significant yield increases (in the case of India see, for example, Sharma and Poleman, 1994). However, despite these gains, the Green Revolution package has come under increasing criticism since the beginning of the 1990s. First, over the long term, the Green Revolution has come to be associated with significant environmental costs. These include falling water tables due to the overuse of tubewells (Bavadam, 2000), waterlogged and saline soils from many large irrigation schemes, declining soil fertility with excessive chemical fertilizer use and water pollution with pesticides (Dhaliwal and Dilawari, 1991; Agarwal, 1995). The Green Revolution has also been associated with the spread of monoculture, which leads to a homogenization of species, greater vulnerability to insect pests and diseases, and to a loss of agro-biodiversity (Thrupp, 2000). Secondly, the sustainability of yield increases has been questioned in view of evidence of diminishing returns in intensive production with new varieties (Conway and Barbier, 1990). Thirdly, the application of the new technique necessitates important investments in seeds, fertilizers, pesticides and irrigation, which are beyond the reach of all but the largest farmers (Joshi, 1992). Indeed, new varieties perform well only when all the necessary inputs are available in sufficient quantities (Conway and Barbier, 1990). Thus, irrigation is often necessary, given that crops may fail if water is not provided in sufficient quantity at the opportune time. Uniformly produced seeds may also not be as well adapted to local conditions as farm-produced seeds. Furthermore, new seeds tend to be much more expensive than farm-saved seeds (Kahama, 1995).

The long-term implications of the Green Revolution for PGR diversity have only been moderately positive, since uniformity or monoculture generally leads to a loss of agricultural biodiversity. It is likely that this situation will be repeated with genetically modified varieties, with the added concerns about unwanted dissemination and contamination of biodiversity by the latter varieties. In both cases, a broader assessment of the impacts of the introduction of new varieties must take into account not only the increased yields, but also the sustainability of the increases and the negative impacts on diversity. The main distinction is that in the case of the GMOs, the legal incentive for their development is provided in large part by IPRs.

3.3.5 Impacts on TK

Intellectual property rights undeniably have important impacts on TK. First, they foster a more commercial approach to the question of TK use. In other words, they foster a new outlook on TK that mainly focuses on the uses of TK that have commercial potential. Secondly, intellectual property rights provide an incentive for registering TK. This is becoming increasingly necessary in view of broader economic changes at the national and international levels which lead to the progressive erosion of TK. Thirdly, the direct and indirect appropriation of TK has increasingly led to the realization that procedures for access are imperative, given the important commercial stakes. As a result, there has been increasing pressure at the international and national levels for the development of a legal regime to regulate access to TK and to allocate the benefits accruing from products derived from TK.

The progressive appropriation of TK through intellectual property rights also has an impact on the ownership of TK. In fact, this is probably the most crucial aspect from a legal point of view. The main impact of the introduction of IPR over TK-related inventions is that TK itself cannot be protected through IPR and is therefore in the public domain. Further, TK generally does not qualify for patent protection because TK itself is not deemed to be 'state-of-the-art'. In general, intellectual property rights over TK-related inventions foster a shift in control from TK holders towards intellectual property rights holders. In fact, it is this shift in property rights in favour of new holders that has led individuals, organizations and governments to challenge patents, such as the turmeric patent (Anuradha, 2001). It is noteworthy that the practical consequences of the turmeric or neem patents for original TK holders are likely to be insignificant, at least in some cases. Thus, in the case of the turmeric patent, the existence of a patent in the USA would not have had any practical consequences for everyday users of turmeric as a healing agent throughout India, where it was not recognized.

3.3.6 Tension between IPR and competition

This section shall introduce the interrelationship between IPR and competition laws and policies. This interaction is relevant when it comes to assessing the scope of protection of IPR in the light of public policies aimed at promoting economic efficiency and the diversity of supply of protected goods and services. Intellectual property forms of protection grant exclusive rights that come close to monopoly rights. This can trigger tensions with competition laws and policies. The interactions between competition and IPR become even more complex on the international level. As illustrated in the case studies above, legal issues concerning TK related to PGR present in general strong cross-border characteristics. It is therefore necessary to deal with these issues not only at the national but also at the regional and global levels. There are a variety of different legal approaches towards competition, ranging from an absence of specific legislation via lax law to very strict and incisive rules with effective sanction mechanisms. In comparison to competition laws and policies, the international intellectual property system appears to be much more harmonized, in particular through the set of TRIPS rules that ensures minimum standards of protection. This higher degree of harmonization of intellectual property rights does not only concern procedural and substantive rules, but also the institutional design that serves to implement them such as public registries, specialized courts and collecting societies. In contrast, procedural and substantive competition rules are mainly national or, as in the case of the European Community, regional.[58] The same applies for the institutional framework.[59] There are initiatives to promote multilateral cooperation in the field of competition in the OECD, the UNCTAD (United Nations Commission for Trade and Development), and more recently in the World Trade Organization (WTO).[60] A strong link between international competition and international trade laws and policies is obvious when anti-competitive private business practices have the effect of replacing traditional tariff and other non-tariff barriers to foreign markets after they have been removed by international trade rules (Kennedy, 2001). Possible approaches in the WTO consist of elaborating harmonized minimum competition policy standards (e.g. the TRIPS approach) or an agreement on core competition principles.[61] However, for the time being, there is no consensus on the question of whether the WTO should deal with this matter, and, if so, in what manner.

As described above, in the long term the Green Revolution has been detrimental to PGR diversity, as it has, in many instances, caused uniformity and monoculture and, thus, a loss of agricultural biodiversity. This development was partly caused by the absence of appropriate competition laws and policies. To address this risk, the impact on the competitive environment must be carefully taken into consideration when new forms of IPR are introduced. Lawmakers are therefore well advised to adapt also the competition laws when using intellectual property to achieve policy goals. Patent laws, for example, trigger contractual practices among the concerned economic players. These contractual arrangements may be reasonable and legitimate for private business purposes, but can be problematic for the achievement of policy goals in the public interest. Concretely, there are contractual practices that are necessary to cope with constraints induced by IPR that can qualify as anti-competitive behaviours. In the case of sequential inventions, for example, patent licence agreements help to overcome certain deadlock situations by structuring the relationship between primary and secondary innovators. The following section will discuss the questions of contractual arrangements that are based on IPR and their impacts on competition.[62] The last section will illustrate in further detail the issues at stake in the light of the example of the seed industry, where the introduction of IPR to promote private investments eventually contributed to building up a highly concentrated industrial sector dominated by a few big players. This example outlines how the instrument of intellectual property forms of protection may fail to reach public policy goals in the absence of appropriate competition laws and policies.

Contract arrangement among sequential innovators[63]

Overly broad patent protection can stunt or slow the pace of future innovation. The scope of patent protection refers to both breadth and height. The *breadth* of a patent defines the range of products that are encompassed by the patent claims, which protect the patent holder against potential imitators or closely similar inventions. The *height* of the patent confers protection against improvements or applications that are easy or trivial. The broader the scope of

[58] For a summary on different national and regional approaches in the area of competition law and policy, see Kennedy (2001, p. 22 ff.).

[59] With respect to the institutional design of US competition institutions and the complexities of their interactions, see Peters (1999, pp. 40–67).

[60] See Kennedy (2001) at p. 97 (overview of OECD guidelines); p. 110 (Draft International Antitrust Code); p. 118 (UNCTAD principles on restrictive business practices). This author outlines the arguments in favour of and against the integration of competition rules into WTO law, and comes to the conclusion that the WTO is not the appropriate forum to deal with this matter, in view of the very different legal approaches that exist between the countries in this field, for the time being.

[61] For further references, see the WTO website: http://www.wto.org/english/tratop_e/comp_e/comp_e.htm

[62] For an overview on policies that may be pursued by antitrust legislation, see First (2002, pp. 175–194).

[63] Author: Gloria Pasadilla.

the patent protection, the higher the degree of monopoly power; but conversely, the narrower the patent, the lower the incentive for the innovator. The question is, what is the optimal level of scope for society? The issue is how to balance the incentives, particularly in the case of *sequential innovations*, between the primary innovator and secondary inventor (those who introduce improvements that may have better industrial applications). The same issue applies to related inventions that somehow infringe upon an initial discovery. Clearly, both deserve some rewards because, for instance, without the initial invention, without its disclosure and knowledge spillover, the secondary innovation would not come about. Conversely, there are primary innovations that initially do not have significant applications, and whose value actually derives from the values of the secondary innovations built upon them. The problem concerns how to design an optimal contract that provides incentives for both.

This is an important issue because all technological advancement is the result of a combination of both primary and subsequent improvements of initial discoveries: hence all innovations require some form of encouragement and incentive at different stages. Having too broad or too narrow a protection can therefore stunt the growth of knowledge, either at subsequent stages or at the very font of discoveries itself. Consider a very broad protection of primary inventions. Barton (1997a) takes the example of a biological receptor, a research tool that is important for the study of schizophrenia, but may not be marketable in itself. If given a broad patent, the inventor of the receptor will have a monopoly over the entire research area, even without necessarily defining any marketable product that is of benefit to the public. From an economic efficiency standpoint, society does not lose out if the initial innovator also has a comparative advantage in the development of all related applications. However, if other innovators exist who have a better capacity to bring about the perfection of its application, then the broad patent that excludes better research companies implies an efficiency loss to society.

Obtaining a licence from the original innovator is one option for secondary innovators. The question arises as to when is the best time to negotiate. If *ex-post*, that is, after the secondary innovator has already invested in research, then his bargaining position is weak, because all his investments are already sunk (i.e. non-recoverable), while the original innovator has the option of refusing to grant any licence. This 'hold-up' problem, where all of the bargaining is on the side of the original innovator, can be solved through the imposition of compulsory licensing, perhaps invoking the 'essential facility doctrine'.[64]

If negotiations instead occur *ex-ante*, that is, before the secondary innovator has incurred sunk investment costs, then there is greater scope for a more balanced outcome. The secondary innovator's bargaining power is improved, making him less susceptible to a hold-up problem. But the nature of these contracts remains subject to possible competition challenges, as will be discussed below.

The joint venture agreement helps in the pooling and sharing of risks. Through joint ventures, researchers are allowed to proceed, and (if successful), rents are shared and divided up accordingly amongst the inventors. However, there is also the danger that joint ventures can function much like a cartel, deterring the entry of third-party innovators.

A 'dependency licence' for follow-on inventors is another possible arrangement (Barton, 1997b). This is akin to the situation pertaining to cross-licensing and grant-backs, whereby the follow-on inventor patent requires the authorization of the holder of the prior patent, and the original holder may not apply the same improvement without authorization by the holder of the follow-on inventor patent. This dependency licence essentially gives the follow-on innovator the right to obtain licences from the holder of the initial patent. It might require that a royalty is

[64] For an overview on the 'essential facility' doctrine under US and EC law, see Meinhardt (1996, pp. 137–160).

paid, but at least it significantly improves the secondary inventor's bargaining power and incentive. The royalties can either be determined in court or through negotiation.

Some basic principles for IPR and competition policy

Considering the number of possible contractual arrangements related to IPRs, not only between primary and secondary innovators, it is worthwhile gaining an overview of competition or antitrust aspects of these arrangements.[65] IPR can be viewed as conferring monopoly rights, whereas competition policy is concerned with the promotion of competition in the market; or it can be seen that competition policy is about short-run allocative efficiency and IPR about long-run dynamic efficiency. This fact appears to create an inherent conflict between governments' IPR and competition policies. It is true that competition policies may at times impose limits on market power of IPRs, but this conflict seems to be more apparent than real. In fact, the two instruments reinforce each other, because innovation is a spur to competition, and competition also acts as a spur to innovation.

Several market arrangements, including a variety of licensing techniques, help defray the short-run misallocations, e.g. from less than optimal use of the innovation, derived from exclusive IPR rights. They play an important role in further dissemination and utilization of the innovation. However, these arrangements can also be made into a front for anti-market activities, i.e. cartel activities. Such activities are examined briefly in the following section.

LICENSING

The granting of a licence allows the utilization of IPR by a non-patentee and promotes further innovation. A patentee may refuse to license, but if the IPR satisfies the 'essential facility doctrine', or if competition is severely threatened (for instance as is the case with mergers or abuse of dominant position), the patentee can be forced to grant a compulsory licence. In most cases the terms of the licence agreement depend upon the relative bargaining positions of the licensor and licensee. In general, licensing poses a greater competition risk if licensors and licensees are actual or potential competitors, or are in a horizontal relationship. Vertical relationships, on the other hand, or when a patent is an input to the production of another product, normally receive a more tolerant treatment by the competition authorities. However, some related agreements tied to licensing agreements have often fallen under competition scrutiny. Examples include licensee price restrictions, exclusive territories, exclusive dealing, grant-backs, reach-through royalties and tying arrangements.[66]

In most jurisdictions, the treatment of many licensing agreements has changed from being *per se* unlawful to one that merits a *rule of reason* approach. For instance, *exclusive territories* – where the licensor grants the licensee a monopoly over a particular territory by agreeing not to sell in the same territory, nor license anyone else who would operate in the same geographical market – may at times be anti-competitive, because this carves out markets among competitors. Yet often it may not be necessary for there to be any production of the licensed product at all. For example, in the *Maize Seed* case, the European court ruled that such an agreement between the INRA research institute in France and the Nungesser company in Germany was indispensable in order for the investment by the German firm to introduce the seed variety in Germany to be economically viable.[67]

[65] See OECD (1998) and Anderson and Gallini (1998).

[66] See OECD (1989) for more detailed discussion on the different IPR and competition policy interaction.

[67] The contract was between Kurt Eisele, a German citizen, and INRA, but he transferred his rights under the agreement to his firm, Nungesser. Eisele agreed to register the varieties and arrange for their marketing in Germany. To encourage his investment, INRA promised Eisele that it would endeavour to prevent exports to West Germany from France (Korah, 1983).

Otherwise, their investment into the introduction and development of the seed variety would not have been worthwhile, if others could free ride on those initial costs by competing in the same market.

Exclusive dealing provides another example. This licensing agreement either prevents the innovator from transferring the innovation to the licensee's competitors, or prevents the licensee from purchasing its supplies from the licensor's competitors. The first can create a problem if the licence is transferred exclusively to a horizontal competitor in a concentrated market, particularly if this eliminates competition. Exclusive transfer to a firm that is vertically related to the innovator,[68] however, may create less of a competition problem. The second scenario has the effect of denying rivals sufficient outlets for exploiting their technologies, and thus has an adverse effect on competition. However, at the same time, this type of exclusive dealing may be the only way for the licensor to have control over the licensees, especially where it is difficult to determine how much a licensee uses up the technology *vis-à-vis* rival technologies. Noting that competition problems can arise on a case-by-case basis, exclusive dealing, like exclusive territories, is analysed under a rule of reason approach, rather than being considered to be a violation *per se*.

In an ordinary competition policy sense, *price restriction* tends to be viewed *per se* as being unlawful. Yet, in the context of IPR, licensee price restrictions may provide the incentive necessary for the innovator to license and thus permit the diffusion of the technology. The idea is that if the profit made through licensing would be less than the innovator's profit if he were to produce the product alone, then he would have no incentive to license. Hence, price restrictions may be considered to be one of the allowable restrictions that reasonably give the patentee the reward he is entitled to secure. But this again is scrutinized on a case-by-case basis, because not all licensing with price restriction is better than no licensing at all. In some cases, the negative allocative effects of price restriction may impose a greater economic cost than the positive benefits of diffusion.[69]

Grant-backs provide a method by which licensors seek to protect themselves against the possibility that licensing will foster the emergence and growth of future competitors. They allow the patentee some rights over improvements made by others upon his innovation. The economic justification behind allowing *bounded grant backs*, i.e. where the original licensor is not given an *exclusive* licence for any improvements to his innovation, is to give the original innovator an incentive to license. Bounded grant backs strike a middle course that ensures that the original licensor is not displaced from the market, while still leaving licensees with a significant motivation to innovate (OECD, 1998).

Reach-through royalties are royalties based on total sales. This is administratively simple, but is effectively similar to exclusive dealing arrangements in that it acts as a disincentive for using competing technology. Just like exclusive dealing, it is approached on a rule of reason basis. *Tying*, or linking the sale of patented products to the purchase of other goods (including goods whose patent protection may have lapsed), also has both pro- and anti-competitive effects. Tying has pro-competitive effects if it is necessary to secure the overall quality of the product, in the same way that car maintenance services tied to car sales can ensure a certain level of quality promised by the manufacturer. Hence, normal competition law applies a rule of reason standard in analysing such cases, but this may be prohibited if tying unnecessarily raises barriers to entry (similar to exclusive dealing agreements).

---

[68] A relationship is vertical when the IPR are complementary inputs (even if the licensor and licensee are otherwise competitors in manufacturing products covered by the IPR).

[69] See Gallini and Trebilcock (1998, p. 43).

MERGERS, ACQUISITIONS, PATENT POOL

Analysis of *mergers* of firms with significant IPR is carried out using standard merger analysis, i.e. the competition authorities consider dominant market positions and the potential impact on competition in the product market. But, in addition, they use the *innovation market approach*, which means that if mergers threaten to significantly reduce competition in the R&D market itself, then the mergers are in danger of being disapproved. This is the case, for instance, if the merged entities have less incentive to proceed quickly with innovation and R&D than is the case if they were separate entities. Some solutions can include compulsory licensing of one or the other of the IPRs, or the forcible spin-off of one R&D group in order to maintain competition in innovation.

Patent *acquisitions* are sometimes used to avoid costly licensing transactions. But if used to accumulate a '*killer patent portfolio*' to preclude any innovation around the main product, then there is scope for competition authorities to intervene. This is particularly true when firms monopolize a technology by not only obtaining patents on products and process which they intend to use and sell, but also patents which they intend to leave idle, thereby amassing a portfolio that it is difficult for competitors to innovate around. The usual solution for this type of acquisition is compulsory licensing.

Patent *pooling* is another useful arrangement, particularly if it puts into the same pool blocking patents. It helps in the efficient utilization of IPR and avoids costly infringement litigations. However, it can also function as a cartel-like arrangement for existing IPR holders, making it difficult for new entrants to innovate around any one existing patent in the pool. Per-use royalty of an IPR package from a patent pool can also increase the marginal cost of production. Patent pools can also act as a disincentive to future research, especially if any improvements to the existing IPR package are automatically shared between all of the patent pool members, thus giving rise to a free-riding problem. The useful rule of thumb to check whether patent pooling is efficiency enhancing is whether cross-licensing implicit in patent-pooling is necessary to compete at maximum efficiency (i.e. one firm cannot use its own technology unless it has a licence to use another IPR in combination with his own). Otherwise, the negative effects of patent pool may outweigh its positive effects.

IPR AND COMPETITION: THE CASE OF THE SEED INDUSTRY

This case study shall illustrate how intellectual property protection plays a role in affecting the market structure of an industry and derives competition implications from the changed competitive structure.

The seed industry market structure. There are about 1500 seed companies (Rabobank International, 2001); however, power is concentrated in only a few, with the top ten seed firms accounting for more than 30% of the roughly US$30 billion dollar commercial seed market. These seed companies specialize in the breeding and production of hybrid and improved crop seeds. Traditionally they have mostly been 'standalone' or independent firms, but with the advent of biotechnology, seed sales have become a crucial direct link for biotech firms, as they embody the input of genetic material into the agricultural production process. This is a fundamental reason behind biotech firms' vertical integration with the seed industry, as discussed below.

Prior to the merger frenzy in the mid-1990s, there was a wave of acquisitions approximately a decade earlier. The 1978–1980 period of mergers coincided with the strengthening of amendments to the US Plant Variety Protection Act. At that time, a number of observers identified a direct causal relationship between the strengthening of intellectual property rights and merger activity, as the IPR triggered expectations of increased earnings in the seed sector. However, whereas many of the acquiring firms in the 1980s' merger round were new entrants to the sector, the 1990s' round involved existing participants and high-profile multinational firms (Lesser, 1998).

Table 3.1. Key global players and their positioning in the seed market (Rabobank International, 2001).

Big league	Minor league	Niche players
DuPont (Pioneer)	Limagrain	Cebeco, Pau Euralis
Pharmacia (Monsanto)	Grupo Pulsar	Ball, Pennington
Novartis (Syngenta)	Sakata	DLF, Svalof Weibul
	Advanta (AstraZeneca)	Saaten Union, Sigma
	KWS	Ragt, DSV, Maisadour
	Delta & Pine Land	Barenbrug
	Dow Agro	
	Aventis	

This wave of consolidation has been thoroughly discussed elsewhere,[70] but what we shall provide here is a summary of the results of this series of acquisitions. It should be noted, however, that some of those acquisitions have been spun off a few years afterwards for a variety of reasons: (i) anticipated synergies might have failed to materialize; (ii) concerns over consumer acceptance of genetically modified organisms and thus, the underperformance of the biotech firms relative to pharmaceuticals, may have led to an increase in pressure from shareholders; and (iii) antitrust scrutiny of mergers might have had a deterrent effect.

Some of the basic features of the 1990s' merger round can, however, be highlighted. First, several large chemical and pharmaceutical firms moved into plant biotechnology, making huge investments in life sciences, and acquiring all of the large national seed firms (e.g. Pioneer, DeKalb, Agracetus, Mycogen, etc.). Some chemical and pharmaceutical firms merged horizontally (e.g. Rhône-Poulenc and Hoechst to form Aventis), then integrated vertically to seed breeding and marketing. The impact of this upon the seed industry is that the large set of small start-up firms which appeared in the 1980s had, by the end of the 1990s, either folded or been acquired by the new agronomic system's giants (Graff *et al.*, 2001).

Thus, in contrast to the diffuse structure in existence during the 1980s, the emergent industry structure now consists of a relatively small number of tightly woven alliances among pharmaceutical firms, biotech research firms and the seed industry. The life sciences industry has solidified to five–seven major firms that are highly vertically integrated and organized around a major life science firm (Table 3.1). These five major gene giants that dominate the life science industry are: Du Pont, Pharmacia (Monsanto), Syngenta, Aventis, and Dow. Together, they account for 60% of global pesticide market, 23% of commercial seed market and virtually 100% of the transgenic seed market (RAFI, September 1999).

As the seed industry became more concentrated, the share of biotechnology patents likewise consolidated on the few major companies. As a result of the wave of buyouts, the purchased firms' intellectual property rights came to be held by its 'mother firm'. Graff *et al.* (2001) found that the top seven seed firms own more than 80% of the total patents in agricultural biotechnology, whilst the three major ones held 55% of the total patents. DuPont and Pharmacia own a majority of all major types of patents: 38% of transformation technology patents, 31% of gene patents and 81% of germplasm patents, the latter merely reflecting the aggressive buyout strategies of these two firms in the seed industry. This

[70] See for instance, Barton (1998), Hayenga and Kalaitzandonakes (1999), Fulton and Giannakas (2001).

pattern raises concerns regarding potential entry difficulties for new firms in the agricultural biotechnology industry, as anyone trying to get in runs the risk of being blocked or considered to be infringing upon any of the biotech patents held by the major firms.

Thus, in both the product and innovation markets, the major firms have cornered a majority share, raising concerns of possible anti-competitive behaviour in the seed market and potential slowdown in the rate of agricultural biotechnology innovations.

Reasons for industry restructuring and the role of IPR. Noting that industry consolidation has led not only to concentration in the product market share (i.e. seeds) but also to concentration in the patents and specialized assets used for research and development in biotech, what was the motivation of giant firms in moving into the seed business? What role did intellectual property rights play in the seed industry transformation?

There are several competing reasons that may explain the restructuring of the industry. Some are unrelated (or only marginally related) to IPR, while others are centred on the intellectual property issue.

Non-IPR reasons. Strong demand complementarily between chemical and biotechnology products is one reason that might have motivated the amalgamation of seed and chemical companies. Consider a single firm producing both insecticides and pesticides and transgenic crops. Such a firm will be more profitable because it can price its products so that the use of the complementary product is encouraged. For instance, Monsanto tried a product-tying strategy when selling Roundup™ (a dominant herbicide with glyphosate as an active ingredient), with Roundup Ready™ crops which are glyphosate tolerant, to maintain their considerable market power in the glyphosate market (Hennessy and Hayes, 2000).

Innovation life cycle is another possible explanation for the consolidation exhibited by the agricultural biotechnology industry (Kalaitzandonakes and Hayenga, 2000). The idea is that it is typical that at the early phase of innovation – the fluid phase – that new entrants gain access, the total number of firms increases, and all of them engage in innovation and experimentation with product designs and operational characteristics. Over time, a specific product becomes the standard, and product innovation subsides, while process innovation may continue at lower cost. Finally, the rate of both product and process innovation dwindles. At each stage of the innovation life cycle, there is a corresponding change in the market structure. The number of firms peaks during the fluid phase and then eventually drops off to a few central players as the dominant design becomes established. The remaining firms emulate the features of the dominant product concept and compete on efficiency.

When applied to the agricultural biotechnology industry, Kalaitzandonakes and Hayenga (2000) note that the number of firms peaked in the early 1980s, as they competed in product innovation and various product forms, including transgenic plants and genetically engineered microorganisms. The dominant design emerged in the early 1990s – transgenic plants with a pesticide action – and consolidation began shortly thereafter.

Yet while the innovation life cycle appears to explain horizontal integration among firms engaged in biotechnology R&D, it does not sufficiently explain the vertical integration of pharmaceutical/chemical firms with seed companies.

IPR-related reasons. While non-IPR-related reasons may provide a partial explanation for the restructuring of the agricultural biotechnology industry, they raise sufficient questions to prompt a search for answers elsewhere. For instance, even as the innovation life cycle can explain horizontal mergers of R&D firms, it falls short of explaining the vertical integration in the life science industry. Thus, others have offered other explanations for the emergent market structure, which are directly related to intellectual property rights.

First, since IPR create monopoly power to its owner, a firm may want to erect barriers to entry for potential competitors. This can be done by leveraging control over key intellectual properties to block potential imitation or minor innovation improvements (Lesser, 1998). By accumulating such blocking patents, the patent owner maintains its monopoly rents within a specific market and for a specific period of time. Thus, industry concentration can be motivated by the desire to control IPR, which results in the maintenance of a firm's monopoly power, and therefore provides an explanation for the industry consolidation.

Another reason why firms may want to accumulate patents by buying companies with IPR is to be able to use them as bargaining chips in negotiations with other firms. That is, knowing the high propensity of patent infringement in the biotechnology industry, having a number of patents give firms the necessary leverage or threat to sue back if they, in turn, are sued for infringement. Patent ownership then protects firms from rival patents or enables them to negotiate for the utilization of certain key technologies on an equitable basis (Joly and de Looze, 1996). Thus, what happens in a concentrated market structure where a few firms own most of the patents is an implicit cross-licensing among the firms (Barton, 1998). Without a sufficient number of potentially infringeable patents, a firm is more vulnerable to being sued for infringement by other companies.

A third IPR-related explanation for industry consolidation is the economies of scope in research or the desire to exploit complementarities in the use of specialized assets in biotech R&D. Graff *et al.* (2001) argue that the mere desire to accumulate patents to block entry would have led to an increase in the sheer number of owned patents, rather than in an increased diversity of patents. Since the increased industry concentration shows that major firms have accumulated not only a greater number of patents but also a more diverse one, an explanation can be found in the mutual complementarities of these assets. For example, the isolation of a gene leading to a gene patent will have a greater value if there are enabling technologies to use these genes; or if the firm owns a large array of elite germplasm into which those genes can be inserted. This explains the vertical integration of many biotechnology firms into the seed sector, as superior germplasm were essential complementary assets for agrobiotechnology.

The question is: why were so many mergers necessary for the exploitation of complementarities, when other possible contractual arrangements, such as licensing or joint ventures with the seed companies, exist? The fourth IP-related explanation relies upon the low appropriability of intellectual property rights in biotechnology and high transaction costs in contractual arrangements to provide an explanation for industrial consolidation in the agrobiotechnology industry.

The high transaction costs in licensing arrangements are due to the value allocation problem from these arrangements. Since firms do not know completely the full potential utilization of the resulting innovation, it is difficult to establish the correct cost and benefit sharing arrangement. Because of the difficulty of arriving at optimal licensing contracts, an acquisition alternative is thus often preferred.

Low appropriability of intellectual property rights and significant patent overlap comes about when firms have similar technology profiles. The weak differentiation of profiles is, in turn, due to the large size of a common knowledge base from academic research and publicly funded research programmes. Thus, it happens many times that different patents are merely based upon different procedures that are aimed at the same applications, e.g. gene insertion on different crops using gene gun technology or microprojectile methods. Consequently, in the face of similar patents, the probability of litigation is strong, and so is the incentive to merge or enter into cross-licensing agreements.

But why integrate vertically into the seed sector? Since crop biotechnologies demonstrate a significant degree of technical imitation and high-quality proprietary germplasm, being a key complementary asset for commercialization, this facilitates a stronger market position than biotechnology know-how and IPR on specific genes. This strategy of vertical integration into the seed business and ownership of germplasm has become an almost necessary strategy for technology firms in the face of contested intellectual property rights (Joly and de Looze, 1996; Hayenga and Kalaitzandonakes, 1999).

To summarize, demand complementary between chemicals and seed, as well as the innovation life cycle, offers possible explanations for the trend of consolidation in the agrochemical/agrobiotech industry. But the existence of intellectual property rights appears to have had much to do with the vertical integration. In particular, firms have had incentives to buy firms with IPR: (i) to block entry of potential competitors; (ii) as a bargaining chip for an equitable use of rival technologies; (iii) because of complementarities of key intellectual assets like transformation technologies, genes and germplasm; or (iv) because of high transaction costs in licensing agreements along with low level of technology differentiation and intellectual property appropriability.

3.4 Conclusions[71]

The intellectual property system as it is reflected in the TRIPS agreement has been developed over centuries, predominantly by, and for the purposes of, industrialized countries. Arguably, most forms of protection are not adapted to the needs of TK holders. This is particularly true for patent protection, which constitutes a complex and costly legal tool, especially with respect to its international component. It requires considerable expertise and financial resources to acquire and manage these intellectual property rights, as well as the will and capacity to commercialize the protected products and processes. Appropriate corporate, institutional and contractual structures are necessary to take full advantage of this system. Furthermore, when patent protection shall contribute to reach certain policy goals such as biodiversity and equity, it needs the implementation of corresponding competition laws and policies. This type of IPR must therefore be radically adapted in order to satisfy the aspirations of TK holders that go beyond the grant of mere 'defensive rights' to prevent the most visible dysfunctions of the system. The same is true, however to a lesser degree, for Plant Breeders' Rights and *sui generis* protection systems. It is therefore questionable whether classical forms of IPR are appropriate for the purposes of TK holders in the light of the economic and non-economic concerns at stake. One may rather explore new concepts such as the '*domaine public payant*' to address both the demand from TK holders and from society at large in order to fulfil policies that are able to balance private and public interests.

[71] Author: Christophe Germann.

Bibliography

Abbott, F.M. (1989) Protecting First World assets in the Third World: intellectual property negotiations in the GATT Multilateral Framework. *Vanderbilt Journal of Transnational Law* 22, 689.

Abbott, F., Cottier, T. and Gurry, F. (1999) *The International Intellectual Property System: Commentary and Material.* The Hague, London, Boston, p. 224 ff.

Agarwal, B. (1995) *Gender, Environment and Poverty Interlinks in Rural India.* UNRISD, Geneva.

Anderson, R. and Gallini, N. (1998) *Intellectual Property Rights and Competition Policy in the Knowledge-Based Economy.* Cambridge University Press, London.

Anuradha, R.V. (1998) Sharing with the Kanis – a case study from Kerala, India. Available at http://www.biodiv.org/doc/case-studies/cs-abs-kanis.pdf (accessed 31 May 2002).

Anuradha, R.V. (2001) IPRs: implications for biodiversity and local and indigenous communities. *Review of European Community and International Environmental Law* 10, 27.

Arrow, K.J. (1962) Economic welfare and the allocation of resources for invention. In: *The Rate and Direction of Inventive Activity: Economic and Social Factors*. National Bureau of Economic Research, New York, p. 609.

Baer, K.W. (1995) A theory of intellectual property and the Biodiversity Treaty. *Syracuse Journal of International Law and Commerce* 21, 259.

Barton, J.H. (1997a) The Biodiversity Convention and the flow of scientific information. In: Hoagland, K.E. and Rossman, A.Y. (eds) *Global Genetic Resources: Access, Ownership and Intellectual Property Rights*. Association of Systematics Collections, Washington, DC, p. 51.

Barton, J.H. (1997b) Patents and antitrust: a rethinking in light of patent breadth and sequential innovation. *Antitrust Law Journal* 65, 449–466.

Barton, J.H. (1998) The impact of contemporary patent law on plant biotechnology research. In: OECD, *Intellectual Property Rights and Competition Policy*. OECD, Paris.

Bavadam, L. (2000) Retrograde development. *Frontline*, 17/11.

Besen, S. and Raskind, L. (1991) An introduction to the law and economics of intellectual property. *Journal of Economic Perspectives* 5(1).

Black's Law Dictionary (1990) *Black's Law Dictionary*, 6th edn. St Paul, Minnesota.

Bosselmann, K. (1996) Plants and politics: the international legal regime concerning biotechnology and biodiversity. *Colombia Journal of International Environmental Law and Policy*, 111.

Clark, P. (1999) Kava case study. South Pacific, paper prepared for WWF.

Commission on Intellectual Property Rights (CIPR) (2002) *Integrating Intellectual Property and Development Policy*. CIPR, London, pp. 11–14.

Conway, G.R. and Barbier, E.B. (1990) *After the Green Revolution – Sustainable Agriculture for Development*. Earthscan, London.

Cottier, T. (2004) *The Agreement on Trade Related Aspects of Intellectual Property Rights* (to be published).

Dhaliwal, G.S. and Dilawari, V.K. (1991) Impact of green revolution on environment. In: Hansra, B.S. and Shukla, A.N. (eds) *Social, Economic and Political Implications of Green Revolution in India*. Classical Publishing, New Delhi.

Downes, D. and Laird, S.A. (1999) *Innovative Mechanisms for Sharing Benefits of Biodiversity and Related Knowledge – Cases Studies on Geographical Indications and Trademarks*. UNCTAD Biotrade Initiative.

Dunleavy, K. and Vinnola, M. (1999) E.U. biotech directive departs from U.S. practices. *The National Law Journal* 24 May 1999, p. C11.

Dutfield, G. (2001) *The International Biotrade, Conservation and Intellectual Property Rights*. Oxford University Press, Oxford, UK.

European Patent 436257 (filed 10 July 1991) Method for Controlling Fungi on Plants by the Aid of a Hydrophobic Extracted Neem Oil.

First, H. (2002) Antitrust's goals: theories of antitrust in the United States and Japan. In: Jones, C.A. and Matsushita, M. (eds) *Competition Policy in the Global Trading System, Perspectives from the EU, Japan and the US*. Kluwer Law International, The Hague, pp. 175–194.

Fulton, M. and Giannakas, K. (2001) Agricultural biotechnology and industry structure. Available at http://www.agbioforum.missouri.edu

Gadgil, M. (1997) A framework for managing India's biodiversity resources in the context of CBD and GATT. *RIS Biotechnology and Development Review* 1/1.

Gallini, N. and Trebilcock, M. (1998) Intellectual property rights and competition policy: a framework for the analysis of economic and legal issues. In: Anderson, R. and Gallini, N. (eds) *Intellectual Property Rights and Competition Policy in the Knowledge-Based Economy*. Cambridge University Press, London.

Germann, C. (2003) Content industries and cultural diversity: the case of motion pictures. *Culturelink*, October 2003.

Girsberger, M.A. (1998) The protection of traditional plant genetic resources for food and agriculture and the related know how by intellectual property rights in international law – the current legal environment. *The Journal of World Intellectual Property* 1(6), quoted 'Legal Environment', p. 1023.

Girsberger, M.A. (1999) *Biodiversity and the Concept of Farmers' Right in International Law, Factual Background and Legal Analysis*. Peter Lang, Berne, Switzerland, p. 329 ff.

Graff, G., Rausser, G. and Small, A. (2001) Agricultural biotechnology's complementary intellectual assets. *Review of Economics and Statistics*.

Greengrass, B. (1991) The 1991 Act of the UPOV Convention. *European Intellectual Property Review* 13, 466.

Gupta, A.K. (2001a) *Case Study 2: India*. World Intellectual Property Organization, Geneva.

Gupta, A.K. (2001b) Creativity, conservation, and compensation: Honey Bee Network approach to augment, sustainably utilize biodiversity, and share the benefits. On file with the authors.

Hayenga, M. and Kalaitzandonakes, N. (1999) Structure and coordination system changes in the US biotech and value-added grain market. Paper prepared for the IAMA World Food and Agribusiness Congress.

Heitz, A. (1991) Genetic resources and Plant Breeders' Rights. In: *Seminar on the Nature of and Rationale for the Protection of Plant Varieties under the UPOV Convention*. UPOV Publication No. 727 [E], pp. 114 and 118.

Hennessy, D.A. and Hayes, D.J. (2000) Competition and tying in agricultural and seed markets. *Review of Agricultural Economics* 22, 389–406.

Joly, P. and deLooze, M. (1996) An analysis of innovation strategies and industrial differentiation through patent applications: the case of plant biotechnology. *Research Policy* 25, 1014–1027.

Joshi, B.H. (1992) *An Analytical Approach to Problems of Indian Agriculture: A Theoretical and System Approach*. B.R. Publishing, New Delhi.

Kahama, C.G. (1995) *Tanzania into the 21st Century*. Tema, Dar es Salaam, Tanzania.

Kalaitzandonakes, N. and Hayenga, M. (2000) Structural change in the biotechnology and seed industrial complex: theory and evidence. In: Lesser, W.H. (ed.) *Transitions in Agbiotech: Economics of Strategy and Policy*. Food Marketing Center, University of Connecticut, Connecticut.

Kennedy, K. (2001) *Competition Law and the World Trade Organization, The Limits of Multilateralism*. Sweet and Maxwell, London, p. 22ff.

Kloppenburg, J.R. Jr and Kleinman, D.L. (1988) Seeds of controversy: national property versus common heritage. In: Kloppenburg, J.R. Jr (ed.) *Seeds and Sovereignty – The Use and Control of Plant Genetic Resources*. Duke University Press, Durham, North Carolina, p. 173.

Korah, V. (1983) Exclusive licenses of patent and plant breeders' rights under EEC law. *The Antitrust Bulletin* 28(3).

Landes, W.M. and Posner, R.A. (1989) An economic analysis of copyright. *Journal of Legal Studies* 18, 325.

Langinier, C. and Moschini, G. (2002) The economics of patents: an overview. Working Paper No. 293. Iowa State University, Center for Agricultural and Rural Development, Ames, Iowa.

Lebot, V. and Lévesque, J. (1989) *The Origin and Distribution of Kava (*Piper methstticum *Forst. f., Piperaceae) A Phytochemical Approach*. National Tropical Botanic Garden, Hawaii, p. 224.

Lebot, V., Merlin, M. and Lindstrom, L. (1992) *Kava the Pacific Drug*. Yale University Press, New Haven, Connecticut, p. 112.

Lele, U. *et al.* (eds) (1999) *Intellectual Property Rights in Agriculture*. World Bank, Washington, DC.

Leskien, D. and Flitner, M. (1997) *Intellectual Property Rights and Plant Genetic Resources: Options for a* Sui Generis *System*. International Plant Genetic Resources Institute, Rome.

Lesser, W. (1998) Intellectual property rights and concentration in agricultural biotechnology. *AgBioForum* 1, 56–61. http://www.agbioforum.org

Machlup, F. (1999) An economic review of the patent system. In: Abbott, F., Cottier, T. and Gurry, F. (eds) *The International Intellectual Property System: Commentary and Material*. Kluwer Law International, The Hague, p. 224ff.

Meinhardt, M. (1996) *Die Beschränkung nationaler Immaterialgüterrechte durch Article 86 EG-Vertrag, unter besonderer Berücksichtigung der Essential Facilities-Doktrin*. Berne, dissertation, pp. 137–160.

Ministry of Environment and Forests – Government of India (1998) Benefit Sharing Model Experimented by Tropical Botanic Garden and Research Institute. Available at http://www.biodiv.org/doc/case-studies/cs-abs-in.pdf (accessed 31 May 2002).

Narayanan, P. (1997) *Patent Law*, 2nd edn. Eastern Law House, Calcutta.

National Research Council (1992) *Neem – A Tree for Solving Global Problems*. National Academy Press, Washington, DC.

Nijar, G.S. (1996) *In Defence of Indigenous Knowledge and Biodiversity: A Conceptual Framework and Essential Elements of a Rights System*. Third World Network, Penang, p. 4.

Odek, J.O. (1994) Bio-piracy: creating proprietary rights in plant genetic resources. *Journal of Intellectual Property Law* 2, 141.

OECD (1998) *Competition Policy and Intellectual Property Right.* OECD, Paris.

OECD (1999) *Competition Policy and Intellectual Property Right.* OECD, Paris.

Pasadilla, G. (2002) Intellectual property rights and competition policy: implications on the seed industry. Working Paper, World Trade Institute, Berne (unpublished).

Peters, B.G. (1999) United States competition policy institutions: structural constraints and opportunities. In: Doern, G.B. and Wilks, S. (eds) *Comparative Competition Policy. National Institutions in a Global Market.* Clarendon Press, Oxford, UK, pp. 40–67.

Puri, K. (2002) How intellectual property rights could work better for developing countries and poor people. Commission on Intellectual Property Rights statement made at the Conference held in February 2002. Available at http://www.iprcommission.org/papers

Rabobank International (2001) *The Indian Seed Industry: On the Threshold of Consolidation.* Food and Agribusiness Research. Rabobank International, The Netherlands.

Rangachari, C.S. and Subbarav, D. (1998) Biodiversity & IPR: ethical dimension. *Economic Times*, 17 March.

Rangnekar, D. (2000) *Intellectual Property Rights and Agriculture: An Analysis of the Economic Impact of Plant Breeders' Rights.* Actionaid, UK, London.

Rural Agricultural Foundation International (RAFI) (September 1999) see http://www.etcgroup.org *Science* 269, 5230.

Scotchmer, S. (1991) Standing on the shoulders of giants: cumulative research and the patent law. *Journal of Economic Perspectives* 5(1).

Sharma, R. and Poleman, T.T. (1994) *The New Economics of India's Green Revolution – Income and Employment Diffusion in Uttar Pradesh.* Vikas, Delhi.

Shelton, D. (1995) Fair play, fair pay: preserving traditional knowledge and biological resources. *Yearbook of International Environmental Law* 5.

Shiva, V. (1994) Farmers' rights and the Convention on Biological Diversity. In: Sanchez, V. and Juma, C. (eds) *Biodiplomacy – Genetic Resources and International Relations.* African Centre for Technology Studies, Nairobi.

TBGRI – Arya Vaidya (1995) Agreement of 10 November 1995 between the Tropical Botanic Garden and Research Institute and Arya Vaidya Pharmacy (on file with the authors).

ten Kate, K. and Laird, S.A. (1999) *The Commercial Use of Biodiversity – Access to Genetic Resources and Benefit-Sharing.* Earthscan, London.

Thrupp, L.A. (2000) Linking agricultural biodiversity and food security: the valuable role of agrobiodiversity for sustainable agriculture. *International Affairs* 76, 265.

US Patent 5,124,349 (filed 31 October 1990) Storage Stable Azadirachtin Formulation.

US Patent 5,827,521 (filed 9 March 1995) Shelf Stable Insect Repellent, Insect Growth Regulator and Insecticidal Formulations Prepared from Technical Azadirachtin Isolated from the Kernel Extract of Azadirachta Indica.

US Patent 5,885,600 (filed 1 April 1997) Natural Insect Repellent Formula and Method of Making Same.

Walden I. (1995) Preserving biodiversity: the role of property rights. In: Swanson, T. (ed.) *Intellectual Property Rights and Biodiversity Conservation.* Cambridge University Press, Cambridge, p. 176.

WIPO (1998–1999) *Report on Fact-Finding Missions (FFM) on Intellectual Property and Traditional Knowledge.* WIPO, Geneva.

Wolfgang, L. (1995) Patents on native technology challenged (patent battle over seeds of the neem tree). 15 September.

Part II

Towards *Sui Generis* Rights

4 Origin and Allocation of Traditional Knowledge and Landraces

Part 1: Origin and Allocation of Traditional Knowledge and Traditional PGRFA: Basic Questions

SUSETTE BIBER-KLEMM

World Trade Institute, University of Berne, and Faculty of Law, University of Basel, Maiengasse 51, CH 4056 Basel, Switzerland

4.1 Introduction

This chapter deals with the question of if and how it is possible to allocate local varieties of PGRFA and traditional knowledge to the communities and/or individuals that maintain and develop these types of informational values. Allocation would be the prerequisite to creating incentives for their maintenance and protection within the framework of trade, such as the sharing of benefits with the communities or the creation of *sui generis* rights.

The objectives would be to find ways to acknowledge and evaluate the creativity, skill and economic input vested by farmers, farming communities and people living in a traditional way, and to maintain and develop their knowledge about the natural resources that sustain their livelihoods.

These issues are to be discussed against the background of the goal of this study, which is to analyse possible (legal) instruments to foster the maintenance and sustainable use of biodiversity and TK in the context of international trade that would promote the goals of equity and fairness, and create the basis for autonomous decision-making and financial return.

Given the complexity of the problem (see Chapter 1), a mix of instruments will be needed, combining rights, instruments and institutions for their implementation. From the economic point of view, measures that increase the returns on sustainable use of biodiversity and maintenance of TK are of specific importance.

Economic theory holds that one goal incentives must fulfil is the closure of the profitability gap between the private and public sectors and/or the social values of biodiversity. To this end it is important that the financial returns accrue at the level where the good – in our case the informational value of PGRFA and TK – is created (see Chapter 1).

One legal option to secure financial compensation to the providers of the information, which is discussed in this context, is the allocation of private rights to PGRFA and/or associated TK (see Chapter 1 and WIPO/GRTKF/IC/3/8). This option is discussed in both this and the following chapter (Chapter 5).

At this point, the focus is on the

conditions that must be met to allow the allocation of such rights to the resource-holders, i.e. to the 'local and indigenous communities embodying traditional lifestyles' (Article 8(j) CBD) and to the farmers and farming communities (Article 9 ITPGRFA).

The following elements are deemed essential to this end: it must be possible to localize the relevant information, i.e. to attribute it to a specific geographical region and/or social entity. To this end it must, first, be possible to define the information that is to be protected and to be able to distinguish it from other, possibly similar information; at the very least, the information must be identifiable. And secondly, it must be possible to identify the holder(s), owner(s) or author(s)[1] of the information (in the case of the farmer-breeders of the landrace) or, as mentioned, to specify a geographical place of origin. The key question in this context is, of course, how the 'legitimate' holder of the information can be determined, and this in turn leads, thirdly, to the necessity to determine the criteria for the allocation of the information to a specific social entity.

This will be analysed in the following discussion with respect to both PGRFA and associated TK.

4.2 Origin and Allocation of Traditional Knowledge

4.2.1 Introduction

As mentioned above, the allocation of traditional knowledge to its holders and stewards is the prerequisite for creating incentives for its maintenance, for its protection in the trade context, and as a basis for the fair and equitable compensation of its use. To this end it is necessary to define the knowledge to be protected, to identify the specific protective needs, and to specify whom – individual or community – is the holder of the knowledge.

In order to answer these questions, it is to be taken into account that TK has specific characteristics and protective needs in comparison with the information generated in the so-called formal creative and innovative processes. In creating a protective legal basis, these have to be taken into consideration, together with the goals of protection.

Definition and relevant characteristics

As mentioned in Chapter 1, at present no generally accepted, uniform definition of TK seems possible. The CBD gives some indications as to the TK falling under its regime. It speaks of 'knowledge, innovations and practices, relevant for the conservation and sustainable use of biological diversity' (Article 8(j)). This would mean for our context that: (i) only TK in connection with biological resources (i.e. in the CBD and IT context) is encompassed; and (ii) the protection would have to be limited to knowledge, innovations and practices which first originate in indigenous and local communities *embodying traditional lifestyles* and which are relevant for the *conservation and sustainable use of biological* diversity.

Three questions arise from this: Is this limitation to local communities living in 'traditional' ways practicable for the designation of owners of TK? What does 'traditional lifestyle' exactly mean? How could the delimitation be put into effect? The second question is closely linked to the first. It occurs in the context of the current debates on the protection of TK, which perceive the topic primarily in a 'North–South' dichotomy. The question would be whether the issue ought to be considered in a more generalized context and in a global perspective.

As to the characteristics, the following viewpoints may illustrate the difference between the 'traditional' and 'formal' approaches to science.

Augustine (1997), a Chief on the Mi'kmaq Grand Council, holding a Bachelor of Arts in anthropology and political science

[1] The 'author' here is understood in the non-technical sense of 'creator' in general, i.e. broader than in copyright.

from Canada, lives in 'both worlds' and has undertaken to analyse and to compare traditional aboriginal knowledge and occidental science.

He bases his analysis on the *Oxford English Dictionary*, which defines science as:

> 1) the state of fact of knowing, a theoretical perception of a truth; 2) knowledge acquired by study, mastery, trained skill; 3) a recognized department of learning ... ; and 4) in a more restricted sense, a branch of study ... (Augustine, 1997, p. 3).

He reckons that each of these definitions can to a greater or lesser extent be applied to indigenous science. He characterizes indigenous science as 'a disciplined approach to knowing and understanding the nature of reality, systems of relationships, and the energies and processes of the universe' (Augustine, 1997, p. 3).

He insists that TK also implies a specific process of learning: 'TK not only acquires knowledge from the distant past, but updates this knowledge according to its own methodologies' (Augustine, 1997, p. 8).

He sees the main difference to occidental science as the fact that '... TK can never be a branch or department of knowledge, but remains inseparable from the cohesive whole, from a way of being and of coming to learning' (Augustine, 1997, p. 3). He identifies as the main differences of the two systems/approaches: (i) the lack of a connection to Earth of the technological, industrial society (p. 6), which could tentatively be translated as a lack of knowledge of the limitation of resources and of a respectful approach to the environment; and (ii) the compartmentalization and fragmentation of the scientific approach in the occidental world, in comparison with the contextual analysis of the knowledge. 'Native traditions teach the whole of nature in a practical, functional macro approach to the environment' (Augustine, 1997, p. 6).

From a 'Western' point of view, van den Daele (2001) characterizes traditional knowledge (in a global sense) as 'embedded knowledge', that is, knowledge that, besides its value in information, also has social and cultural meaning, and as 'embodied knowledge', being knowledge that cannot be represented adequately in explicated rules or textbooks but is ingrained in people through socialization. Accordingly, he characterizes 'Western' scientific knowledge as disembedded and disembodied, as knowledge as 'information' which is global and impersonal, in contrast to knowledge as 'culture' which is local and personal.

Shortcomings of the formal IP system

These characteristics may explain some of the shortcomings of the formal system of intellectual property, which has been designed for the 'Western' type of science systems. It is submitted that for the creation of mechanisms to protect TK it is important to take these shortcomings into account. There is, first, the problem of the duration of the right. TK dates back for generations (or even time immemorial) and is handed down to future generations. Aboriginal people and communities therefore may want to protect their traditional knowledge for an unlimited time.[2] Formal IPR, in particular patents, protect only for a restricted period of time. Secondly, there might be a problem of control over the use made of the knowledge: communities want some control over the use of knowledge that originates from their culture. Formal IPR as a rule (except trade secrets) do not allow for such control, as there is an obligation to make the information available for the public, and as, in any case, the protected information falls back into the public domain after the expiry of the IPR. Thirdly, the scope of the formal IPR might be too narrow. Indigenous peoples want to protect their (entire) culture and not only one iso-

[2] See WBCSD (2003) Dialogue: Indigenous peoples' right of control over their knowledge should last as long as the community use of that knowledge is active and efforts are made to keep it confidential within the concerned group of holders of traditional knowledge.

lated manifestation.[3] Further, there is the problem of the enforcement of the rights. The enforcement of intellectual property rights is generally the responsibility of the rights holders. This requires knowledge of the IPR scene, finances, technical means, capacity and capability. Therefore aboriginal IP holders may be at a disadvantage when defending their IP rights if they do not have access to these resources (Indian and Northern Affairs Canada, 1999, p. 10).

4.2.2 Types of traditional knowledge

The above discussion indicates that different types of traditional knowledge exist. The following grouping is considered to be relevant as a basis for discussing legal solutions to the problem:

1. Traditional knowledge can be *associated* with a biological resource, which is the case for information on the effects of medicinal plants or on the specific qualities of a crop. However, it can also be *integrated* into a biological resource, as is the case of cultivated crop varieties and domesticated animals, which are the product of the skills of generations of farmers and breeders.
2. Traditional knowledge can be freely accessible within a community and be known by everybody, as is the case with folk remedies such as curcuma or neem. On the other hand, access and usage can be regulated and restricted within the community, such as plants used for ritual purposes in sacred ceremonies like *Ayahuasca*/*Yagé*.
3. Traditional knowledge can be allocated to a specific, clearly defined group within a community, such as to shamans or other healers acting as stewards of the knowledge; or to a community as a whole, as is the case with the *Hoodia* cactus that is used by the San people in southern Africa. Conversely, it can be integrated in the culture of a society in general.
4. Differences may also exist in the way that a community or a people deal with certain types of knowledge: it can wish to keep it secret within a community, or only transfer it as a gift, in that its spiritual character is opposed to marketability. Or it can wish to market the information, in which case insisting upon fairness in the transaction, including the sharing of the benefits.
5. Traditional knowledge can be documented in a written form, either by the holders of the traditional knowledge themselves, as is the case in some systems of traditional medicine such as Ayurveda and Siddha, or by others such as scientists or historians.

So a difference can be made in the degree of publicity and thus in the accessibility of the knowledge. There is knowledge that is strictly secret and not open to use by outsiders, even not by all members of the group. In turn there is knowledge that is generally known within a specific community: it might be openly accessible for all members of the community, or be protected by customary laws, or maintained and managed by specific stakeholders. And, finally, there is knowledge widely known within a society.

These examples illustrate that, in analogy with formal IPR, there may exist different protective needs, such as:

- Absolute protection and secrecy. This can, for instance, be the case for sacred knowledge.
- Autonomy to decide if and how the information is used, for instance, for medicinal knowledge belonging to a specific indigenous community or to several specific communities.
- Or the relatively free access to knowledge, but with compensation for the holders and/or sharing of benefits resulting from its use. This could be the case for knowledge that is generally known.

Therefore the option should be evaluated to create different categories of instruments, according to the protective needs.

[3] For a more detailed discussion see Chapter 6.

4.2.3 Degree of publicity: the 'public domain' discussion

In contrast to formal IPR, which prevent information becoming public domain knowledge, the goal of rights to TK might be to take TK out of the 'public domain' in the formal sense (WBCSD, 2003). This issue, as with formal IPR, involves different interests:

1. The interests of the holders of the information or, in the case of IPR, of the holders of the rights. In the IPR these interests are mainly defined as economic interests; in TK there might be wider and/or different interests involved, such as respect, autonomy and control. However, this is the interest in the private good character of the information.
2. The interest of the public in general to be able to make use of the information, e.g. to make use of a medicinal plant. This interest corresponds to the public good character of the information.
3. Finally, the interest of the public, which I would describe as a 'future interest', e.g. in maintaining creativity, or in maintaining traditional knowledge, as an important element for conservation and sustainable use of biogenetic resources in the future (which is the optional value, public good character).

It is submitted that in designing protective instruments for TK, a balance between these different interests must be found. In particular, the public and private interests should be carefully balanced. The private interest of holders of TK could be for example to keep the information secret and/or to respect taboos, or to maintain the autonomy to decide about (all) future uses. On the other hand, the private/public interest could consist of making use of TK as a starting point for further creative processes, or in the open exchange of crop varieties.

4.2.4 The question of ownership and origin

Origin

Origin can be defined from both geographical and social viewpoints.

The geographical definition of origin may be differentiated as follows. It might be possible to trace the knowledge to the specific area where it originated. An example could be the knowledge about kava, which originated in the Pacific Islands; St John's wort, which is a typical European TK; or Ayahuasca, which originated in the Amazon.

This geographical designation can vary in size from a village, where a healer has specific knowledge; to a region, where a specific community lives, or where specific knowledge is maintained, such as the knowledge about the *Hoodia* cactus in South Africa or the maintenance of potato varieties in the Peruvian Andes; to a country, such as neem and curcuma in India; or to parts of or an entire continent.

This leads to various types of ownership of the knowledge. TK can be owned by an individual holder, in his own right, as seems to be the case for African shamans (Nwokeabia, 2001). More frequently, however, TK is owned collectively and can be traced back to a specific community. In this case the prerequisites for protection can be further differentiated according to whether the community still exists; the knowledge is still in use in the community; or whether a time limit exists for knowledge that has only been in the open since a certain period.[4] One of the problems might be that several aboriginal groups claim ownership over the same or similar knowledge and may differ as to how this knowledge should be protected or shared. Further, knowledge that is so widely held may be considered public knowledge in a specific region. So the geographical, and possibly historical, origin might be a further connecting point, and the information might be allocated to a state.

[4] See, for instance, the Peruvian law on TK that clearly defines the public domain. If knowledge is in this public domain for no longer than 20 years, a part of the benefit resulting from its use has to be paid into the Fund for the Development of Indigenous Peoples (Article 13; see Chapter 2, this volume).

Registration schemes are proposed as an instrument to allocate TK for both privately owned and public TK (see second part of this chapter and Chapter 6).

4.2.5. Summary

The following questions are to be asked in relation to the creation of protective instruments:

First, which types of information should be protected, and what would be the objective of the protection? Would the goal be to grant autonomy to decide on the use made of the TK, or would it rather be to create a clear basis for its marketing? Secondly, which elements of ownership ought to be protected? Is it full ownership or a mere right to compensation for its use? And to what extent should be the scope of the protection in view of the subject matter, time frame and planned utilization? For instance, is the protection limited to industrial utilization?

What about an absolute right in the sense of duration and in the sense of vetoing its use at each stage of its development or analysis? Would this be compatible with a possible public interest in the broad commercialization of the knowledge, such as the development of new medicines?

These questions have to be answered in view of the goals of protection, such as the public interest in maintaining TK, and the private interests of owners of TK, such as respect for their 'taboos', autonomy to decide upon the use of their knowledge, and related economic interests; additionally, to make use of TK as a starting point for further creative processes.

It is submitted that different types of protective means might have to be created, which allow for a differentiated definition of 'public domain', and that regarding the protection of traditional knowledge, the public and private domain need to be carefully balanced.

4.3 The Concept of Origin of PGRFA

4.3.1 Background and terminology

As described above, the goal of legal instruments to protect the informational value contained in PGRFA would be the support and maintenance of *in situ* on-farm breeding. Accordingly, our focus will be the varieties of plant genetic resources as maintained and developed by local farmers or farming communities – the so-called landraces or farmers' varieties.

Within these traditional seed supply systems (which frequently are characterized as 'informal'[5]), different types can be distinguished (Correa, 2000, p. 13). First, the traditional system, based on the use and continuous improvement of farmers' varieties, characterized by farmers engaged in selecting and saving seeds, and bartering with neighbouring farmers or farmers in different villages. Secondly, a system in which some farmers specialize in the production of improved seeds for the local (regional) market[6] or even for the utilization in intensified farming systems.[7]

The two systems differ with regard to the characteristics relevant for the creation of *sui generis* rights. Whereas in the second case it is possible to clearly identify both breeders and bred varieties,[8] the crucial question is whether this is also possible in the first case.

Accordingly, the focus will be on the decentralized, traditional (informal) on-farm *in situ* breeding of farmers' varieties. The issue is whether in these systems it is possible to establish where a landrace comes from, and to establish this origin with enough precision that allows its allocation to a specific social entity.

[5] Alternative terminology: decentralized seed supply systems or farmers' seed supply systems, in contrast to the industrialized, centralized production of seeds.

[6] Personal information: Sanjaya Gyawali, Li-Bird, Pokhara, Nepal.

[7] Personal information: Renato Salazar, PEDIGREA, Philippines. See also the initiative by SATIVA (http://www.sativa.org) in Switzerland.

[8] Or at least the variety for the time being, as it is bound to change over time (and space).

To this end, the term 'origin' has to be analysed more closely. On the one hand, the term 'origin of landraces' has to be examined. Is it to be understood in a geographical sense only, or does it contain elements of breeding in the technical sense too? And, if so, what would these elements be?

On the other hand, 'origin' is also a technical term, used in the CBD to define the ownership of the sovereign state over its biological resources. Therefore, the meaning of this terminology and its worth in the context of the allocation of landraces has to be explored.

To answer the first question, it is essential to clarify beforehand the notion of PGRFA and of farmers' varieties/landraces, to give some background information on their specific characteristics, and to describe the relevant characteristics of the *in situ* on-farm breeding of landraces.

Plant genetic resources for food and agriculture: landraces

The International Treaty on Plant Genetic Resources for Food and Agriculture defines PGRFA as 'any genetic material of plant origin of actual or potential value for food and agriculture'.[9] 'Genetic material' in turn is defined as 'any material of plant origin, including reproductive and vegetative propagating material, containing functional units of heredity'.

Landraces or farmers' varieties thus are a specific type of PGRFA. From the legal point of view, they can be negatively defined as varieties that cannot be protected by PBR[10] because they do not fulfil the prerequisites of uniformity, stability and distinctness. The question is whether a positive description or definition of farmers' varieties is possible.

The scientific literature offers a great variety of definitions of landraces, which, in addition, have changed over time (see in detail below. For instance, Jarvis and coworkers give a rather general definition. They characterize a landrace as a crop variety that is bred and cultivated by farmers and adapted to local environmental conditions (Jarvis *et al.*, 2000, p. 8). Harlan (1975) gives a more differentiated characterization of *populations* of landraces, implying their identifiability on a local level:

> Landrace populations are often highly variable in appearance, but they are each identifiable and usually have local names. A landrace has particular properties or characteristics ... Each has a reputation for adaptation to particular soil types ... They also may be classified according to expected usage ... All components of the population are adapted to local climatic conditions, cultural practices, and disease and pests. (cited in Jarvis *et al.*, 2000, p. 9)

Zeven (1998, p. 137), in turn, concludes that as landraces have a rather complex nature, it is not possible to give an all-embracing definition. He differentiates between autochthonous and allochthonous[11] landraces, and proposes to define an autochthonous landrace as a variety with a high capacity to tolerate biotic and abiotic stress, resulting in a high yield stability and an intermediate yield level under a low-input agricultural system (adapted from Mansholt, 1909).

In defining the difference between autochthonous and allochthonous varieties, Zeven introduces elements of time and variability. He defines an *allochthonous* landrace as an *autochthonous* landrace of a foreign region that has recently been introduced into the region

[9] This excludes the plant genetic resources which are of importance for their biochemical qualities in the context of, for example, pharmaceuticals and health care, or for their specific 'material' qualities (e.g. the non-wood forest products used for industrial, building and housing purposes) or 'industrial crops' (e.g. rubber, oil palm). They can be 'wild' or domesticated.

[10] As defined by the Convention of the International Union for the Protection of New Varieties of Plants (UPOV) (see Chapter 2, this volume).

[11] 'Autochthonous' means inhabiting a place or region from earliest known times; synonymously: aboriginal, derived from within a system, endemic, indigenous, native. Conversely, 'allochthonous' means originating from outside a system.

concerned.[12] After its introduction it may become contaminated with genotypes of the autochthonous landraces. Depending on the number of generations of after-growth and on the frequency of seed change, it may become an autochthonous landrace.

In sum, characteristics appearing in all definitions of landraces are their capacity to adapt to local environmental conditions and their high stress tolerance, and therefore yield stability. In the given situation of subsistence farming, these qualities appear as highly advantageous. The identifiability of varieties, in turn, is not as generally acknowledged.

In situ *on-farm conservation*[13]

Traditional farming systems, in maintaining and further developing farmers' varieties/landraces, essentially contribute to the diversity of agricultural biological resources. *In situ* on-farm conservation of agrobiodiversity thus is increasingly recognized as an important supplement to *ex situ* gene banks.

In situ conservation of PGRFA allows for the maintenance of mechanisms that are important for the further evolution of traditional PGRFA, such as the hybridization within and between populations of wild, weedy and cultivated plants, the competition among genotypes, the natural and conscious selection by farmers at the local level and the exchange of different genotypes among farmers and farms (Brush, 1994; see also Brush, 2000).

Farmers play an important role in this system, as the maintenance of agroecosystems, the further evolution of PGRFA, and human intervention are mutually interdependent. On the one hand, they contribute their breeding skills and their knowledge about farmers' breeding technologies. On the other hand they maintain the on-farm crop evolution system. This includes: (i) the maintenance and intergenerational tradition of the existing crop variety and relevant, additional information; (ii) the maintenance and intergenerational tradition of the innovatory breeding skills; and (iii) the maintenance of the ecosystem services provided by the non-industrial farming systems, allowing the intercropping between domesticated landraces and their wild and weedy relatives.

Zeven (1998, p. 136) underlines the influence of human selection on the development of landraces. The intensity of this on-farm selection process – between conscious or unconscious selection – varies. Unconscious selection is generated by the farmer in changing the growing conditions. Conscious selection is understood, for example, as choosing the best plant/seed for propagating or experimenting with new material or old material in new conditions.[14]

The question is what this means for the definition of origin and/or authorship of landraces and their allocation to specified social entities. Would it be possible and appropriate to honour the creative input into a specific landrace variety by defining criteria identifying its farmer-breeders? What would be these criteria? And would it be appropriate and possible to distinguish between the creators and origin, where 'creator' or 'breeder' indicates the result of

[12] A key issue in plant breeding is the open access to the resources and the exchange of crop varieties and related information at local, regional and international levels. At the local level, in traditional societies and in subsistence farming systems, sharing and exchange of varieties is traditional and of great importance for the evolution of local varieties, and to assure food security. This exchange takes place in a reciprocal relationship.

[13] For more details see Biber-Klemm (2002).

[14] In the legal context, the issue of conservation and maintenance of traditional PGRFA *in situ* on farm is discussed under the notion of Farmers' Rights. Farmers' Rights are the answer to the 'enormous contribution that the local and indigenous communities and farmers of all regions of the world, particularly those in the centres of origin and crop diversity, have made and will continue to make for the conservation and development of plant genetic resources which constitute the basis of food and agriculture production throughout the world' (Article 9.1 ITPGRFA; see Chapter 6).

conscious breeding, and 'origin' designates the localization of an 'autochthonous' landrace?[15] Or is the distinction between origin and creator simply one of degree rather than of principle?

From this it can be concluded that, as basic elements for the allocation of a landrace to a specific entity, the landrace must be distinguishable/identifiable and/or fulfil other criteria, such as the possibility of ascribing it to a specific social entity or geographical region, i.e. it must be possible to identify the social and/or geographical origin of the landrace.

The question is whether the notion of origin of plant genetic resources as used in the scientific debate on PGRFA and/or as integrated in the CBD is suitable to serve as a basis for creating rights to traditional PGRFA. This question will be treated in the next paragraph.

4.3.2 Origin

'Origin' of PGRFA seems to have different meanings in legal and scientific contexts.

The scientific understanding of origin

The notion of 'origin' of PGR in the scientific context is closely linked to the theory of N.I. Vavilov, postulating the existence of so-called 'centres of origin' of domesticated crops.

Vavilov found that there were certain areas in the world where crop plant diversity was extremely intense, i.e. regions containing a high level of diversity of a number of crops (see Fowler, 2000). The areas of greatest diversity were believed by him to represent the centres where the crops were originally domesticated (Hawkes, 1983, p. 52 ff.).

Besides these (primary) centres of diversity, so-called secondary centres of diversity are recognized for many crops. These are due to the movement and exchange of crops throughout history. The high degree of diversity in these secondary centres is due to a long history of cultivation of a crop, combined with environmental and social factors supporting diversification (Raymond and Fowler, 2001, p. 4).

However, the high degree of diversity in Vavilov's 'centres of origin' does not refer primarily to the diversity of individual crop varieties nor to distinctive properties, but rather to diversity in general. Vavilov's theory was that areas of high diversity correspond to the areas of origin, i.e. where the crops were originally domesticated. So, his notion of origin has nothing to do with the origin of individual varieties with distinctive properties (Fowler, 2000).

The distinction between primary and secondary areas of diversity deals with sequences in the development of agriculture. 'Vavilov, Zeven and de Wet and others, in their employment of the concept of "centres" were more interested in a crop's diversity or origins, which may not be quite the same as a species' (CGIAR, 2001, p. 1).

Thus, science looks not primarily at the diversity within a specific crop variety nor at its individual origin, but at the centres of crop diversity in general, being interested in where the greatest amount of diversity can be found. Integrated in this question is the enquiry after the process of domestication and diversification of a crop, a question that also eminently includes a historical element.

Yet, in our context it is important to be aware that 'diversity has no fixed address. New forms and combinations ... can arise wherever a crop is grown, regardless of where it was domesticated' (CGIAR, 2001, p. 2).

Country of origin according to the CBD

In the CBD, the notion of 'origin' is the connecting point for the identification of biological resources falling under the sovereignty of the state. Only genetic

[15] Zeven (1998) defines 'autochthonous' as a landrace grown for a long period in the farming system concerned. As the environment changes annually and as the landrace becomes 'contaminated' – purposely or not – with a few genotypes of other landraces or cultivars, it will continuously adapt itself.

resources of which the providing state is the 'country of origin' (or which have been acquired in accordance with the CBD) are covered by the access and benefit-sharing regulation of the CBD (Article 15.3).

The CBD defines the country of origin as the country that possesses the genetic resources in *in situ* conditions. *In situ* conditions for domesticated and cultivated species[16] are 'the surroundings where they have developed their distinctive properties' (Article 2.12).[17] In view of this definition, several details need to be considered.

First, the wording 'have developed' implies a time element. The question is whether this term points to the development of specific properties due to ecological factors only, or whether it also includes evolution furthered by indirect or direct human intervention. The CBD definition of 'domesticated or cultivated' species as 'species in which the evolutionary process has been influenced by humans to meet their needs' (Article 2.7) points to the latter interpretation. Secondly, the term 'distinctive properties of species' indicates a qualitative element. The primary question is, of course, what is meant by this criterion. This question is complicated by the fact that it is not clear on which taxonomic level the comparison has to take place. The CBD speaks of distinctive properties *of species*. Is this wording to be read as the difference *between* species,[18] or as the difference *within* species; that is, between varieties?[19]

As is apparent, this makes an enormous difference to the definition of 'origin'. If interpreted on the species level, the 'country of origin' would be where a specific crop species has been domesticated or basic traits have been developed in historical times. The interpretation of the level of variety, in turn, would allow a more current and narrow designation of origin.

Country of origin in the International Treaty on PGRFA

The International Treaty confirms the national sovereignty of states over their *own* PGRFA (Article 10), without defining criteria for ownership in detail.

The International Treaty speaks of 'centre of origin' and 'centre of crop diversity', referring to these notions in the context of farmers' rights (Article 9). According to its interpretation, 'centre of origin' means a 'geographical area, where a plant species, either domesticated or wild, first developed its distinctive properties'. 'Centre of crop diversity' refers to a geographic area too. It is defined as an area that contains a high level of genetic diversity in *in situ* conditions, i.e. in 'surroundings where they [crop species] have developed their distinctive properties' (see Article 2).

Thus, the International Treaty uses the same criteria (distinctive properties of species) as the CBD.

Discussion

The current discussions on the definition of origin in a legal context focus on the question with a view to the *state sovereignty* over domesticated genetic resources.[20] In this context the CBD speaks of distinctive properties of *species*. According to the wording of the Convention this can be read as differences *between species* (or possibly subspecies), but not a priori *within* species.

It is important to be aware that the questions asked by the CBD differ essentially from those asked by science. Science,

[16] Defined as 'species in which the evolutionary process has been influenced by humans to meet their needs' (Article 2 CBD).

[17] *In situ* conditions for wild species are defined as the ecosystem where they exist in natural surroundings, such as ecosystems and habitats (Glowka *et al.*, 1994, p. 22). Thus the definition differs for wild and for domesticated or cultivated species.

[18] Or possibly subspecies such as the potato *Solanum tuberosum*; but not, for instance, *S. andigenum* or *S. juzepczukii* (see Tapia and De la Torre, 1997, p. 12).

[19] Likewise Fowler (2000).

[20] Compare, for example, Fowler (2000) and Hardon *et al.* (1994).

in inquiring into 'origin', asks for the (historical) centres of crop diversity, being interested in where the *greatest amount of diversity* can be found. The legal interest, in turn, is to establish *ownership*. Science asks where a crop species has been domesticated, and enquires after the (historical) *process* of domestication and diversification: the CBD asks for the *place* specific properties of species have been developed.

Accordingly, in scientific and legal literature the line of argument in the interpretation of the notion of origin varies. Difficulties arise through the CBD's link between 'origin' and '*in situ* conditions'.

Glowka *et al.* (1994), in an early interpretation of the CBD, seem to understand the CBD definition as referring to varieties, in particular landraces. For the interpretation of '*in situ* conditions' they refer to the known formula of 'surroundings where they have developed their distinctive characteristics', which – in connection with *in situ* conservation – they interpret as 'those areas where humans have created agricultural systems in which they have, in turn, developed identifiable plant varieties (known as landraces) ... This is independent of the (geographical) relation to the wild populations from which they originated' (p. 22).

In contrast, in the scientific literature treating origin in connection with the CBD's assertion of state sovereignty over genetic resources, origin seems in most cases to be interpreted in an historical (Vavilovian) sense.[21] Fowler (2000) argues at the historical and species level and concludes that the definition of origin is technically and financially not feasible. He maintains that the CBD definition is to be understood in an historical sense, as most crops and certainly most of their properties originated long ago (p. 4, although in his examples he mixes species and varieties).[22] He argues,

> many properties come in infinite gradations. Many ... will have been developed over time, and over a wide expanse of territory, encompassing more than one country. Some properties might have multiple origins both in time and place ... Proving the historical origin – pinpointing both the time and place of each – is well beyond the grasp of today's science or science budgets (Fowler, 2000, p. 5).

In contrast, Hardon *et al.* (1994, p. 12), differentiate between 'origin' (apparently understood in an historical sense) and the current geographical distribution in referring to 'landrace groups which can be distinguished within geographical regions on the basis of a complex of name, morphology or usage'. They conclude that it might be theoretically possible to trace the origin of samples to their original site or origin of the population/landrace. However, it requires the actual sample to be a true representation of its *original source*. The problem of tracing origin becomes even more intractable if the combinations of genes and genotypes in a sample has been altered by selection and recombination and/or regeneration (Hardon *et al.*, 1994, p. 15).

In the literature (for instance, Hardon *et al.*, 1994; Girsberger, 1999; Fowler, 2000) the viewpoint prevails that the origin (in the CBD sense) of domesticated crops or, more precisely, crop species, cannot be clearly defined given: (i) the evolution of the distinctive properties since time immemorial; (ii) the breeding history of the crops, involving an uncountable number of parental lines; and (iii) the culture/custom

[21] This might stem from the fact that the 'origin' discussion was, or is, inspired by the discussion on the creation of a multilateral system of access and benefit-sharing in the IU revision process, the lines of arguments being built up to prove that the clear definition of origin is not practical or feasible and therefore other solutions must be found. This line of argument has to be seen given the background of the strong rejection of models of ownership and property, from the apprehension that such models could prevent the free flow of germplasm that is vital for the maintenance and evolution of crops.

[22] 'Vavilov's theory focused on *crops* and on *regions;* the CBD's on *properties* and countries, a definition requiring a considerably higher level of precision. Nevertheless both depend on a detailed knowledge of history that, for the most part, must reach back beyond the founding of the nation state itself' (Fowler, 2000, p. 4).

of open exchange of crops and their (intercontinental) flow (see Brush in Girsberger, 1999, p. 60).

These findings and conclusions as to the CBD definition of 'origin of domesticated *species*' somehow contrast with the results of ethnobotanical research, which – at least at the level of crop *varieties* – describe distinctive local breeds.[23]

Thus, in the following discussion it is proposed to examine whether the reference to this historical understanding of 'origin' is compelling and useful in our context. It is suggested to investigate whether the question of origin could be considered on two different levels: (i) the level of differences *between species, and subspecies*, which are relevant to establishing origin in view of state sovereignty; and (ii) the differences *within* species, between varieties, which could be relevant to establishing origin on a lower geographical level.[24]

Conclusions

The key question as to the definition of origin is how the notion of 'distinctive properties' of species and/or of varieties is to be understood.

As a working hypothesis it is put forward that, for the sake of the allocation of the genetic information as contained in landraces, the legal (CBD) definition of origin is not differentiated enough. As it is positioned on the species level (apparently on the basis of the more historical scientific understanding of the term), further interpretative steps are necessary.

Thus, in the context of the question of the allocation of landraces to specific social or geographical entities, the necessity of a *functional* interpretation of the term 'origin' is proposed, answering to the specific characteristics of landraces.

From this, the following questions arise. First, can specific characteristics be established that allow the distinction of different varieties of landraces; and secondly, do these characteristics allow the allocation of a variety to a geographical entity on a relatively low level (such as community or farming family)?[25]

4.3.3 A functional interpretation of origin

In general

The goal of the CBD's definition of 'country of origin' is to allocate genetic resources to the state. The notion of 'country of origin' is the connecting point for the establishment of the ownership of the state over its genetic resources. So, the definition refers to the level of the nation states in defining the origin of the genetic information.

However, as elaborated above, for the creation of effective incentives for the conservation of traditional PGRFA, the allocation to a lower level, in a geographic or socio-political sense, would be preferable.

The question is – what would be the relevant criteria? Could the CBD criteria for the definition of 'origin' be adapted for this purpose? As mentioned above, the CBD definition of origin of domesticated species encompasses several elements: geographical (being the surroundings); biological (distinctness); social (influence of the evolutionary process by human needs); and temporal (that it has developed).

Among these criteria, distinctness seems

[23] See, e.g. Schneider, 1995; Asfaw, 1999; Gonzales, 1999; Pionetti and Suresh, 2002.

[24] This question is to be answered first from a botanical/taxonomic point of view. Does the allocation of species/subspecies to a 'regional/national' (i.e. higher), and of varieties to a 'subnational' (i.e. lower), geographical level make sense from the botanical/taxonomic point of view? Or are domesticated plants always species (as, for instance, the potato *Solanum tuberosum*; but consider *S. andigenum* and *S. juzepczukii* (see Tapia and De la Torre, 1997, p. 12).

[25] This question has to be distinguished from the protective criteria, which must be defined according to the goals the protection is intended to fulfil. From this incentive point of view, in the case of PGRFA, the maintenance of the crop diversity and its evolutional potential (genetic diversity and adaptability to exogenous factors, in contrast to the UPOV criteria of uniformity and stability), as well as the conscious innovative breeding, are all relevant (see Zeven, 1998).

to be a basic element for the identification of a variety. The central point is that a variety must be distinct enough to be able to identify it in comparison with other varieties. If not, the problems of multiple protection of one variety, and an infinite duration of rights over a variety, would arise (IPGRI, 1999).

Further elements could be found in the characteristics of a landrace. Among the elements to describe landraces discussed by Zeven (1998, pp. 35–36),[26] the criteria of the breeding history, in particular in connection with human selection, and the element of the common appearance (integrity, combined with diversity), seem, in analogy with the above-mentioned CBD criteria, to be promising for the identification of specific landraces.

The question is whether one or a combination of these criteria could provide connecting points for the definition of the origin of a landrace.[27]

The criterion of distinctness

Zeven (1998), citing Kiessling (1912), states that a landrace is a mixture of phenotypes, but that (the majority of) these phenotypes have a common appearance, which makes them at least somewhat different from another landrace of the same crop.

Thus, one element in identifying a landrace could be its appearance, as even if the landrace is genetically diverse, the genetic diversity may not hold for all (expressed) characters.

Distinctness, besides stability and uniformity, is one of the criteria for the allocation of Plant Breeders' Rights. It is true that – as a rule – landraces cannot be protected by PBR, because by definition they do not correspond to the criteria of stability and uniformity. On the contrary, the opposite qualities of genetic diversity within a specific landrace and its resulting capability of adapting to exogenous (detrimental) factors are specific characteristics that, according to scientific literature, should be encouraged.

However, keeping in mind these specific qualities, the question might be asked whether the criterion of *distinctness*, as elaborated in the application of the UPOV regulations, might be helpful for the identification of landraces and possibly adopted to this end.

THE UPOV CRITERION OF DISTINCTNESS

UPOV defines 'variety' as a plant grouping within a single botanical taxon of the lowest known rank.[28] Such a variety is distinct if it is clearly distinguishable from any other variety whose existence is commonly known at the time of the filing of the application (see UPOV, 1991, Article 1(vi)). The question is how the criterion of 'clear distinguishability' is implemented in practice.

TESTING DISTINCTNESS

UPOV has created guidelines for testing distinctness, homogeneity and stability of new varieties of plants.[29]

According to the Revised General Introduction to the Guidelines, the characteristics used for the identification of crop varieties must be capable of precise recog-

[26] Breeding history, diversity/integrity, adaptation, yield stability/lower yield, resistance/tolerance and human selection.

[27] These criteria need not correspond to the requirements that a *sui generis* right would have to comply with. These would have to be defined taking account of the activities to be supported, taking care to avoid any detrimental effects.

[28] This means distinct types at the lowest level taking account of the botanical nomenclature such as species or a subspecies: UPOV (International Convention for the Protection of New Varieties of Plants) as revised (1991) Article 1.vi.

[29] This consists of a general introduction and guidelines for individual crop species: UPOV Revised General Introduction to the Guidelines for the Conduct of Tests for Distinctness, Homogeneity and Stability of New Varieties of Plants. TG/1/2, 1979–11–14. UPOV Guidelines for the Conduct of Tests for Distinctness, Homogeneity and Stability: Potato *Solanum tuberosum* L.. TG/23/5, 1986–11 21. A list of test Guidelines can be found at http://www.upov.org/eng/publctns/pdf/testguid.pdf

nition and description. The varieties with which a tested variety has to be compared are those whose existence is a matter of common knowledge. The first basis of comparison is normally those varieties that are considered to be similar to the tested variety (Revised Introduction s.19).

Testing is then done according to qualitative and quantitative criteria. A *qualitative*[30] difference exists when 'the respective characteristics show expressions which fall into two different so-called "states of expression"', that is, different characteristics as described in the species guidelines. For potato leaves this would, for example, be the size (very small to very large), the width (narrow to broad), the waviness of margin (low to high), and the depth of veins (shallow to deep). *Quantitative* characteristics are those that are measurable on a one-dimensional scale and show continuous variation from one extreme to the other.

It seems that the tests are conducted on the basis of testing the expressions of genetic variety, i.e. on the basis of morphological and physiological[31] criteria only, not (yet?) taking account of recent techniques to enable the characterization of the *genetic material*, such as, for example, DNA sequencing (for details see Hardon *et al.*, 1994, pp. 9–11).

With a view to the identification of landraces, Hardon *et al.* (1994, p. 7) maintain that the recognition of the contribution of farmers in maintaining and enhancing genetic diversity requires reliable and objective *genetic* definition of the identity of landraces (and populations of wild species). However, taking account of the UPOV testing procedures, the question is whether, if not based on UPOV, another system for identifying landraces based rather on *morphological* and *physiological* analyses might be developed.

To this end, it must be possible to distinguish landraces of the same variety from other, similar varieties. This leads to the questions of first, to what extent a variety has to be distinct; secondly, to which characteristics this criterion ought to apply; and thirdly, how the 'state of the art', that is, the stock of landraces the examined variety would have to be compared with, is to be defined.

ADDITIONAL INDICATIONS FOR DISTINCTNESS OF LANDRACES

Additional indications for distinctness of a landrace could be: (i) its use being current in a specific, clearly demarcated geographical area to which, in turn, a specific social community can be ascribed; and (ii) its specific name (if more than one landrace of the same species is grown in the same region). The name might indicate specific distinctive characteristics (e.g. colour, region of provenance, size of plant or grains) (Zeven, 1998, p. 135).

The problem here is that (morphologically or genetically?) identical varieties can have different names in different regions and that due to the custom of open exchange, the allocation to a specific region could be difficult (although findings in the IPGRI paper in part 2 of this chapter suggest this is not the case).

The criteria of uniformity and identifiability

The criterion of uniformity is closely linked to the criterion of distinctness. It serves, first, to define clearly the subject matter to be protected. It means that the characteristics that are required to define distinctness must (in principle) be uniform for the variety: uniformity is used to distinguish between varieties.

The criterion of uniformity required by the UPOV plant variety protection regime is presently controversial. The concern is that this criterion reinforces trends towards

[30] Qualitative characteristics should be those that show discrete discontinuous states with no arbitrary limit on the number of states (Revised General Introduction, pp. 10–11).

[31] Plant anatomy, or 'morphology', refers to the description of the structure and parts of a plant. The 'phenotype' is the 'outward physical manifestation' of the organism, anything that is part of the observable structure, function or behaviour of a living organism (in contrast with the genotype, which is the internally coded, inheritable information).

genetic uniformity and thus has a negative effect on agrobiodiversity.

Therefore it is proposed to replace 'uniformity' with 'identifiability', the latter term leaving more flexibility to include landraces and thus to encourage heterogeneity (IPGRI, 1999, p. 15).

The element of time

As mentioned above, the breeding history can be another element for defining the functional origin of a landrace. It can be imagined that a specific variety has a long history in a specific region, or even that it traditionally belongs to a specific farming family. Accordingly, it is submitted that the breeding history, the creative input by farmers into a specified landrace (in analogy with the input of the formal breeder), and the social aspect inherent in the *in situ* breeding process, could be taken account of by a time criterion.

The problem is, however, to establish the relevant time period and to create supporting evidence.

4.3.4 Conclusions

The question at the outset was whether it is possible to establish a specific type of 'farmer-breeders' or, more appropriately, a functional definition of origin of landraces. This, in turn, leads to the question as to what would be the criteria for defining such a functional notion of origin.

Identified as possible criteria were distinctness and/or identifiability (by morphological criteria, possibly supported by the name and/or a delimited geographical origin); time as an indicator for the breeding history; and the social element.[32]

Accordingly, the question arises as to whether it would be possible to establish these facts in the field, and whether this kind of information has already been, or could be, stored in registration schemes containing the passport data of collected landrace-accessions. In the following part of this chapter, the authors conduct such a test of the data included in the CGIAR System-wide Information Network for Genetic Resources (SINGER), not with the intention of actually facilitating IPR claims, but just to test one of the biggest PGR databases in the world. If this is not the case, the subsequent question is whether it would be possible to create corresponding documentation schemes in the future.

4.3.5 Questions for further research

Given the above, several questions may be deduced for further research regarding the feasibility of allocating traditional PGRFA (and traditional knowledge) by registration systems. The basis would be to evaluate the described options by analysing an existing registration system. The basic question is: which of the above-described criteria (being time frame/breeding history; distinctness/identifiability) ought to be considered as a connecting point and could be made operational for the allocation of landraces?

This leads to the more scientific, technical and practical questions, such as:

- To what degree are landraces distinct between communities or regional entities? Is it possible, and frequently occurring, that specific villages, farming families or regions have specific (distinguishable) landraces with specific characteristics?
- Is it possible, theoretically and practically, to determine the number of years a specific landrace has been managed by a specific social unit? (5 years, 10 years, 20 years?) How long would this period have to be in order to allow for a legitimate allocation?
- And would it be possible in practice to allocate on the basis of these data specific landraces to specific (legally rele-

[32] Other characteristics such as adaptation, yield stability/lower yield, resistance/tolerance are criteria for distinguishing landraces from formally bred varieties. They are the elements that ought to be considered in establishing protective criteria.

vant) entities such as communities or families?

In evaluating the SINGER database, the question is whether in its present set-up it can be used as an instrument of allocation to specific communities on the basis of the *passport data* of the accessions stored in the system. Do the passport data for the accessions contain the relevant categories? Would an allocation be possible on the basis of the *existing entries*; or, if not, would it be possible and practicable to create additional passport data to this end?

Part 2: Farmers, Landraces and Property Rights: Challenges to Allocating *Sui Generis* Intellectual Property Rights to Communities over their Varieties

M. Halewood, J.J. Cherfas, J.M.M. Engels, Th. Hazekamp, T. Hodgkin and J. Robinson[33]

International Plant Genetic Resources Institute (IPGRI), Via dei Tre Denari 472/a, 00057 Maccarese (Fiumicino), Rome, Italy

4.4 Introduction

In this chapter, we analyse a number of issues associated with the possibility of allocating *sui generis* intellectual property rights to farmers over their landraces. (We generally prefer to use the term 'farmers' varieties', and, in the context of this chapter, the most accurate term might be 'plant groupings' or 'populations'. However, to maintain consistency within this volume, we will refer to 'landraces'. In those instances when we do refer to 'varieties' in this paper, we are not using it to mean varieties potentially protected by plant variety protection laws, i.e. those that are distinct, uniform and stable.) In Section 4.5, we examine the historical evolution of the concept of 'landraces'. In this context, we emphasize the relatively recent trend towards appreciating the dynamic role farmers play in developing, using and conserving landraces. We also highlight a few generally agreed upon characteristics of landraces: for example, their diverse, heterogeneous nature and the fact that in marginal environments they produce relatively stable yields (compared to 'modern' varieties grown under similar diverse circumstances). In Section 4.6 we examine the interplay of three key factors – plant reproductive systems, environment and human uses – that contribute to the incidence, duration and dissemination of distinct landraces (or, more accurately, distinct plant populations). In Section 4.7, we consider challenges associated with developing *sui generis* intellectual property laws to create legally enforceable property rights for farmers over landraces. In this context, we consider both technical legal options (for example, conditions for the protection of landraces pursuant to some form of *sui generis* intellectual property law), and the significance of the introduction of such laws in light of the importance of farmers' existing patterns of use, conservation, and exchange of landraces.

In Section 4.8, we depart from the progression of analysis in the previous parts and examine the quality and quantity of data that exist from past collecting missions that could be used to create a link between existing gene bank accessions and the communities from which, or near to which, the material was collected. In this context, our

[33] The authors would like to thank the following people for their comments on earlier drafts: Susette Biber-Klemm, Thomas Cottier, Peter Button, Carlos Correa, Jurg Schneider, Mohammed Sadiki, Leslie Lipper, Conny Almekinders, Rene Salazar, Daniela Solieri, Danielle Manzella, Kirit Patel, Devendra Gauchan, Pablo Eyzaguirre and Emile Frison. Any errors or misrepresentations are the responsibility of the authors alone.

primary source of data is the Consultative Group on International Agricultural Research (CGIAR) System-Wide Information Network for Genetic Resources (SINGER). In Section 4.9, we analyse the range of legal options that could be used to confer rights and benefits upon similarly situated communities in the future.

We conclude, in Section 4.10, that due to: (i) the highly diverse and constantly changing nature of landraces generally, and (ii) the dynamic interaction between farmers and their crops, it would frequently be difficult, from a legal technical perspective, to create *sui generis* intellectual property laws that vest rights of control (*qua* ownership) in farmers and farming communities over their varieties. Furthermore, in those cases where it may be legally technically feasible, it seems likely that to take advantage of such protections, farmers would need to alter their behaviour, particularly their patterns of exchange and their use of regenerative materials. Farmers farm the way they do in order to maximize their food security and to improve their livelihoods. There is a significant risk that the gains that might be realized through the introduction of intellectual property incentives to alter their existing practices would be outweighed by the losses realized as a result of these same changes. Nevertheless, we do not rule out the possibility that, under certain circumstances, it may be possible to establish a strong enough connection between a farmer, community or group of communities and a distinct landrace in order to justify recognizing that the farmer or community should enjoy some form of property right (or some other form of control that may not be a property right *per se*) over the landrace.

4.5 Historical Evolution of the Concept of Landrace

4.5.1 From nature to nurture: the anthropologization of the landrace

There is no taxon equivalent to 'landrace' in the International Code of Nomenclature for Cultivated Plants (Trehane *et al.*, 1995). It is not a cultivar, which is by definition distinct uniform, and stable. Although landraces are, in some sense, distinct (see discussion later in this chapter), they are neither necessarily uniform nor regularly stable (that is, demonstrating true breeding) in the absence of selection.[34] While various attempts have been made to define and describe landraces, as we shall establish in the following paragraphs, none of them offers an operational method that allows a sample to be categorized as a landrace in the absence of information about its use, cultivation and management.

The use of 'landrace' to denote a type of biological entity first emerged in the 1890s. Zeven (1998) provides a review of definitions since that time. In its first incarnation, a landrace was seen by breeders and others (though clearly not by the farmers themselves) simply as a source of material for scientific breeding and therefore worth conserving. There was recognition that landraces were often genetically variable populations (one reason they were valuable to breeders), and that yields were often lower but more stable from year to year than those of improved varieties. Beyond this, not much effort was put into defining the landrace, except to distinguish it from wild species and from the products of scientific breeding.

One common attribute of landraces in the early definitions was that they were autochthonous or endemic. Under this view, a landrace was associated with a particular place. Exactly how long the association had to exist was, however, a matter of

[34] In this case, we are referring to the stability of the genetic structure of landraces, which is different from the stability of yield or production or other performances of landraces. Indeed, farmers actually select indirectly for yield stability, particularly in adverse environmental conditions. They certainly do not select for genetic stability.

disagreement. Early authors talk about 'time immemorial'. Later, the idea of a farmer's generation or two (that is, 30–60 years) in consistent use was considered enough to denote a landrace. This evolution in the definition of landraces occurred over time as people progressively gained more interest in local populations and as data on local diversity built up.

Yield stability, which is often explicitly associated with wide adaptability, is another attribute that several authors use to describe landraces – generally to distinguish them from the products of scientific breeding. For some (for example, Harlan, 1975b), yield stability is a consequence of, and secondary to, genetic diversity within the landrace. This stability can arise in two ways. First, wide adaptability, as represented by genetic heterogeneity, will enable a population to yield under a wide range of environmental conditions. Secondly, environmental conditions that fluctuate from year to year will tend to favour different genotypes in different years. This characteristic is particularly important in marginal environments.

Early authors downplayed the role of human activity in maintaining the link between landrace and locality. Indeed, von Rümker (1908, cited by Zeven, 1998) said that no human selection is carried out; the qualities of a particular landrace are the result of an unselected adaptation to its growing conditions. Von Rümker added that a landrace would preserve its distinguishing characteristics even when grown outside its native region. Natural selection alone adapts the landrace to frost, drought, temperature and other environmental parameters. This view, which seems risible today, remained common through the 1940s, when Banga (1944, cited by Zeven, 1998) wrote that landraces survive 'without or with only little mass selection'.

Harlan (1975a) and Brown (1978, cited by Zeven, 1998) were among the first to state that landraces have come to rely on cultivation for their survival, which suggests that at least some form of artificial selection was taking place. Within a decade, views on the maintenance of landraces had shifted dramatically. Hodgkin *et al.* (1993) state that 'the most important feature' of landraces is that human intervention is needed to create and maintain them. Brush (1995), too, claims that landraces owe their existence to farmer selection.[35] Louette (2000) (Case Study 4.1) demonstrates that farmers' selection is critical to the maintenance of local maize landraces in Cuzalapa, Mexico. Teshome *et al.* (1999) show that farmers in Ethiopia are consistent in recognizing and distinguishing sorghum (*Sorghum bicolor*) landraces and that their accuracy 'approximates the accuracy of standard scientific taxonomic approaches'; further, local farmers are relied upon to maintain them. Fernandez (1995) provides an extensive list of examples of multiplication and conservation practices associated with indigenous varieties (that is, landraces), many of which (for example, planting mixtures, blending, sowing wild relatives, conscious hybridizing and the allowance of clones to flower) would inevitably lead to changes in the genetic, and thus agromorphological, constitution of the populations. Prain and Campilan

[35] Zeven (1998) concludes his review by adapting a definition (or is it a description?) that goes back to the early years of the 20th century. He writes that 'an autochthonous landrace is a variety with a high capacity to tolerate biotic and abiotic stress resulting in a high yield stability with an intermediate yield under a low input agricultural system'. A plant population, under this view, can only be called a landrace if one also knows about its agronomic performance and place of origin. It is curious that after providing such a full account of the development of the appreciation of farmers' roles in landrace development, conservation and use, Zeven's own new definition underplays the element of dynamic farmer selection/maintenance, which would reflect the shift from a 'naturalist' view that has come out clearly in recent research. Of course, as this comment on Zeven's definition makes clear, not all researchers are equally convinced of the central, consciously dynamic role of farmers. By way of another example, Hawkes (1983) states that the adaptation of landraces to their environment is the result of selection 'largely of an unconscious nature'.

(1997–98) demonstrate the 'cultural saliency' of the role of particular sweet potato landraces (calling them 'cultivars') in land consecration and first-planting rituals in upland Irian Jaya. In the absence of this cultural value, these particular sweet potato landraces would have vanished long ago.

Case study 4.1: Farmers' key role in selecting and perpetuating maize landrace traits

Louette (2000) investigated the approach of maize farmers in the Cuzalapa watershed of Jalisco state on the Pacific coast of Mexico. Cuzalapa farmers grow several kinds of maize in the presence of one another and of teosinte and other wild relatives of *Zea*. Since this species is a more-or-less obligate out-breeder, it would thus require considerable selection to maintain the genetic integrity of a sample under these circumstances. Louette defined a seed lot as the set of kernels of a specific type that a farmer selects and sows 'to reproduce that particular maize type'. In contrast, she defined a 'variety' or 'cultivar' as a set of farmers' seed lots 'that bear the same name and are considered to form a homogeneous set'. The seed lot was a physical entity; the 'cultivar' or 'variety', on the other hand, was associated with a name and was a semantic entity. A variety was considered 'local' if it had been planted for at least one farmer's generation (that is, approximately 30 years), or 'exotic' if it had been either recently introduced or was planted only episodically. An exotic variety could be a landrace that was local somewhere else or a commercially improved variety. Varieties could be a farmer's own seed, a local variety from another farmer or an introduction. Each seed lot was independently classified as 'own seed' (from the farmer's previous harvest, even if the parent generation was exotic).

Over six seasons (3 years) farmers grew 26 named varieties. Only six were local, but they occupied 80% of the land planted to maize and were grown by everyone. The other 20 were exotic, each cultivated by only a few farmers. Of the seed lots, slightly more than half (53%) were own seed; more than a third (36%) were from other farmers in Cuzalapa, and 11% from outside the region. In terms of area, 45% was sown to own seed, 40% to seed from Cuzalapa and 15% to exotic introductions

Some of the most widely grown varieties, which were a major source of household food, were local, and yet seed was often brought in from beyond the watershed. Farmers attached no great importance to their own seed, which they considered equivalent to the seed of that variety from another farmer. They also believed in the need to change seed sources regularly to maintain productivity. Thus, 'it appears unlikely that any farmer in Cuzalapa sows seed derived from a stock bequeathed directly from his parents'.

Gene flow among all the varieties planted in the area was assessed by looking for purple grains within white or yellow ears, which indicates pollination by the variety Negro. Farmers made no attempt to isolate varieties, and the observed gene flow was as high as 10–20% in the outside rows. This number fell to 1% by 2 or 3 m within the plot, but as Negro is not itself homozygous for grain colour, these figures underestimate the level of crossing. A further influence on the genetic make-up of a variety is drift, which is the result of using too few ears to produce a seed lot. More than 30% of the seed lots sown are drawn from fewer than 40 ears, but fields sown with less representative seed lots do not differ in their isoenzyme diversity from those sown with more representative seed lots. Despite the limitations of isoenzyme diversity as a metric, it is clear that gene flow sustains the diversity and productivity of the varieties.

Farmers maintained the characteristics of named varieties by selecting ears at harvest to use for seed lots. They chose well-filled kernels from healthy ears, sometimes discarding the top and bottom kernels in the row. The mean weight of ears selected for seed was 30% higher than for ears drawn at random from the harvest. By selecting ears after harvest, however, farmers did not exclude outside rows, and yet selection still reduced contamination with foreign pollen. Deliberately contaminating

Negro with (recessive) white or yellow varieties resulted in an increase in non-purple kernels from 7.5% to 16.5% when seeds were selected at random, but this number stayed constant in farmer-selected seed.

Louette concluded that the way in which farmers actually managed their seed 'calls into question the genetic definition of a landrace'. (The authors note, however, that Louette did not offer such a genetic definition, save to say that one of the concerns expressed about the *in situ* conservation of landraces is the danger of genetic contamination from other varieties.[36]) She also concluded that 'the assumption that traditional systems are closed and isolated with respect to the flow of genetic material is clearly contradicted by the results of this study'. She pointed out that seed lots would disappear if a farmer does not save seed from a particular planting, but that farmers are not concerned about that. They are also happy to use seed lots of local varieties from outside the watershed. Landraces in Cuzalapa are distinct and diverse, but they are subject to intense selection and rely on constant infusions from outside the locality.

4.5.2 Details of farmers' innovation: improving landraces' contribution to livelihoods is not necessarily linked to promoting their distinctness, stability and uniformity

Not surprisingly, the older attempts to define and characterize landraces were often carried out without the benefit of experience with farmers in their fields. More recent research, however, has relied on turning to farmers and asking them how they manage the genetic resources they use. For example, Louette (2000) introduced her study of maize farmers in Cuzalapa by stating that the 'terms and concepts used in this work are based on farmers' own practices and concepts'. Accordingly, as already set out earlier, she defines 'varieties' as sets of seed lots selected and named by farmers. Looking at how the people who grow the crop actually deal with such issues as genetic identity, gene flow and seed source has resulted in the emergence of a new kind of sub-specific category called the Farmer's Unit of Diversity Management (FUDM). The FUDM is a way for scientists to measure crop diversity based on the way the farmer perceives diversity. Diversity could be reflected in seed quality, forage biomass, colour, plant height, cooking value and so on, depending on how the farmer categorizes the diversity he or she is using (Jarvis, personal communication). The FUDM is a relatively new concept in as much as it looks at defining plant groupings through reference to farmers' use values and/or physical traits, rather than by local names alone, as names have been found not to adequately reflect diversity. Local names have also been found to refer to different varieties of the same crop[37] (Busso *et al.*, 2000; Grum, personal communication).

Bringing anthropological and biological analytical approaches together has revealed that in many instances farmers have a detailed understanding of the dynamics of the varieties with which they work. In Morocco, for example, Sadiki *et al.* (2001) show that the units of diversity in farmers' names for faba bean (*Vicia faba* L.) correspond clearly to the units identified by a statistical analysis of standard morphological measurements. A more detailed study of lucerne (*Medicago sativa*) in Morocco identifies separate FUDMs, each associated with a single-seed producer in a specific village (Bouizgaren *et al.*, 2001). The authors conclude that the farmers' names do not consistently distinguish the FUDM, and that the FUDM should be associated with the farmers' description (usu-

[36] The authors wish to comment that this may not in fact always be true. Contamination that would introduce useful characters such as disease resistance would help *in situ* conservation and diversity.

[37] Studies of sorghum in Mali and Zimbabwe reveal that the name a farmer gives to a variety is an imperfect piece of information, if we want to know something about the variety's genetic make-up. Varieties that share a name can differ markedly from one another, while apparently identical varieties can have different names.

ally a morphological description) of these units. Bouizgaren *et al.* (2001) state unequivocally (albeit without reference to the issues raised later in this chapter) that the different FUDMs 'belong' to different villages and are used by the farmers who are neighbours of the seed producer. The use of farmers' names to identify FUDMs without corroborating biological data is, however, prone to misinterpretation – and not only because the names themselves are incomplete pieces of information.

A wide-ranging study of taro (*Colocasia esculenta*), barley (*Hordeum vulgare*) and rice (*Oryza sativa*) in mountain communities in Nepal showed that the ability to link names with genetic discrimination depends crucially on the species in question. Bajracharya *et al.* (2001) used molecular tools to assess genetic variability in the three species. In rice, there was very little difference between landraces with the same name maintained by different farmers. The authors conclude that the name farmers give a variety, based largely on the appearance of the panicle and grain, is a poor indicator of its genetic make-up, which at these high altitudes is far more powerfully influenced by selection for resistance to cold. Furthermore, the lack of variability among all of the samples of rice suggests that these samples could all have been selected from the locally common landrace called Jumla Marshi. Barley is different. Each of the four named varieties considered in the study differs significantly from all of the others, and all four are somewhat diverse within the landrace, even though some landraces are grown over a much wider agroecological range than others. This fact may reflect the existence of seed-exchange systems. Finally, taro's position is somewhat intermediate. The varieties, as named by farmers, show some differences between each other, but statistical analysis placed each variety into one of only two distinct groups.

The concept of FUDM as a means of identifying the relationship between farmers and landraces is gaining acceptance. One drawback is that it generally involves a fair amount of work to collect the material, gather the anthropological, sociological and agricultural information about it, and then to marry these data to a detailed biological analysis. On the other hand, this work can be carried out for a relatively low cost (Jarvis *et al.*, 2001).

The fact that farmers recognize and select for certain traits does not signify, however, that farmers purposefully work towards the development, conservation and use of 'distinct' landraces. Jarvis *et al.* (2001) point out that '[f]armers' preferred or valued characteristics for a given variety may be distinct from those used to distinguish or name a variety' and that the 'characteristics used to distinguish a particular variety may differ from those characteristics that the farmer actively selects in the next generation of the plant population'. For example, in Morocco, farmers prefer 'fortas', a durum wheat landrace, for the taste of the couscous made out of its flour, but it is recognized and named using characteristics of the spikes (for example, colour, length and plant height) (Sadiki, personal commentary). In a study of sorghum landrace diversity and loss over a 20–year period in Zimbabwe and Mali, Grum (personal communication) demonstrated that seed exchange and selection practices can result in the preservation of desired agromorphological traits over time within a constantly changing constellation of plant groupings. While the names of between 30 and 70% of varieties (in the dry north and humid south of Mali) have been lost, a preliminary analysis of agromorphological traits indicated that a significantly lower number of traits had been lost (Mikkel Grum, personal commentary). Louette (2000) demonstrated that maize farmers mix large proportions of seeds of different maize varieties/landraces/populations in with a considerable range of local material in their fields each year, allowing them to cross-pollinate. Against this backdrop of rapidly introgressing foreign and local material, the farmers select similar seed lots from year to year, thereby maintaining a recognizable population that Louette defined as a landrace. The study was relatively short, analysing only 3 years and six generations of cultivation. Yet even over this time frame, Louette established that there is considerable genetic difference between the seed lots selected by the farmers, despite the fact that the selected-for traits remained similar. Even if selection for traits by different farmers remains similar, selection would take place under different management practices and environmental conditions (for example, different plots with different soil conditions and so on), so that the same selection practices would lead to variations among the populations within the same landrace (Sadiki, personal commentary). Louette made a strong case that the 'landrace' of maize in her study area was one unit in a highly diverse metapopulation. One of her conclusions was that:

> [t]he traditional management of maize in Cuzalapa contributes more to the conservation of a general level of diversity than to the conservation of genetically stable and distinct maize populations. A landrace is far from a stable, distinct, and uniform unit. Its diversity is linked to the diversity of the material sown in the area, and then related to the diversity of the introduced varieties.

Well-rehearsed facts, which are certainly worth repeating in this context, are that farmers' systems of informal exchange are crucial to: (i) their ability to constantly introduce new material into their cropping systems; (ii) maintaining high levels of diversity; and (iii) maintaining relatively stable yields. In Nepal, 90% of the seed planted by farmers is accessed through informal sources (that is, from their own farms or obtained through exchanges with other farmers). In Cuzalapa, Louette (2000) found that out of the 27 varieties grown by the studied farmers over the course of 3 years, only six were considered to be local – that is, grown in the region for more than 30 years. All the others were introduced from outside the region. Furthermore, even the seed lots for local varieties were frequently brought in from outside the region. As a consequence, no 'introduced seed lot that is morphologically similar to a local variety would be distinguished, so no exotic variety with characteristics similar to those of local varieties would be recognized as a distinct cultivar'. This is an important fact in as much as it means that '[a] "local variety" is therefore not constituted by seed lots of exclusively local origin'. All told, each year just under 50% of the seed that the farmers planted was from other farmers or from outside the region. The farmers did not know the actual origin of the exotic material – they knew where they got it, but nothing beyond that fact. In a similar vein, Prain and Campilan (1997–98) reported an extraordinary turnover of sweet potato landraces by farmers practising shifting cultivation in Wawawaga, Central Irian Jaya. The authors reported that in the two years compared – 1994 and 1998 – a total of 47 landraces were cultivated each year by the farmers studied. However, in 1998, only 27 (roughly 58%) of the landraces were the same. Forty-two per cent had been replaced by other sweet potato landraces. Castello (2002) also confirmed the importance of farmer exchange and selection for maintaining diversity. He compared genetic diversity in maize collected from farmers' fields with maize stored in a gene bank that was collected from the same fields 5 years earlier. He found that 5 years after the collecting mission, there was more genetic diversity in the material currently in the farmers' fields than was contained in the *ex situ* materials.

In formal systems, the rights holder, or his or her assignees, generally sells protected varieties to farmers who may be spread over a large region, and even globally. In the case of landraces, which are adapted to specific environmental conditions, the seed is produced by the community of farmers for their own needs. If introduced into another region, such material will undergo a new selection process by the farmers. So, over time, the issue is mainly access to these landraces as sources of diversity rather than as 'cultivars' *per se*. Farmers are selecting characteristics and traits rather than the structure of the landrace, because it is these traits and characteristics that respond to their needs (yield, product harvested and so on). They do not select for distinctiveness or uniqueness *per se*,[38] but they do select for what responds to and fits their needs in terms of production (that is, security and stability of production, yield and quality). Therefore, landraces play the role of a source of diversity from which each farmer selects and adapts in his or her own way to meet his or her own needs over time.

4.6 How the Interplay of Plant Reproduction Systems, Human Uses and the Environment Affect the Incidence and Duration of Distinctions between Landraces

In the context of analysing the historical evolution of the term 'landrace' in the previous section, we focused primarily on the importance of human influences on landrace use, conservation and improvement. In this section, we shift our analytical focus to an historical analysis of a wider combination of factors – reproductive systems, environment and human uses – that affect the incidence, longevity and dissemination of distinct landraces.

The three major breeding systems in plants are cross-pollination, self-pollination and clonal propagation. It is important to recognize that these systems are not always absolutely distinct. There is a continuum from completely clonal, with no evidence for seed set, through to obligate out-breeding, which is ensured by mechanisms to create self-sterility. Looked at simplistically, clonally maintained species (including potatoes, cassava, dates and olives) ought most likely to retain their distinctive features over time, as their reproductive systems permit only limited or no gene flow. In-breeders (including beans, rice, wheat and barley), which usually self-pollinate, would change relatively slowly. Finally, out-breeders, cross-pollinated for the most part (including maize and pearl millet), would change most rapidly. Conversely, and equally simplistically, out-breeders would be potentially the fastest to develop new distinctive traits, clones would be the slowest, and in-breeders would lie somewhere in the middle in the absence of selection forces. However, the following paragraphs will demonstrate that environment and human uses can end up affecting entirely different outcomes *vis-à-vis* distinctness between plant groups than one might predict when looking at the breeding systems of plants alone.

Commerce and trade will spread varieties and genes, while germplasm associated with closed, isolated cultures will probably remain distinct longer. Non-biological components can also influence the distinctness of a landrace. These are essentially the features of the environment, including climate, soil and geographical features such as altitude and topography. For example, crops growing in isolated environments, such as high, deep valleys and dense tropical forests, are likely to change less than those growing in open environments, windswept plains and areas well served by road networks. Likewise, crops growing in regions where climate and other factors are very different from year to year might diverge more rapidly than those in more stable environments.

4.6.1 In-breeders

Few in-breeders are absolutely self-pollinating. For the rest, a small amount of introgression, from wild and weedy relatives or other varieties growing nearby, can represent an important source of novel genes that adds an element of dynamic change to the population's genome. A landrace is more likely to be affected over the years than a modern variety, for which fresh seed is often purchased anew each season.

For example, the wild rice *Oryza barthii* is a staple of southern and central Sudan. It is truly wild but represents a very important food source and is a valuable resource for breeders of *O. glaberrima* (Wood and Lenné, 2001). More importantly, in this instance, although domesti-

[38] We do not mean to suggest that commercial breeders select for distinctness, uniformity and stability (DUS) exclusively.

cated rice (*O. sativa*) is a predominantly in-breeding species, when the two species grow in proximity, wild alleles introgress into the *O. glaberrima* genome. Similarly, in Asia, genes from *O. rufipogon* introgress into the domesticated rice genome as a result of their growing in close proximity (Xiao *et al.*, 1996). In Pahang, Malaysia, the Pesagi swamp farmers cultivate vitreous rice and glutinous rice together, harvesting ear by ear to take advantage of introgression (Lambert, 1985). Harvesting seed ear by ear enables farmers to select desired types and to rogue off-types for the next season's rice seed. The individual cultivars are continuously improved through seed selection and the seeds of different households, although having the same name, are genetically very different. Moreover, farmers experiment with off-types and newly received varieties, thus contributing to a very dynamic gene pool. The Apau Ping Dayaks of East Kalimantan cultivate upland rice in shifting cultivation systems (Setyawati, 1996). Each farmer manages about five different varieties, but there is a continual search for new varieties. Interestingly, the varieties cannot be selected based on adaptation to soil conditions as seed is saved before new land is opened up. Thus, it is necessary to have either a range of varieties, expecting that some will be adapted to the conditions likely to be encountered, or to cultivate varieties that are known to be broadly adapted to a range of conditions.

Seed exchange among communities will also affect gene flow, generally leading to a reduced distinctiveness of varieties in an area, in the absence of selection. In Nepal, for example, it is common to exchange barley and wheat seed every 3 years among groups of villages, which leads to similarity of types over the district (Iijima, 1964). Parzies *et al.* (2001) measured gene flow for two different crops, barley landraces from northern Syria and pearl millet (*Pennisetum glaucum* ssp. *glaucum*) landraces from Rajhastan, India. They predicted that lower gene flow at greater geographic distance would be associated with restricted seed-exchange systems. Gene flow for barley was lower than that for pearl millet, and the seed exchange of barley was indeed restricted while pearl millet landraces were exchanged over far greater distances. Pearl millet is more of an out-breeder than barley and so might be expected to enjoy greater gene flow, but Parzies *et al.* concluded that the greater gene flow seen in pearl millet was more the result of differing seed-exchange strategies than of the differing mating systems of the two crops. This is a clear indication that social (seed exchange) systems and biological systems are intertwined.

Common beans (*Phaseolus vulgaris*) are perhaps one of the most predominant in-breeding crops, although they can cross-pollinate to a limited extent under conducive conditions. In Malawi, as in many other countries, beans are grown in heterogeneous mixtures of genotypes that form local landraces.[39] These genotypes differ in seed colour and canopy characteristics as well as in several other traits. Farmers consciously adjust the mixtures to preserve heterogeneity and maintain the characteristics of the landrace (Martin and Adams,

[39] A landrace of an out-breeding species will be made up of individuals that are heterozygous at most alleles, while a landrace of an in-breeding species will tend to be made up of several different, but largely homozygous, types of individuals. This, in itself, can be a source of confusion, particularly in the case of an in-breeder such as *Phaseolus vulgaris*. Some authors (Martin and Adams, 1971, cited by Zeven, 1998) describe the mix of diverse genotypes grown by one farmer in one field as a landrace. Others (Voss, 1992, cited by Zeven, 1998) say that the farmer is growing a mixture of several landraces. Certainly farmers often distinguish the seed-colour forms in a bean mixture with different names, but it must be remembered that a single-coat colour could nevertheless be associated with several different genotypes and that an otherwise uniform genotype could be associated with two or more seed-coat alleles. Taking this argument further, uniform appearance can easily mask genotypic diversity. Zeven (1998) offers the example of an Austrian wheat landrace that appeared genetically uniform. Nevertheless, an investigation of the storage proteins in 50 randomly chosen grains revealed the presence of 31 different phenotypes.

1987a, b). One benefit of these mixtures is a kind of within-population buffering that allows the crop to better withstand pressures from pests and diseases and to compete for resources, including moisture and light. The response of the population as a whole to all of these pressures will affect the frequencies of each (in-bred) genotype from year to year, but if the alterations in population structure are not in line with the farmers' requirements, they will actively intervene. For example, Adams and Martin (1988) reported that beans with a white seed coat were more susceptible to disease than were other forms but that farmers consciously added them to the sowing mixture when their numbers dropped below a certain level.

Case Study 4.2: Gene flow between in-breeders; a case study of sorghum

Sorghum (*Sorghum bicolor*) is a staple cereal crop in Africa across a wide belt of the Sahel. Gene flow among landraces and between landraces and wild relatives plays a large role in the evolution of sorghum landrace genomes.

Sorghum is largely an in-breeder, but a significant amount of out-crossing takes place. For example, there is substantial gene flow when sorghum is sown across large tracts of open savannah. *S. verticilliflorum* is the progenitor of *S. bicolor* and grows over extensive areas of tall grass savannah in Sudan (Wood and Lenné, 2001). It is harvested and eaten and represents a valuable genetic resource for breeders. It is also a significant source of wild alleles for the cultivated crop. This demonstrates the permanent influx of genes from the wild into cultivated material. Ethiopian farmers use the same strategy by interplanting (or tolerating) domesticated sorghum with weedy material from the same species. The Nuba Mountains area of Sudan is relatively small, yet populations of cultivated sorghum and sesame (*Sesamum indicum*, which is largely, but not exclusively, self-pollinated) are very variable (Bedigian and Harlan, 1983). The Nuba people are geographically isolated and culturally diverse in many respects, including their agricultural practices and crop plant rituals. Bedigian and Harlan suggested that this cultural diversity contributes to cultivar diversity. Landraces of sesame are sown in mixtures with sorghum landraces, and several wild and weedy relatives of both species grow in the cultivated fields and undoubtedly cross with the cultivated varieties. Although seed selection to maintain desired characteristics would remove any obviously undesirable gene combinations, McGuire (2002) suggested that gene exchange has kept the landrace cultivars of sorghum dynamic in Harerghe, Ethiopia.

4.6.2 Out-breeders

Pearl millet and maize are examples of out-breeders. Pearl millet, although self-fertilizing, is protogynous (that is, the female part of the flower matures before the male), making it largely an out-breeder. This characteristic promotes gene flow within landraces and between landraces and wild relatives. Busso *et al.* (2000) used molecular markers to investigate the diversity of pearl millet landraces in two Nigerian villages. They found that there was a 'greater level of similarity between different landraces grown on the same farm than between identically named landraces grown by different farmers in the same village'. This finding cautions against using local landrace names as indicators of genetic identity, because a single name can embrace several populations that differ greatly at the genetic level. This is clearly demonstrated for several species. Within landraces, variation, expressed as differences between populations of the same landrace, arises as a result of cultivation and selection of seed lots by different farmers under different management and plot environmental conditions. (This variation usually affects traits selected by farmers less than it affects others.) Names can only be used reliably within a given geographic area or with a set group of users (for example, at village, community, or provincial levels). It is also a testimony to the effect of the active cross-breeding reproductive system of pearl millet – the 'different' landraces and wild relatives maintained by individual farmers introgressed to the point that they were more similar to each other than to the landraces by the same name outside the geographical reach of their own pollen. The geographical isolation of the farmers' plots from one another reduced the possibility of cross-pollination between farms. This isolation was maintained by the fact that these farmers, unlike those in the Cuzalapa valley described by Louette (2000), rarely exchanged seeds, thereby maintaining the isolation of their crops. Consequently, although the landraces themselves were genetically very similar, the overall mix of landraces on each farm was potentially unique. This outcome too is very different from that which was obtained with respect to maize in the Cuzalapa valley, where seed for varieties considered to be 'local' were frequently obtained from outside the region. In the final analysis, however, one important similarity exists between the results obtained from the Louette and Busso studies. Both emphasize that the landraces (or 'varieties' as they are called in Louette's study) are best described not so much as distinct, uniform and stable varieties, but rather as sub-units of the diversity of materials constantly shifting within the region in which they were grown (Louette refers to 'metapopulations'). The geographical area of significance is much smaller *vis-à-vis* the Nigerian farmers growing pearl millet, but the overall conceptual link between local landraces and larger, shifting distributions of diversity is common to both.

4.6.3 Clones

Many extremely ancient clones of a range of crops mentioned in Roman literature are still being cultivated (Robinson, 1996). The Dottato fig, for example, which is mentioned by Pliny (AD 23–79), is still grown in Italy. The Sari Lop fig has been grown in Turkey for 2000 years, and the ancient cultivar Verdone has been cultivated in the Adriatic for centuries. Aroids, date palms (some clones of which may date back to Neolithic times), and olives (some little changed since Roman times) are other long-lived clones. Golding hop has been grown in Britain for 250 years. No wild forms of ginger exist, although there are many distinct types in cultivation. Garlic, grapes (10,000 separate clones of which are known), saffron, sisal, vanilla and black pepper are among the many vegetatively propagated crops that exist as ancient clones. Although some can be bred (as opposed to selected), modern breeders have made little progress compared to the accomplishments of early cultivators. Consequently, the genetic constitution of these landrace genomes has changed predominantly as a result of mutation and the clonal degeneration associated with disease susceptibility. This situation contrasts with that of the in-breeders and out-breeders discussed earlier, whose genomes exist in relatively dynamic states over time. Yet while some clonal landraces have changed little, they have often become widely distributed, often precisely because they are such good selections.

Many of the most ancient clones are long-lived perennial species and are cultivated as such. One can draw a further distinction between those crops that people generally choose to maintain clonally on an annual cycle, such as potato and cassava; those that lack all ability to reproduce sexually, such as garlic and most bananas; and those that are kept for long periods, such as fruit trees. There is generally more opportunity for the annually maintained species to change, although the products of sexual reproduction among crops that are typically propagated clonally make important contributions to the diversity and identity of these crops (for example, olives, dates, potatoes and so on).

Many thousands of potato landraces are cultivated in the Andean region, including seven distinct species of various ploidy levels (that is, having different numbers of chromosomes) and numerous wild relatives. Brush *et al.* (1981), commenting on potato cultivation in Peru, argued that considerable hybridization occurs because the people sow mixtures of potato species and genotypes, including modern varieties. Thus, novel genes are constantly being introduced into a crop through introgression (for example, by chance cross-pollination, as discussed later in this chapter), which might be considered to be genetically constant for long periods as it is reproduced clonally by replanting harvested tubers. Brush *et al.* also identified the considerable exchange of tubers among communities, which also serves to keep the landrace populations of potato dynamic. Johns and Keen (1986) reported a similar situation in the Altiplano of western Bolivia, with new potato types constantly appearing in farmers' fields as a result of chance cross-pollination and unconscious tuber harvesting. Potatoes are particularly prone to generating volunteer plants, as some tubers always remain unharvested.

Most landraces of domesticated potato from the southern Andes of Peru are extremely versatile and broadly adapted to a range of ecogeographical conditions (Zimmerer, 1998). Many landraces have become cosmopolitan through germplasm exchange that has taken place through the extensive networks of barter, gift giving and trade that exist among the relatively well-linked communities of the area. The adaptability of these varieties results from a variety of factors, including selection for tolerance to a range of biotic and abiotic stresses characteristic of a range of altitudes and climates, gene flow among numerous domesticated and wild potato species, agricultural practices of crop rotation (which leave tubers in the field), and seed-exchange systems. The result is that the genetic structure of potato landraces is not related to differences in habitat (Zimmerer, 1998). In essence, landraces can be locally characteristic, but they are generally components of large, open, interlinked genetic systems.

Cassava (*Manihot esculenta*) is another crop that people generally reproduce clonally from stem cuttings. In nature, however, the species is a perennial out-breeder. In the tropical rainforest of Peru, the Jivaro tribe maintains a dynamic collection of landraces. As old cultivars are lost, new ones appear as volunteers, possibly hybrids from crosses with wild relatives, and these are nurtured to maturity (Boster, 1984). However, as a result of the relative isolation of cassava farming communities in the Amazon, the exchange of planting material is restricted, and the local cassava populations remain moderately differentiated (Salick *et al.*, 1997). Moreover, unlike the highly versatile potato populations cultivated in the Andes, populations of cassava from the Amazon basin are more narrowly adapted to their environment. However, on a field scale, people grow extremely diverse mixtures of germplasm (Boster, 1984; Salick *et al.*, 1997).

Schneider (1995) reported that for another clonally propagated crop, sweet potato (*Ipomea batatas*), considerable indigenous knowledge has been associated with volunteer seedlings. Like cassava, spontaneous seedlings, often generated through hybridization, were observed in New Guinea and could be adopted as new varieties or, perhaps more accurately, be incorporated into existing varieties. The future fate of these volunteers was determined at harvest, when they were either perpetuated through cuttings or disposed of. The conclusion is that even landraces that are clonally reproduced show a considerable degree of variability. The influence of breeding systems on variety diversity is gradual.

This albeit simplified examination of the part played by breeding systems and farming practices in the development and preservation of distinctiveness serves to accentuate the conclusions of the first section. Although it may be possible to identify populations that we can choose to call landraces (or FUDM), gene flow and selection, which are influenced by breeding systems, physical environment, farmers' cultivation practices, seed exchange and trade, will result in changes in distinctness over time, with the result that landraces are components of large, interconnected, dynamic genetic networks that often defy strict definition according to the formal system as it is currently constructed.

4.7 Assessing Potential Intellectual Property Policy Responses

It is difficult to trace back the exact processes by which specific human communities shaped crop development. There are globally important crops such as taro and sorghum for which it is difficult even today to delineate the boundaries between wild and domesticated types, landraces, local cultivars and modern varieties. While even the biological boundaries are not always clear, the distinctions between landraces and modern cultivars are largely social – that is, the modern variety is the product of a formal crop-breeding process undertaken with clear breeding objectives set out at the beginning of the process. Landraces are the result of selection pressures arising from interactions between cultivars, the natural environment and human communities that husband the crop. With respect to human influence, the landrace is the result of an ongoing process over a much longer period of time in which one or more human communities are involved. The degree to which this process can be isolated over time and space enables one to associate specific landraces with specific communities living within a specific environment. Some crops, such as teff, are clearly associated with the cultures and communities of the Ethiopian highlands. Prior to 1970, no other peoples grew the crop outside this area, and those that now grow teff outside the Ethiopian highlands obtained their genetic material from the communities in those areas. The same can be said for fonio in the Guinea savannah and the Sahelian zones of West Africa, and quinoa in the Andean highlands, among other examples of neglected and under-utilized crops. For globally distributed crops, such as maize, wheat, sweet potato, potato and so on, this association is much more difficult, even when we know the origins of the crop and how it has spread. The landraces are the result of exchanges and contributions by many cultures and communities that carried the crop to new environments, used it in different ways, and managed it according to the preferences embedded in their cultures and food habits. The ability to differentiate this people and cultivar association over time and in a particular space is the *sine qua non* for any assignation of rights to a landrace.

4.7.1 Challenges

Any attempt to create legal rights (on behalf of the communities) and obligations (on the part of third parties) with respect to landraces will require the creation of criteria for assessing the material concerned and its relationship to the subject individual, community or communities on a case-by-case basis. Research on this topic to date can be divided into explorations of three alternative legal mechanisms: (i) *sui generis* intellectual property laws that vest property rights in communities as (a) owners of landraces *per se* and/or (b) owners of rights to market the landraces in association with the name of the community or territory where it is grown; (ii) access laws, which provide communities with the right to refuse, or to set conditions upon, third parties collecting landraces grown on their land; and (iii) recognition (and elevation to levels of national and international recognition) of communities' own traditions, customs and laws regarding the use, exchange and conservation of landraces.

In the following paragraphs, we will focus on *sui generis* intellectual property laws that would vest rights of ownership in farmers over their landraces *per se* (that is, the first category of mechanisms listed above). In so doing, we are not suggesting that this form of intellectual property law has the most potential to 'protect' farmers' interests in their landraces, or that they would be more effective than access laws or the recognition of communities' customary laws and practices. In fact, we tend towards the opposite conclusion. That said, we have chosen to focus on this form of intellectual property law for a number of reasons. First, there is currently a great deal of discussion in community, national and international fora about the potential of *sui generis* intellectual property laws to protect farmers' interests in their landraces by creating exclusive forms of control over those landraces and vesting them in farmers and farming communities. We hope that we can make a contribution by introducing more directly technical, social and biological considerations into these discussions than has been done in the past. Secondly, the preceding analysis in this chapter regarding landraces does not have much bearing on whether or not access laws can be effectively used to provide communities with a means to protect their interests in their landraces. If there is an access law in place that provides communities with the right to consent to, or to refuse, collecting on their land (or land in which they have interests), it does not matter if the material in question can be identified as a landrace or not. The community concerned would have the right to control access to it, no matter what it is, *as long as it is located on their lands.*[40] Thirdly, a study of communities' own traditions, customs and laws, and how they may (or may not) be recognized within national and international legal frameworks, is well beyond the scope of this chapter. We feel a much more robust appreciation of local communities' practices, customs and laws is ultimately a necessary element in developing a truly effective system for the protection of landraces. However, this subject is also beyond the scope of this study. Finally, we want to reiterate that in this chapter, we do not focus on the possibility of the appellation of origin laws as a means to provide benefits to farmers and farming communities. Instead, we focus almost exclusively on legal mechanisms to create *sui generis* legal forms of control of third parties' uses of the landraces *per se*, which is more akin to existing plant variety protection laws than *appellation d'origin*.

While we certainly do not want to limit ourselves to analysing the criteria for protection that are part of the 1978 and 1991 International Convention for the Protection of New Varieties of Plants (UPOV Convention),[41] they do provide an obvious point of departure for analysing potential intellectual property protections for landraces.[42] It is widely agreed that the conditions for protection established within the UPOV Convention – that plant varieties must be distinct, uniform and stable – generally preclude the extension of protection to landraces (though there may be exceptions). This agreement is founded upon a widespread appreciation of many of the facts set out in the preceding analysis – that generally speaking, landraces are not uniform nor are they sufficiently stable over time.[43] Furthermore, as we shall discuss in some detail later in this chapter, questions linger about the degree of distinctness of landraces. However, it is still worthwhile to attempt to locate where, within the UPOV framework of protection for different kinds

[40] In this case, we are assuming that the national access law would recognize the right of individuals and communities to refuse (or allow) access to genetic resources located on their lands.

[41] International Convention for the Protection of New Varieties of Plants (1991) [hereinafter UPOV Convention].

[42] Leskien and Flitner (1997), IPGRI (1999) and the Crucible II Group (2001) have analysed options for the domestic implementation of 'effective *sui generis* protections' for plant varieties, in furtherance of Article 27(3)(b) of the WTO Agreement on Trade-Related Aspects of Intellectual Property Rights, 15 April 1994, in the Marrakesh Agreement Establishing the World Trade Organization, 15 April 1994, 33 I.L.M. 15 (1994) [hereinafter TRIPS Agreement]. In all of these three studies, among other things, the authors investigated variations on the distinctness, uniformity and stability [hereinafter DUS] criterion that are included in the UPOV Convention, *supra* note 10. They did so in the context of attempting to lessen the purported effect of the DUS criterion to create incentives for plant breeders to develop varieties with narrow genetic bases. In this context, they considered that distinct (D) and identifiable (I), or just distinct alone, might be preferable conditions for protection. However, the authors did not have protection for farmers' varieties firmly in mind when considering these criteria; nor did they engage in analyses of data regarding landraces *per se*.

[43] Again, stability in this case refers to the structure of the landrace and not to the stability of the protection. The structure of landraces is dynamic, changing over time in response to fluctuations in growing conditions, including farmers' management and environmental factors. These dynamic changes lead to the stability of the landraces and their stable performance over time.

of material, landraces would come closest to 'fitting in' and being protectable (at least by way of analogy). In regard to their structure, landraces are most comparable to population varieties. There is, however, a very significant difference between modern varieties, including population varieties, and landraces. Modern varieties are continuously reproduced (i.e. maintained) from the same known and stable parents, so the system for their production is 'closed' with no new genes being introduced. As the earlier analysis has emphasized, the development and use of landraces takes place entirely in systems that are radically 'open' in comparison. Landraces generally are, therefore, as stated earlier, components of large, interconnected, dynamic genetic networks that defy strict definition.

Of course, landraces possess some degree of distinctness. Otherwise, from a purely logical perspective, it would be impossible even to discuss the possibility of different landraces and it would not be possible for farmers to engage in any of the selection practices that we describe earlier. If there is nothing to favourably distinguish a sub-population when farmers are selecting seed, they will pass it over in favour of other populations (Boster, 1985). Of course, as we have stated, farmers generally do not care about distinctness *per se*. Yet by choosing and selecting traits that meet their needs, they end up creating populations the structure of which corresponds to specific phenotypes. However, several questions still remain. Is the incidence of the traits selected by farmers consistent enough over space and time to be able to say that the landrace is distinct enough, or, to use a different term, identifiable enough, that it can be protected by a *sui generis* intellectual property system?[44] Is such a system technically practical?[45] Is it worthwhile?

---

[44] The International Union for the Protection of New Varieties of Plants [hereinafter UPOV] has issued guidelines for determining distinctness (UPOV, TG/1/3, 2002), pursuant to which 'a variety may considered to be clearly distinguishable if the difference in characteristics is: (a) consistent, and (b) clear'. UPOV further issues species guidelines concerning traits to be considered for distinctness and the thresholds for the levels of those traits within the variety concerned. One possibility for a *sui generis* system of intellectual property protection for landraces would be to attempt to define alternative (lower) minimum threshold levels for the incidence of the particular trait or traits in question on a trait-by-trait, crop-by-crop basis. This standard-setting exercise could be done in advance of applications on a generic basis, or it could be done, at least at first, on a more reactive basis, with governments stating what crops they would be willing to consider claims on, and then creating expert committees to consider whether or not a minimum threshold of incidence for the particular trait or traits is a satisfactory way of defining the variety in question, and, furthermore, whether the variety in question meets this standard. Clearly, there is potentially a considerable cost associated with establishing such an alternative system of standards for landraces, and countries would have to be careful to consider the costs and benefits before embracing it outright. Another possible problem is that such protections for widely heterogeneous landraces will represent 'grabs' over significant ranges of diversity overall, leaving less and less 'room' for subsequent claims, on a much more rapid basis than is currently the case where varieties with comparatively narrower genetic and morphological bases are the objects of plant variety protection laws.

[45] The European Union (EU) has recently recognized the need to permit the marketing of 'landraces and varieties that are naturally adapted to the local and regional conditions' under existing seed legislation (Commission of the European Communities, 2001). While it fails to offer any criteria for the definition of landraces, the Biodiversity Action Plan for Agriculture treats landraces and traditional varieties as two separate categories and suggests that each type could be admitted to national lists as a 'conservation variety' if it meets the requirements of seed marketing directives (EEC Directive 70/457 and EEC Directive 70/458).

In anticipation of this regulation coming into force, a Finnish project (Onnela, 1999) examines several alternatives for the registration of landraces. The project concludes that there is not sufficient information about the characteristics of landraces and the amounts of genetic variation in their populations to ensure that individual landraces could be identified for efficient conservation without duplication. While acknowledging that molecular methods [restriction fragment length polymorphisms (RFLPs), random polymorphic DNAs (RAPDs)] could give insights into diversity at the DNA level, the report concludes that these methods are too expensive for routine use. Gross morphology and phenology would still be required, not least of all because the assessments of DUS required by UPOV are still based on morphology.

There is no simple answer to any of these questions, at least not yet. Research in the field is still relatively new, and, as the earlier analysis demonstrates, the data being collected are extraordinarily complex and subject to multiple interpretations. A few general trends in research results, however, allow us to make the following observations. First, somewhat ironically, the relatively recent rise in appreciation for the dynamic and often innovative role of farmers in conserving and improving landraces represents an additional challenge to creating *sui generis* intellectual property protections. On the one hand, farmers' innovation provides a justification for the granting of intellectual property rights in the first place.[46] On the other hand, the recognition of farmers' innovation is accompanied by the recognition that landraces themselves are changed and dispersed more rapidly than previously appreciated. This dynamism undermines in some ways the possibility of distinct or identifiable landraces coming into being, or at least makes them harder to locate as fixed entities over space and time. (Rather, it contributes to their being generally considered not uniform or stable enough for UPOV-style protection.) Older concepts of landraces that show them as being solely the products of natural adaptation to local environment – concepts that downplayed the innovative role of farmers – were accompanied by implicit assumptions about the longevity of connection between distinct landraces with distinct geographic locations (characteristics that would have made them easier to identify for the purposes of making them the objects of intellectual property protection).

Secondly, farmer-distinguished traits or characteristics are not necessarily distinct in the sense of being unique. Recall that the 'local' maize varieties in Cuzalapa were frequently planted from seed lots brought in from outside the region. Trade and exchange dispersed these varieties to the point where no community could claim a special connection to them. While distinct among the maize landraces planted in the valley, they were not unique to the valley. The high rate of trade of landraces between farmers, and their availability through local markets, ensures that many landraces are dispersed over wide areas, including across international borders. Indeed, the entire history of agriculture, both as practised by farmers and by scientific breeders, is unthinkable without a wide-ranging movement of plants around the globe.[47]

Note 45 continued

The report concludes that registration and support in Finland would be available to, *inter alia*, 'landraces which can be morphologically distinguished from other varieties'. Distinctness is essential, but the authors appear to feel there is no need to meet the UPOV criterion of uniformity. However, the report also concedes that a variety could be registered as a landrace simply if it has been cultivated for a long time on the same farm and belongs to a diverse varietal complex 'although it is not proven to be completely distinct'. For cereals, in particular, which are mostly in-breeders, the report suggests that data be collected on the characteristics of the different types observed in a sample sent for registration and that their proportions in the population are added to the varietal description. In the second season of assessment, different types found from an analysis of 100 individual plants in the first season may be grown as ear rows to further assess their genetic uniformity.

[46] Here we are making reference to the standard neo-liberal Lockean justification for property ownership: mixing labour with raw materials to create something new being the basis for recognizing ownership over the thing produced. Other equally convincing justifications for or against property ownership may also come into play *vis-à-vis* the same set of actions and relationships.

[47] The well-established history of the movement of crop germplasm and their interdependence (and importance to food security) are the factual building blocks upon which the International Treaty on Plant Genetic Resources for Food and Agriculture [hereinafter International Treaty] is based. The criteria for the inclusion of crops and forages in the multilateral system of access and benefit sharing created by the International Treaty is that they are important to food security and that the members of the world community are interdependent in their reliance upon them. Article 11.1 states:

Thirdly, farmer-distinguishable traits frequently may not be associated with characteristics that identify the landrace. Farmers may preserve traits across landraces over time. This leads to the frequently made conclusion that the diversity of landraces is linked to the diversity of the material sown in the area and the diversity of introduced varieties, which may be the single most important factor distinguishing landraces from the products of institutional 'modern' plant breeding. More distinct than any particular landrace, therefore, is the whole mix of material that is being used by farmers in a geographical area. Linking rights to such a wide-ranging mixture would be very challenging indeed!

Fourthly, removed from the environment in which it was adapted and the farmers who maintain it, the key characteristic that distinguishes landraces (and some other taxa) from modern-bred varieties is their genetic diversity. (This factor does not apply as much to clonal and truly self-pollinating landraces, as is discussed earlier.) This genetic diversity is the foundation of their wide adaptability, but it makes 'proving' the existence of a landrace very difficult. Any attempt to define a seed sample as a landrace by measuring its genetic diversity must fail because it cannot capture changes in this diversity over time. Thus, to add a note about the frequencies of particular genotypes or phenotypes in the population to a description of a landrace is informative only about the landrace as it exists after a particular growing season in a particular place. The normally wide diversity of a landrace almost guarantees that frequencies of alleles will change from season to season. What, then, is the 'correct' frequency distribution? Two different landraces might by chance converge on the same frequency distribution in a season. As a further complication, if a landrace is moved from one area to another so that frequency distributions, averaged over a suitable period, have changed, is it now a different landrace?

We do not want to be understood to say that it is impossible, in all cases, to trace a connection between a particular landrace and community or communities. We do want to stress, however, that most local farming systems do not function in ways that place importance on the generation of distinct landraces for the sake of distinctness. Farmers' planting, selection and exchange of seeds influence modifications in the landraces and the combination of landraces that they use. Their exchange and trading practices, in particular, ensure the dispersion of landraces over large geographical areas, frequently without regard for cultural or political boundaries. As a consequence, identifying any one particular landrace as distinct from all other related landraces and establishing that it has a unique connection with a particular community or group of communities will be difficult.[48]

In this context, it is worth considering what Salazar *et al.* (2003) refer to as the 'new farmers' varieties'. In particular, the authors refer to two relatively recent means by which farmers are producing new materials: by making selections from existing modern cultivars and through participatory plant breeding (with an emphasis on 'those varieties resulting from selection in segregating materials received from plant breeders'). They write that:

> [m]ost of the new farmers' varieties do not generally reach the level of distinctness and especially the level of uniformity and stability produced by plant breeding institutions and commercial seed companies. In conclusion, these new farmers' varieties are

Note 47 continued

'In furtherance of the objectives of conservation and sustainable use of plant genetic resources for food and agriculture and the fair and equitable sharing of benefits arising out of their use, as stated in Article 1, the Multilateral System shall cover the plant genetic resources for food and agriculture listed in Annex I, established according to criteria of food security and interdependence'.

[48] In this context, it is important to note that we are not focusing on the appellation of origin law (or any close analogue to it), pursuant to which it would not be so crucial to be able to define the boundaries of the landrace that would be the subject of the protection.

therefore more distinct, uniform and stable than landraces, but often less distinct, uniform and stable than modern varieties of the formal system.

On the one hand, these 'new farmers' varieties' might 'have an upper hand' in terms of satisfying looser conditions for protection than the landraces we have discussed so far. On the other hand, as Salazar *et al.* note, one could argue that they are no longer landraces *per se*. In the context of the future analysis of possible *sui generis* intellectual property rights being created for farmers' varieties, it would be clearly worthwhile to focus more on the products and processes of participatory plant breeding. In fact, there have been cases of such products satisfying the distinctness, uniformity and stability (DUS) requirements for existing plant variety protection laws.

4.7.2 Hypothetical case

Of course, there will always be exceptional cases. For the purposes of analysis, let us consider a hypothetical situation concerning a known landrace: chianquiahuitl maize, which Louette (2000) says is of unknown origin and probably not cultivated any longer outside the Cuzalapa valley. This information provides us with a 'head start' in establishing that the landrace is unique to that valley, not having been traded or exchanged with farmers outside the valley for some time and not being grown by farmers outside the valley. (One can see already that once legal benefits are associated with sole cultivation, the onus of proof will be relatively high in terms of proving that no one else is growing the variety elsewhere.) In this case, therefore, we are already beyond a significant hurdle, assuming that Louette is correct. The next question would be whether the chianquiahuitl is distinct from other landraces both inside and outside the valley. If challenged through legal proceedings (with the possibility of needing to call expert evidence in a courtroom), could one defend the claim that chianquiahuitl exists as a distinct landrace, or would one be forced to acknowledge that the line of demarcation between it and other landraces/wild relatives is sometimes too blurry to discern with accuracy? Proving its distinctness by reference to its genetic diversity is not possible, and is probably too expensive in any case. Thus, proof will have to turn on phenotypic characteristics. Farmers faced with the cost of litigating such a case would want to be very sure that, despite the high degree of introgression between maize landraces in the valley, they could defend their claim that chianquiahuitl is distinct.

If one could be relatively confident that chianquiahuitl was indeed distinct, the next issue (of course, these issues can be dealt with in different order) would be whether the connection between the community and the landrace was of a nature to justify its being granted *sui generis* intellectual property rights. This would of course depend upon the legislation in place. Existing plant variety protection (PVP) laws can protect varieties that plant breeders breed themselves or which they (or others) discover and develop. It appears that the farmers of the Cuzalupa valley did neither. On these bases, their claim would fail. However, a *sui generis* landrace protection law could have alternative criteria to suit the special nature of landraces and farmers' relationships to them. For example, it could recognize seasonal selection of seed as a form of breeding (a possibility recognized in existing plant variety protection laws, provided of course the new variety is distinct, uniform and stable.) Perhaps the law could be structured to take into account the fact that in the absence of farmer intercession, most landraces would cease to exist. Based on this fact, the law could include a presumption that, based on proof of the fact that a farmer or farmers were using a landrace over several generations (of the plant), they were, *de facto*, a kind of *sui generis* plant breeder, actively innovating to maintain the existence of the variety. If such a presumption were recognized in the law, one could 'argue backwards' that, based on the co-existence of landrace and farmers, in situations where they have not received the material from outside the region for a specified period of time, they can be deemed to be the 'breeder' of the landrace. This conclusion would of course depend entirely upon the national law of the country concerned or, by extension, upon the international laws harmonizing national laws.

Another issue that has to be addressed is the identity of the party claiming to be the owner of the landrace. Is it all of the inhabitants of the valley; only the farmers; or only the farmers actually growing chianquiahuitl at the time of the suit? How would this issue be determined? In cases where there was a perfect coincidence between the geographical distribution of the landrace and the borders of the local political administrative unit (or the borders of a population of people who considered themselves to belong to a common cultural unit), the issue would be easier to deal with, at least in principle. In other situations, it may be hard to forge a consensus about who should share in the joint or co-ownership of the landrace. Again, the national law in the country in question would have to be clear about the possible identities of the parties who could claim 'ownership' and associated benefits. In such a case the local communities should have a key role in resolving the dispute.

This example illustrates a few of the challenges that would exist in proving: (i) the existence of a distinct (that is, unique) landrace; and (ii) a historical connection between the landrace and a community or communities sufficient to justify 'ownership'. They are not necessarily insurmountable in every case, but one needs to keep the difficulties in mind in assessing whether or not it is worthwhile to create a national, sub-regional or international system of *sui generis* intellectual property protection for landraces *per se*, which is analogous to existing plant variety protection laws but less demanding in terms of the conditions of protection.[49]

Up to this point, we have not discussed the possible range of rights that could be conferred on farmers in connection with such a system of protection. One possibility would be to award the 'owners' with exclusive and semi-exclusive bundles of rights such as those created by the UPOV Convention and most countries' PVP laws. In such a case, third parties would need to obtain permission from farming communities before using their protected landraces for a variety of proscribed purposes. This necessity would provide a basis for farmers to negotiate for compensation for third parties' uses of their landraces. Another possibility would be to grant exclusive labelling rights (that is, a hybrid form of appellation of origin) or trademark protection connected with the marketing of the landraces. All of these options raise additional issues to be considered. In this chapter, we stop short of such considerations, focusing as much as possible on the preliminary question of attributing distinct landraces to particular communities.

4.7.3 Cautionary note regarding perverse incentives

The utilitarian justification for patent, PVP and copyright laws is that they will provide incentives for inventors, plant breeders and authors, respectively, to innovate where they might not otherwise do so. The short-term grant to those people of time-limited, semi-exclusive property rights over their

[49] We have not actually addressed a number of associated challenges, such as how a farming community actually learns that material over which it has rights is being used by others; the prohibitive cost of litigation and difficulties associated with actually getting the farmers organized to launch a legal action and possibly travel to, and participate in, court proceedings far from their farms, possibly in other countries; the technicalities of launching international law suits where national laws regarding landrace protection is not harmonized with (or existing in any form with) other countries; and, ultimately, whether or not such an enterprise can be justified on the basis of the returns farming communities might realize through the creation of a *sui generis* intellectual property regime.

creations is the price that society pays for the wider benefits generally associated with higher levels of innovation.

Farmers, however, are already engaging in innovative activity *vis-à-vis* the evolution, maintenance and use of landraces. One has to be careful, then, about introducing a system of incentive and reward, and must make sure that it reinforces, rather than alters, their behaviour. We have demonstrated, earlier in this chapter, that farmers' patterns of innovative uses of landraces frequently contribute to their being in a state of flux, metamorphosing over time in response to farmers' selection priorities, planting patterns, seed exchanges, decisions about trading priorities and so on. As we have stated above, it would be rare for farmers' varieties to be distinct, uniform and stable. However, when their varieties satisfied those criteria, farmers could seek protection under existing plant variety protection laws. In this chapter, we are focusing, *ab initio*, on landraces that we have said generally are not distinct, uniform and stable. Indeed, to create and maintain distinct, uniform and stable varieties, farmers would generally have to change their ways of farming, selecting, exchanging and trading seeds. Focusing on distinctness alone as an alternative condition of protection might lessen the pressure on farmers to alter their behaviour in order to obtain whatever benefits might flow from intellectual property protection, but it would none the less create such incentives. For instance, in order to ensure the distinctness of chianquiahuitl, the farmers of the Cuzalapa valley could feel the need to grow it in geographically isolated plots so it would not be subject to introgression from other maize landraces. This technique would interrupt the dynamic interaction of selection and gene flow that Louette describes. Furthermore, the farmers would have to deny any neighbouring farmers' requests for chianquiahuitl seed in order to ensure their claim to be the only *de facto* breeders and conservers of the landrace. However, Louette (2000) has demonstrated that the farmers, and maize diversity generally, experience high degrees of introgression between landraces as well as from the exchanges of materials between farmers.

The issue of the possibility of perverse incentives raises what is perhaps a more fundamental question about the underlying reason for creating intellectual property protections for landraces in the first place. From an economist's point of view, policy makers assign intellectual property rights as a means of correcting market failures and of creating incentives for innovation by achieving a more efficient allocation of resources. The rational for creating *sui generis* intellectual property rights in landraces would therefore be based on the assumption that there is not enough innovation by farmers and farming communities *vis-à-vis* their landraces. The fact is, however, that very few people appear to be arguing that farmers are not being innovative enough in this regard. Instead, the arguments for creating intellectual property rights for farmers appear to be more commonly rooted in equity concerns, and these concerns could possibly be addressed by other means that are less complicated from a legal technical perspective; specifically, means that would not include focusing so much on the subject matter of the protection — that is, the landrace itself (L. Lipper, personal communication). For example, equity concerns can be addressed through distributing benefits to farming communities that are derived through the use of materials accessed on their lands (whether or not they can be defined as distinct or identifiable landraces) in the development of commercialized products, through taxes on commercial seed sales, or even through forms of intellectual property laws, such as appellation of origin laws, which do not focus so much on the creation of new materials but rather on the relationship between producers in a region and their 'traditional' products.

4.7.4 Converse situations: when *sui generis* protections could be feasible

We do not want to sound too negative. The preceding recitation of difficulties also provides insights into when it might be possible to protect a landrace through some form of *sui generis* intellectual property mechanism. There are instances where farmers are already not exchanging seed with their neighbours (Busso *et al.*, 2000) and the possibility of introducing an incentive whose negative effects might outweigh its positive effects would not be an issue. There are also instances where varieties have been grown exclusively in contained geographical areas (Louette, 2000). There are surely instances where (unlike chianquiahuitl in the Cuzalapa valley) the exclusively grown landrace was developed in the same area by the same farming communities that are there now. There are probably also instances where the sole farmers using the landrace are all part of a local political unit (who could represent them) or a cultural group (who could be represented more or less expeditiously through some form of cultural organization). There may also be instances where there is no question that a landrace is clearly distinct from all other related landraces. When all, or most, of these conditions intersect, *sui generis* intellectual property protections would be a real possibility. Policy makers at different levels will ultimately have to decide if, in their jurisdictions, these conditions coincide with sufficient frequency and with the promise of substantial enough gain to dedicate the resources to creating a *sui generis* intellectual property protection regime and subsequently administer and enforce it. In this context, Salazar *et al.* (2003) note that the products of participatory plant breeding are potential candidates for *sui generis* forms of intellectual property protection.

It is also critical to recognize that, to date, there simply is not enough research regarding many of these key issues to be able to make an accurate prediction as to the likely degree of coincidence of these factors. The International Plant Genetic Resources Institute (IPGRI) is currently conducting research regarding the correlation of farmers' perceptions to population structures. This work has involved, among other things, studying the strength of the link between farmers' selected-for traits and scientifically measured phenotypes under experimental conditions.

4.8 Mining Data for Connections between Communities and *Ex Situ* Collections

In this section, we examine the extent to which data gathered during plant genetic resources collecting missions could be used to establish a link between gene bank accessions and their originating communities. The primary source of data for this exercise was the System-Wide Information Network for Genetic Resources (SINGER), which is the genetic resources information exchange network of the International Agricultural Research Centres of the Consultative Group on International Agricultural Research (CGIAR). It provides access to information on all of the collections of genetic resources held by the CGIAR centres in their gene banks. Together, these collections comprise more than half a million samples of crop, forage and tree germplasm of major importance for food and agriculture. We also looked at data from the IPGRI and the International Board for Plant Genetic Resources (IBPGR) collecting missions. It is important to note that most of the accessions referred to in this chapter have been 'designated' pursuant to the 1994 Agreement between the International Agricultural Research Centres of the CGIAR and the Food and Agricultural Organization (FAO), placing these collections of plant germplasm under the auspices of the FAO 'in trust for the benefit of the international community'.[50] The object of the exercise was simply to test an existing large database for the kinds of data it held that might be relevant. It was certainly not to build a case for the recharacterization of those materials.

[50] SGRP (2003) *Booklet of CGIAR Centre Policy Instruments, Guidelines and Statements on Genetic Resources, Biotechnology and Intellectual Property Rights,* available online at SINGER webite: http://singer.cgiar.org/booklet.pdf (last accessed 22 December 2004).

The primary focus of the investigation was to ask how far the passport data gathered in SINGER could be used to link accessions to communities. The IPGRI collecting data were used to support this analysis, and we also asked whether any historical trends in the kind of information recorded on collecting missions could be discerned.

4.8.1 Data selection

SINGER maintains a number of central databases on genetic resources in the CGIAR. These central databases hold a selection of data from the contributing CGIAR gene banks, which allow SINGER to function as a gateway to all germplasm accessions in the CGIAR. This gateway function relies on the identification or passport data that these accessions have in common. We selected several data fields that might be able to establish a linkage between a collected accession and its originating community. These fields included latitude, longitude and the elevation of the collecting site; the country of original collection; the sub-country location data; the source of the collected material; and the year and month of collection. Since the central SINGER databases use only a selection of data to fulfil their gateway function, we considered it essential to establish what additional types of information gene banks might have about landraces and their origin. We decided to use data from the IBPGR/IPGRI collecting missions as good examples of the type of information that would have been collected. As these missions cover a period of 20 years, the survey would try to detect any historical trends in documentation practices in time.

4.8.2 Results

The total number of accessions in SINGER on 11 October 2001 was 519,804, of which 115,377 (22%) were recorded as landraces. In order to establish a link between the collected landraces and their originating communities, it is essential to have at least some information on when, where and from whom the germplasm was collected. It would also be useful, as indicated earlier, to have data on community uses of these materials and the ways in which the communities may have altered these materials over time.

With respect to the date when the accession was collected, most collecting forms included years, months and days of collecting activities. Where these were not included in the forms, the collecting mission codes provided approximations. We estimated that with some additional effort the year of collection could be established for more than 90% of the accessions.

To determine where the accessions were collected, place names and geographical coordinates were both used. All landraces in SINGER have either geographical coordinates or at least the country of origin recorded, and 96% of the landraces have more detailed location data on the collecting site. In most cases, it will include an indication of the nearest village or town. Often the direction and distance from the collecting site to these settlements are indicated (for example, 'Kalmang Bala. 40 km NW from Mansehra. NWFP').

About one-third of the landraces recorded in SINGER have geographical coordinates. Of the remaining two-thirds without geographical coordinates, most were collected in the late 1970s and early 1980s, a period of extensive germplasm collecting worldwide. Many of these collecting missions were considered emergency missions to salvage genetic diversity in areas where the risk of genetic erosion was high. It is possible that the huge influx of new accessions had an influence on the quality of the computerized documentation.

To determine from whom the landrace material was collected, SINGER itself does not provide many entry points. The descriptor 'source of the collected material' describes the source in very broad categories such as farm store, farmer's field, garden, local market, town shop and so on. Additional data such as local (landrace) names, use, farmer or farm name, indication of ethnic group and so on, which could be used to strengthen the linkage between a landrace and a particular community, are not part of the central SINGER databases, which were designed to hold only as much data as needed to give the user an overview of all accessions in the CGIAR. More detailed accession-related data can be found in the databases of the contributing centres, and these data can often be accessed directly from SINGER.

Turning to the IBPGR/IPGRI collection missions, we randomly selected 50 samples and scored them on the presence or absence of important data types. For geographical coordinates, the IBPGR/IPGRI collecting mission data seem to suggest a much higher incidence than SINGER (70% *versus* 33%).[51] Data on the environment (32 out of 50) and ethnobotanical data (31 out of 50) were relatively common. The environmental data usually included descriptors such as topography and soil characteristics. For ethnobotanical data, the most common items listed were local names and use. More detailed information on the communities involved, such as names of farmers/farms, ethnic groups and so on, were very scarce (5 out of 50). Occasionally, a collector would describe how local markets play an important role in dispersing genetic diversity through rural communities or indicate how long farmers have been cultivating a particular landrace, but, overall, we encountered very few comments of this nature.

We did not discover any historical trends in the ways the information was recorded. The relatively small sample size (50 out of approximately 500) and the short period covered would make it difficult to find any trends, with the exception of the very obvious ones. The assumption that the quality of the collecting reports would improve over time could not be confirmed. Throughout the period 1976–95, we encountered very detailed and good reports as well as some disappointing ones. Quality seemed to depend much more on the effort made by the individual collector than on any determination to raise standards over time.

4.8.3 Data-mining conclusions

For most of the landraces collected and maintained by gene banks, it is possible to trace back the geographical origin of the sample with a reasonable amount of certainty. The data from SINGER suggest that 32% of the landraces have actual geographical coordinates. Ninety-six per cent have location data on the collecting site at the sub-country level. Likewise, the date of collecting can be determined for a large part. SINGER has the actual year and month of collecting for about half of its landraces, but one may expect that this percentage could be significantly boosted, at least to determine the collecting year, by inference from the collecting mission and collector codes. To assist with the linkage between certain landraces and communities, the most useful information to find in the data reviewed are local names and uses.

4.8.4 Collecting forms, ethnobotanical descriptors

By way of follow-up to our analysis of data available through SINGER and the IBPGR/IPGRI collecting forms, we conducted an analysis of gene banks' current collecting forms, to determine if, in the future at least, more and better ethnobotanical information might be collected in the course of collecting missions. This information would be useful to forge a legally relevant connection between the peoples concerned and the collected materials (Hazekamp, 2002).

In order to obtain sample germplasm collecting forms, an e-mail was sent to ten

[51] The reason for the apparent discrepancy was not further investigated, partly because even with the lower figures in the System-Wide Information Network for Genetic Resources it had already been established that the original location where a landrace was collected could be relatively easily determined using geographical coordinates and the other available location data.

Table 4.1. Gene banks requested to participate in the study.

CGIAR gene banks	National gene banks
CIAT (Colombia)	Asian Vegetable Research Development Centre (Taiwan)
CIMMYT (Mexico)	Australian Medicago Gen. Res. Centre (Australia)
CIP (Peru)	Australian Winter Cereals Collection Agricultural Research Centre (Australia)
ICARDA (Syria)	Institute for Biodiversity Conservation and Research (Ethiopia)
ICRAF (Kenya)	Institute for Plant Genetics and Crop Plant Research, Gatersleben (Germany)
ICRISAT (India)	N.I. Vavilov Research Institute of Plant Industry (Russia)
IITA (Nigeria)	National Gene Bank of Kenya (Kenya)
ILRI (Ethiopia)	National Genetic Resources Laboratory (USA)
INIBAP (France)	Royal Botanical Gardens Kew (UK)
IRRI (Philippines)	SADC Plant Genetic Resources Centre (Zambia)
WARDA (Ivory Coast)	

national or regional gene banks and 11 gene banks of the CGIAR. The selection of the gene banks (see Table 4.1) took into account the extent of their collecting activities and/or the expectation that they would have a particular interest in ethnobotanical descriptors. This review was based on the responses (including germplasm collecting forms) from eight national or regional gene banks and six CGIAR centres.

The review of the gene bank collecting forms undertaken in this study indicates that descriptors regarding the collecting location and the time of collection are common and standard throughout the gene banks. When it comes to documenting information on the originators of traditional cultivars, the situation is variable. Although most gene banks have a number of basic descriptors, including collecting source, collecting type and local name, only five out of the 14 gene banks included in this review have developed forms with an extensive ethnobotanical component. These are the Asian Vegetable Research Development Centre (Taiwan), the Institute for Biodiversity Conservation and Research (IBCR-Ethiopia), the South African Development Community (SADC) Plant Genetic Resources Centre (SPGRC-Zambia), the International Crops Research Institute for the Semi-Arid Tropics (ICRISAT) (India) and the International Rice Research Institute (IRRI) (Philippines). These five gene banks have had fairly recent development take place since the mid-to-late 1990s. The diversity in approaches and ethnobotanical descriptors used by these five is an indication that this area is still very much in development, and no firm standards have yet been agreed upon.

When looking in more detail at these five forms, we notice several trends:

- The donors of the material are described in different levels of detail. Basic descriptors include name, ethnic group and language used for the local name. Age and gender issues occur less frequently. In particular, the SPGRC provides much more detail, such as human settlement history, division of labour at the household level, land ownership and tenure, and the source of information (key informant designation/social status and an indicator of reliability of information). This provides much more information on the social context.
- Although the source of the material is usually described in some detail (type of collecting site, environmental traits), the actual origin of the material is described in variable degrees of detail. The ICRISAT form provides the most complete descriptors in this regard, indicating not only how and from whom the material was obtained but also how long this material has been in the possession of the donor. The SPGRC captures data

on the consumption of the plant material (local, selling at market), which provides some indication on the exchange or flow of the particular germplasm between communities.

- By far the greatest attempt on the part of the ethnobotanical descriptors to capture indigenous knowledge was by using descriptions of special traits (performance, tolerance and resistance), cultivation and selection criteria. Besides a common descriptor on plant use, only the IBCR and the SPGRC specifically document the cultural significance of the crop. The IBCR form introduces many descriptors used to rank the specific preferences for local varieties as a way to describe biodiversity. The SPGRC is the only one to indicate whether access to indigenous knowledge is restricted in any way or not.

Although the gene banks included in this review have good documentation on where and when germplasm was collected, only a few have developed extensive sets of ethnobotanical descriptors. The basic data that most gene banks do have are recordings of local names. However, past experience with recorded local name data (for example, in the rationalization of germplasm collections) has shown them to be of variable use, often needing substantial cross-checking before their validity can be established. They are therefore regarded as being indicative at most.

Considering that the volume of global germplasm collecting dropped sharply at the beginning of the 1990s, one can presume that the amount of ethnobotanical data collected by gene banks, in a structured manner, is fairly limited. For example, even though the ICRISAT's collecting form, which includes considerable ethnobotanical descriptors, has existed since 1997, only 100 accessions have been collected using this form as of late 1999 (Hazekamp, 2002). The future collecting of ethnobotanical data by gene banks would benefit from the development of an international standard for ethnobotanical descriptors.

4.9 Assessing Possible Policy Responses to the Database Analyses

Our analysis of SINGER databases and gene bank collecting forms reveals that, of the wide range of information that would be potentially relevant in establishing a legally relevant link between collected materials and communities (for example, community uses, cultural significance, history of use, details about distribution and exchange, evidence that the material may be unique and so on), only information related to the location of collected material is consistently provided. We know in which country 99% of the material was collected. For 96% of the accessions, there is more fairly general information about the locations of the collected samples within the country concerned (for example, 'collected 40 km north-west of [village X]'). In 30% of the cases, where longitudinal and latitudinal coordinates accurate to the seconds-level are listed, it is possible to know where the sample was collected within a few kilometres (Hazekamp, 2001).

It must be further borne in mind that there are no available data concerning where any of the collected information is *not* found – that is to say, whether the accession is unique in any way. The fact that it was collected at longitude X and latitude Y does not mean that the same material could not also have been collected at X ± 1 and Y ± 1, or anywhere else in the world for that matter.

Location and the time of collecting on their own are not sufficient data upon which to found a claim for a *sui generis* UPOV-style plant variety protection right. Nor does it appear to provide a sufficient factual basis for a strong claim for other forms of intellectual property protection, such as geographical indications protection or trademark protection. Geographic indications protection can apply to goods originating from within a particular territory, 'where the given quality, reputation or other characteristic or the good is essentially attributable to its geographical origin'.[52] Our raw location data do not provide any indication about which of any accession's traits can be attributed to an adaptation to the local environment. Again, working from the older conception of landraces – that landraces were simply crops adapted (unaided by conscious farmer selection) to the local environments – it might have been possible to 'argue backwards' that the very existence of the landrace in a particular location necessarily means that it has developed some or many of its characteristics as a result of natural selection pressures.

Location data and dates of collecting may, however, be useful as *supporting evidence* in more broadly substantiated claims for intellectual property rights. For example, current claims based in part on longevity of association between a community and a landrace could use the collecting location data to establish that the landrace in question existed in the vicinity of the claimants for at least X number of years. The location data could also be useful as a means of defeating claims for intellectual property rights over identical material by other parties. The World Intellectual Property Organization (WIPO), in a project endorsed by the Intergovernmental Committee on Intellectual Property and Genetic Resources, Traditional Knowledge and Folklore, has recently launched a test website to be included in the prior art searches of intellectual property offices around the world. The IPGRI, on behalf of the System-Wide Program on Genetic Resources (SGRP) and SINGER, has made an agreement with WIPO to make a link from their pilot website to SINGER for the purposes of facilitating searches of SINGER by national patent offices doing prior art searches.[53]

Finally, as alluded to earlier in this chapter, location data are really all that one needs in situations where a domestic access law provides individuals and communities with the right to accede to, or deny, requests for access to materials located on their lands. Of course, most of the material listed in the SINGER database was collected throughout the 1970s, 1980s and, less so, 1990s. Presumably, prior to 1993, when the Convention on Biological Diversity (CBD)[54] came into force, in the absence of any explicit national access law that created such obligations, it was not necessary to get the permission of either governments or constituent communities when collecting materials and taking them out of the country for storage in gene banks. Certainly, since 1993, as far as internationally situated institutions such as the International Agricultural Research Centres (IARCs) are concerned, it is incumbent upon them to comply not only with national access laws but also with the spirit of the CBD, which requires collectors to obtain the prior informed consent of national governments (and arguably local communities).

4.10 Conclusions

Throughout this chapter, we have discussed a number of factors that make the project of creating workable *sui generis* intellectual property protections for landraces that are analogous to (but considerably different from) plant variety protection laws particularly challenging. Are these challenges insurmountable? In many cases, we think they may be. In others, we are not so sure. Most of the data we have to rely on regarding farmer–landrace interactions were not generated in the context of seeking to establish cases for intellectual property protection. Instead, they tend to have been collected in the context of research on farmers' decision-making and management practices and how they impact upon the conservation of agrobiodiversity. Researchers

[52] TRIPS Agreement, see note 41.

[53] See the World Intellectual Property Organization's [hereinafter WIPO] Portal of Online Databases and Registries of Traditional Knowledge and Genetic Resources, available online at WIPO: http://www.wipo.int/tk/en/databases/tkportal/index.html (last accessed 22 December 2004).

[54] Convention on Biological Diversity, 5 June 1992, 31 I.L.M. 818 (1992).

doing this kind of work are content with the evidence that farmers are distinguishing between plant populations on the basis of any number of traits as evidence of their management priorities and competence. The legal analyst, however, wants to know if the material concerned is distinct from all other landraces as evidence of the basis of a claim for exclusive interests in the same material. Certainly, much more detailed data than we have referred to in this chapter would be necessary on a case-by-case basis to 'prove' that intellectual property protection rights should be granted.

Farmers and farming communities have been making invaluable contributions to crop development and agrobiodiversity creation and conservation for millennia. It is critically important that their enormous contributions are recognized and rewarded. This chapter has focused on the interface between the biological and social factors that contribute to the development of distinct and identifiable characteristics in landraces, and one approach to creating *sui generis* intellectual property protections for these landraces. Our conclusion, with respect to this particular approach – one that is analogous to existing plant variety protection laws, but with looser conditions for protection – is agnostic, running to cautiously negative. There may be conditions under which it would be possible to identify and protect a landrace pursuant to the looser *sui generis* conditions for protection than we have discussed in this chapter. On the other hand, it appears that often it would not be possible. To be certain, it would be necessary to conduct research on a crop-by-crop, community-by-community, process-by-process (for example, participatory plant breeding) basis. Such pilot studies are under way, and we are waiting for their results. In the meantime, we recommend that it would perhaps be more fruitful, for those interested in pursuing intellectual property-related approaches, to investigate *sui generis* options that are more closely analogous to appellation of origin laws, where history and patterns of use and production by the people of the region are more the focus than the biological limits of a particular plant population.

Bibliography

Part 1

Augustine, S.J. (1997) Traditional Aboriginal Knowledge and Science Versus Occidental Science. Paper prepared for the Biodiversity Convention Office of Environment, Canada.

Asfaw, Z. (1999) The barleys of Ethiopia. In: Brush, S.B. (ed.) *Genes in the Field. On-farm Conservation of Crop Diversity.* International Plant Genetic Resources Institute, Rome; International Development Research Institute, Ottawa, Canada; Lewis Publishers, Boca Raton, Florida, pp. 77–107.

Biber-Klemm, S. (2002) The community-based conservation and management of genetic diversity in agroecosystems: the role and function of law. In: Engels, J.M.M. *et al.* (eds) *Managing Plant Genetic Diversity.* CAB International, Wallingford, UK.

Brush, S.B. (1994) Providing Farmers' Rights through in-situ conservation of crop genetic resources. Commission on Plant Genetic Resources, First Extraordinary Session, Rome 1994. Background Study Paper No. 3.

Brush, S.B. (2000) The issues of *in situ* conservation of crop genetic resources. In: Brush, S.B. (ed.) *Genes in the Field. On farm Conservation of Crop Diversity.* IPGRI, Rome; IDRC, Ottawa; Lewis, Boca Raton.

CGIAR (Commission on Genetic Resources for Food and Agriculture) (2001) Crops proposed for the multilateral system; centres of diversity, locations of *ex-situ* collections and major production countries. Background Study Paper No. 12.

Correa, C. (2000) Options for the implementation of farmers' rights at the national level. Trade-Related Agenda, Development and Equity (T.R.A.D.E.) Working Papers No. 8; South Centre.

Fowler, G. (2000) *Implementing Access and Benefit-Sharing Procedures under the Convention on Biological Diversity: The Dilemma of Crop Genetic Resources and their Origins.* Global Forum on Agricultural Research, Dresden, Germany, 21–23 May.

Girsberger, M.A. (1999) *Biodiversity and the Concept of Farmers' Rights in International Law. Factual Background and Legal Analysis.* Peter Lang, Berne, Switzerland.

Glowka, L., Burhenne-Guilmin, F. and Synge, H. (1994) *A Guide to the Convention on Biological Diversity.* Environmental Policy and Law Paper No. 30, IUCN.

Gonzales, T.A. (1999) The cultures of the seed in the Peruvian Andes. In: Brush, S.B. (ed.) *Genes in the Field. On-farm Conservation of Crop Diversity.* International Plant Genetic Resources Institute, Rome; International Development Research Institute, Ottawa, Canada; Lewis Publishers, Boca Raton, Florida, pp. 193–216.

Hardon, J.J., Vosman, B. and van Hintum, Th.J.L. (1994) *Identifying Genetic Resources and their Origin: The Capabilities and Limitations of Modern Biochemical and Legal Systems.* Commission on Plant Genetic Resources, Background Study Paper No. 4, Rome.

Harlan, J. (1975) Our vanishing genetic resources. *Science* 188, 618–621.

Hawkes, J.G. (1983) *The Diversity of Crop Plants.* Cambridge University Press, Cambridge.

Indian and Northern Affairs Canada (1999) *Intellectual Property and Aboriginal People: A Working Paper.* Indian and Northern Affairs Canada, Research and Analysis Directorate. Industry Canada, Intellectual Property Policy Directorate.

IPGRI (1999) *Key Questions for Decision Makers. Protection of Plant Varieties under the WTO Agreement on Trade-Related Aspects of Intellectual Property Rights.* IPGRI, Rome.

Jarvis, D.I., Myer, L., Klemick, H., Guarino, L., Smale, M., Brown, A.H.D., Sadiki, M., Sthapit, B. and Hodgkin, T. (2000) *A Training Guide for In Situ Conservation On Farm.* Version 1. International Plant Genetic Resources Institute, Rome, Italy.

Nwokeabia, H. (2002) Why industrial revolution missed Africa. A 'traditional knowledge' perspective. Economic Commission for Africa, Doc. ECA/ESPD/WPS/01/02.

Pionetti, C. and Suresh, R. (2002) Diversity on the Deccan Plateau. *Seedling,* April 2002.

Raymond, R. and Fowler, C. (2001) *Sharing the Non-monetary Benefits of Agricultural Biodiversity.* Issues in Genetic Resources No. 5. IPGRI, Rome.

Schneider, J. (1995) Farmer practices and sweetpotato diversity in Highland New Guinea. In: Schneider, J. (ed.) *Indigenous Knowledge in Conservation of Crop Genetic Resources.* CIP-ESEAP/CRIFC, Bogor, Indonesia, pp. 63-70.

Tapia, M.E. and De la Torre, A. (1997) *Women Farmers and Andean Seeds.* IPGRI and FAO, Rome.

van den Daele, W. (2001) Modern science and traditional knowledge in Western societies. Wissenschaftszentrum Berlin, Berlin (unpublished presentation, on file with author).

World Business Council for Sustainable Development (WCBSD) (ed.) (2003) *Intellectual Property Rights in Biotechnology and Healthcare – Results of a Stakeholder Dialogue.* Summary Report. Earthprint, Stevenage, UK. Available at website http://www.wz-berlin.de/ipr-dialogue/

Zeven, A.C. (1998) Landraces: A review of definitions and classifications. *Euphytica* 104, 127–139.

Part 2

Adams, M.W. and Martin, G.B. (1988) Genetic structure of bean landraces in Malawi. In: Gepts, P. (ed.) *Genetic Resources of Phaseolus Beans.* Kluwer Academic, Dordrecht, The Netherlands, pp. 355–374.

Bajracharya, J., Rijal, D.K., Tiwari, P.R., Shakya, D.S., Baniya, B.K., Steele, K.A., Witcombe, J.R., Sthapit, B.R., Brown, A.H.D., Hodgkin, T. and Jarvis, D.I. (2001) Understanding genetic diversity and farmer named varieties in taro (*Colocasia esculenta* L. Schott), barley (*Hordeum vulgare* L.), and rice (*Oryza sativa* L.) in the mid and high hills of Nepal through isozyme and microsatellite DNA analysis. Presented to the International Symposium on Managing Biodiversity in Agricultural Ecosystems, Montreal, Canada, November 2001.

Bedigian, D. and Harlan, J.R. (1983) Nuba agriculture and ethnobotany, with particular reference to sesame and sorghum. *Economic Botany* 37, 384–395.

Boster, J.S. (1984) Classification, cultivation and selection of Aguaruna cultivars of *Manihot esculenta* (Euphorbiaceae). *Advances in Economic Botany* 1, 34–47.
Boster, J.S. (1985) Selection for perpetual distinctiveness: evidence from Aguaruna cultivars of *Manihot esculenta. Economic Botany* 39, 310–325.
Bouizgaren, A., Birouk, A., Kewrfal, S., Hmama, H. and Jarvis, D. (2001) On farm conservation of alfalfa farmer's units of diversity (FUD) in Morocco. Presented to the International Symposium on Managing Biodiversity in Agricultural Ecosystems, Montreal, Canada, November 2001.
Brush, S.B. (1995) *In situ* conservation of landraces in centers of crop diversity. *Crop Science* 35, 346–354.
Brush, S.B., Carney, H.J. and Huaman, Z. (1981) Dynamics of Andean potato agriculture. *Economic Botany* 35, 70–88.
Busso, C.S., Devos, K.M., Ross, G., Mortimore, M., Adams, W.M., Ambrose, M.J., Aldrick, S. and Gale, M.D. (2000) Genetic diversity within and among landraces of pearl millet (*Pennisetum glaucum*) under farmer management in West Africa. *Genetic Resources and Crop Evolution* 47, 561–568.
Castillo, F., Herrera, E., Moreno, V., Romero, J., Nuñez, I., Ballesteros, V., Sanchez, J., Ortega, R., Ramirez, P., Kato, A., Goodman, M.M., Smith, M.E., Ramirez, A., Qualset, C.O. and Espejel, F. (2002) Potential of local diversity for the improvement of maize production in Mexico. In: Chavez-Servia, J.L., Arias-Reyes, L.M., Jarvis, D.I., Tuxill, J., Lope-Alzina, D. and Eyzaguirre, C. (eds) *Proceedings of a symposium: Managing Crop Diversity in Traditional Agroecosystems.* 13–16 February 2002, Merida, Mexico. International Plant Genetic Resources Institute, Rome, pp. 55–56.
Crucible II Group (2001) *Seeding solutions. Vol. 2, Options for National Laws Governing Access to and Control over Genetic Resources.* IPGRI, IDRC, DHF, Rome.
Fernandez, P.G. (1995) Seed systems, indigenous knowledge systems and genetic diversity. In: Schneider, J. (ed.) *Indigenous Knowledge in Conservation of Crop Genetic Resources.* Proceedings of an international workshop, Cisarua, Bogor, Indonesia, 30 January–3 February 1995, CIP-ESAP/CRIFC, pp. 11–17.
Harlan, J.R. (1975a) *Crops and Man.* American Society of Agronomy, Madison, Wisconsin.
Harlan, J.R. (1975b) Our vanishing genetic resources. *Science* 188, 618–621.
Hawkes, J.G. (1983) *The Diversity of Crop Plants.* Cambridge University Press, Cambridge, UK.
Hazekamp, T. (2001) Report on a five day consultancy for GRST, in the framework of IPGRI contribution to the Swiss Agency for Development and Cooperation (SDC) project on Indigenous Knowledge, 8–15 October 2001.
Hazekamp, T. (2002) Analysis of the ICARDA, IRRI, ICRISAT germplasm databases for digitalized ethnobotanical data of barley, rice, and sorghum landraces. Consultancy report, August 2002.
Hodgkin, T.H., Ramanatha Rao, V. and Riley, K.W. (1993) Current issues in conserving crop landraces *in situ*. Paper presented at the FAO-IBPGR On-farm Conservation Workshop, 6–8 December 1993, Bogor, Indonesia.
Iijima, S. (1964) Ecology, economy and social systems in the Nepal Himalaya. *Development Economics* 2, 91–105.
IPGRI (1999) *Key Questions for Decision Makers: Protection of Plant Varieties under the WTO Agreement on Trade-Related Aspects of Intellectual Property Rights.* IPGRI, Rome.
IPGRI (2001) *NUS Strategy.* IPGRI, Rome.
Jarvis, D.I., Myer, L., Klemick, H., Guarino, L., Smale, M., Brown, A.D.H., Sadiki, M., Sthapit, B. and Hodgkin, T. (2001) *A training Guide for* in situ *Conservation on the Farm,* Version 1. IPGRI, Rome.
Johns, T. and Keen, S.L. (1986) On-going evolution of the potato on the Altiplano of Western Bolivia. *Economic Botany* 40, 409–424.
Lambert, D.H. (1985) *Swamp Rice Farming: The Indigenous Pahang Malay Agriculture System.* Westview Press, Boulder, Colorado.
Leskien, D. and Flitner, M. (1997) *Intellectual Property Rights and Plant Genetic Resources: Options for a* Sui Generis *System.* Issues in Genetic Resources No. 6. IPGRI, Rome.
Louette, D. (2000) Traditional management of seed and genetic diversity: what is a landrace? In: Brush, S.B. (ed.) *Genes in the Field. On-farm Conservation of Crop Diversity.* Lewis Publishers, Boca Raton, Florida.
Martin, G. and Adams, M.W. (1987a) Landraces of *Phaseolus vulgaris* (Fabaceae) in northern Malawi. I. Regional variation. *Economic Botany* 41, 190–203.

Martin, G. and Adams, M.W. (1987b) Landraces of *Phaseolus vulgaris* (Fabaceae) in northern Malawi. II. Generation and maintenance of variability. *Economic Botany* 41, 204–215.

McGuire, S. (2002) Farmers' views and management of sorghum diversity in western Hareghe, Ethiopia: implications for collaboration with formal breeding. In: Cleveland, D. and Soleri, D. (eds) *Collaborative Plant Breeding: Integrating Farmers' and Plant Breeders' Knowledge and Practice.* CAB International, Wallingford, UK.

Onnela, J. (1999) *Landraces in Finland: Proposal for Varietal Research, Registration and Maintenance System of Cereal, Forage Grass and Legume Landraces and Old Commercial Cultivars.* Plant Production Inspection Centre, Loimaa, Finland.

Parzies, H.K., Brocke, K.V., Spoor, W. and Geiger, H.H. (2001) Contrasting seed management practices for landraces of barley and pearl millet in Rajhastan, India, inferred from gene flow data. Abstract from the twenty-sixth EUCARPIA Congress, Plant Breeding: Sustaining the Future, Edinburgh, UK, 10–14 September 2001.

Prain, G. and Campilan, D. (1997–98) Farmer maintenance of sweet potato diversity in Asia: dominant cultivars and implications for *in situ* conservation. CIP program report.

Robinson, R.A. (1996) *Return to Resistance: Breeding Crops to Reduce Pesticide Dependence.* agAccess, Davis, California.

Sadiki, M., Belqadi, L., Mahdi, M. and Jarvis, D. (2001) Diversity of farmer-named faba bean (*Vicia faba* L.) varieties in Morocco: a scientific basis for *in situ* conservation on-farm in local ecosystems. Presented to International Symposium on Managing Biodiversity in Agricultural Ecosystems, Montreal, Canada, November 2001.

Salazar, R., Louwaars, N. and Visser, B. (2003) Protecting farmers' new varieties: Some thoughts [unpublished].

Salick, J., Celinnnese, N. and Knapp, S. (1997) Indigenous diversity of cassava: generation, maintenance, use and loss among the Amuesha, Peruvian Upper Amazon. *Economic Botany* 51, 6–19.

Schneider, J. (1995) Farmer practices and sweet potato diversity in highland Papua New Guinea. In: Schneider, J. (ed.) *Indigenous Knowledge in Conservation of Crop Genetic Resources.* Proceedings of an international workshop, Cisarua, Bogor, Indonesia 30 January–3 February 1995, CIP-ESAP/CRIFC, pp. 63–70.

Setyawati, I. (1996) Environmental variability, IK and the use of rice varieties. *Indigenous Knowledge and Development Monitor* 4, 11–13.

SGRP (2003) Booklet of CGIAR Centre Policy Instruments, Guidelines and Statements on Genetic Resources, Biotechnology and Intellectual Property Rights, available online at SINGER website: http://singer.cgiar.org/booklet.pdf (last accessed 22 December 2004).

Teshome, A., Fahrig, L., Torrance, J.D., Arnason, T.J. and Baum, B.R. (1999) Maintenance of sorghum (*Sorghum bicolor*, Poaceae) landrace diversity by farmers' selection in Ethiopia. *Economic Botany* 53, 79–88.

Trehane, P., Brickell, C.D., Baum, B.R., Hetterscheid, W.L.A., Leslie, A.C., McNeill, J., Spongberg, S.A. and Vrugtman, F. (eds) (1995) *International Code of Nomenclature for Cultivated Plants – 1995 Adopted by the International Commission for the Nomenclature of Cultivated Plants. Regnum Vegetabile 133.* Quarterjack Publishing, Wimborne, UK.

Wood, D. and Lenné, J. (2001) Nature's fields: a neglected model for increasing food production. *Outlook on Agriculture* 30, 161–170.

Xiao, J., Grandillo, S., Ahn, S.D., McCouch, S.R., Tanksley, S.D., Li, J. and Yuan, L. (1996) Genes from wild rice improve yield. *Nature* 384, 223–224.

Zeven, A.C. (1998) Landraces: a review of definitions and classifications. *Euphytica* 104, 127–139.

Zimmerer, K.S. (1998) The ecogeography of Andean potatoes. *Bioscience* 48, 445–454.

5 A New Generation of IPR for the Protection of Traditional Knowledge in PGR for Food, Agricultural and Pharmaceutical Uses[1]

Thomas Cottier and Marion Panizzon
World Trade Institute, University of Berne, Hallerstrasse 6/8, CH 3012 Berne, Switzerland

5.1 Introduction

The goals of sustainable social and economic development of rural areas, and of fairness and equity in international economic relations, require a multitude of strategies in different walks of life, whether social, economic or legal. Validation of traditional lifestyles and related knowledge, ranging from agricultural know-how and food production to medical treatment, is one of the strategies. It supports global efforts towards sustainability and the preservation and use of biodiversity. It assists in bringing about decentralized habitation, sustainable livelihood and well-being of rural areas around the world. It helps to protect biodiversity and support an ecological balance, distributive justice, a sustained nutritional basis and food safety for all in an age of biotechnology. This chapter seeks to further investigate the potential of validating traditional knowledge by legal means, in particular intellectual property rights. It seeks to address one of the most difficult, controversial and thorny issues in the current debate on the legal status of PGRs. In the pursuit of the abovementioned goals and values, it explores to what extent a new generation of intellectual property rights could complement current policies.

Existing forms of intellectual property are of paramount importance in many areas related to traditional knowledge. In particular, biotechnology fully qualifies for intellectual property protection and thus exclusive rights, while traditional knowledge generally forms part of the public domain, accessible to the public at large. The imbalance of legal rights and protection is at the heart of the problem. It is therefore essential to explore the potential of IPR protection for traditional knowledge with a view to redressing current inequities.

[1] The authors thank Dr Susette Biber-Klemm and Professor Jerome H. Reichman, Barbara Ledda and Michael Halewood for valuable comments on earlier versions of this chapter.

Chapter 5 highlights the complexity and difficulty, if not impossibility, of such a venture in the field of PGR. Yet, the basic function of rights compels further exploration of the possibilities of this avenue: more than any other instrument or an IPR could empower the rural community or the individual to defend their interests. It could provide a basis for a better deal, for linking knowledge up with the industrialized world and for putting knowledge and information to work generating much-needed income. It could assist in bridging traditions and the modern world and by doing so contribute to conservation through use of economically valuable knowledge and information to the benefit of rural populations. Can it be done? To whatever extent we succeed, we also succeed in implementing basic values enshrined in human rights: the rights to economic well-being and development, and more specifically a right to overcome subsistence farming and earn the income for a decent life in rural areas.

This chapter first sets out the reasons why intellectual property protection for traditional knowledge would be beneficial to the overall IP system, redressing current imbalances. It reports current efforts to bring about enhanced protection in different international organizations and in domestic legislation. It then turns to develop the legal foundations, scope and organization of what we call Traditional Intellectual Property Rights (TIP Rights).

5.2 The Strategic Case for Intellectual Property Protection

5.2.1 The legitimacy and rebalancing of the IPR system

IPR, in the short and medium term, mainly serves the interests of industrialized countries and emerging economies. Strategic defenders of the current IPR system, which are countries depending on effective and world-wide protection of IPR, should therefore be interested in seeking ways to advance the system and to concretize its equitable components. They should share an interest in rendering it equally useful for rural populations. They should support tools that allow interfacing the life of those people with the information society and world of modern technology – this time as a provider, and not merely as a receiver in the process of transfer of information.

Strategic defenders of the established IPR system need to face the fact that many developing countries, in building their economy, will, for decades to come, essentially continue to depend on promoting and advancing sustainable agriculture, leaving subsistence farming behind. Strategic thinking in the field of IPR policies needs to take this into account. It needs to recognize that TK has often been the trigger point for new product development, especially in sectors of speciality food and beverages, horticulture, pharmaceuticals, personal care and cosmetics. With the advances of biotechnology, a renewed interest in ethnobotany, and the increasing demand for herbal remedies (see Hansen and van Fleet, 2003, p. 4), the erosion of TK might be halted. Building upon TK presents a unique opportunity for developing countries to access the global markets, in the fields of both agriculture and medical treatment.[2] Appropriate incentives and rewards for creators and caretakers of TK need to be created so that income generated from proceeds reaches these persons and at least is shared with them. Currently, outsiders often take away TK from its custodians without proper, or indeed any, compensation or reward being given to them (see, e.g., Downes, 2000, p. 255). Efforts to remedy this situation should also be explored within the system of intellectual property protection in an effort to rebalance the IP system and make it more attractive to agricultural economies and rural populations in developing countries.

[2] Cf. http://www.dfait-maeci.gc.ca/aboriginalplanet/ 'Traditional Indigenous Knowledge and Intellectual Property Rights', Canada, Department of Foreign Affairs and International Trade, last accessed 11 January 2006.

5.2.2 The need to clarify the connection between traditional knowledge and existing IPR

Intellectual property protection in its different forms is about protection of commercially valuable information. Valuable information, created by human efforts and investment, is at the heart of protecting human creativity and investment by means of patents, plant variety protection, copyright, trademarks, geographical indications, lay-out designs or undisclosed information (trade and commercial secrets). It would be inconsistent to consider IPR the principal instrument for protecting 'information' while completely denying such protection to TK despite its potential economic value.

The relationship of conventional IPR and TK needs to be clarified. It needs to be rendered compatible with the overall need to square and interface the worlds of traditional wisdom and the science of biotechnology. The current relation and state of affairs clearly is not satisfactory for traditional knowledge holders and has caused much anger, mostly in those developing countries, which are unequipped with the means for effective patent litigation – again to the detriment of the proprietary concept underlying the IPR system. There have been at least two showcase patent cases where the rights were not granted to the communities from where they originated, but to subsequent innovations building upon traditional information and use. The United States Patent Office (USPTO) granted patent protection to industrial 'inventors' for the neem (pesticide, skin condition treatment) (see Roht-Arriaza, 1996, pp. 921–922; Ramani, 2001, p. 1148, fn. 4),[3] turmeric (wound healing) (see Ramani, 2001, p. 1148, fn. 5; see also Chapter 3, 3.3.1, this volume)[4] and ayahuasca vine (anti-depressant),[5] without checking for 'prior art'. This not only caused outcries of anger in India and Brazil respectively, but also demonstrated an unsettled relationship of competing claims to commercially valuable information (see Chapter 6, 6.2.1, this volume).[6] At the heart of the problem is that TK has been considered as a matter in the public domain, and thus by accessible and exploitable by the public at large, and thus by any company interested in the information.

5.2.3 Competing interests and conflicting legal concepts

Approaching potential legal protection requires analysing competing interests in the field. We first turn to the structure of the seed industry, strongly dependent on IP protection, and then turn to the legacy of international law, leaving TK entirely in the public domain. We are faced with a variety of different and competing interests that set the stage for potential legal protection of TK and related products. We are faced with the

[3] See, for example, US Patent 6,623,766, assigned to the Council of Scientific and Industrial Research, New Delhi, IN, on 23 September 2003, concerning the insecticidal formulation of neem controlling malarial vector, mosquitoes: http://patft.uspto.gov/netacgi/nph-Parser?Sect1=PTO2&Sect2=HITOFF&p=1&u=/netahtml/search- bool.html&r=1&f=G&l=50&co1=AND&d=ptxt&s1=neem&OS=neem&RS=neem; all other US patents involving neem are available under: http://www.uspto.org (accessed 11 January 2006).

[4] See US Patent 5,401,504 of 28 March 1995 in which 'the use of turmeric in wound healing' was assigned to the University of Mississippi Medical Center, http://patft.uspto.gov/netacgi/nph-Parser?Sect1=PTO2&Sect2=HITOFF&p=1&u=/netahtml/search-bool.html&r=7&f=G&l=50&co1=AND&d=ptxt&s1=turmeric.TTL.&OS=TTL/turmeric&RS=TTL/turmeric (accessed 11 January 2006).

[5] See US Patent 5,256,533 of 26 October 1993, a patent accorded to beta-carboline, to which literature refers to as being found in ayahuasca plants. The patent's assignee is the Board of Regents, The University of Texas System, available at: http://patft.uspto.gov/netacgi/nph-Parser?Sect1=PTO2&Sect2=HITOFF&p=1&u=/netahtml/search-bool.html&r=1&f=G&l=50&co1=AND&d=ptxt&s1=ayahuasca.PPDB.&OS=SPEC/ayahuasca&RS=SPEC/ayahuasca (accessed 11 January 2006). The ayahuasca plant's use by the indigenous people of Amazonian Peru, Ecuador, Bolivia, Colombia and Western Brazil was described as early as the 1850s probably by Harvard University ethnobotanist Richard Schultes who may have referred to English naturalist Richard Spruce: see http://www.ethnopharmacology.com/ayahuasca (accessed 11 January 2006).

[6] See WTO Document IP/C/W/370, p. 5.

typical tensions between appropriation of information and leaving it in the public domain.

The structure of the seed industry

A large part of research and development in agricultural and pharmaceutical biotechnology is with the private sector. Today, more than 70% of cutting-edge agricultural biotech is produced and controlled by the private sector (see IPGRI, 2000, p. 103). IPR relating to PGRFA and PGM, as well as to the TK associated with agricultural or pharmaceutical uses of genetic resources, will remain indispensable tools to finance and secure R&D operations in the long term. Strategies designed to realize the social, nutritional and ecological functions of TK need to take these tools into account. The establishment of an equitable balance between interests and rights of industry and those of TK holders therefore cannot be achieved by simply dismantling IPR protection. And to the extent that IPR are ignored and policies to conserve and use TK exclusively rely upon the public sector and the taxpayer, direct legal linkages between the industry and the custodians of TK are likely to remain weak.

Existing IPR, in particular relating to genetic engineering, tend to promote further concentration and monopolies in the area of the seed industry (see Chapter 3, 3.3.6, this volume). IPR need to be redesigned in order to meet the challenges TK presents for the R&D industry. TK cannot be simply ignored unless governments are ready to fully finance all research in the field. Given governments' limited resources, it is not advisable to return to public monopolies.

New IP policies therefore should work to promote the sharing of technologies within the IPR system by building upon different titles of protection. TK protection could be an important element to this effect, besides competition policies that act against the abuse of dominant positions and cartels.

PGRFA in the public domain

Traditional knowledge, in particular relating to plant genetic resources for food and agriculture (PGRFA), is customary knowledge and therefore has been part of the public domain. Exceptions exist where knowledge is limited to the holder, in particular in medicinal and spiritual practices. Any attempt to protect traditional knowledge by means of rights faces the difficulty of withdrawing TK from the public domain and of appropriating its uses and commercialization. Chapter 5 describes the difficulties not only of identifying and localizing potential right holders, but also the cultural and societal implications such a move entails. Moreover, it is argued that the creation of *sui generis* rights creates adverse incentives and is not beneficial in most cases (see Chapter 7, 7.3.3, this volume). Keeping PGR for food and agriculture in the public domain is also one of the main goals of the International Treaty on Plant Genetic Resources for Food and Agriculture (ITPGRFA), discussed shortly, to include as many crops as possible under a list securing mutually guaranteed free access to the resources for the purposes of breeding and cultivation. Interests underlying the agreement seek to discourage recourse to and use of intellectual property protection for TK with a view to enhancing the free flow of genes. The question arises as to what extent the two philosophies can be intertwined and rendered compatible.

Linkages to global public goods

Finally, TK can be conceived as being part of global public goods, i.e. a heritage the protection of which is a concern to the world community at large, and not limited to the states exercising sovereign rights over the resources concerned. Providing the basis for long-term biodiversity and global food security, they enjoy a particular quality which is difficult to capture in contemporary legal terms. While the Doha Development Agenda (DDA) is explicitly committed to promoting environmental goods, it does not specifically relate to TK-produced goods.[7] Global public goods are a concept for ensuring that the WTO is not reduced to a 'supply-driven agenda, ignoring developments in actual environmental markets' (Vaughan, 2003, p. 2). It is not related to common heritage concepts of public domain. While changes in the classification schemes of the Harmonized System (HS) and lower tariffs promote the market access for public goods (see Vaughan, 2003 and Chapter 8), it is unclear as to how this should play out in terms of IP protection. Would it be beneficial, or rather detrimental, to the protection of such goods? To what extent can IP protection contribute to enhanced commercial use, thus conservation through use, of such goods? To what extent should promotion of such goods be left to the tools of trade regulation, addressed in Chapter 8 of this volume? Attempts to address the legal status of TK in different international and national instruments reflect the difficulty and inadequacy of dealing with the problem so far.

5.3 Current State of the Law and Efforts Made to Date

5.3.1 International instruments

Convention on Biological Diversity (CBD) and benefit sharing

With the adoption of the CBD in 1992, legal protection in the realm of PGRFA was introduced for a 'product of nature' (Hawtin, 2003, pp. 429–430). The CBD responded to emerging inequities stemming from private industry's use of PGRFA by reasserting national sovereignty and jurisdiction over such resources. None the less, the Global Biodiversity Forum (GBF) of the CBD contracting parties maintains the dialogue with the business sector and secures financially crucial partnerships.[8] In order to keep international trade in PGR free of protectionism under national intellectual property (IP) laws, the CBD imposed ABS principles and devised the instruments of 'mutually agreed terms' and 'prior informed consent' (PIC) (see also Chapters 2 and 6; Bragdon, 2003, pp. 442–443).[9] However, these flanking policies proved to be too cumbersome to sustain continued investment by transnational corporations. They also proved ineffective in developing countries that lacked the means to install national registration systems, obtain patents or oppose the patents of others on relevant TK (see Gupta, 1999). They fail, in the final analysis, to define those entitled to access benefit-sharing and leave the matter in a realm of legal uncertainty.

Plant Variety Protection (UPOV)

In its 1978 and 1991 Acts, the UPOV established a *sui generis* system of plant-related IPR for plant breeders (see Bragdon, 2003). This effort to invest breeders with private rights to their improved varieties obliged even the FAO to weaken the pure public domain approach that had been embodied in the International Undertaking (IU).[10]

Three interpretative resolutions pertaining to the IU confirm that plant breeders' rights are FAO compatible. Developing countries, which were then still relying on national sovereignty and the principle of benefit sharing to protect their plant genetic resources from bio-piracy, found themselves at an even greater disadvantage when confronted by assertions of plant breeders' rights. Some responded by developing hybrid regimes that combined

---

[7] See Doha Declaration 14 November 2001, paras. 31–33.

[8] See http://www.gbf.ch (accessed 11 January 2006).

[9] CBD Article 17.3.

[10] International Undertaking on Plant Genetic Resources, UN Food and Agriculture Organization, 22nd Session, para. 285, Annex, U.N. Doc. C/83/REP (1983) [hereinafter IU]; FAO Conference Resolution 4/89 acknowledges that UPOV's plant breeders' rights are not at odds with the IU's objectives; FAO Conference Resolution 5/89 recognizes and defines farmers' rights; and FAO Conference Resolution 3/91, finally, recognizes national sovereignty over genetic resources.

UPOV rights with CBD duties (see Chapter 2; Kongolo, 2001), while many joined UPOV, particularly under pressure from the EU and the USA.

International Treaty on Plant Genetic Resources for Food and Agriculture (ITPGRFA)

The United Nations' FAO deserves credit for raising international awareness of genetic diversity as 'one of humanity's most important resources' (Hawtin, 2003, p. 430). The IU of 1983 recognized plant genetic resources as the 'common heritage of humanity', and its underlying policies are rather hostile to the idea of protecting these resources by means of IPR.[11] It is feared that private intellectual property rights could undermine the public domain status of plant genetic resources for food and agriculture.

Consistent with this outlook, the ITPGRFA, adopted in 2001, reluctantly recognized that plant breeders (protected by UPOV) using plant genetic materials available from the gene banks of CGIAR Centres may 'take out IPR' on new products (Article 12.3(d)).[12] Moreover, Articles 12.3 (f) and (g) of this treaty does not prevent private parties from claiming IPR on modifications of plant genetic materials or on materials for which access (under the CBD) had been legally obtained (see de Carvalho, 2003, p. 97).

Otherwise, Article 12.3 is opposed to *introducing* IPR on traditional knowledge and on PGRFA. Article 12.3(d) prohibits IPR that limit the facilitated access to the PGRFA/genetic parts/components in the form received from the multilateral system (MS). The Treaty emulates the CBD's goals of 'conservation and sustainable use of plant genetic resources for food and agriculture and the fair and equitable sharing of benefits derived from their use'.[13] To this purpose, the ITPGRFA Contracting Parties agreed to list 35 crops and 29 forages found to be in the public domain as pertaining to the public domain and thus open for free exchange (Annex I ITPGRFA). This legal instrument, devised to 'facilitate access to PGRFA and to share the benefits of its utilization' (Article 10.2 ITPGRFA), is termed 'multilateral system' (MS). None the less, the Treaty does not attenuate the competition between holders of traditional knowledge and plant breeders and farmers over the ownership and use of TK-related material. Any new intellectual property right in traditional knowledge must accordingly take account of the relationship between TK protection and existing rights of breeders and farmers (see Girsberger, 1999).[14]

The WTO Agreement on Trade-Related Aspects of Intellectual Property Rights (TRIPS Agreement) and the Paris Convention for the Protection of Industrial Property

With the adoption of the TRIPS Agreement in 1994 (Cottier, 2005), the patenting of plant genetic resources for food and agriculture benefited from a major step towards global recognition and enforcement.[15] Article 27.3(b) of the TRIPS Agreement

---

[11] FAO's Global Plan of Action for the Conservation and Sustainable Utilization of Plant Genetic Resources for Food and Agriculture, available at http://www.fao.org/ag/cgrfa/PGR.htm, was introduced by the Leipzig Declaration in June 1996 and is monitored by the Commission on Genetic Resources for Food and Agriculture. It addresses foremost the *ex situ* and *in situ* preservation of plant genetic resources. It promotes 'core collections' of germplasm through national governments in cooperation with IGOs and NGOs and warns against the seed industry's use of gene banks as an excuse to neglect collecting biodiversity-rich crops, under-utilized crops and local varieties.

[12] Compare Article 12.3(d) with Articles 12.3(e) and (f) of the ITPGRFA.

[13] ITPGRFA, Article 1.1.

[14] FAO Conference Resolution C5/89 (1989).

[15] Agreement on Trade-Related Aspects of Intellectual Property Rights, Annex 1c of the Marrakesh Agreement Establishing the World Trade Organization, signed on 15 April 1994, The Results of the Uruguay Round of Multilateral Trade Negotiations, The Legal Texts, WTO Secretariat, Geneva 1994 [hereinafter TRIPS Agreement].

specifically requires Members to provide either patents or a *sui generis* regime for novel and stable plant genetic resources. In addition, Article 27.3(b) narrows the scope of exclusions from patentability. In practice, the flexibility inherent in the *sui generis* option has reportedly created more problems than solutions, and it has also proved ineffective as an instrument for addressing the protection of traditional knowledge as such (see Bragdon, 2003, pp. 446–448; Llewelyn, 2003, pp. 305–308). Indeed, the TRIPS Agreement makes no express reference to traditional knowledge (see Dutfield, 2002b).

The protection of traditional knowledge was not a subject under discussion during the Uruguay Round of multilateral trade negotiations. However, the inclusion of Article 10bis of the Paris Convention for the Protection of Industrial Property into the TRIPS Agreement brought about a first basis for addressing misappropriation of traditional knowledge by taking recourse to principles of unfair competition. Article 10bis of the Paris Convention obliges Members to the Convention to assure protection against any act of competition 'contrary to honest practices in industrial or commercial matters'. The provision sets forth an illustrative list of such practices, including the creation of confusion, discrediting and misleading the public. It does not explicitly address misappropriation of efforts, achievement and work of others, but allows taking this into account in further specifying the principle of honest business practices. Importantly, Article 10bis of the Paris Convention is subject to WTO dispute settlement and enforcement under the Dispute Settlement Understanding (DSU) of the WTO (Chapter 2).

Explicit reference to TK was introduced in paragraph 19 of the Doha Agenda adopted in 2001.[16] The topic, however, has not made progress in the meantime. In July 2004 it was dropped from further work, along with efforts stalled to further develop protection of geographical indications[17] (Panizzon and Cottier, 2005). The decision was taken in an effort to facilitate negotiations, which are difficult enough.[18] Given the importance of the topic for trade in agriculture and market access – the key issues of the Doha Agenda, the matter is likely to return on the agenda, sooner or later.

DISCUSSIONS IN THE TRIPS COUNCIL

At the Council for TRIPS, efforts were undertaken to make the CBD and its Bonn Guidelines mutually compatible with TRIPS. So far to no avail, the initiative would require IPR under TRIPS to be given an interpretation consistent with the principle of 'access and benefit sharing', and conversely, the CBD would have to be viewed as consistent with TRIPS-imposed IP obligations (see Dutfield, 2002a; see also Chambers, 2003). We note that, so far, no discussions have taken place on the basis of rights and obligations under Article 10bis of the Paris Convention as incorporated into the TRIPS Agreement.

General discussions held in the TRIPS Council revealed a willingness to deal with the matter and to take the quest for TK protection seriously, until the matter was stalled in July 2004. A number of arguments were made in support of TK protection. Since the *raison d'être* of the TRIPS Agreement is to provide IP protection for a broad range of subject matters, including new ones such as plant varieties, biological materials, lay-out designs and computer software, it was said that 'it is only equi-

[16] See Doha Ministerial Declaration (DMD), adopted 14 November 2001, WTO Document WT/MIN(01)/DEC/1 of 20 November 2001. The DMD provides that WTO members revise Article 71.1 of TRIPS relating to Articles 24.1 and 2 for GIs, and Article 27.3(b) for TK, whereby any reform must take into account the development dimension.
[17] TRIPS Agreement Articles 22 to 24.
[18] Cf. WTO Doha Work Programme, Decision adopted by the General Council on 1 August 2004, WTO Document WT/L/579 of 2 August 2004 [hereinafter WTO, Doha Work Programme]. The Doha Work Programme is also called the 'General Council's post-Cancún decision'.

table that TK should be given legal recognition'.[19] Moreover, 'an egalitarian system of IPR is one which does not exclude a priori any section of society'.[20] It was further stated, 'nothing in the TRIPS Agreement prevents the WTO from adopting a specific protection regime for TK'.[21] TRIPS Article 8.1 would seem to encourage the creation of legal protection for TK as a measure 'necessary to protect public health and nutrition and to promote the public interest in sectors of vital importance to [...] socio-economic and technological development'. However, Article 8 TRIPS only relates to national IPR legislation, which in our view bears the risk of a fragmented protection of TK. It was argued that 'granting of patents on TK already in the public domain amounts to unauthorized appropriation of the knowledge' and such patents should be revoked.[22]

Switzerland, which made the most far-reaching proposal of a disclosure requirement constitutive to patent application, immediately softened the potential impact of the constitutive disclosure requirement, by stating that TK should primarily be dealt with at WIPO, leaving the TRIPS Council only to 'benefit' from WIPO's work.[23] The EC seemed ready to accept a disclosure of origin criteria, albeit not a constitutive one, and advocated that it be held apart from patent application criteria. Under the EC proposal, a failure to disclose the origin would trigger criminal sanctions and/or civil liability, but not affect the validity of the patent.[24] Brazil welcomed the EC's proposal that it would be important to disclose the source of the country of origin of the biological resource and of TK used in an invention.[25] Other suggestions proposed are bilateral contracts between the TK holder and the person or company wishing to access the patent (but also pointing to the difficulty of 'negotiations between unequal parties and the problem of prior informed consent') and a *sui generis* system of protection (protection of proprietary rights that serve to ensure fairness and equity).[26]

DISCUSSIONS IN THE COMMITTEE ON TRADE AND ENVIRONMENT

The CBD is an MEA and not an IP-related treaty; thus, constructing a PIC requirement contingent on a patent application is not so much a challenge of linking the CBD to the WTO (which has, since *US-Shrimp*, embraced MEAs), but rather one between the CBD and the TRIPS specifically (see Prakash, 2000, pp. 155–157). The question is whether the issue of TK in the CBD will ever be discussed at the TRIPS Council. Additional committees will need to deal with the matter of TK once flanking policies are included. This is particularly true for agriculture, tariffs and TBT/SPS concerns.

World Intellectual Property Organization (WIPO)

WIPO has dedicated a considerable amount of work to the problem of TK protection. It has designed model laws, discussed shortly, compiled databases to assist national legislative efforts to protect TK, and it has devoted considerable effort to seeking a consensual definition of TK. Initially, WIPO essentially proposed a bottom-up approach under which developing countries would first assess how existing national mechanisms of intellectual prop-

[19] WTO Document IP/C/W/370, p. 3.
[20] WTO Document IP/C/W/370, p. 3.
[21] WTO Document IP/C/W/370, p. 11.
[22] WTO Document IP/C/W/370, p. 11.
[23] WTO Document IP/C/W/400, paras 25–27.
[24] See *Report of the Meeting held on 8 October 2002,* Note by the Secretariat, WTO Document WT/CTE/M/31, 2 December 2002, para. 74.
[25] See *Report of the Meeting held on 8 October 2002,* Note by the Secretariat, WTO Document WT/CTE/M/31, 2 December 2002, para. 75.
[26] See *Report of the Meeting held on 8 October 2002,* Note by the Secretariat, WTO Document WT/CTE/M/31, 2 December 2002, para. 75.

erty could be more effectively used to protect TK before introducing protection at the international level.[27] Work in WIPO has progressed considerably in recent years, integrating both domestic and international levels of protection. Important policy papers prepared by the WIPO secretariat in 2004 propose a number of basic objectives and principles, based upon which TK protection should be developed both in domestic and international law:[28]

- Responsiveness to the needs, aspirations and expectations of TK holders.
- Understandable, effective, appropriate and accessible protection.
- Flexibility and comprehensiveness.
- Equity and benefit-sharing.
- Consistency and supportiveness with existing IP systems.
- Recognition of rights.
- Complementarity of defensive and positive protection.

Specific principles suggested for discussion (reproduced verbatim) include the need to:[29]

- Prevent the misappropriation of TK.
- Define acts of misappropriation as acquisition and appropriation of TK by unfair means.
- Provide examples of misappropriation, such as acquisition of TK by theft, bribery, coercion, breach or inducement to breach of contract, confidence, fiduciary duties, acquisition of TK without prior informed consent if such consent is required, etc.
- Protect TK against other acts of unfair competition.
- Ensure maximum flexibility for national authorities to implement the principles in a legal form appropriate to their own legal systems, national priorities and stakeholder needs.
- Apply the principle of prior informed consent to TK.
- Ensure equitable benefit-sharing from industrial and commercial uses made with gainful intent, if the TK conveys a technology-based advantage in industry and commerce to the user.

WIPO submits what are called *general aspects of the guiding principles*, which TK could take into account:[30]

1. Respond to the need for greater respect and recognition of TK holders and their greater involvement in legal and policy processes aimed at promoting their interests.
2. Seek protection that is balanced and proportionate, given the broader public interests and the specific interests of TK-holder communities, recognizing fair use or fair dealing and the legitimate public domain, while repressing truly inequitable conduct.
3. Target the misappropriation of TK rather than the mandatory creation of exclusive property rights (although leaving this option open where national policymakers, in consultation with TK holders, elect to take this policy choice as a means of economic development for TK holders).
4. Aim to ensure that benefits arising from commercial or industrial uses of TK are distributed equitably.
5. Apply the principle of prior informed consent to TK in harmony with existing legal systems at the national and international levels.

[27] See *Elements of a Sui generis System for the Protection of Traditional Knowledge*, Report of the Third Session of the Intergovernmental Committee on Intellectual Property and Genetic Resources, Traditional Knowledge and Folklore, WIPO Document WIPO/GRTKF/IC/3/8, 29 March 2002 [hereinafter WIPO, *Elements of a Sui generis System, Third Session*] paras 4, 5, 8, 17–57; but see paras 8, 42–43.
[28] *Protection of Traditional Knowledge: Overview of Policy Objectives and Core Principles*, Intergovernmental Committee on Intellectual Property and Genetic Resources, Traditional Knowledge and Folklore, Document prepared by the Secretariat, 7th Session, Geneva, 1–5 November 2005 para. 16, WIPO Doc. WIPO/GRTKF/IC/7/5 (20 August 2004) (reproduced verbatim). On the integration of domestic and international levels in the work of the Committee see in particular para. 36.
[29] WIPO/GRTKF/IC/7/5 para. 21 (reproduced verbatim).
[30] WIPO/GRTKF/IC/7/5 para. 23 (reproduced verbatim, footnotes omitted).

6. Leave open flexibility for national authorities to give effect to the principles in a manner compatible with their own legal systems, national policies and stakeholder needs.
7. Aim for consistency with the diverse range of existing TK laws described to the Committee, and be supportive of the various policy and legal choices already taken to protect TK.
8. Are compatible with existing international intellectual property principles, doctrines and standards, and with international non-IP instruments, processes and fora which address different aspects of traditional knowledge protection, such as the CBD, FAO, the International Forum on Forests (IFF), the ILO, UNCCD, UNCTAD, UNEP, the UN Permanent Forum on Indigenous Issues, and the WHO and WTO. The principles pay close attention to being fully compatible with, and supportive of, those processes and instruments, but they do not seek to enter into areas addressed by them, leaving the relevant issues to those instruments and processes.
9. Address TK in a manner complementary to the protection of TCEs/expressions of folklore (see WIPO/GRTKF/IC/7/3), and combining both positive and defensive protection.
10. Address the characteristics of TK and needs of TK holders through customized mechanisms, which draw on and build further upon the existing principles of IP law.
11. Address the substance of TK protection at the international level, leaving open the precise legal form and status for decision by Committee members and WIPO Member States as appropriate. This approach reflects the insight that greater understanding of the substantive principles of TK protection may facilitate the development of appropriate international vehicles. The suggested draft principles do not prejudge any decisions on the legal form and status, but are so drafted as to serve as input to any of the particular approaches to the international dimension set out in WIPO/GRTKF/IC/6/6.
12. Apply the combined approach to TK protection which was generally supported at the Committee's sixth session, using the policy tools of unfair competition, prior informed consent, exclusive rights and compensatory liability, while referring to customary laws and protocols to the extent possible and appropriate.
13. Do not require any registration or compilation of TK in databases, in view of the widely voiced concerns about the potential downside of such databases. The principles do not suggest protection should be predicated on the use of TK databases at the local, national or international levels. They do not preclude the optional and voluntary use of registries or databases for those communities and countries that wish to do so, since several communities or countries have already established such databases or registries, and they choose to continue working with them.
14. Address the international dimension in line with the requirements of the Committee at its sixth session.
15. Derive their structure and contents from the framework document submitted by the African Group, as widely supported by Committee members.
16. Do not prejudice the application of existing IP systems to TK subject matter and should not affect the functioning of existing systems, other than providing improved defensive protection of TK.

Operational legal principles suggested by the WIPO Secretariat essentially rely upon the concept of prohibiting unfair competition and protection against misappropriation:[31]

1. Traditional knowledge shall be protected against misappropriation.
2. Any acquisition or appropriation of traditional knowledge by unfair or illicit means constitutes an act of misappropriation. Misappropriation may also include deriving commercial benefit from the acquisition or appropriation of traditional knowledge when the person using that knowledge knows, or is grossly negligent in

[31] WIPO/GRTKF/IC/7/5 Annex I, p. 5 para. B (reproduced verbatim).

failing to know, that it was acquired or appropriated by unfair means; and other commercial activities contrary to honest practices that gain inequitable benefit from traditional knowledge.

3. In particular, legal means should be available to suppress:

(i) acquisition of traditional knowledge by theft, bribery, coercion, fraud, trespass, breach or inducement of breach of contract, breach or inducement of breach of confidence or confidentiality, breach of fiduciary obligations or other relations of trust, deception, misrepresentation, the provision of misleading information when obtaining prior informed consent for access to traditional knowledge, or other unfair or dishonest means;

(ii) acquisition of traditional knowledge or exercising control over it in violation of legal measures that require prior informed consent as a condition of access to the knowledge, and use of traditional knowledge that violates terms that were mutually agreed as a condition of prior informed consent concerning access to that knowledge;

(iii) false claims or assertions of ownership or control over traditional knowledge, including acquiring, claiming or asserting intellectual property rights over traditional knowledge-related subject matter by a person who knew that the intellectual property rights were not validly held in the light of that traditional knowledge and any conditions relating to its access; and

(iv) commercial or industrial use of traditional knowledge without just and appropriate compensation to the recognized holders of the knowledge, when such use has gainful intent and confers a technological or commercial advantage on its user, and when compensation would be consistent with fairness and equity in relation to the holders of the knowledge in view of the circumstances in which the user acquired the knowledge.

4. Traditional knowledge holders should also be effectively protected against other acts of unfair competition, including acts specified in Article 10bis of the Paris Convention. This includes false or misleading representations that a product or service is produced or provided with the involvement or endorsement of traditional knowledge holders, or that the commercial exploitation of products or services benefits holders of traditional knowledge.

According to WIPO, the legal form of protection may be realized through a special law on traditional knowledge, the laws of intellectual property, including unfair competition law and unjust enrichment, and other areas of civil and panel and administrative law.[32] WIPO suggests that the

> form of protection need not be through exclusive property rights, although such rights may be made available, as appropriate for the holder of traditional knowledge, including through existing or adopted intellectual property systems, in accordance with the needs and the choices of the holders of the knowledge, national laws and policies, and international obligations.[33]

These suggestions were further supported by additional documents and studies undertaken by the Secretariat.[34]

Finally, we note that the United Nations Conference on Trade and Develop-

[32] WIPO/GRTKF/IC/7/5 Annex I, p. 6.

[33] WIPO/GRTKF/IC/7/5 Annex I, p. 6.

[34] See The Protection of Traditional Knowledge: Outline of Policy Options and Legal Elements, Intergovernmental Committee on Intellectual Property and Genetic Resources, Traditional Knowledge and Folkore, WIPO/GRTKF//C/7/6 (27 August 2004); The Protection of Traditional Cultural Expressions/Expressions of Folklore: Overview of Policy Objectives and Core Principles, ibid., WIPO/GRTKF/IC/7/3 (20 August 2004); Recognition of Traditional Knowledge within the Patent System, ibid., WIPO/CRTKF/IC/7/8 (23 July 2004); Genetic Resources: Draft Intellectual Property Guidelines for Access and Equitable Benefit-Sharing, ibid., WIPO/GRTKF/IC/7/9 (30 July 2004).

ment (UNCTAD) proposed to work on the basis of agreed minimum standards to support an internationally recognized *sui generis* system.[35]

5.3.2 Recommendations by international bodies and national legislation

Model laws

As of 31 August 2004, the NGO GRAIN lists at least 14 countries, including regional organizations, which had made, or were in the process of making available, specific legal protection for TK-related subject matter.[36] As for international organizations, WIPO and UNESCO have put together model provisions for national laws on the protection of folklore (see Dutfield, 2002b, pp. 32–34).[37] However, they neglected ensuring that states search for, and document, the expressions of folklore (see Gopalakrishnan, 2002, pp. 726–727). WIPO is building databases for sustaining national legislative efforts to protect TK. It has also devoted much effort to finding a working definition of TK, but cannot yet claim sufficient consensus for proposing an operational WIPO definition. National model legislation reacts to the lack of definition at the international level by each creating their own definitions instead of introducing operational language and effective protection.[38]

WIPO has defined some requirements for a national *sui generis* system to protect TK.[39] It proposes a bottom-up approach, suggesting first seeing how existing national mechanisms of intellectual property can be more effectively used to protect TK before introducing protection at the international level. The Organization of African States (OAU) adopted Model Legislation for the Protection of the Rights of Local Communities, Farmers and Breeders in 1998. It provides for a 'hybrid system' at the centre of which figure the interests of farmers, while acknowledging breeders' rights and 'community intellectual property rights'. In two respects the model goes beyond ABS, PIC and UPOV: Part I (a) of the legislation recognizes and protects 'the inalienable right of local communities including farming communities over their biological resources and crop varieties, knowledge and technologies'. Article 23(1) recommends states ensure that at least 50% of benefits obtained through sharing resources are channelled to 'the concerned local community or communities in a manner which treats men and women equitably' (Kongolo, 2001, pp. 364–366, 368; see Chapter 2).

Domestic legislation

A review of existing efforts shows that a number of developing countries have made efforts to protect TK beyond existing IPR and to develop *sui generis* systems. India in 2000 created IPR protection for TK by means of the Indian Biological Diversity Bill, being the first legislation of this kind in any of the 12 megadiverse countries. The legislation sets up a National Biodiversity

[35] See GRAIN, Biodiversity Rights Legislation (BRL), Traditional Knowledge: The latest IPR trap, available at: http://www.grain.org/brl/?typeid=30 (last accessed 11 January 2006). The African Union, Andean Community, ASEAN, Bangladesh, Bolivia, Brazil, Costa Rica, India, Morocco, Free Trade Area of the Americas (FTAA), Pacific Forum, Pakistan, Panama, Peru, are the countries GRAIN lists as having installed or drafted biodiversity rights legislation specifically aimed at protecting traditional knowledge.

[36] See WIPO, *Traditional Knowledge – Operational Terms* para. 36.

[37] See http://www.wipo.int/tk/en/folklore (accessed 11 January 2006).

[38] See, for example, the Draft African Model Legislation for the Protection of the Rights of Local Communities, Farmers and Breeders which defines Traditional Knowledge thus: 'community knowledge or indigenous knowledge is the accumulated knowledge *that is vital for conservation and sustainable use of biological resources* and/or which is of socio-economic value, and which has been developed over the years in indigenous/local communities', cited in: WIPO, *Traditional Knowledge – Operational Terms*, para. 37.

[39] See WIPO, *Elements of a Sui generis System, Third Session* paras 1–3.

Authority (NBA), whose approval will be required when applying for intellectual property rights to protect an invention based on a biological resource from India or on TK located in India. When granting its approval, the NBA can impose benefit-sharing conditions. These may include fees or royalties, joint ownership, transfer of technology or a venture capital fund for benefit claimers, or the payment of monetary or other compensation (see Cullet, 2001, pp. 212–230; see Chapter 2, 2.4.1, this volume).[40]

The Republic of the Philippines recognizes the ownership of TK by the communities, and empowers them to manage it in accordance with their customary laws.[41] The Brazilian approach recognizes indigenous and local communities' right to prevent unauthorized third parties from using, exploiting, experimenting, disclosing, transmitting and re-transmitting data and information that integrate or constitute associated TK. Protection is construed by contracts of access (bilateral approach), which are aimed at the sharing of benefits arising from the use of genetic resources and associated TK.[42]

Peru's system protects TK in a manner such that only collectives, but not individual holders, are eligible for protection. An important feature of this system is that it requires the free exchange of knowledge between the different communities holding the same TK. PIC is a focal point for the users of the TK-held knowledge. The PIC is to be managed by an administrative body, which represents the community. Moreover, the system offers, in addition to IPR, instruments against unfair competition, in order to ensure an equitable appropriation of TK by third parties. The IPR component of the system consists of licensing, which is only granted upon written consent of a knowledge-holding community (see Eugui, 2002, p. 16, and Chapter 2, Section 2.3.1, this volume).

5.3.3 Conclusions

International instruments lay open mutual tensions and a lack of coherence, reflecting competing constituencies and fora. National legislation partly follows international recommendations, but offers diverging solutions. Often, as the laws of the Philippines, Brazil and India suggest, TK is protected through hybrid systems of ABS and defences against misappropriation by third parties, although the latter instrument seems to be closer to a civil law remedy than to true IPR. *Sui generis* protection, developed in response to Article 27 of the TRIPS Agreement, partly relies upon IP concepts in granting exclusive rights. Partly it is limited to providing the basis for contractually arranged benefit-sharing agreements.

All national legislation is, by definition and however skilled, limited in effect to the territory of the state and not able to manage complex international constellations. Apart from the CBD, there is no international instrument in place protecting TK effectively and internationally. Nevertheless, today there is still more scepticism than consensus regarding a new global IP protection for TK. It is submitted that such scepticism is partly based upon the absence of viable concepts. It is partly based upon the rejection of making recourse to IPR and partly on sound analysis of technical possibilities. Recent work undertaken in WIPO, discussed above, has made substantial contributions to the way of coming to grips with TK protection in legal terms. Thus, we need to examine the objec-

---

[40] See also http://envfor.nic.in/ (accessed 11 January 2006).

[41] See The Indigenous People's Right Act, 1997, Act No. 8371; and the Rules and Regulations Implementing the Act, Order No. 1, 1998.

[42] Quoted in: *Review of Existing Intellectual Property Protection for Traditional Knowledge,* Report of the Third Session of the Intergovernmental Committee on Intellectual Property and Genetic Resources, Traditional Knowledge and Folklore, WIPO Document WIPO/GRTKF/IC/3/7, 6 May 2002 [hereinafter WIPO, *Review, Third Session*], para. 16.

tions brought against the IP system and analyse to what extent internationally defined *sui generis* exclusive rights could contribute to harmonizing the international legal system and whether this offers a viable option to redress the inequities addressed above.

5.4 Basic Objections to IPR Protection

It is no secret that the TRIPS Agreement in its present form and scope primarily responds to the needs of industrialized countries (see among others, Downes, 2000, p. 255). It does not offer much to farming communities. It was acceptable to developing countries mainly because it formed part of an overall package and single undertaking, which included a pledge to liberalize market access in agriculture and to textiles. These goals materialize only slowly. The Agreement on Textiles was abolished on 1 January 2005, leaving trade in textiles under regular GATT rules. These efforts, however, have not yet remedied a basic imbalance. Given this background, the idea of introducing new types of IPR is strongly contested, *inter alia* for the following reasons specifically relating to TK:

- First, it is argued that an IPR protection of TK is impossible for systemic reasons underlying the rationale of IPR. Under this view, TK does not live up to the criteria of innovation and novelty, which preconditions any IPR protection.
- Secondly, and from a systemic point of view, it is argued that, as a practical matter, it is impossible to discern and individualize intellectual ownership of these resources and encapsulate them into private rights.
- Thirdly, the North–North, R&D, argument is that the protection of traditional knowledge with IPR is hostile to the goals of developing a truly open system of crop exchange and accessibility for the purposes of breeding and research.
- Fourthly, IPR conflict with the other concepts of legal protection for biodiversity. The CBD respects international agreements with an IP protection component relating to biodiversity (such as Art. 27(3)b TRIPS) only to the extent that the 'exercise of those (other treaties') rights and obligations do not cause a serious damage or threat to biological diversity'.[43]
- Finally, it is argued either by the knowledge holders themselves or by NGOs that IPR, as all private law concepts, are alien to the informal sector and the communities' customary laws reflecting it, and thus should not be used. IPR will destroy traditions of free exchange and mutual communal support. It is argued that the creation of IPR creates perverse incentives and destroys traditional methods of breeding.

We turn to these objections first.

5.4.1 Innovation *vs* traditions

While patents, plant variety protection, copyright, industrial designs and integrated circuits require novelty as a precondition for granting an exclusive right, an innovative step is not contingent for trademarks, geographical indications, appellations of origin, data protection and trade secrets. Trademarks serve the purposes of distinguishing products on the market and to protect marketing efforts. Protection of TK through trademarks resembles economic assistance to a collective.[44] Geographical indications and appellations of origin protect specific qualities of a product, which draw their defining feature from a particular region and TK developed therein.[45] However, the extension in the framework of TRIPS of the provision on geographical

[43] Article 22 CBD.

[44] See WIPO *Review, Third Session*, para. 26, as WIPO says: '[T]raditional knowledge holders enjoy special assistance very much in line as the assistance provided in many countries to small and medium enterprises.'

[45] See, for example, http://www.bangkokpost.com As of writing this contribution, Thailand is debating whether and how Jasmine rice might fall under its forthcoming legislation on geographical indications.

indications to products other than wine and spirits in Article 23 of TRIPS has not been successful to date (see Nascimento, Chapter 6, this volume; Addor and Grazioli, 1997, pp. 865 ff.). The protection of undisclosed information is unrelated to the novelty of the product. It may well have emerged from the use of TK, which is kept to a small number of people.[46]

The intellectual property system therefore is not alien to the protection of existing products: it only partly focuses on innovation. An invention is not required. Neither the TRIPS, nor the Paris Agreement nor the European Patent Treaty or the US Patent law (35 USC 101) define the term 'invention' (see Leskien and Flitner, 1997). What all rights therefore share is the goal of protecting human labour and investment of financial and human resources from unlawful appropriation by third parties. This is at the core of the system. This is what justifies in the end the grant of exclusive rights, in order to deter infringements and combat counterfeiting and piracy within a market economy. It is true that food production in developed countries will rely more heavily on IPR to protect a certain cultivation method, or an environmentally friendly production process, because this is where the money lies in developed countries' food production. In contrast, IPR protection is not at the forefront of developing countries' concerns (see Smale *et al.*, 2001, p. 2).

5.4.2 Assignable resources and communitarian rights

Developing countries dismiss IPR because they fear that the exchange and availability of TK-related products would become more difficult, if not impossible (South–South refusal). These are important concerns, mainly voiced within the FAO and CGIAR system. It is no secret that even in industrialized countries with adequate resources to thoroughly investigate patent claims, registration and patent litigation are costly. *A fortiori*, the examination process in developing countries has been wanting as the *India-Patents* case before the WTO has shown, and many bad patents are regularly issued (see Chapter 3; Dutfield, 2002a, pp. 910–911). Yet, it is not a valid argument to deny indigenous communities patent protection because they cannot afford it. Efforts at building capacity are under way and supported and may equally benefit TK protection.

The idea of allocating private rights to TK depends upon the possibility of identifying potential holders of IPR on a local or regional basis. The factual study conducted by IPGRI on behalf of this project (Chapter 4) concluded that assignability is not possible for a large share of landraces, since their origins cannot be properly traced (see IPGRI, 2000, p. 71; see also Chapter 4, 4.3.2, this volume). The study does not exclude that it is impossible to trace a connection between a particular landrace and a particular community in all cases, but emphasizes that distinction is no trait in traditional breeding practices and not a concern of farmers. The result is based upon an analysis of SINGER as well as IBPGRI/IPGRI collection missions. The report notes that all the accessions indicate either the geographical coordinates or at least the country of origin recorded; 96% of them detailed location of the collecting site in SINGER. In IBPGRI/IPGRI Collecting Missions data, ethnobotanical information on local names and use is relatively common, while detailed indications on farms or ethnic groups remain very scarce. Only five of 14 gene banks developed collecting forms with extensive ethnobotanical information, most of them established in recent years. Where they exist, they contain information on special traits (performance, tolerance, resistance), cultivation and selection criteria and local names which, however, are only indicative and inconclusive. The report also says that location data do not indicate where any of the collected data are *not* found: 'Our raw data does not provide any indication about which any

[46] See UNCTAD Document, TD/B/Com.1/EM.13/2, pp. 11–13.

accession's trait can be attributed to adaptation to the local environment'. It is thus considered insufficient for the claim of *sui generis* rights. At the same time, data are considered sufficient to provide supporting evidence to defeat claims by others and to demonstrate prior art in the context of claims to IPR. Equally, it is held sufficient to provide the basis for rights to approve or deny access to resources located on their lands. The study concludes on an 'agnostic to cautiously negative' note as to the creation of *sui generis* rights for landraces.

> There may be conditions under which it would be possible to identify and protect a landrace pursuant to the looser sui generis conditions for protection that we have discussed in this paper. On the other hand, it appears that it often would not be possible. To be certain, it would be necessary to conduct research on a crop-by-crop, community-by community, process-by-process (e.g. participatory plant breeding) basis. Some studies are underway, and we are waiting for their results. (Chapter 4, Section 4.10, this volume)

The analysis discussed *sui generis* rights in terms of potential protection for landraces, essentially building upon the tradition of plant variety protection. Protection of traditional knowledge, however, primarily focuses on *information*. Knowledge about appropriate use of resources under contextual circumstances is at the heart of protection, rather than the particular landrace. The latter essentially operates as evidence and a product derived from information. And too little is known about ethnobotanical information documenting such information and its origins. Data are still fragmentary and consist of circumstantial evidence. The analysis of accession data further shows that identification of origin of information underlying the breeding process has not been a criterion in describing accessions under the traditional public domain approach. It does not exclude that future efforts and studies may provide enhanced and more systematic evidence on the assignability of knowledge relating to PGR. Current raw data are considered a sufficient basis for the establishment of prior art and the frustration of IPR claims by others. Such evidence therefore is considered sufficiently precise in order to demonstrate that a particular practice and knowledge, as evidenced by a particular accession, has been in use in a particular community or region, albeit not in an exclusive manner. It thus may well serve as a basis for further research into the positive assignability of TK.

The ability to assign knowledge relating to PGR may be further facilitated if the relevant time span under consideration for granting IPR is limited, and only takes into account the labour of farming communities and regions of recent decades. Evidently, the longer the relevant period, the less likely it is possible to determine such allocation. The determination of this period will depend on the final results of further factual studies, and on political will. Assuming at this stage that a period of 50 years was considered as a relevant time span for defining assignability, many of the imponderabilities could be reduced and eliminated. Finally, we note that the intellectual property is open to communities and sharing among a great number of people and limited to individuals. The current international IPR system is accustomed to defining collective rights (See, for example, Nwokeabia, 2003, p. 1).[47] Trademark law developed collective marks. Communitarian rights of particular regions *per se*

[47] See also Nwokeabia, at nwokeabia@yahoo.com, who states that national IPR or *sui generis* systems, even those relating to TK, might not recognize collective ownership. Customary laws in Africa, for example, 'recognize only communal ownership of TK, despite its vital role (in Africa, up to 80% of the development in agriculture and medicine originate from TK) with detrimental consequences for individual innovation, as the consequence is either secrecy in high-income sectors, such as medicine and indifference in low-income sectors, such as agriculture'. Thus, neither an individual-only recognition nor a collective-only TK protection is favourable to the promotion of TK, but a combination of both is desirable.

define geographical indications and appellations of origins. The enforcement of copyright and related rights has a long tradition of operating on the basis of collecting societies to which authors and artists belong and where proceeds are allocated on the basis of income-sharing, rather than as individualized gains. Formed on the basis of private law, these associations may easily transgress national boundaries and even comprise producers and owners unrelated in territorial terms (Germann, Chapter 6). Assignability under IPR could therefore encompass fairly large groupings and communities, even transgressing national borders.

We conclude for such reasons that a final determination of the potential of future assignability of TK associated with PGR is premature at this stage, given the data and research available. The conceptual door should not be closed prematurely.

5.4.3 Free flow and crop exchange

Thirdly, there is North–North opposition to IP protection for TK. Industry fears operating within a much narrower research exemption or being condemned to paying off the added value of TK it had previously obtained for free, but had capitalized upon. It is likely and reasonable to assume that plant breeders fear that IPR protection of TK will critically narrow down the research exception under which they have been operating.

States are free to operate restrictions on the free flow of PGRFA in the interests of protecting the TK of their communities, but only within the boundaries of international treaties. The ITPGRFA is such an important limitation, enjoining a limited number of specific crops from being subjected to exclusivity, licensing requirements and thus remuneration. To what extent this list will be extended in a process of progressive liberalization will depend on financing the international system by way of taxpayers' contributions. Since states are unwilling to part with a source of income and/or to raise taxes, in order to reimburse those who have lost revenue with the TK now in public domain, a number of important crops have not been included in the ITPGRFA. The ITPGRFA concept is too removed from the market economy for the TK to yield any profit for the knowledge bearer. Instead, an IPR will subject the seed under IPR protection, whereby the knowledge bearer as a right holder will be able to appropriate the profit and the competitors will have the incentive to build upon the TK, in order to become TIP-right holders themselves.

5.4.4 Destruction of traditional ways of life

This argument is about the fundamental opposition to IPR themselves. Sometimes this is based on evidence, but often it is not. Most non-governmental organizations (NGOs) – with the exception of the Centre for International Environmental Law (CIEL) – have from the outset been opposed to aggrandizing the scope of IP protection at the international level (See, for example, Downes, 1997). Their fixed stance stems from their categorical opposition to the WTO Agreement on Trade-related Aspects of Intellectual Property Rights (TRIPS), which they see as a betrayal of the developing countries.[48] Their aim is to render TRIPS ineffective. Most NGOs accuse the international community of deepening the divide between developed and developing countries, when obliging the developing countries to establish an IPR system of protection.[49] These groups view the expanse of geographical indications or the protection of TK as merely a fig leaf, which leaves the basic inequality uncovered. It is interesting

[48] Agreement on Trade-Related Aspects of Intellectual Property Rights, Annex 1c of the Marrakesh Agreement Establishing the World Trade Organization, signed on 15 April 1994, The Results of the Uruguay Round of Multilateral Trade Negotiations, The Legal Texts, WTO Secretariat, Geneva 1994 [hereinafter TRIPS Agreement].

[49] See, among others, NGOs specialized in the field of agricultural diversity: GRAIN (http://www.grain.org) and RAFI (http://www.rafi.org), GAIA Foundation, Actionaid (mail@actionaid.org.uk).

to observe that those challenging the basic inequities of the IPR system for developing countries often refuse to cooperate in improving the system for the holding countries' benefit.

True, private proprietary rights are not necessarily better suited for preserving the biodiversity of a region than is public ownership of the land, for instance in traditional pastoral collectives (see *The Economist*, 2002, pp. 48–50). Chapter 6 points to the risk of perverse incentives by IPR. Instead of fostering innovation, *sui generis* rights may induce farmers to change traditional patterns of breeding, seed exchange and decisions about priorities in order to create and maintain distinct, uniform and stable landraces, thereby losing traditional dynamic interaction of gene flow (6.2.3). Yet, while private ownership may not be the best solution for indigenous people, this is not true for settled farming in general. Farmers are perfectly accustomed to notions of property and ownership of land, and much of human history is about assessing such rights. It is at the root and heart of modern legal systems as they emerged upon agricultural settlement by formerly migrating and hunting tribes. And today, the struggle for land reform in many countries where the colonial heritage is still present through large land ownership has often not been completed. Many still fail to dispose of rights that would allow the mortgaging of property and thus access to investment. In addition, the younger generation in developing countries is reported not to believe any longer that traditional ways of using TK can result in economic returns (see Balick, 2003). Their career choices increasingly adapt to the model of globalization. They believe that modernizing is the only way to render their economies marketable.

For such reasons, and in the long run, in order to choose to preserve TK, nomadic pastoralists (estimated 190 million worldwide) will need a legal tool and to see economic returns in TK. The current and increasing loss of traditional knowledge, induced by changing patterns of land use and farming, indicates that traditional ways of dynamic breeding may not be able to halt the process of erosion. For TK not to be lost, for future innovations to be based upon TK, and by the indigenous communities themselves, and for the younger generation to be able to build upon their own identity and to have the freedom to define themselves by their own roots, it is crucial that IPR empower TK (see Nwokeabia, 2001, p. iii). Finally, it is important to note that shaping IPR would allow limiting their use on interaction and intercourse with the industrial sector, in particular biotechnology, while leaving traditional breeding patterns among farmers and farming communities outside the reach of protection. We recall that IPR are mainly sought and necessary for bringing about more equitable power relations with the formal sector.

5.4.5 Conclusion

In conclusion, we do not think that the arguments made against attempts to develop and create new types of IPR in order to address the problem of erosion and equity, with a view to create better conditions for sustainable development and the empowerment of the rural population, justify halting our inquiry into the potential of intellectual property protection of TK. We should face the challenge.

5.5 A New IP Right in Traditional Knowledge

5.5.1 Subject matter

The goal of identifying specific communitarian IPR rights expressing the information on development, conservation and use of PGR can be pursued under different names. WIPO has explored the adoption of a working definition for TK.[50] Narrowing it down

[50] See Intergovernmental Committee on Intellectual Property and Genetic Resources, Traditional Knowledge and Folklore, Fourth Session, Geneva, 9 –17 December 2002, Brief Summary of Working Documents, WIPO Document WIPO/GRTKF/IC/4/INF/6, 6 December 2002, p. 3.

from a general perception, it adopted a strict working definition: 'the content or substance of traditional know-how, innovation, information, practices, skills and learning, rather than to the form of its expression'.[51] The WHO agreed on a working definition for TK-related medicines.[52] Other authors have explored the most fitting terminology for TK (Dutfield, 2002b, pp. 8–11; see also Posey and Dutfield, 1996, pp. 12–13; Chapter I, Crucible II Group (2001) Seeding Solutions Vol. II).

At the outset it should be noted that international IP instruments have a tendency to 'leave specific determinations of the boundaries of protectable subject matter up to domestic authorities'.[53] With respect to TK relating to PGRFA, it would be possible to refer to the concept of farmers' rights and further develop this concept. As of today, these 'rights' were designed to counter plant-breeding rights (see Girsberger, 1999, p. 150). However, they have remained without the qualities of true rights. Farmers' rights may serve as a title to obtain public funding and support. They are not more than a label, and are moreover restricted under the International Treaty to domestic support, bar any international dimension. We therefore advise not to refer to this concept as it merely adds to the confusion. Also, we do not think it useful to denominate TK protection in terms of broad human rights. While TK protection serves the ultimate goal of enhancing human dignity and development, we prefer, in effecting this goal, to take recourse to the terms of intellectual property rights (see Cullet, 2001, p. 228).

An activity-specific approach referring to 'traditional medicinal knowledge', 'traditional ecological knowledge', 'traditional and local technology knowledge', 'know-how and practices' is both difficult to maintain in an increasingly interdependent reality and detrimental to the formulation and enforcement of a legal protection. A more general term – albeit open to the new and changing forms of TK – is preferred.

This study suggests the use of the term of Traditional Intellectual Property Rights ('TIP Rights') for a *sui generis*, private law entitlement vested in communities and which covers data and information about plant genetic resources (PGR both for the purposes of food and agriculture (PGFRA) and for medical purposes). The concept encompasses a given community's traditional knowledge of methods and skills of use of PGR for these purposes and includes usage of the plant that embodies such knowledge. It may be objected that using the term of 'customary' IPR is more accurate than 'traditional', because it includes the 'contemporary indigenous practices and beliefs, which form inherent part of any TK definition'.[54] We use the term 'traditional' only to signal the nature of the customary process of gathering the information by means of experience, observation, experimentation and creation, as opposed to bioprospecting and other modern forms of collecting and developing formal, science-based knowledge.

5.5.2 Right holders

We believe that a TIP right should be inclusive, rather than exclusive, with respect to prospective beneficiaries of protection. It should encompass the spectrum of all knowledge-bearers touched by such a law, ranging from individuals to communities,

[51] Intergovernmental Committee on Intellectual Property and Genetic Resources, Traditional Knowledge and Folklore, Seventh Session, Geneva, 1–5 November 2004, Protection of Traditional Knowledge: Overview of Policy Objectives and Core Principles, WIPO Document WIPO/GRTFK/IC/7/5 [hereinafter WIPO Document WIPO/GRTFK/IC/7/5], para. 31.

[52] See The WHO Traditional Medicine Strategy 2002–2005, available at http://www.who.int (accessed 11 January 2006).

[53] WIPO, *Traditional Knowledge – Operational Terms and Definitions*, paras 4–14, citing the WTO TRIPS Agreement as an example that does not give a definition for 'invention', nor defines how much innovation is necessary for the protection by international IP laws.

[54] WIPO, *Traditional Knowledge – Operational Terms and Definitions*, para. 19.

associations and cooperatives. There is no formal limitation to indigenous or traditional communities. Right holders include all custodians of TK, in particular farming communities around the world.[55] All groupings engaged in creating, conserving and using TK in their daily working and cultural lives are eligible. Given the strong intergenerational component of TK, it will be difficult to identify a single holder. Because TK is often held by families and even by lineages, TIP rights should partake of a communitarian nature (see Downes, 2000, p. 255). In the event that individuals add value to traditional knowledge, they could qualify for conventional modes of IP protection, such as trade secrets (undisclosed information) utility models or petty patents, plant variety protection, copyrights or database rights in value-adding collections of information. That the TIP right as such is reserved for a collective entity hardly presents a new phenomenon in IP law. Consider, for example, that even a modern patent law deals more with teams of inventors than single persons and 'most intellectual property assets are owned by collective entities, which in many cases represent large and diffuse groups of individuals'.[56]

Different communities in different regions of the world may hold the same TK simultaneously.[57] Given that they may assert parallel rights to the same traditional knowledge, cooperative efforts in assigning or transferring such knowledge to firms for industrial application would be in their mutual interests. Also beneficial might be agreements setting prices for such purposes, short of collusion running foul of anti-trust laws. In this regard, states participating in the proposed IP scheme would need to adopt appropriate exceptions for collective administration of TIP rights, so as to avoid pitting different communities against one another in a way that would drive down the prices paid for their traditional knowledge. It would undermine a primary goal of establishing such rights if antitrust law denies traditional communities an exceptional use of monopolistic market power.[58]

5.5.3 Conditions, content and scope of rights

Defining the content and scope of TIP rights in traditional knowledge poses the most difficult issue before us. This topic above all others is particularly subject to negotiations and compromise. The scope of protection may range from fully developed property rights to a limited entitlement focusing on compensation without entailing powers to prevent others from using materials for commercial purposes. However, it should be clear at the outset that TIP rights extend only to *commercial* activities undertaken by public or private entities. The very rationale for such rights is to introduce a new level of economic benefits that alters the balance of power between those who possess traditional knowledge and commercial actors who seek to exploit and build upon such knowledge with a view to marketing new products. In keeping with general principles of intellectual property law, activities partaking of private use

[55] See also WIPO/GRTKF/IC/7/5 para. 34: '[I]t was stated at the Committee's sixth session that, while indigenous peoples are important stakeholders in this discussion, not all TK belonged to indigenous peoples, and that it was necessary also to consider non-indigenous holders of TK, such as farming communities. In addition, the narrow sense of "ownership" may be inapplicable in some circumstances, the relationship between a community and its TK often being seen in terms of custodianship or responsibility.'

[56] WTO Document IP/C/W/347/Add.1, para. 34, the WTO cites the example of General Motors, which owns intellectual property rights on behalf of a community of shareholders that is much larger and more diffuse than most identified traditional communities.

[57] See WTO Document IP/C/W/347/Add.1.

[58] See *Elements of a Sui generis System for the Protection of Traditional Knowledge*, Report of the Fourth Session of the Intergovernmental Committee on Intellectual Property and Genetic Resources, Traditional Knowledge and Folklore, WIPO Document WIPO/GRTKF/IC/4/8, 30 September 2002 [hereinafter WIPO, *Elements of a Sui generis System, Fourth Session*] p. 26.

should not be covered. Nothing therefore would bar individuals from using TK generated by others for private purposes short of engaging in commercial activities.

Intellectual activity

While there is no need to require novelty as a prerequisite of protection, a *certain level of intellectual activity* should accordingly be required for IP protection. TK not eligible for IP protection thus encompasses 'items not resulting from intellectual activity in the industrial, scientific, literary or artistic fields, such as human remains, languages in general, and other similar elements of "heritage" in the broad sense'.[59] TIP rights covering information on plant genetic resources in particular should protect TK relating to nutrition, health prevention, healing and other human activities. In order to identify TK protectable by IP rights, and to render it eligible for legal protection, the specific subject matter needs to be recognized or documented in some codified knowledge system.[60] A tension arises from the need to distinguish such commercially valuable information from other components of TK that are interwoven with the processes of everyday life and merely reflect the community's values.[61] It is argued that properly designed TIP right would only protect commercially valuable information that was not interwoven with the processes of everyday life.[62] The distinction, however, will be difficult to draw. In particular, breeding practices by farmers certainly are 'interwoven with the processes of everyday life', but at the same time may form part of a codified knowledge system.

The object and content of the right

The principal object and content of proposed TIP rights is knowledge, and thus information. It is neither the product ensuing from or based upon TK, such as a landrace or plant variety, nor the raw material (PGRFA, plant genetic material (PGM)) at its basis. The proposed IPR protection of TK basically is of limited use in protecting information relating to PGRFA and PGM. A TIP right rewards the intellectual effort of identifying the PGR's propensity for human use. It leaves the PGRFA and PGM untouched by private rights. Where the TK is *associated* with the particular use of a biological resource, as when it pertains to the effects of medicinal plants or the specific quality of a crop, the TIP right encompasses certain uses one can elicit from that plant.

We reiterate that any TIP right pertaining to PGRFA should also encompass use of the PGR, but should not protect the genetic information contained in the seed. This would establish an equitable relationship between plant breeders and the TK holders. Whether the relevant TK is only associated with plant genetic material (e.g. pharmaceutical plants) or forms part of the PGRFA (e.g. seed, domesticated animals), the right

[59] WIPO, *Elements of a Sui generis System, Fourth Session,* at para. 25: 'human remains' are part of the 'heritage of indigenous people'. In a broad sense, heritage includes the result of any kind of 'intellectual activity' protected by IPRs. But heritage also contains artistic creations, as well as human objects, sites, removed from intellectual activity, and which IPRs are not designed to protect. Both types of 'heritage of indigenous people' fall under the UNESCO preservation regime pursuant to the 'Principles and Guidelines for the Protection of the Heritage of Indigenous Peoples' developed under the Auspices of the United Nations (UN) Working Group on Indigenous Populations, UN document number E/CN.4/Sub.2/1994/31, cited in ibid., Annex II, at 9.

[60] WIPO, *Elements of a Sui generis System, Fourth Session* at para. 32.

[61] See WIPO, *Review, Third Session,* para. 33.

[62] Cf. WIPO, *Traditional Knowledge – Operational Terms,* paras 33, 59, suggesting that one needs to be cautious when drawing the line. The *sui generis* system of Panama [Article 1 Law No. 20, of 26 June 2000] distinguishes TK that has commercial utility from other TK. Such a distinction has, to a certain extent, the effect of disintegrating the holistic aspect of TK in which commercial use and spiritual components are intertwined. WIPO suggests allocating to customary law the function of managing the sacred sources of TK, leaving it to international and national laws to regulate the commercially relevant TK.

should protect both the traditional knowledge and the product. As to the underlying genetic information, the desirable compromise with research needs to preserve the status quo of the public domain approach. Research parallel to or in combination with TK relies on genetic data from the public domain.

The concept raises a number of difficult issues that need to be addressed. We concede that it may sometimes prove difficult to distinguish the product that TK identified and that the TIP right properly protects – such as a plant – from what counts as the underlying genetic information. We suggest a process-oriented rather than a product-oriented approach to making such distinctions. The key question is whether merely identifying the genetic code of a plant discovered by some indigenous communities should trigger a duty to license the use of the plant (from the community concerned, assuming it is protected by TIP rights?) or whether identifying the genetic code is of such value to all human beings and research that it should be regarded as an activity falling within the public domain.

Another question is whether using the genetic code to create a new or improved product amounts to using the 'plant' in a restricted sense that should be licensed by the TK holder. A research-friendly approach might subject only those products that were created without interfering with the genetic material to qualify for licensing, such as extracting or synthesizing substances, while piecing together strands of genetic information (whether identified or not) to develop a new product would remain unaffected by the TIP right's scope.

An open question is whether the genetic information encapsulated within the plant (or seed) is exempted from the scope of the TIP right because the communities have usually neither identified the genetic sequences making up the seed nor have they matched the function of any given seed to a certain strand of the genetic code, and for contemporary international treaty law, genetic information falls under the research exemption.[63]

A recurring question will be whether a bio-prospector ameliorating or *a fortiori* 'discovering' a new use for the substance originally associated with beneficial qualities by the TK holder would also be obliged to obtain a licence from the TK-holding community or whether the new use discovered would exempt the bio-prospector from redeeming the original holder. The PGRFA and PGR themselves will most likely be governed by material transfer agreements (MTAs) or ABS-contracts. MTAs could enable patent protection to be obtained on genes isolated from the material in exchange for the obligation to make available such rights at a reasonable royalty for commercial developing markets and for zero royalty to subsistence farming in the countries of origin (see Kush, 2003). Thus, it is imaginable that a recipient under an MTA, usually the pharmaceutical industry, will have to apply for a licence with the traditional community who identified the beneficial value of a plant, while the traditional community under the MTA will have to apply for a licence to use the genetic code, once the industry has been able to isolate it from the plant genetic resource. It is imaginable that there will be in effect an exchange of patent rights.

A difficult question also arises from the need to adequately compensate holders of rights in TK when a TK component is claimed as part of a 'combination patent' in jurisdictions that recognize such patents. Here the object is to prevent would-be patentees from avoiding the need to license the traditional knowledge or otherwise to compensate its holders simply because they file for a combination patent (see Hansen and van Fleet, 2003, p. 11). For example, the *Maca* and Velvet Deer Antler USPTO patent demonstrates that industry can easily circumvent the recognition of TK under cur-

[63] Whether the identification of genetic information as such should ever qualify as a patentable invention or should be left in the public domain is a much-debated question that lies beyond the scope of this contribution.

rent law, simply by combining two substances into a novel product, whereby one of the two substances making up the product had originally been identified by TK holders, '[w]e have found that a combination of the Andean tuber, *maca*, and antler, such as from deer or elk, is useful in raising testosterone levels in men with low normal or low testosterone levels, while enhancing vitality'.[64] Industry can rely on the support of the USPTO, which will register the patent, despite prior art. In the *maca* case, 'the combination of these two plants, despite individually documented traditional uses, was not published in written form; therefore, the USPTO was legally bound to approve the patent' (American Association for the Advancement of Science, AAAS, 2002). The scientists claiming patent protection for *maca* as 'treatment to increase testosterone levels and to provide a sense of enhanced well-being in men' admit prior art as to the *maca* component.[65] However, the admission is only partial, as the scientists emphasize the nutritional, wound-healing and antidepressant uses of *maca* documented to exist in the Andean region, and downplay the traditional knowledge as quasi-unreliable prior art, 'anecdotal information suggest[s] that *maca* alleviates [...] menopausal syndromes [and] male erectile dysfunction'.[66] Even worse, there exists as well prior art for the second component of the mixture to be patent registered as an invention, the elk antler. The only difference is that it is Chinese medicine and not Peruvian knowledge holders that have been using elk antler 'in a variety of remedies'. Again the patent applicants take pains to ignore any prior art: 'In Chinese medicine, deer antler is used in a variety of remedies. However, there is little documentation in the medical literature to support this usage. The long term safety has not been studied.'[67]

The use of either of the single individual components (PGR) by traditional knowledge holders in Peru (*maca*) or China (elk antler) would not infringe the US patent on the combination, nor would the documentation (prior art) of a single component render the combined product ineligible for patent protection, even though the patenting of the single components taken individually should not be possible because of the existing prior art (see Hansen and van Fleet, 2003, p. 11).[68] In our view, this method of combining substances in order to circumvent compensation of the TK holder should be closely monitored under good faith standards by the registering patent office.

Rights conferred

We reiterate that the purpose of establishing TIP rights is to define ownership and the obligation that results from commercial use of information that takes the form of traditional knowledge. The normal scope of any IPR is to prevent a third party from using information for commercial purposes without the consent of the right holder. The restricted usage is further defined by spe-

[64] See US Patent 6,093,421 of 25 July 2000, available at: http://patft.uspto.gov/netacgi/nph-Parser?Sect1=PTO2&Sect2=HITOFF&p=1&u=/netahtml/search-bool.html&r=1&f=G&l=50&co1=AND&d=ptxt&s1=*maca*&s2='deer+antler'&OS=*maca*+AND (accessed 11 January 2006), Summary of the Invention; the inventors are US scientists, the assignee is the Biotics Research Corporation (Rosenberg, Texas).

[65] US Patent 6,093,421 of 25 July 2000, Objects of the Invention.

[66] US Patent 6,093,421 of 25 July 2000, Background of the Invention.

[67] US Patent 6,093,421 of 25 July 2000, Background of the Invention.

[68] US Patent 6, 093, 421, July 25 2000, available at: http://www.uspto.gov/patft/index.html For the individual components (*maca* and velvet deer antler) of this fertility drug, the USPTO refused patent protection since prior art by the Andean population had been shown. However, for the combination of both substances, no prior art existed, which made the recipe eligible for patent protection. Thus, the US patent was granted to Biotics Research Corporation, a Texan industry. Reference to TK in the claims of the patent is very unspecific – 'used in folk medical traditions as a mood enhancer, as an antidepressant, and wound healer' – and leaves the reader somewhat in doubt as to whether or not the Andean region and/or Chinese medicine cited might have used a combination of the substances for increasing fertility in men.

cific operations, such as putting the protected item on the market. Beyond such typical concerns, policy makers must carefully delimit the extent to which TK rights should be subject to a research exception, under which use of the otherwise protected information would not require the right holders' consent. Similarly, policy makers must decide whether to allow potential licensees to test their products and obtain market approval prior to obtaining the consent of the right holders. The modalities of protection remain open to debate and negotiations. Rights may be defined to entail the full range of powers normally associated with intellectual property rights: the right to prevent others from commercially using the information for specific purposes and the right to control the commercialization of related products. Rights, however, are limited to compensation, obliging the right holder under a legal licence to allow commercial utilization against compensation. Holders of TK are likely to seek strong rights while users, especially those in developed countries, may prefer to limit the right to compensation or to adopt the compensatory liability regime (CLR) suggested by Professor Reichman in 2000 (see Reichman and Lewis, 2005).

Between an exclusive right, entailing the right to prevent and stop others from commercial use of protected information for specific purposes and rights to control commercialization of related products, and a clear entitlement to compensatory liability lies the possibility of adopting a misappropriation regime (see Correa, 2002; and Dutfield, 2002b, p. 29 referring to Correa). Such a regime seems consistent with a broader unfair competition approach, which some have proposed (see Correa, 2002, p. 30). The African Group within the Council for TRIPS endorsed the misappropriation approach in 2003.[69] Proposals made by WIPO, discussed above, are essentially based upon unfair competition, albeit they do not exclude the establishment of IPR properly speaking.

An approach rooted in unfair competition can benefit from Article 10bis of the Paris Convention for the Protection of Industrial Property, forming part of the multilateral trading system of the WTO under the TRIPS Agreement. Just as undisclosed information has been protected on this basis by Article 39 of the TRIPS Agreement, which established a new IPR category in international law,[70] so it can be argued that to use information and knowledge generated by others and expressed in a specific product is unfair to the extent that it serves to facilitate copies or a derived product without the consent of the creator. Recourse to protection of unfair competition essentially relies upon a broad principle seeking to avoid misappropriations. There are many advantages to this approach. There is no need to define right holders, as the instrument is essentially defensive. There is no need to define the scope and content of rights in specific terms. It essentially depends on a case-by-case approach, failing to provide legal security at the outset. Protection by means of unfair competition, and thus existing obligations under Article 10bis Paris Convention, as incorporated in the TRIPS Agreement, offers an interesting starting point, upon which – as in other areas – rights could eventually build.

All of these proposals need to be carefully evaluated in light of their potential impact on the free flow of genetic material envisaged by the FAO's International Treaty and their human rights implications (see Cullet, 2001, p. 228 referring to Chapman, 2000;[71] and Chapman, 2002, pp. 861–882) We suggest, however, to continue

[69] *Taking Forward the Review of Article 27.3(b) of the TRIPS Agreement – Joint Communication from the African Group,* Council for Trade-Related Aspects of Intellectual Property Rights, WTO Document IP/C/W/404, 26 June 2003.

[70] See TRIPS Agreement, Article 39 (embodying the protection of confidential information within the framework of Article 10bis of the Paris Convention).

[71] Audrey R. Chapman, Approaching Intellectual Property as a Human Right, Obligations Related to Article 15(1)(c), UN Doc. E/C.12/2000/12, 2000.

evaluating the potential of positive rights and proactive protection. The concept of unfair competition, strictly speaking, does not entail positive rights. All forms of IPR are rooted in unfair competition rules, but go beyond both in form and substance. For example, using someone's invention or product name is protected under unfair competition, in the absence of patent or trademark protection. Such protection, however, is not proactive and there is no legal security as to the scope of rights. Unfair competition, in other words, is both a starting point and a remedy of last resort. Even if broad principles can be defined more precisely, adjusting to the needs of TK protection,[72] they cannot form the basis and legal entitlement upon which the use of rights in terms of negotiations and licensing can be built. Proactive protection by means of TIP rights therefore may be beneficial from this point of view. For such reasons, we also look into the problem of duration, creation and registration of rights.

5.5.4 Duration of rights

IPR that stimulate innovation and creation are generally limited in time. After a period in which innovators reap their rewards, innovation enters the public domain of human knowledge and becomes freely available. With the idea of TIP rights, the process is reversed. Traditional knowledge that has resided in the public domain without being privatized by law becomes, as of a particular date, available and subject to an intellectual property right owned by a community.

Active use of TK should constitute both a prerequisite and limit of protection. Rights to TK should only arise so long as the process or information exists within, and is being used by, a particular community. Once it fades into the past, it is no longer of commercial interest to the community and should no longer be granted protection. When the use of active TK is protected, however, the period of liability for specific uses will have to be determined by law. Regardless of whether an exclusive rights regime or a compensatory liability regime is adopted, the question arises as to how long the licensee must remain liable to the licensor (Reichman and Hasenzahl, 2002).[73] The problem here is that a new product will often only be placed on the market after a number of years of research and testing, and the amount of equitable sharing of benefits cannot be properly defined during the first years of licensing.

One solution would be to differentiate the licensing fees according to a menu of options, with one charge for a period of R&D, and additional compensation to become due once the product reaches the market. Such compensation would then be limited for a certain period of time sufficient to generate adequate benefit sharing. At some point, e.g. after 10 years, the duty to compensate would cease and the use of TK by that licensee would no longer trigger a duty to remunerate the right holder.

We stress, however, that the underlying TIP rights in traditional knowledge as such should not lapse and would again become operational if the TK and related materials were used by another economic operator or for different purposes. At this point, the process of licensing or compensatory liability would revive and additional revenues would be generated. In effect, one could envisage that TK protection – not unlike trade secret or trademark protection – could last indefinitely, so long as active use is made of the relevant information, and this solution would be consonant with the intergenerational and incremental nature of traditional knowledge.[74]

[72] See proposals by WIPO, WIPO Document WIPO/GRTFK/IC/7/5, supra note 51.

[73] These authors note that the TRIPS Agreement, Article 31, marks an important step for developing countries, because it recognizes compulsory licensing on grounds other than the abuse of rights and thus permits the use of such licensing to promote public policy goals.

[74] But see WIPO, *Elements of a Sui generis System, Fourth Session*, para. 78 (seeming to say that once TK becomes commercially viable, it should be limited to a specified period of protection, beginning with the point in time of commercial exploitation).

5.6 The Creation and Registration of Rights

5.6.1 Registration requirements

The creation of TIP rights may follow, as in IPR law in general, formal and informal genesis. Cultural rights following patterns of copyrights are likely to exist without formal registration. The same holds true for protection based upon unfair competition. Rights relating closely to the traditions of industrial property could rely upon registration with a government or international agency (see Cottier, 1998, pp. 578–581). Registration itself can take either the form of declaratory or constitutive registration, with the difference that a declaratory registration would not formally establish a legal title but essentially serve as evidence and documentation of prior art for the purpose of strengthening claims of traditional communities against infringement prior to the vesting of the formal legal title.[75] To the extent that it can be shown that the technical rule under patent application lacks novelty and has been used before in the informal sector, the patent claim lacks an important and constitutive ingredient.

Registration of TK is controversial.[76] Yet, it is already under way, mainly with a view to demonstrate prior art in the context of granting patents. Generally speaking, registration amounts to a detailed description and accession of the information and its traditional usage and applications. The examination can either be reduced to a formal one as Portugal seems to have opted for, or it can be a technical one, which as provided for in Panama's and Thailand's law includes the post of a patent examiner solely for IP rights of indigenous people.[77]

Functions under both constitutive and declaratory registration can only be assumed if the information is collected in a systematic manner. This typifies the problem of resources for the adequate enforcement of traditional intellectual property rights, because TK is often unsystematically related from generation to generation. Evidence in the form of databases can most effectively be done on a national level by governmental agencies, NGOs or universities.[78] It can take place right from the beginning at the international level by an international organization or international NGOs. As recording information is considered 'the primary vehicle for the preservation of the information' (Balick, 2003), a TIP right with its necessary registration requirement contributes to the most basic form of preserving knowledge for future generations. The IP community may support these efforts. In the US, for example, a good-will initiative by IP firms, the Public Interest Intellectual Property Advisors (PIIPA), helps to fill the void of lacking international law requirements for TK, by arranging immediate support for TK holders. PIIPA is 'an independent international service and referral organization that can help fill the need for assistance by making the know-how of intellectual property professionals available to developing countries' usually based upon the legal requirement that all lawyers shall engage in *pro bono* work (Gollin, 2003, 1).[79] Even if their services only cover a small portion of the problems encountered by developing countries, it is a valuable incentive to alleviate the cost of patent litigation.

Costly registration procedures can be avoided by building upon existing inventories of knowledge. Biodiversity databases for PGRFA, such as the SINGER, could be expanded to contain TK-relevant informa-

[75] Cf. WIPO, *Elements of a Sui generis System, Fourth Session*, p. 30.
[76] See IPGRI, 2000.
[77] See WIPO, *Elements of a Sui generis System, Fourth Session*, p. 29.
[78] See Alaska Traditional Knowledge and Native Foods Database, which is a joint effort of the US Environmental Protection Agency, the University of Alaska and the Alaska Native Science Commission, available at: http://www.nativeknowledge.com, or the Honey-Bee Network Innovation Database, available at: http://www.sristi.org/honeybee.html (accessed 11 January 2006).
[79] See http://www.piipa.org (accessed 11 January 2006).

tion, albeit they were not designed for such purposes in the first place. Another option is to construct a TK-specific biodiversity registry specifically for such a purpose (see Chapter 3, Section 3.3.2, this volume). IT technology offers an enormous potential for gathering and interlinking information on the subject and for creating a worldwide inventory. While the subject matter will be protected under TK protection, the information will thus be made available to the public at large as a trade-off common to the principles of transparency in intellectual property protection. First-hand national experience is building up in the creation of such registries. India separates the TK biodiversity registry into one for TK relating to traditional medicine and one relating to Ayurvedic medicine. China has only a single TK database relating to medicinal knowledge, while the World Bank seems to combine TK relating to PGRFA with TK relating to medicinal PGM into a single registration system.[80]

Ancillary to a registration requirement of the patent holder, there are proposals before the WTO TRIPS Council to condition the grant of a patent for TK to the disclosure of the origin of the genetic resources. A more relaxed stance would qualify national patent laws with disclosure of origin as a declaratory requirement with civil and/or penal sanctions. Such a disclosure of origin requirement would spur the development of national or international documentation centres (biodiversity registers), where the TK were to be catalogued. Once such a database is established, the TK-holding community applying for IPR protection would no longer have to go through the costly documentation and evidence-gathering process. Once the TK is publicly registered, the TIP right of the identified TK-holding community shall be presumed. Inversely, plant breeding or pharmaceutical industry benefits from the presumption that where the TK biodiversity database refers to public TK without an identified holder, industry can be presumed to be the first holder, unless and until a TK-holding community or individual proves that he/she is the genuine holder. Critical WTO literature points out that a disclosure requirement is incompatible with the current text of the TRIPS Agreement (see de Carvalho, 2000, p. 371).

Different proposals in the TRIPS Council plead for amending the TRIPS to make disclosure an obligation. While the EC proposal limits the legal consequences to those outside patent law, the Swiss propose that a failure to disclose would delay a patent being granted or affect its validity.[81] According to de Carvalho (2000), the disclosure requirement only applies when the genetic resource concerned is preserved *in situ*, because when its active components are isolated from those resources or even synthesized, the link between the invention and the resources may become too weak to be of any significance. Even if the disclosure requirement does not apply to trade secrets, and certificates or plant varieties or plant patents obtained by breeding plant genetic resources because of the extensive application of Article 15(1)(ii) UPOV, it nevertheless applies to human creative efforts such as landraces and natural genetic resources (de Carvalho, 2000, p. 371). There are two statutes that have incorporated the requirement of prior disclosure. These are the Andean decision No. 391 of 16 August 1996 and the Biodiversity Law of Costa Rica (de Carvalho, 2000, p. 371).

---

[80] See generally http://www.wipo.int/tk/en/databases/tkportal/index.html (accessed 11 January 2006).

[81] See EU (Council for Trade-Related Aspects of Intellectual Property Rights, Communication from the European Communities and their Member States, WTO Document IP/C/W/383, 17 October 2002), Switzerland (Council for Trade-Related Aspects of Intellectual Property Rights, Article 27.3(b), the Relationship between the TRIPS Agreement and the Convention on Biological Diversity [...], Communication from Switzerland, WTO Document IP/C/W/400, 28 May 2003, Council for Trade-Related Aspects of Intellectual Property Rights – Article 27.3(b), the Relationship Between the TRIPS Agreement and the Convention on Biological Diversity [...] (from Switzerland – Revision), Brazil, Cuba, Ecuador, India, Peru, Thailand, Venezuela (Council for Trade-Related Aspects of Intellectual Property Rights, The Relationship between the TRIPS Agreement and the Convention on Biological Diversity [...] India, Peru, Thailand, Venezuela, WTO Document IP/C/W/403, 24 June 2003), The African Group (WTO Document IP/C/W/404).

One option, which would not require the usually costly registration by the collective or the individual, is the idea to construct the TIP right as a presumption. In order for the TK holder to be protected against unlawful use by third parties, a third party user would have the burden of disclosing that the origin of the biological or genetic resource is a different one than that registered in the database, if the TK holders are identified in the catalogued TK. The third party user has the additional burden of showing that if the TK is under national sovereignty, or belongs to a specific/specified knowledge holder (community, shaman) the user has prior informed consent (PIC) to use the TK.

If the TK biodiversity registry does not identify a specific TK holder, it is to be discussed whether or not the TK is still presumed to be held by an identifiable, but not yet catalogued, TK holder. A first option, favouring TK, is that also in the case of uncatalogued TK holders, industry would have to show that there is no TK holding community or individual, before it can use the TK. The second and more industry-friendly option is to propose that if the biodiversity registry does not identify a TK holder, the TK is 'public traditional knowledge'. Under the latter option, a TK holder (as opposed to the industry) would have the burden of proof to come forward with the affirmative evidence (as opposed to industry's negative burden of proof that there is no holder of knowledge) that he or she is the holder of the knowledge.[82]

As authors have pointed out, disclosure and prior informed consent requirements for TK have, in contrast to the PGRFA, not yet been included in international patent treaties (see Philippe Cullet, Chapter 3). Yet, conditioning the grant of a patent on such a requirement would successfully link the objectives of the CBD ABS system to that of IPR protection of TK.[83] If the PGRFA is shown as in a constructed research commons (Annex I ITPGRFA or in a CGIAR database), there could be a presumption in favour of the user, that no TK related to the genetic resources exists, and that industry will not have to prove PIC or apply for ABS.

5.6.2 Opposition procedures and judicial review

Registration would be subject to opposition procedures and eventually to legal challenge. Upon a published application, other claimants may object to the exclusive authorship of TK-related information and use and argue that they are equally entitled. They may argue that a particular activity and practice was previously or at least simultaneously used. They may claim exclusive authorship or joint authorship. Objections would result in negotiations and consultations, with a view to sharing property rights. This may result in joint associations relating to particular TK-related information. These associations of shared ownership may include communities and regions around the world. There is no need for contiguity. The international organization in charge would be in a position to administer these rights.

Finally, TK rights, as any IP right, will be subject to judicial review. Users may argue that the registration was unlawful because the information was no longer actively used or did not really exist in the first place. Right holders, on the other hand, may use court proceedings to stop unlawful use of TK-related information and ban commercialization of products due to lack of consent or due to lack of adequate compensation. Disputes may arise in interpreting and applying licensing agreements. Again, it is possible to create uniform and centralized global authorities to deal with these

[82] For a detailed discussion see IPGRI, 2000, pp. 95–101.

[83] Linking PIC and a disclosure requirement to the presumption of a TIP right empowers the local communities, because it adjusts to the reality that TK holders will not be able to carry the costs of registration, thus, it realizes the goals of equity and distributional justice. The downside of the presumption system is that it does not necessarily stimulate innovation, and thus, it might prejudice intergenerational equity.

kinds of claims. The example of domain name protection dispute settlement at WIPO shows a possible path which may be followed. Alternatively, it is possible to use traditional avenues of national court systems. It is evident that the inherently international dimensions will be more difficult to realize.

5.7 The Interface with Other IP Rights

The protection of TK aims at closing an important gap within the international system of intellectual property rights, essentially protecting the use of PGR with associated knowledge. It would apply in addition to other rights. All, in their own ways, play an important role relating to *commercial products* based upon plant genetic resources. We merely recall again, at this stage, the importance of plant patents and plant variety rights protecting new and stable varieties. We recall trademarks, including collective marks and geographical indications and appellations of origin, both frequently used for agricultural products. Finally, we recall the relevance of trade secrets and the protection of undisclosed information and the principles protecting against unfair competition. All of them form part of the international IPR system (Abbott *et al.*, 1999). Geographical indications and plant variety protection should be briefly emphasized as they show close links to the concept of TK protection.

5.7.1 The complementary function of geographical indications

Geographical indications and appellations or origin relate the qualities of a product to the qualities of a particular region and can be used for products 'where a given quality, reputation or other characteristics of the good is essentially attributable to its geographical origin' (Article 22.1 TRIPS Agreement). Geographical indications (GIs) and appellations or origin are protected in a number of international agreements beyond the TRIPS Agreement, and have been used mainly in Europe. They either take the form of registered rights or are protected by means of prohibition of unfair competition. The TRIPS Agreement limits protection to a minimal standard of unfair competition. As a practical matter, it does not exclude the use of geographical names, as long as the true origin of the product is made known to the consumer. Absolute protection is currently granted only for wines and spirits. Efforts are under way under the Doha Agenda of the WTO to expand absolute protection to agricultural products.[84] While they have been facing resistance in particular from agricultural exporters of the New World, developing countries increasingly realize the potential of GIs for protecting niche products in relation to TK (Panizzon and Cottier, 2005). Indeed, GIs offer an important avenue for protecting specialities produced on the basis of TK from imitation and use of denominations expressing a particular quality of, and expectations in, the product. Many countries are currently undertaking efforts to explore the export potential of such products and develop an interest in expanding protection. It will be a matter of finding an appropriate balance between absolute and relative protection in light of currently somewhat excessive standards under European law.[85] It will allow the focusing of TK protection on knowledge, while products derived

[84] See Addor and Grazioli (2002). See also Dwijen Rangnekar, *A Review of Proposals at the TRIPS Council: Extending Art. 23 to Products Other than Wine and Spirits*, UNCTAD/ICTSD Capacity Building Project on Intellectual Property Rights and Sustainable Development, 12–15, 44–45 (May 2003) available at http://www.iprsonline.org

[85] See Council Regulation (EEC) No. 2081/92 of 14 July 1992 on the protection of geographical indications and designations of origin for agricultural products and foodstuffs (OJ No. L 208 of 24 July 1992, p. 1) and Commission Regulation (EC) No. 2400/96 of 17 December 1996 on the entry of certain names in the 'Register of protected designation of origin and protected geographical indications' provided for in Council Regulation (EEC) No. 2081/92 on the protection of geographical indications and designations of origin for agricultural products and foodstuffs, OJ No. L 327 of 18 December 1996, pp. 0011–0012.

therefrom can be addressed by GI protection, in addition to trademarks and collective marks. We cannot see major tensions between TK and GI protection. They will, in general, run in parallel. This is different in respect to plant variety protection rights.

5.7.2 Relation to plant breeders' rights

As matters stand, well-funded plant breeders relying on their R&D capacities can locate plant genetic resources for food and agriculture and exploit them without any legal or compensatory obligations. Once a TIP right was established, instead, it would oblige plant breeders to search for a potential right holder to whom some legal obligations had to be discharged. One beneficial side effect of such an obligation could be that traditional knowledge, which developing countries lack the resources to lay open, might become properly documented and safeguarded to facilitate dealings between the two sets of rights holders.

Under a TIP right, protection for the communities' traditional knowledge would kick in at an earlier stage than the protection of plant breeders' rights. This follows because the TIP right would confer proprietary protection on the relevant traditional knowledge before the plant breeder had taken any innovative step to modify plant genetic resources. Once a TIP right was established at the international level, it would thus saddle future plant breeders with an international legal obligation to seek a licence to use any such resources that were linked to protected traditional knowledge and to compensate the right holder.

The coexistence of these two regimes raises many other difficult questions that can only be hinted at here. For example, if the plant breeding industry isolates a genetic sequence from traditional material, does the genetic sequence also count as information emanating from TK or simply as genetic material barred from qualifying for any IP protection? Does the process of extracting genetic information – absenting any modification of that genetic material – amount to an innovation, which would confer an IP right on the plant breeder? Should the international system reward a process that required intellectual creativity, even if the material itself, the genetic sequence laid open, is not (yet) modified? The answer to this question is in turn linked to the ethical and political debates about the patentability of life forms (see Chapter 3).

We recognize that critics may argue that opening the door to the protection of traditional knowledge by proprietary rights could elicit further claims that genetic research results should be similarly protected. Without staking out a position on these more abstract issues, we believe that a workable solution, allowing science sufficient leeway for research on plant genetic resources for food and agriculture, would result from applying the TIP right to protect both the traditional knowledge and the seed that it produces, but not the underlying genetic information as such.

A system to provide TIP rights in traditional knowledge would have to be carefully aligned with the existing regime that protects plant breeders' rights under the UPOV treaties.[86] In particular, a TIP right would need to be defined against the plant breeder's privilege (see Chapter 3, Section 3.2.3, this volume).[87] How to reconcile the conflicting interests of plant breeders and TK rights holders is crucial to determining how the benefits from TK will be shared and how rights to TK can be exchanged for technical support. In our view, the line of demarcation between the two regimes will require resort to the dispute resolution system of an intergovernmental organization, which would dispose of impartial sci-

[86] Cf. Article 17, International Convention for the Protection of New Varieties of Plants (UPOV), Paris, 2 December 1961, as revised at Geneva on 19 March 1991, UPOV Doc. 221(E), 1996; also available at http://www.upov.org

[87] Cf. UPOV Article 17.

entists and experts who could allocate relevant TK to its legitimate assignees.[88]

5.8 The Fora for Future Negotiations

Which fora are appropriate for negotiating TIP rights (see Cottier, 1998, pp. 581–584; contra Dutfield, 2002b, pp. 3–8)? Developing TK protection has been, and will continue to be, a matter to be pursued in different organizations. The question pertains in a broader sense to the linkages of global environmental regulation with international economic law, the latter ranging from trade goals of best allocation of resources, to competition law, and to IPR. Specifically at issue is the linkage of the WTO TRIPS Agreement with related Non-WTO (specifically, Non-IP treaties) such as the FAO ITPGRFA and the CBD. Treaty linkages have so far only been discussed for the relationship of the General Agreement on Tariffs and Trade (GATT) and MEAs, where conflicts were solved by evolutionary treaty interpretation or via recourse to Article 31(3)(c) of the Vienna Convention on the Law of Treaties (VCLT) (see Marceau, 1999, pp. 120–123; Howse, 2000, pp. 55–62; and Pauwelyn, 2001, pp. 542–550). At issue is also the relationship of WTO and WIPO in developing intellectual property rights.

We submit that the matter ultimately needs to be settled by WTO negotiations for the following reasons. First, the TRIPS Agreement owes a duty to developing countries. Secondly, disputes arising under the TRIPS Agreement fall under the jurisdiction of the well-functioning dispute settlement mechanisms of the WTO. Thirdly, the matter is related to the patenting of life forms and any reform of Article 27(3)(b) of TRIPS will need to be addressed in relation and balanced with TK protection. Fourthly, TK protection shall be dealt with consistently with the function and extension of geographical indications. Fifthly, TRIPS is not the only avenue within the WTO to address the problem. Compared to all other fora, it is the only one that has the potential to promote trade in TK-related goods by fostering non-discriminatory market access and market conditions (national treatment) (see Cottier and Panizzon, Chapter 8, this volume). Dealing with the matter within the multilateral trading system of the WTO, however, is not meant to exclude work in other fora, in particular WIPO, UNEP and FAO. Essential building blocks, including defining rights and registration systems, should be developed in such fora and may eventually be linked to WTO rights and obligations by way of reference or incorporation. Conceptually, WIPO is well ahead in defining TK protection, and a possible agreement resulting from the efforts of the Committee on Intellectual Property and Genetic Resources, Traditional Knowledge and Folklore may well be eventually integrated into the multilateral trading system. The task ahead is a matter of bringing about appropriate coordination and coherence in policy making.

5.9 Conclusions

It is often argued that creating a new international *sui generis* system of TIP rights addressing knowledge in relation mainly to the use, but also genesis, of PGR is not suitable for farming communities. How could they have access to international institutions? How could they trust these institutions, thousands of miles away? Why should they rely upon information-age technologies in order to protect and advance tradition and heritage? More profoundly, how should they trust a property rights approach in the light of the fact that PGRFA and PGM enjoy a long tradition of use in the public domain? We do not believe that these obstacles, while serious, cannot be overcome.

This is primarily a learning process. As

[88] Choosing the right forum for future negotiations is thus particularly important, and in our view, the WTO is the preferred forum for a number of reasons. For reasons of space, this topic lies beyond the scope of this contribution.

much as farmers have adopted concepts of real and movable property rights, they may adopt concepts of intellectual property rights, and in some cases it is clear that indigenous and farming communities already have done so. The potential to enhance their standing, bargaining powers, valuation and reward is likely to attract attention and support. In terms of organizations, it will rarely be a single farmer or small community that enjoys these rights. Entitlement will be regional, national or even international associations, not unlike collective societies, which administer rights and licensing and equitable distribution of proceeds. Existing institutions defending farmers' interests may absorb these functions. They may be defended by organizations to be created. There is, of course, a risk of creating additional and costly bureaucratic structures. But this in itself is not sufficient to deny the approach, as many other and unchallenged institutions would need to be discarded on the same grounds. It will be a matter of creating conditions of good governance in order to make sure that these institutions are workable.

The structure of international relations and commerce has considerably changed in the process of globalization. No longer are states the only actors in the system. Business associations, large corporations and NGOs operate along with specialized international intergovernmental organizations. The creation of structures dealing with TK in these different fora will render farming communities in developing countries more powerful by giving them a voice heard, a tool to act and representation in a world of concentrating economic actors, in particular in the field of PGRFA and, increasingly, also of biotechnology. Rules on TIP rights therefore need to entail the substantive and procedural and institutional requirements to interact with these industries. We maintain that ultimately it is only international trade regulation within the WTO that has the power to halt the devolution of TK, as it will only be the symbiosis of global trading and worldwide IP protection that will provide the precondition for TK to become empowered to participate in global trade. In the age of globalization, monetary valuation creates the necessary power, and only trade that transcends borders. TK associated to PGRFA, its holders, and the seed at the basis, will not become 'players' unless their identity becomes a tradable asset.

Neither the approach of the CBD, expressed by its ABS system, nor the public goods approach of the FAO IU, the FAO ITPGRFA and the FAO Agreement with CGIAR, manifested by its public domain/in-trust approach, are by themselves in a position to sufficiently and effectively empower the collectives which have identified, collected, developed and conserved PGRFA with their TK. Different avenues and options exist to express TK rights, either by means of establishing positive rights, or making available defensive rights. We suggest continuing to work on positive TIP rights. These rights, if properly defined and registered, will offer legal security and registration of knowledge. They will contribute to a global inventory of TK and assist in preserving it for posterity. Defensive approaches, based upon the concept of prohibition of unfair competition and misappropriation of knowledge, are equally of great interest, but perhaps a second-best solution if the task of establishing TIP rights should fail.

For assignable and identifiable information relating to the use of PGR, we therefore propose the creation of TIP rights, the scope of which will need to be defined in negotiations, but which at least should entail a right to compensatory liability. We also support the requirements of prior informed consent and disclosure of the origin to be preconditions to a patent application. At the same time, we recall that IPR and TIP rights are not exclusive. In particular, enhanced protection of geographical indications will increase the potential for development of local and international niche markets for TK-based products. All this will need to be part of an overall package of trade policy measures, further discussed in Chapter 8 in support of developing viable agricultural structures in developing countries.

Discussions held in different fora, in

particular UNEP, FAO, WIPO and WTO, demonstrate the difficulty of dealing with complex and multifaceted issues such as TK in separate international organizations, even separate committees of a single institution. Addressing TK experiences the full exposure to a highly fragmented structure of international organizations, even if limited to intellectual property issues. Even within the WTO, different committees have been concerned. Without advancement of coherence in international fora, and better policy coordination, the goals of sustainable development cannot be achieved. The constitutional questions of policy processes and decision making beyond the bounds of traditional international cooperation go hand in hand with complex substantive issues. More adequate structure–substance pairings are imperative.

Yet, whether or not a new generation of rights is prepared and negotiated within UNEP, WIPO or FAO, it would eventually need to become part of the global trading system and thus of WTO law. International disputes on TIP rights would therefore be subject to WTO dispute settlement, and a trade-related enforcement mechanism would support the full implementation of the treaty. This will be of particular interest to least developed countries (LDCs). A major contribution in rebalancing the international intellectual property system and the WTO TRIPS Agreement would be achieved.

Bibliography

AAAS (2002) *Patents, Traditional Knowledge, and the USPTO, Clearing Up Some Misconceptions in Order to Stop Biopiracy.* Report on Science and Human Rights, Volume XXII, No. 2, available at: http://shr.aaas.org/report/xxii/2_patent.htm

Abbott, F., Cottier, T. and Gurry, F. (1999) *The International Intellectual Property System: Commentary and Materials*, 2 volumes. Kluwer Law International, The Hague.

Addor, F. and Grazioli, A. (2002) Geographical indications beyond wines and spirits – a roadmap for a better protection for geographical indications in the WTO TRIPS Agreement. *Journal of World Intellectual Property* 5(6), 865–897.

Balick, M.J. (2003) Traditional knowledge: lessons from the past, lessons for the future. Discussion Draft, Conference on Biodiversity, Biotechnology and the Protection of Traditional Knowledge, St Louis, MO, 4–5 April 2003, available at: http://law.wustl.edu/centeris/Confpapers/, on file with the authors.

Bragdon, S. (2003) Major legal regimes affecting plant genetic resources. In: Cottier, T. and Mavroidis, P.C. (eds) *Intellectual Property: Trade, Competition, and Sustainable Development*, World Trade Forum Vol. III. University of Michigan Press, Ann Arbor, Michigan, pp. 441–453.

Carneira da Cunha, M. and Almeida, M. (2000) Indigenous people, traditional people and conservation in the Amazon in Brazil. *Daedalus, Journal of the American Academy of Sciences* 129(2), 315–338.

Chambers, B. W. (2003) Emerging international rules on the commercialization of genetic resources – the FAO International Plant Genetic Treaty and CBD Bonn Guidelines. *Journal of World Intellectual Property* 6(4), 311 ff.

Chapman, A.R. (2000) *Approaching Intellectual Property as a Human Right, Obligations Related to Art. 15(1)(c).* UN Doc. E/C.12/2000/12.

Chapman, A.R. (2002) The human rights implications of intellectual property protection. *Journal of International Economic Law* 5(4), 861–882.

Correa, C.M. (2002) Protection and promotion of traditional medicine, implications for public health in developing countries. Study, August 2002, on file with the authors.

Cottier, T. (1998) The protection of genetic resources and traditional knowledge: towards more specific rights and obligations in world trade law. *Journal of International Economic Law* 5(4), 555–584.

Cottier, T. (2005) The Agreement on Trade Related Aspects of Intellectual Property Rights. In: MacRory, P.F.J., Appleton, A.E. and Plummer, M.G. (eds) *The World Trade Organization: Legal, Economic and Political Analysis.* Springer, Berlin, pp. 1017–1089.

Crucible II Group (2001) *Seeding Solutions. Vol. II: Options for National Laws Governing Control over Genetic Resources and Biological Innovations.* IPGRI, Rome.

Cullet, P. (2001) Property rights over biological resources – India's proposed legislative framework. *Journal of World Intellectual Property* 4(2), 211–230.

de Carvalho, N.P. (2000) Patent Law and Policy Symposium: Re-engineering patent law: the challenge of new technologies. *Washington University Journal of Law and Policy* 2, 371–401.

de Carvalho, N.P. (2003) From the shaman's hut to the patent office: in search of effective protection for traditional knowledge. Discussion Draft, Conference on Biodiversity, Biotechnology and the Protection of Traditional Knowledge, St Louis, MO, 4–5 April 2003, available at: http://law.wustl.edu/centeris/Confpapers/, on file with the authors.

Downes, D. (1997) Using intellectual property as a tool to protect traditional knowledge: recommendations for next steps. CIEL Discussion Paper, available at: http://www.ciel.org/Publications/, on file with the authors.

Downes, D.R. (2000) How intellectual property could be a tool to protect traditional knowledge. *Columbia Journal of Environmental Law* 25, 253–282.

Dutfield, G. (2002a) Sharing the benefits of biodiversity – is there a role for the patent system? *Journal of World Intellectual Property* 5(6), 899–931.

Dutfield, G. (2002b) *Protecting Traditional Knowledge and Folklore: A Review of Progress in Diplomacy and Policy Formulation*, UNCTAD/ICTSD Capacity Building Project on Intellectual Property Rights and Sustainable Development, Draft (October 2002), available at: http://www.ictsd.org/unctadictsd/docs

Eugui, D.V. (2002) Intellectual property in the FTAA: new imbalances and small achievements. *Bridges* 6(8), 18–22, available at: http://www.ictsd.org

Girsberger, M.A. (1999) Biodiversity and the concept of farmers' rights in international law. *Studies in Global Economic Law*, Vol. 1. Peter Lang Publishing Group, Berne, Switzerland.

Gollin, M.A. (2003) Answering the call: public interest intellectual property advisors. Paper prepared for Conference on Biodiversity, Biotechnology and the Protection of Traditional Knowledge, St Louis, MO, 4–5 April 2003, available at: http://law.wustl.edu/centeris/Confpapers/, on file with the authors.

Gopalakrishnan, N.S. (2002) Protection of traditional knowledge, the need for a *Sui generis* law in India. *Journal of World Intellectual Property* 5(5), 725–742.

Gupta, A.K. (1999) Rewarding creativity for conserving diversity in the Third World: can IPR regime serve the needs of contemporary and traditional knowledge experts and communities in the Third World? In: Cottier, T., Widmer, P. and Schindler, K. (eds) *Strategic Issues of IP Management in a Globalizing Economy*. Hart Publishing, Oxford, UK, pp. 119–129.

Hansen, S.A. and van Fleet, J.W. (2003) *A Handbook on Issues and Options for Traditional Knowledge Holders in Protecting their Intellectual Property and Maintaining Biological Diversity*. American Association for the Advancement of Science (AAAS), Washington, DC, available at: http://shr.aaas.org/tek/handbook/

Hawtin, G. (2003) Management of plant genetic resources in the CGIAR: problems, prospects and the quest for equity. In: Cottier, T. and Mavroidis, P.C. (eds) *Intellectual Property: Trade, Competition, and Sustainable Development*. The World Trade Forum Vol. III. University of Michigan Press, Ann Arbor, Michigan, pp. 425–440.

Howse, R. (2000) Adjudicative legitimacy and treaty interpretation in international trade law: the early years of WTO jurisprudence. In: Weiler, J.H.H. (ed.) *The EU, the WTO, and the NAFTA, Towards a Common Law of International Trade*. Oxford University Press, Oxford, UK, pp. 35–69.

International Plant Genetic Resources Institute (IPGRI) (2000) The Crucible II Group. *Seeding Solutions*, 71, Vol. I. International Development Research Centre, Dag Hammerskjöld Foundation, Rome.

Kongolo, T. (2001) New options for African countries regarding protection for new varieties of plants. *Journal of World Intellectual Property* 4(3), 350–371.

Kush, G.S. (2003) Biotechnology: public private partnership and IPR in the context of developing countries. Paper prepared for Conference on Biodiversity, Biotechnology and the Protection of Traditional Knowledge, St Louis, MO, 4–5 April 2003, available at: http://law.wustl.edu/centeris/Confpapers

Leskien, D. and Flitner, M. (1997) Intellectual property rights and plant genetic resources: options for a sui generis system. IPGRI, Issues in Genetic Resources No. 6, June 1997, available at: http://www.gtz.de/biotech/dokumente/ipgri_e.pdf (last accessed 18 April 2004).

Marceau, G. (1999) A call for coherence in international law. *Journal of World Trade* 33(5), 87–152.

Mgbeoji, I. (2001) Patents and traditional knowledge of the uses of plants: is a communal patent regime part of the solution to the scourge of biopiracy? *Indiana Journal of Global Legal Studies* 9, 163.

Nwokeabia, H. (2001) *Why Industrial Revolution Missed Africa: A 'Traditional Knowledge Perspective'*. Economic Commission for Africa (UNECA), ECA/ESPD/WPS/01/02, on file with the authors.

Nwokeabia, H. (2003) *The Economics of African Indigenous Knowledge*, World Bank IK Notes, No. 53, available at: http://www.worldbank.org/afr/ik/iknt53.pdf

Panizzon, M. and Cottier, T. (2005) Developing country interests and negotiating positions on protection of geographical indications and traditional knowledge. In: Petersmann, E.-U. (ed.) *Developing Countries in the Doha Round*. Robert Schuman Centre for Advanced Studies, European University Institute, Florence, Italy, pp. 227–268.

Pauwelyn, J. (2001) The role of public international law in the WTO: how far can we go? *American Journal of International Law* 95(3), 535–578.

Posey, D.A. and Dutfield, G. (1996) *Beyond Intellectual Property: Toward Traditional Resource Rights for Indigenous Peoples and Local Communities*. International Development Research Centre, Ottawa.

Prakash, S. (2000) WTO rules – Do they conserve or threaten biodiversity? *Journal of World Intellectual Property* 3(1), 155–165.

Ramani, R. (2001) Note and Comment: Market realities v.indigenous equities. *Brooklyn Journal of International Law* 26, 1147–1176.

Reichman, J.H. and Hasenzahl, C. (2002) Non-voluntary licensing of patented inventions, historical perspective, legal framework under TRIPS, and an overview of the practice in Canada and the United States. UNCTAD/ICTSD Capacity Building Project on Intellectual Property Rights and Sustainable Development, Draft, September 2002, available at: http://www.ictsd.org/iprsonline

Reichman, J.H. and Lewis, T. (2005) Using liability rules to stimulate local innovation in developing countries: application to traditional knowledge. In: Maskus, K. and Reichman, J.H. (eds) *International Public Goods and Transfer of Technology Under a Globalized Intellectual Property Regime*. Cambridge University Press, Cambridge, UK, pp. 337–366.

Roht-Arriaza, N. (1996) Of seeds and shamans: the appropriation of the scientific and technical knowledge of indigenous and local communities. *Michigan Journal of International Law* 17(4), 919–965.

Smale, M., Day-Rubinstein, K., Zohrabian, A. and Hodgkin, T. (2001) The demand for crop genetic resources: international use of the US plant germplasm system. EPDT Discussion Paper No. 82, p. 2.

The Economist (2002) Nomadism in Mongolia. *The Economist*, 21 December 2002.

Vaughan, S. (2003) Trade preferences and environmental goods. *Carnegie Endowment for International Peace, Trade, Equity, and Development*, Issue 5, February.

6 Flanking Policies in National and International Law

Susette Biber-Klemm,[a] Philippe Cullet,[b] Christophe Germann,[c] Andrea Nascimento Müller[d] and Jonathan Curci[e]

[a]*World Trade Institute, University of Berne, and Faculty of Law, University of Basel, Maiengasse 51, CH 4056 Basel, Switzerland;* [b]*School of Law, School of Oriental and African Studies, University of London, Thornhaugh Street, London WC1 0XG, UK;* [c]*Institute of European and International Economic Law, University of Berne, Hallerstrasse 6, CH 3012 Berne, Switzerland;* [d]*Legal Advisor, Rue de Lyon 48, CH 1203 Geneva, Switzerland;* [e]*Brigham Young University, 620N 1066E Provo, Utah, USA*

6.1 Introduction

Chapters 4 and 5 have examined ways in which intellectual property can address the protection of traditional knowledge for individually assignable knowledge. Beyond the protection of TK through TIPs or farmers' rights, states have a number of instruments within the intellectual property system and beyond that they can use to address specific issues related to the protection of traditional knowledge. This chapter examines some avenues in which the current international intellectual property regime could be used or adapted in the context of TK. The first section focuses in particular on possible adaptations to existing patent procedures, such as the introduction of a requirement of prior informed consent. It also addresses ways in which some other forms of intellectual property rights, such as geographical indications, could be used in the context of TK protection and the relevance of labelling as a tool to foster better recognition of traditional knowledge. The second section looks at the management and enforcement of existing intellectual property rights in the context of traditional knowledge protection. It focuses in particular on the contribution that the collecting society's model could make to traditional knowledge protection, the question of registration and the possibility of establishing a clearing-house mechanism.

6.2 Existing Intellectual Property Rights: Avenues for Further Development[1]

The intellectual property rights system has found it difficult to adapt itself to the new challenges brought about by the development of genetic engineering in industrialized countries, and issues related to the use of plant genetic resources and traditional knowledge as the basis for products which can be protected under the formal intellectual property rights system. The current system is not well adapted for dealing with issues related to plant genetic resources and traditional knowledge, but some adjustments can nevertheless be proposed. This section examines some of the ways in which intellectual property rights could be

[1] Author: Philippe Cullet.

 Rights to Plant Genetic Resources and Traditional Knowledge (eds S. Biber-Klemm and T. Cottier, assoc. ed. D. Szymura Berglas)

made more relevant in situations where protected knowledge is linked to or derives from traditional knowledge. This includes diverse elements from adaptation to patent procedures to the further development of the protection concerning geographical indications.

6.2.1 Novel uses of the patent model

Patents are only one of several types of intellectual property rights relevant in the context of traditional knowledge over plant genetic resources. However, patents are not only widely used but are also at the centre of most international debates and therefore deserve special attention. There are a number of levels at which the procedures for obtaining and enforcing patent rights can have significant impacts on the management of traditional knowledge linked to plant genetic resources. These include the introduction of novel ways to address the issue of prior art, the introduction of new conditions for patentability imposing an obligation to seek prior and informed consent from traditional knowledge holders, the increased use of the notion of joint inventorship to cover situations involving the use of traditional knowledge and the redefinition of existing conditions for patentability to make them more directly traditional knowledge relevant, for instance, by allowing the grant of petty patents.

Prior art, disclosure and prior informed consent

PRIOR ART

Prior art is information that is accessible to the public in written form or through an oral presentation, or, depending on the country, in any other form.[2] In the context of patent law, prior art is information that is accessible to the public before the filing date of the application or before the priority date of the application, depending on national rules. The level of ease with which access to information can be obtained is treated differently in different countries, but it is generally accepted that information can only be deemed to be accessible as long as there are no factors that severely restrict access.[3]

The question of prior art has been the object of significant debates between developed and developing countries. Questions that surround prior art are linked to the procedures that patent offices normally put in place to verify the accuracy of the claims embodied in the patent application with regard to the criterion of novelty.[4] Novelty is judged against existing knowledge in the public domain. Different types of problems can arise in the context of traditional knowledge. Problems surface where public domain knowledge is not easily accessible through the normal search procedures that patent offices use. This has been of particular concern in relation to traditional knowledge in the case of patent applications in the USA because oral publication in a foreign country does not count as proof of prior art within the USA. In other words, issues surrounding prior art are in part linked to the territoriality of patents. Different strategies can be used to address existing problems. First, where a patent claim is made on knowledge that is in fact in the public domain in another country and documented in written form, it is possible to oppose the grant of a patent. This implies that even if the patents office is not aware of the written source in a foreign country and grants the patent, this does not bar subsequent opposition. As noted in Chapter 3, this can be a cumbersome and expensive strategy because it forces the opposing party to challenge the patent in the jurisdiction where it was granted and show that

[2] Cf. Article 8 of the Draft Substantive Patent Law Treaty, WIPO Doc. SCP/9/2 (2003).

[3] See, for example, Standing Committee on the Law of Patents, Information Provided by Members of the Standing Committee on the Law of Patents (SCP) Concerning the Definition of Prior Art, WIPO Doc. SCP/6/INF/2 (2001).

[4] See, for example, Article 15 of the Patent Cooperation Treaty 1970 and Rule 33 of the related Regulations.

each and every claim made in the patent application is invalid because it is anticipated in prior art. It is, however, an option that some countries like Venezuela are seeking to pursue.[5] Secondly, patent offices in OECD countries are often likely to miss out on relevant information because they limit their search procedures to specific information sources.

One way in which patent offices can be assisted in avoiding the grant of patents on inventions anticipated in traditional prior art is through the development of specific traditional knowledge databases (see below and Chapter 3). The need for such databases stems in part from the fact that patent examiners cannot be expected to have access to all the relevant traditional knowledge if it has not been compiled and systematically arranged. In this context, efforts undertaken by WIPO to provide online access to traditional knowledge databases are significant.[6] Databases can fulfil different functions in the context of the granting of intellectual property rights on inventions derived from traditional knowledge. In the sense of defensive protection, databases improve the availability of traditional knowledge as prior art which can be much more easily searched by patent authorities. The uses they fulfil also include the provision of a common terminology, translation from local languages and in some cases bibliographic references. Databases can also help traditional knowledge holders by enhancing their access to information concerning holders of intellectual property rights derived from traditional knowledge. In situations where traditional knowledge holders themselves control databases, this also strengthens local control over the documentation and use of knowledge.[7] On the whole, traditional knowledge documentation through databases provides an important tool to ensure that public domain knowledge is not privately appropriated in any jurisdiction around the world. Some countries have, however, decided to supplement efforts at documenting knowledge with specific provisions in their domestic laws. Peru has, for instance, adopted a law that includes as one of its specific objectives the need to avoid granting patents on inventions based on traditional knowledge without taking traditional knowledge into account in the examination of novelty and the level of inventiveness disclosed in the invention.[8] This or similar steps constitute one way of ensuring that the issue of prior art is not reduced to the simple accessibility of information. This is important in a context where increasingly low inventive steps are deemed sufficient for patentability. From the point of view of the protection of traditional knowledge, it is necessary to ensure that better access to traditional knowledge through databases does not contribute to a further erosion of the criterion of inventiveness.

Apart from concerns in the context of national patent regimes, the question of prior art has also surfaced in the context of the use of germplasm held in CGIAR gene banks and in the context of benefit-sharing

[5] See Article 83, Venezuela: Ley de diversidad biológica, 27 October 1999, which makes it a duty of the National Biodiversity Authority to review intellectual property rights granted abroad to determine if national genetic resources have been used and in the affirmative to either seek the annulment of the patent or a share of the benefits derived from the invention.

[6] At present, the trial run by WIPO seeks to assess the needs of patent granting authorities. It focuses in particular on determining whether there is a need for specific classification systems for traditional knowledge and on the integration of the data into existing intellectual property information systems. WIPO also seeks to understand the needs of information providers and the conditions under which traditional knowledge documentation can be carried out. For further information, see http://www.wipo.int/globalissues/databases/tk/index.html

[7] Concerning the uses of databases, see, for example, Recommendations of the Workshop on Technical Aspects of Databases and Registries of Traditional Knowledge and Associated Biological/Genetic Resources, Cochin, India, 11–13 November 2002 (on file with author).

[8] Article 5, Peru: Régimen de Protección de los Conocimientos Colectivos de los Pueblos y Comunidades Indígenas Vinculados a los Recursos Biológicos (2002) (see Chapter 2).

schemes. The case of a wild rice variety from Mali with specific resistance to bacterial rice blight is of special interest in this context. The special trait of *Oryza longistaminata* was first identified by Indian researchers who then transferred it to the International Rice Research Institute (IRRI). Research at IRRI led to the development of cultivated varieties with blight resistance. Subsequent research at UC Davis led to the mapping of the Xa21 gene responsible for this trait. The University applied for a patent but recognized the need for benefit sharing.[9] This case throws open a number of difficult questions concerning prior art. First, farmers in Mali generally did not have specific knowledge of the specific trait identified by researchers and this variety was often considered a weed.[10] However, a landless community relying on wild varieties of rice, including *O. longistaminata*, had detailed knowledge of the special resistance of this variety. This indicates that the identification of prior art in specific cases can be even more of a challenge than commonly realized, as illustrated in a situation where a given variety is considered a weed by some farmers and is intimately used by other communities. In other words, the identification of prior art itself requires significant efforts to ascertain the exact extent of local knowledge or lack thereof. In this case, the outcome was the realization that, against normal expectations, ethnobotanical knowledge of local PGRFA is not necessarily and exclusively held by local landowning farmers. Secondly, this case is also noteworthy because it involved a CGIAR Centre. This led to a significant controversy because the patent was seen as compromising the efforts of the IRRI and actors in rice-producing regions affected by the bacterial blight problem insofar as the patent was granted in a country not suffering from bacterial blight while restricting the exports of bacterial blight-resistant rice to the USA. In this case, the difficulties linked to the grant of the patent were recognized by UC Davis, which decided to allow non-commercial researchers access to the gene and to allow the IRRI to develop new varieties incorporating the patented gene (Blakeney, 2001b).

DISCLOSURE OF ORIGIN AND PRIOR INFORMED CONSENT

International patent treaties do not include requirements concerning the disclosure of the origin of the resources or knowledge from which the protected invention is derived. The origin of the biological or genetic resources that are used in developing a patented invention has, however, acquired greater importance in the context of the use of traditional knowledge as one basis for research in genetic engineering. First, as noted in the previous section, it has become increasingly important to establish the origin of knowledge used in a patent application to determine whether it is part of prior art or not. Secondly, the origin of knowledge used in an invention indicates whether the patented invention has benefited from the existence of previous knowledge in the public domain. A disclosure requirement in patent applications would have the following benefits:

- It would legally force patent applicants to double-check prior art in their field before applying for a patent.
- It would provide an avenue for claims of benefit sharing or for claims of joint ownership.
- It would provide a legally binding mechanism forcing patent applicants to show that the resources or knowledge they used as a basis for their invention were acquired with the consent of the individual or group concerned.

The issue of prior informed consent is likely to remain contentious at the international level for the foreseeable future. First, a disclosure requirement would shift the burden of proof from the party opposing

[9] US Patent No. 5859339, Nucleic Acids, from *Oryza sativa*, which Encode Leucine-Rich Repeat Polypeptides and Enhance Xanthomonas Resistance in Plants, issued 12 January 1999.
[10] For a detailed account of this case, see WIPO/UNEP (2002).

the grant of a patent to the patent applicant. Secondly, the requirement would possibly involve not only permission to use specific knowledge granted by traditional knowledge holders but also permission from the state concerned for accessing the desired biological or genetic resource as per the requirements of the Biodiversity Convention and the PGRFA Treaty. In this sense, the implementation of a disclosure requirement is of great interest because it constitutes one situation in which countries must take into account international obligations arising from different treaties at the same time.

The question of a disclosure and prior informed consent requirement must be looked at from the point of view of patent applications as well as from the point of view of access and benefit-sharing mechanisms under the relevant treaties. The requirement can be conceived either as an element extraneous to the patent application or as a procedural or substantive condition of patentability. In the first case, the disclosure requirement can simply act as a bridge between the different international obligations of member states without having any direct impact on patent applications and patent rights.[11] In the second case, the requirement can be conceived either as an indirect or direct condition of patentability. The former concerns situations where the validity of a patent that has been granted can be challenged in case the disclosure requirement condition has not been fulfilled.[12] The latter goes further and provides that the granting of the patent can be made conditional on the fulfilment of the disclosure requirement. This last option has been the object of significant debate as it has been argued that if the requirement was made a condition of patentability, this may require an amendment of the TRIPS Agreement (de Carvalho, 2000).

From the point of view of traditional knowledge protection, it is likely that a disclosure and prior informed consent requirement introduced as a condition of patentability would be the most effective instrument to ensure that traditional knowledge is fully taken into account and acknowledged by patent applicants. It would have the advantage of forcing patent offices to examine the question of disclosure and prior informed consent at the outset. The legality of such a requirement under TRIPS has been questioned because Article 27 provides a finite list of substantive conditions that can be imposed on patent applicants. However, there is no need to conceive the requirement as a substantive condition of patentability. Article 62 of the TRIPS Agreement expressly provides that member states can introduce 'reasonable procedures and formalities' as long as they are consistent with the provisions of the Agreement.[13] In any case, the requirement to disclose prior art is part of the grant of patent rights as recognized under Article 29 of the TRIPS Agreement. The inclusion of traditional knowledge prior art should not pose any problem and may in fact be required under existing rules. The only specific issue concerns the question of the disclosure of the geographical origin of knowledge. This, however, need not be seen as a substantive condition for patentability. It constitutes a procedural requirement that can fall into the scope of Article 62 of the TRIPS Agreement.

The consequences of a breach of a disclosure and prior informed consent requirement could range from a financial penalty imposed on the patent holder to the sharing of patent rights under a joint ownership scheme or simply to the revocation of the patent by putting the disclosure requirement on a par with other patentability requirements like novelty (Ragavan, 2001). This would be conceptually close to the proposal to use the notion of unjust enrichment to provide an alternative protection to traditional knowledge holders whose

[11] This is the option adopted by Directive 98/44/EC of the European Parliament and of the Council on the Legal Protection of Biotechnological Inventions, 6 July 1998, 1998 OJ L 213.
[12] For further details concerning the 'unclean hands doctrine', see de Carvalho (2005).
[13] Contra de Carvalho (2005).

knowledge is appropriated without their consent (Gervais, 2002).

The question of disclosure and prior informed consent can be looked at from an intellectual property perspective, but only a broader perspective taking into account the requirements of international environmental treaties can provide a comprehensive picture. The Biodiversity Convention is, for instance, noteworthy for its emphasis on the question of access to genetic resources and the necessity for prior informed consent. Access refers under the Convention to the rights of states to regulate the flow of genetic resources towards other countries. This adds an important dimension to the issues considered insofar as it indicates that the grant of intellectual property rights is dependent not only on the conditions laid down by the state in intellectual property laws, but also on conditions laid down in other laws which are not necessarily directly concerned with intellectual property such as a biodiversity legislation. This has been confirmed by the Bonn Guidelines on access and benefit-sharing (Bonn Guidelines, 2002). Overall, the impact of the Biodiversity Convention on the grant of intellectual property rights may be to require that inventions based on genetic resources or associated knowledge should be denied patent protection if they have not been acquired in a manner conforming to the principle of prior informed consent. This is significant because it covers both the resources and the associated knowledge (see Commission on Intellectual Property Rights, 2002).

The requirements of prior informed consent and disclosure have already been incorporated in some legislation of selected developing countries. The Costa Rican Biodiversity Act, for instance, clearly provides that intellectual property rights on inventions using biological resources can only be granted if a certificate of origin and a statement on prior informed consent are provided to the organs instituted under the Biodiversity Act.[14] The amended Indian Patents Act includes a new requirement concerning the disclosure of the geographical origin of biological materials used in the invention. This requirement is limited to the physical resources and does not specifically involve a prior informed consent requirement (India – Patents (Amendment) Act, 2002). The Philippines have adopted an even stricter framework in their Indigenous Peoples Rights Act. The Act provides that access to biological resources or associated knowledge is allowed only with prior and free consent from the communities. The Act specifically indicates that free and prior consent involves a consensus of the indigenous peoples concerned which must be 'free from any external manipulation, interference coercion, and obtained after fully disclosing the intent and scope of the activity, in a language and process understandable to the community'.[15] Further, the Act also recognizes the rights of indigenous peoples to the restitution of their intellectual property in case it has been acquired without prior informed consent or in violation of local laws or customs.[16] The Venezuelan Biodiversity Act adopts the same approach and takes it to its most stringent conclusion. In effect, the Act provides that no intellectual property rights will be recognized unless the rules for accessing biological resources have been followed. This shifts the burden entirely to the patent applicant who must follow the rules or face annulment of intellectual property rights.[17]

Joint inventorship

Patent law has long acknowledged that several individuals or legal entities could apply jointly for a single patent. In the case of traditional knowledge, the question which is most likely to surface is not whether a biotechnology firm is entitled to apply jointly with traditional knowledge

[14] Costa Rica: Biodiversity Law (1998).
[15] Section 3.g, Philippines: The Indigenous Peoples Rights Act of 1997.
[16] Section 32, Philippines: The Indigenous Peoples Rights Act of 1997.
[17] Article 82, Venezuela: Ley de diversidad biológica, 27 October 1999.

holders for a patent right but rather whether the contribution of the traditional knowledge holders is sufficient to allow them to claim joint inventorship.

The notion of joint inventorship has been the subject of disputes even before the development of inventions based on traditional knowledge. As a result, it may be possible to derive interesting ideas from previous practice for use in the context of traditional knowledge. In the USA, for instance, the statute recognizes the notion of joint inventorship.[18] The main requirements imposed for joint inventorship are that each of the joint inventors must contribute to the inventive element of the protected invention and that they must collaborate with each other. More specifically, the joint inventors must produce an invention by their aggregate efforts and must work on the same subject matter.[19] Further, while it may be difficult to distinguish prior art from the distinct contributions to a single invention, the statute recognizes that joint inventors do not have to work together at the same time. It is sufficient that the contribution of the joint inventor that works first in time constitutes an integral part of the process of joint invention.[20] Finally, the joint inventors do not need to collaborate on each of the claims made in a patent application and do not need to have made the same type of contribution or contributed equally to the invention.[21]

The notion of joint inventorship has been developed in conditions that are completely different from the conditions that a patent office would face in the case of an invention to which a traditional knowledge holder and a genetic engineering firm both contribute. However, it appears that the principles developed could be applied in the case of traditional knowledge. In fact, the only case where joint inventorship seems clearly barred is in the case where the joint inventors do not know of each other's work. In the case of traditional knowledge obtained with prior informed consent (which itself includes a statement by people accessing the knowledge concerning the intended use), normal patent rules can be relatively easily used to provide joint inventorship to traditional knowledge holders if their contribution is clearly linked and relevant to the final product. Overall, the determining test may be that the contribution of the joint inventor must be essential in distinguishing the invention from prior art (Huft, 1995). While the merits of each case would have to be examined individually, there are a number of cases where the contribution of traditional knowledge holders to a given invention may constitute an integral part of the protected invention. However, while it may be relatively easy to determine the joint nature of an invention based on the traditional knowledge of a healer who does not share knowledge with other members of her/his community, this would be much more difficult in a case like the turmeric patent, where the contribution to the overall invention would have been the inventive effort of a whole nation and not that of easily identifiable individual(s).[22]

Traditional knowledge and existing conditions for patentability

The criteria that applicants for patent rights must fulfil were defined with certain types of invention in mind. As highlighted in Chapter 3, the protection of traditional knowledge does not generally fit well within the normal patent model. However, there are ways in which existing criteria for patentability could be used for the benefit of traditional knowledge holders. First, the notion of inventive step poses significant

---

[18] 35 United States Code 116.

[19] *Monsanto Co. vs Kamp*, 269 F. Supp. 818 (District of Columbia, District Court, 15 June 1967).

[20] *Shields vs Halliburton Co.*, 667 F.2d 1232 (US Court of Appeals, 5th Circuit, 19 February 1982).

[21] See, for example, *Ethicon, Inc. vs United States Surgical Corporation*, 937 F. Supp. 1015 (United States District Court, D. Connecticut, 9 September 1996).

[22] On the turmeric patent, see above at Chapter 3.

difficulties in the case of traditional knowledge because patent law distinguishes inventions from discoveries and from public domain knowledge. Traditional knowledge is usually deemed to be part of the state-of-the-art when it is in the public domain. These traditional interpretations could be modified to reflect the specificities of traditional knowledge. Knowledge may be state-of-the-art and yet be novel because it comes from a different knowledge system unrelated and independent from developments in the formal knowledge system. In other words, different scientific systems should not be prejudiced because they do not follow the western model (Dutfield, 2001). Further, in recent years the notion of inventive step has been changing fast with the development of genetic engineering. In fact, the clear distinction that used to exist between invention and discovery is increasingly blurred. In this sense, if patent rights can be obtained on purified natural substances without being challenged according to the concept of discovery, traditional knowledge should also be patentable under the same criteria (Mgbeoji, 2001).

Apart from new interpretations of existing criteria for patentability, it may also be possible to bring traditional knowledge within the formal intellectual property rights system by defining new sub-species of rights. This has already been attempted for different reasons with the development of second-tier patents (petty patents) that could constitute one avenue for the protection of traditional knowledge within the patents system. Second-tier patents were first developed in the 19th century to address some of the perceived shortcomings of the concept of novelty. The UK Utility Design Act of 1843 was the first legislative development in this field. Even though this was enacted to remedy some problems of the patents system, it addressed designs and in practice sought to protect the form of a product, not its function (Janis, 1999). In effect, the developments of second-tier patents was linked to the need to find legal instruments that could provide protection to inventions that were not deemed significant enough to warrant the grant of a patent but were genuine innovations. As a result, the idea behind a petty patent is a protection which is short-term and which is granted on the basis of diminished standards of inventiveness compared to patent law. Another important factor accounting for its development is the need to find a form of intellectual property right that could be obtained faster and more cheaply than patents. In general, second-tier patent systems provided a way to reward functional improvements in three-dimensional shapes of tools or similar developments which neither patent nor trade secret law effectively protected. In other words, second-tier patents rewarded enhanced technical proficiency of tools but did not protect the underlying idea or the manufacturing process (Reichman, 1994).

Second-tier patenting has followed mixed fortunes over the past two centuries but has been revived periodically. It has, for instance, been criticized for not bringing about the intended cost reduction for the benefit of small and medium enterprises, although proposals for a sub-system linked but distinct from the patent system have periodically resurfaced.[23] In the context of traditional knowledge, proposals for the protection of smaller fragments of innovative contributions that do not necessarily fulfil the conditions for patentability are noteworthy.

The protection of relatively insignificant innovations can have interesting consequences. On the one hand, the relaxation of the rule concerning the inventive step could lead to the recognition that a locally developed plant variety is worthy of protection. This would allow a given farmer or farmers to acquire a measure of control over follow-up innovations derived from their variety. On the other hand, states could

[23] See, for example, Amended proposal for a European Parliament and Council Directive approximating the legal arrangements for the protection of inventions by utility model, Official Journal C 248E, 29/08/2000 p. 56.

decide to deny protection of the 'inventive step' for a local variety but simply provide that copying the product should be illegal. In this way, where innovation is deemed insufficient to qualify for a patent, protection through a right akin to copyright would provide the most limited form of protection possible (Reichman, 2000a).

Overall, second-tier patenting in the context of traditional knowledge is of interest because it highlights that debates over the relevance of patents in certain fields has long been disputed. The responses that have been given over time are noteworthy because they could to a certain extent be applied directly to traditional knowledge. These include attempts to redefine the novelty criterion, the search for intellectual property rights that are affordable for small enterprises and whose maintenance costs are low, and rights that can be granted more easily than patents with minimal examination by the competent authority (Stern, 1994). However, while second-tier patenting indicates some further ways in which the patent system could open itself up to traditional knowledge, the responses it provides remain limited by the framework within which they are conceived. Further, petty patents are subject to another type of criticism. In a situation where the scope of patentability is increasingly seen as being too broad, with the potential for harming further scientific development, it becomes even more difficult to strike an appropriate balance between the need for protection and the need to foster overall scientific and technological development (Suthersanen, 2001).

6.2.2 Adaptation of the conditions for granting of plant breeders' rights

Conditions for the granting of patents and plant breeders' rights are related. However, there are enough differences to warrant separate consideration of the conditions for the granting of plant breeders' rights. One of the conditions for protection under the UPOV Convention is that the variety must be uniform. This has two main impacts from the point of view of traditional knowledge. It stops traditional knowledge holders from applying for plant breeders' rights because their varieties are usually unable to fulfil this condition. Further, the condition of uniformity must be put in the broader context of agro-biodiversity conservation. In fact, while the plant breeders' rights model fosters the development of uniform varieties commercialized on a large scale, existing treaties in the field of agrobiodiversity management generally seek to avoid uniformity and monocultures, which constitute a significant cause of loss of biodiversity. Adapting the condition of uniformity would therefore be beneficial from the point of view of traditional knowledge holders and could be said to constitute a requirement from the point of view of environmental treaties. The UPOV Convention, unlike the TRIPS Agreement, does not specifically allow member states to take into account their other obligations in the field of plant variety management and conservation. However, UPOV member states that are also parties to the Biodiversity Convention must, even without formal acknowledgement in UPOV, implement all their obligations jointly. This implies that member states should have at least some limited scope in adapting the conditions of protection to take into account all their international obligations.

The condition of distinctness is another basic criterion for plant variety protection under UPOV. Distinctness is the closest that plant breeders' rights come to a concept of novelty as proposed in the patents regime.[24] The progressive acceptance of protection for essentially derived varieties, in other words varieties that do not differ significantly from the protected variety, has gradually rendered the criterion of distinctness quite ineffective as a

[24] While novelty is one of the basic criteria for protection under UPOV, it differs radically from its counterpart under patent law since it relates to the first commercialization of a variety and not to its inherent characteristics. See Article 6 of the UPOV Convention.

way to measure novelty. From an environmental point of view, the possibility of protecting essentially derived varieties constitutes an incentive for the commercialization of increasingly similar varieties, a factor contributing to crop uniformity and loss of agro-biodiversity.

Different changes to existing plant breeding criteria can be envisaged. The introduction of a prior informed consent requirement would, for instance, constitute an important change that could significantly strengthen the position of farmers' varieties within the UPOV system. This would provide a clear mechanism for ascertaining the origin of the genetic material, the extent of distinctness as judged within the UPOV system and would strengthen the claims of traditional knowledge holders for benefit-sharing. For member states of the UPOV Convention, the introduction of such a requirement may require an amendment to the Convention as it provides a finite list of conditions of protection and a finite list of reasons for cancellation and nullity of breeders' rights.[25] Some countries that have based their legislation on the UPOV system have adopted prior informed consent requirements, indicating that there is scope for further development in this field. The Indian Act, for instance, imposes on commercial breeders the obligation to certify that the genetic or parental material has been lawfully acquired.[26]

Other conditions could be imposed on the granting of plant breeders' rights. Environmental concerns (biosafety) can be built into the conditions for granting rights. The Thai Plant Variety Act provides, for instance, that a new plant variety cannot be registered if it has 'severely adverse impact, directly or indirectly, on environment, health or public welfare'.[27] In other words, plant breeders' rights could be made conditional upon a safety appraisal procedure. Further, food security and food safety concerns can provide the basis for the imposition of further conditions on the granting of the rights to ensure the maintenance of agricultural biodiversity, the promotion of health and the enhancement of national food security.

6.2.3 Geographical indications[28]

Geographical indications were for a long time seen as a supplementary means of intellectual property protection for specific products, with a significant emphasis on wines and spirits.[29] This perception has been changing in the aftermath of the adoption of the TRIPS Agreement. This is due to two main related factors. First, economic globalization and increased international trade makes the international protection of geographical indications much more important for countries and actors claiming such rights. Secondly, a number of developing and developed countries have progressively discovered the commercial potential of some of their geographical indications and are pushing for more extensive protection in this field. A number of geographical indications are closely derived from traditional knowledge. As a result, the possibility of strengthening their protection is significant from the point of view of the protection of traditional knowledge.

Geographical indications are an atypical form of intellectual property rights that do not protect novel elements but rather an accumulated goodwill built up over the years. This goodwill is the outcome of a recognized or perceived link between a product and a geographical area. The purpose of geographical indications is to identify products but not to provide protection to the product as such. Further, geographical indications are also atypical insofar as they are a collective right that only grant producers in a given area the right to use the

[25] For UPOV-1991, see Articles 5, 21, 22.
[26] Article 18, India: Protection of Plant Varieties and Farmers' Rights Act (2001).
[27] Section 13, Thailand: Plant Varieties Protection Act, B.E. 2542 (1999).
[28] Authors: Philippe Cullet and Andrea Nascimento Müller.
[29] See above, Chapter 2, Section 2.3.3.

indication for products of a specific geographical area and that is unlimited as long as the specific conditions for the granting of the geographical indication remain in place. Two main forms of geographical indications have been developed in international law: the first, indications of source, simply denotes the fact that a given product derives from a certain geographical area. The second, appellations of origin, refers to products whose quality or other characteristics are essentially attributable to their geographical origin, either because of natural or human factors. Under the TRIPS Agreement, both forms have been subsumed under the concept of geographical indications.

Geographical indications can take the form of words, phrases, symbols and iconic emblems. Under TRIPS, indications do not necessarily have to be the name of a geographical place on Earth and can therefore include names that relate to a specific geographical area, such as Basmati in the case of rice in certain areas of Pakistan and northern India. However, goods that are protected must originate in the region with which they are associated, which implies that licences for the production of a protected good outside its region of origin cannot be protected (Blakeney, 2001a). Different countries use different criteria to demarcate areas that can be covered by an indication. These can range from a geographical unit linked to a political classification to *ad hoc* definitions such as where a specific wine-growing area is granted a right (WTO, 2001).

Geographical indications can be protected by different means. One of the most common means of protection is through laws that prohibit business practices that may involve the misuse of indications. This includes the repression of unfair competition and consumer protection, for instance, with regard to product labels and food safety. This negative protection imposes on legitimate users the duty to prove that there have been illegitimate uses of an indication.[30] Stronger protection is available in the case of wine and spirits under Article 23 of the TRIPS Agreement. In this case, legitimate users are better protected insofar as competitors in other regions of the world are more restricted in the use they can make of the protected indications. Further, legitimate users are given the right to challenge other users simply on the basis that a product has not been produced in the area identified by the indication without having to prove that there has been unfair competition or that this misleads the public.

Geographical indications and trademarks

Geographical indications have a number of links with trademarks,[31] some of which are of direct relevance in the context of traditional knowledge. In general, trademarks provide two types of protection for geographical indications. First, they can provide protection against the registration and use of an indication as a trademark. In this sense, trademarks provide a tool for preventing third parties from appropriating signs, symbols or names that belong to traditional communities with a view to exploiting them commercially (de Carvalho, 2005). This corresponds closely to the traditional distinction between trademarks and geographical indications. While the former seek to distinguish a specific product or service and cannot be descriptive, the latter are usually descriptive.

[30] See Articles 10, 10bis, 10ter of the Paris Convention for the Protection of Industrial Property, 20 March 1883 (as revised and amended).

[31] A trademark is any word, name, symbol, device, slogan, package design or combination of these that serves to identify and distinguishes a specific product from others in the market place or in trade (International Trademark Association Glossary). Normally a mark for goods appears on the product or its packaging, while a service mark appears in advertising for the services. Depending on the national laws, a trademark may be registered (®) or unregistered (™). There are different types of marks: e.g. trademarks, promoting products, service marks, promoting services, certification marks, which cover characteristics such as product quality, method of manufacturing, and collective marks, which are used by members of a group or organization.

Secondly, trademarks can in some circumstances provide protection for geographical indications against unauthorized use by third parties in cases where an indication has been recognized as a trademark.

The relationship between trademarks and geographical indications and the problems concerning the possible overlaps between the two kinds of rights have been addressed in different ways. The TRIPS Agreement provides as a general rule that registration of a trademark that contains or consists of an indication is not allowed if the goods do not originate from the area covered by the indication and if such registration is likely to mislead the public as to the true place of origin. This is supplemented by a stronger regime in the case of wines and spirits, in which case trademark registration is to be refused or invalidated on the simple basis that the wines and spirits do not originate from the area covered by the indication. Finally, the TRIPS Agreement makes an exception in situations where trademarks have been applied, registered or acquired in good faith.[32]

On the whole, while trademarks may be relevant in the context of the protection of traditional knowledge where specific products derived from traditional knowledge can be so protected, geographical indications are more relevant and interesting from a conceptual point of view. First, where trademarks seek to provide signs that distinguish products from a given entity from their competitors, geographical indications are freely enjoyed by all product manufacturers and traders in the specific area and protect all of them from inappropriate use of the indication. Secondly, unlike trademarks, geographical indications cannot be transferred, a significant advantage in all situations where transferability involves a risk for the weaker party, as is likely to be the case with traditional knowledge holders. Thirdly, the difficulty in registering a geographical indication as a trademark is likely to reduce the risk of conflicts where one trader or producer seeks to appropriate the indication for her/himself through a trademark. Fourthly, despite exceptions, trademarks are usually individual rights, while geographical indications are by definition collective rights and in this sense may be more amenable to adaptation to traditional knowledge protection.[33]

Further development of geographical indications

The incomplete and differentiated protection for geographical indications provided in the TRIPS Agreement has led a number of WTO member states to seek changes in this field. Two main issues are currently being discussed in the TRIPS Council. As called for in the TRIPS Agreement itself and confirmed at the Doha ministerial conference,[34] states are negotiating a multilateral system of notification and registration for wines and spirits. The Doha Ministerial Declaration specifically addressed two distinct issues. First, it confirmed the willingness to negotiate the establishment of the multilateral system just mentioned. Secondly, it acknowledged the need for the TRIPS Council to consider issues related to the extension of the special protection for wines and spirits to other goods.[35]

Some members mentioned different interpretations of this mandate, in documents circulated at the Ministerial Conference (O'Connor and Company, 2003). Negotiations proposals have pitted states with significant interests in the protection of different types of geographical indications, including those relating to traditional knowledge, against states that have generally been reluctant to accept protection

[32] For further details, see Article 24.5 of the TRIPS Agreement.

[33] For an example of a legal regime recognizing collective trademarks, see, for example, Andean Community, Decision 486 – Common Intellectual Property Regime, 14 September 2000.

[34] Article 23.4 of the TRIPS Agreement and para. 18 of the Ministerial Declaration, Doha, 14 November 2001, Doc. WT/MIN(01)/DEC/1.

[35] WTO Doha Ministerial Declaration, para. 18, adopted on 14 November 2001, WT/MIN(01)/DEC/1.

through geographical indications. Thus, Argentina has, for instance, stated that it sees no mandate to negotiate the extension of the protection of GIs to other products.[36] On the opposite side, a number of developed and developing countries have affirmed that there is a clear mandate to launch negotiations on the extension of additional protection for GIs.[37]

The pros and cons of GI extension essentially can be divided into three topics. The first argument relates to the negotiating balance accomplished in the Uruguay Round. The second concerns the deficiency of the scope of protection available in Article 22. Thirdly, there is the potential impact of GI extension on trade, consumers and the TRIPS obligations, of which the biggest would be related to costs (Rangnekar, 2002). Members opposing GI extension, such as Australia, Paraguay, Canada, New Zealand, the USA and Argentina, contend that the obligations entail excessive costs, much of which would fall on developing countries.[38] Those in favour of the extension of GIs, such as Bulgaria, the Czech Republic, Iceland, India, Switzerland and Turkey, argue that there is potential for commercial use, and that such extension would benefit trade and development.[39] The proposal to extend the level of protection afforded to wine and spirits' GIs to other products would include agricultural products, processed foods and handicrafts.

With regard to the use of GIs as a tool for the protection of traditional knowledge, limited discussions have taken place in the context of the TRIPS Council. Some states, such as Switzerland, have indicated that GIs provide a possible tool for the protection of traditional knowledge and genetic resources, for example in the case of kava, neem, Mexican enola beans, Peruvian yacon and Andean nuna beans. The extension would guide the use of GIs on equitable conditions for all products and would promote traditional methods of production and processing, thereby contributing to economic development.[40] It has also been stated that the extension of GIs could constitute an incentive for producers to market their goods internationally, thus promoting international trade.[41] At the Council for TRIPS meeting held in November 2002, Australia argued that GIs could constitute one way of protecting traditional knowledge in a less severe or monopolistic way, although nowhere near as strongly as patents or copyrights.[42]

Geographical indications and traditional knowledge

From the perspective of traditional knowledge, geographical indications are of specific interest for several reasons.

---

[36] WTO Doc. WT/MIN(01)/W/8, 'Communication from Argentina regarding paras 18 and 12 of the Draft Ministerial Declaration' of 12 November 2001.

[37] WTO Doc. WT/MIN(01)/W/11, 'Communication from Bulgaria, Kenya, India and Sri Lanka regarding paras. 18 and 12 of the Draft Ministerial Declaration', 13 November 2001. See also WTO Doc. WT/MIN(01)/W/9, 'Communication from Bulgaria, The Czech Republic, The European Communities and its member states, Hungary, Liechtenstein, Kenya, Mauritius, Nigeria, Pakistan, the Slovak Republic, Slovenia, Sri Lanka, Switzerland, Thailand and Turkey regarding paras. 18 and 12 of the Draft Ministerial Declaration', of 14 November 2001.

[38] WTO, Doc. IP/C/M/38, supra note 16, para. 125.

[39] WTO, Doc. IP/C/W/204, 'Communication from Bulgaria, The Czech Republic, Iceland, India, Liechtenstein, Slovenia, Sri Lanka, Switzerland and Turkey', of 18 September 2000, para. 2.

[40] WTO, Doc. IP/C/M/35, 'Council for Trade-Related Aspects of Intellectual Property Rights – Minutes of Meeting – Held in the Centre William Rappard on 5–7 March 2002', para. 162.

[41] WTO, Doc. IP/C/M/38, 'Council for Trade-Related Aspects of Intellectual Property Rights – Minutes of Meeting – Held in the Centre William Rappard on 25–27 and 29 November, and 20 December 2002', para. 69.

[42] WTO, Doc. IP/C/M/38, 'Council for Trade-Related Aspects of Intellectual Property Rights – Minutes of Meeting – Held in the Centre William Rappard on 25–27 and 29 November, and 20 December 2002', para. 125.

1. They differ from other types of intellectual property rights insofar as they are clearly collective in scope. Geographical indications do not grant a single holder the right to benefit from the protection but rather limit the protection to a specific area. They provide a collective right to use the indication. In other words, they offer an exclusive protection against outsiders to an indeterminate number of people within the region of protection. Protection through geographical indications may therefore provide an interesting avenue to foster protection for products manufactured within a specific area while not restricting the number of rights holders within the area.

2. Geographical indications do not impose any tests of novelty like the patent system. In fact, they can be used specifically to protect traditional products as long as the particular characteristics of these products can be attributed to a specific geographical origin (Commission on Intellectual Property Rights, 2002).

3. Protection through geographical indications does not relate to one specific method of production of a given product. This allows not only different production methods to be covered under a given indication but also for production methods to change over time within the scope of the protection offered (Downes, 2000). In other words, geographical indications intrinsically recognize one of the essential characteristics associated with traditional knowledge, which is its evolution over time.

4. Geographical indications do not imply monopoly control over the knowledge that is embedded in the protected indication. In fact, this knowledge remains in the public domain. This presents advantages and disadvantages. On the positive side, from the point of view of the existing intellectual property rights system, the absence of protection of the knowledge is counterbalanced by recognition in perpetuity as long as the link between the geographical place and the good is maintained. On the negative side, the lack of protection implies that traditional knowledge can be misappropriated (Dutfield, 2000). This is similar to the broader concern over biopiracy in the context of patents.

5. The impossibility of transferring geographical indications outside their region of protection constitutes a major advantage in the context of traditional knowledge.

6. Geographical indications present an advantage over other forms of intellectual property rights for traditional knowledge holders insofar as protection may extend not only to indications that are currently in existence, but also to indications likely to be used in the future.[43] In other words, an indication may constitute a ground for denying appropriation, for instance, through a trademark, in cases where the indication is in use but also in cases where it may be used by traditional knowledge holders for commercial aims in the future.

Certain caveats should also be entered with regard to the use of geographical indications to foster the protection of traditional knowledge. First, they can only be used to protect an indication and cannot constitute a tool for protecting the underlying knowledge. Secondly, geographical indications that are deemed to have become generic lose all their usefulness from the point of view of traditional knowledge protection. Thus, if Basmati rice were to be recognized as a generic name, Indian and Pakistani rice growers would lose all claims on the indication. Thirdly, a number of technical issues may limit the relevance of geographical indications in the protection of traditional knowledge. A system of protection for the benefit of traditional knowledge holders would have to impose strict limitations on individuals or companies eligible to seek registration of an indication. This limitation notwithstanding, outsiders may relatively effortlessly claim the indication if they buy a company producing within the area or decide to manufacture a product within the protected area. This would probably be detrimental for

[43] Cf. *Windsurfing Chiemsee vs Huber*, Cases C-108/97, C-109/97, Judgment of the European Court of Justice, 4 May 1999, [1999] ECR I-2779.

traditional producers. Overall, geographical indications provide some scope for protection of traditional knowledge, but the protection remains limited insofar as indications are conceived as marketing tools and do not protected the knowledge related to the product.

Case study: Potential benefits of geographical indications: the kava case in the South Pacific[44]

Many developing countries that are commercial producers of agricultural goods are interested in increasing their commercialization. Geographical indications might support this goal. One example could be the production of kava kava (*Piper methysticum*) (see also Chapter 3).

Kava products, based on traditional knowledge, originally were produced and consumed in the South Pacific Islands. As noted in Chapter 3, different kava-derived products have been progressively commercialized, leading to a considerable export potential in regional and world markets (Lebot *et al.*, 1992).

However, kava is also produced in other regions (Hawaii, Central America), the production of the South Pacific Islands seemingly being the most traditional and ancient. GIs could generate a competitive advantage either in the form of the indication of the South Pacific Islands as the original geographical source of kava, or in the form of an appellation of origin, which in addition includes information on a specific quality that is essentially attributable to its origin. As this specific quality can be caused by natural or human factors, the traditional knowledge about how to best plant, grow and produce kava could thus be protected.

Therefore, in an attempt to find a solution for the protection of kava's traditional knowledge and products in the South Pacific, GIs could play an important role. There are two basic elements that link TK and GIs: (i) the lack of individual intellectual property rights, the rights therefore being acquired by a group; and (ii) the fact that the knowledge is transferred from one generation to the next. This is supported by collective traditions and rewarding traditions, while allowing for evolution.

When one considers the commercialization of kava, there is a stark contrast between the price of kava, sold to the USA at around US$5–10 per pound or US$11–22 per kilo, and the price of a kava product in the USA, which is between US$12 and US$60 for 60 grams (Downes and Laird, 1999). Kava producers could gain additional value through processing the commodity into end-use products.

On the whole, the kava case shows that GIs can create a competitive advantage for the marketing of a product and, if the marketing is successful, contribute to the economies of developing countries. Furthermore, trademarks and collective trademarks and/or certification marks could also protect kava. The establishment of such marks would enable producers to set up shared standards for sustainable development, as well as the monitoring and enforcement of standards. This would maximize the protection and sustainability of the resources by reducing the number of producers who wish to make short-term profits and. in many cases, do not have a long-term strategy.

6.3 Management and Enforcement of IPR and Traditional PGR: Institutional Design

Granting private rights in order to contribute to fulfilling public policies only makes sense in practice if the right holders can benefit from the economic and non-economic advantages that such rights are supposed to provide. This truism requires that the elaboration of substantive rules of law must be accompanied by the development of a corresponding institutional framework allowing the implementation, management and enforcement of the legal means at stake. In our context, the institutional design must

[44] Author: Andrea Nascimento Müller.

cover three main elements, i.e. the gathering, archiving and the dissemination of relevant information and knowledge, the management and the enforcement of rights referring to such information and knowledge. In various fields of protection by classic intellectual property rights (e.g. patents, trademarks, industrial design, geographical indications, copyright, etc.) registries play an essential role in gathering and archiving information and/or knowledge. For certain fields of protection such as patent and trademarks, the registration generates the exclusive rights, whereas in other fields such as copyright the optional registration has simply a declaratory or informational character that may facilitate the fact-finding procedures for dispute resolution purposes. Models of registration and documentation facilities that may serve law- and policy makers as a source of inspiration in setting up a system that would meet the specific characteristics of maintaining, developing and protecting traditional knowledge are addressed below. Some of these tasks, especially with respect to the dissemination and exchange of information and knowledge, may also be addressed by the clearing house mechanisms described later. Management of the rights by, or on behalf of their holders, as well as their enforcement, are complex undertakings. Reasons of efficiency may lead right holders to take care of these tasks in a collective rather than an individual way. In this case, the model of collecting societies, as addressed below, may provide stimulating inputs to law- and policy makers. This model is based on a 'bottom-up' approach as opposed to the one inspired by the clearing house mechanisms that may also fulfil management and enforcement activities, but relies rather on a 'top-down' approach. Both models may be envisaged to inspire the elaboration of a system that would be suitable for the management and enforcement of rights related to traditional knowledge. However, the collecting societies model, which is based on a spirit of advocacy rather than of arbitrage, may be more suitable to addressing various conflicting interests than the clearing house scheme. This is especially true with respect to enforcement of rights. In addition, the management of rights is likely to be more efficient if it is 'demand driven', i.e. driven by the right holders' demand of rights management services, as in the case of collecting societies, as opposed to the clearing houses feature of functioning in a 'supply-driven' way, i.e. driven by an international bureaucracy's supply. On the other hand, registries and clearing houses may also serve as 'matchmaker' facilities that bring together right holders and users. In this way, these institutions may provide a very valuable contribution to marketing traditional knowledge.

6.3.1 Documentation and registration[45]

Introduction and questions

As described previously (Chapter 1), one of the characteristics of TK is that it is largely handed down orally, from generation to generation. Local laws, customs and traditions define its use and tradition. This specific feature leads to the loss of knowledge at the community level once the cultural tradition chain is interrupted or destroyed. At the international level, this leads to problems of legal security in various respects. This occurs in particular in connection with the access to TK and PGRFA and their utilization outside their traditional area of uses, such as in industrial R&D processes. The problems may consist of the following:

- The difficulties faced by the design holders and guardians/custodians of the knowledge, because, for instance, these are considered not to exist any longer, as was the case with the Kung bushmen and their knowledge about the Hoodia cactus;[46] or because several communities are custodians of similar knowledge, or

[45] Author: Susette Biber-Klemm. I thank Shakeel Bhatti for inspiring comments and valuable input.
[46] GRAIN, BIO-IPR docserver, 17.06.01.

because within the community it is not clear if and by whom information can be traded.

- The fact that the examination of pre-existing information in the patent procedure is difficult or even impossible, as information on the TK in question is not easily accessible. In some systems such an examination is not included in the procedure. The same may be true for the proof of prior art in challenging a patent, in systems where a written proof is needed.
- The delimitation of the public and private domain. Beyond the reach of local laws and customs, from the IPR point of view, the information generally belongs to the public domain and can be used freely without authorization or compensation.

This occurrence has led to a variety of initiatives to document TK. On the national level, systems for documenting TK are created for the sake of preventing its loss, that is, to *preserve* the TK (for instance, Biozulua in Venezuela), and/or to create a basis for proving prior art, so as to *protect* TK against inappropriate or unauthorized use by others.[47] Frequently, such documentation and registration schemes are integrated in national legislation on access and benefit sharing.

But initiatives to document TK also exist as bottom-up initiatives at the local and regional level, to organize and preserve the local knowledge basis, and to serve as a basis for taking decisions on resource management for the benefit of the community.[48] There also exist initiatives to create documentation and registries on the international level, which aim for example at creating networks between different players[49] in order to mainstream indigenous/traditional knowledge into the activities of development partners, and to optimize the benefits of development assistance, especially to the poor,[50] or to serve as a documentation of prior art.[51]

In the preceding chapters some thoughts were given on documentation of TK and PGRFA in connection with its allocation to its creators and/or holders (Chapter 4) and its registration in connection with the creation of TIP rights (Chapter 5). Here the question is: what could be the function of TK documentation/registries at the interface with trade in general, and as a basis for the above-described flanking measures? Thus, according to the problems and goals described in Chapter 1, it has to be asked whether documentation and registration is or could become a viable means to achieve the following objectives:

- To empower the holders of TK and traditional PGRFA to facilitate their autonomous decision about the use of their information. The question in this context is first, whether documentation can be used as a means of allocating information to its holders/custodians or authors (see the above discussion in Chapter 4); secondly, what would be the necessary elements for documentation and the procedural needs for verifying the claims, or possibly to establish common property if the information is shared by several communities; and thirdly, how the decision of the holders of the information on the use made of it can be secured. This question has been discussed above in connection with the

[47] For the terminology, see WIPO/GRTKF/IC/5/12, No. 17.

[48] E.g. the Inuit of Nunavik database, which contains information on their knowledge and use of the land base and resources: see Downes and Laird (1999, p. 5); or the example of the Nicola Tribal Association in British Columbia, which has its traditional knowledge stored in a database, but only the tribal elders have the password to access the information (Dalton, 2002).

[49] See the World Banks IK database on http://www.worldbank.org/afr/ik/

[50] See the examples in Downes and Laird (1999) and the list of databases compiled by WIPO's Intergovernmental Committee on Intellectual Property and Genetic Resources, Traditional Knowledge and Folklore (IGC) at http://www.wipo.int/globalissues/databases/tkportal/index.html

[51] See the WIPO project to compile TK, which is in the public domain in order to provide an organized and searchable instrument to research prior art for patent examiners.

option to create traditional intellectual property rights (Chapter 5).

- To protect against the misappropriation (piracy) of TK and PGRFA. This would be the question as to the method and extent of documentation in order to prove prior art and ownership in the case of publicly inaccessible information, and prior art in the case of public domain information.
- To facilitate the marketing of, and trade in, TK and traditional PGRFA in order to create incentives for its maintenance. This would entail that in order to advertise which information is meant to be marketed, it must at least in part be made publicly accessible. The challenge in this context is to find a solution for the dichotomy between advertising the existence of the information, and to keep the basic information (if it cannot be otherwise protected) secret at the same time.

The subsequent reflections will focus on the market aspect, taking account of the fact that all the issues mentioned are interrelated. This perspective sheds light on the close connection between registration and access and benefit sharing. In particular the reflections on the streamlining of the procedures in order to promote access (discussed in Chapter 6) have to be taken into account.

In the following, the current discussions and positions of the various involved stakeholders will be described and evaluated. Then, on the basis of two examples, the elements necessary to fulfil stakeholder expectations and, in particular, to make use of registers to promote trade, will be analysed and evaluated.

Definitions and terminology

The terms 'documentation' or 'inventory' and 'registration' are often used interchangeably in the current debate on TK. However, it is necessary to differentiate between them.

According to the Oxford English Dictionary, *documentation* is 'the accumulation, classification, and dissemination of information on the material so collected'. Thus, documentation is stocktaking, an inventory of information, recording it in a systematic way. TRIPS defines databases as 'compilations of data or other material, whether in machine readable or other form, which by reason of the selection or arrangement of their contents constitute intellectual creations' (Article 10).[52]

In the context of the documentation of TK, as a rule electronic databases are used. The goal would be to record as much TK answering to the factual criteria specified for the given database.[53]

Registration, on the other hand, means to formally set down specific information in writing in a precise manner. A register is thus an ordered collection, repository or list of information that has an official status. The registration of information in a registry puts that information 'on the record'. It records the fact that the registrant asserts a claim to that information, creation, innovation or object. It confers some legal status on the record, which, however, needs to be established in pertinent legislation that specifies the requirements, procedures for application and dispute settlement.[54]

At this point it is necessary to clearly make the distinction between the registers designed for the 'formal' intellectual property rights – such as patents – and the reg-

[52] Having, for the sake of protection, its focus on the creative aspects implied in the design of databases. There are two different aspects of the IPR protection of databases: (i) the protection of the value of intellectual creation represented by the system of the database itself; and (ii) the protection of the content of the database, i.e. the information stored in it. Only the second is taken account of in this context.

[53] Consider the distinction between TK recorded in a database and codified TK. Codified TK, in particular in the field of traditional medicine, is TK which has been disclosed in writing in ancient scriptures and is fully in the public domain, in contrast to non-codified TK which has not been fixed in writing, often remains undisclosed by its holders and is passed on in oral traditions (WIPO/GRTKF/IC/3/6).

[54] See WIPO/GRTKF/IC/2/3 1 July 2001 para. 118; Downes and Laird (1999, p. 5).

isters designed to record traditional knowledge and/or related *sui generis* intellectual property rights.

In the case of patents, the disclosure of the invention and, accordingly, the general accessibility of the pertinent information, is part of the system (see Chapter 3). In the case of traditional knowledge this must not be the case, in particular if the registration has primarily a defensive character in the sense of proof of prior art, and *sui generis* intellectual property rights do not protect the information. But even in the latter case, given the problems of controlling the utilization of TK and PGRFA in industrial R&D, in order to avoid misappropriation, it might be advisable not to publish the details.[55]

It follows from this that in the case of TK it is also necessary to carefully balance the interests and benefits of the holders of the TK against the interests and benefits of society as a whole.

Current discussions

THE POSITION OF DIFFERENT STAKEHOLDERS

Donor countries. The initiatives to document PGR and associated TK by the donor countries may have different purposes (see, for instance, WIPO, 2001; Ruiz, 2003).

One aspect which is at the forefront of many initiatives is the preservation of knowledge, which otherwise could get lost because of various pressures on traditional culture. This goal encompasses two different aspects: on the one hand, documentation is meant to safeguard the transmission of knowledge to subsequent generations (see, for instance, the Biozulua register of Venezuela[56]). On the other hand, documentation is meant to facilitate the transmission of TK, and to promote the sustainable use of biological resources within the communities.[57]

A second, frequently prevailing, goal is the prevention of the illicit acquisition of intellectual property rights over TK by third parties. Documentation facilitates the proof of prior art or obviousness, incorporating procedures to oppose patents already granted.[58] However, there are also initiatives to make documentation accessible for patent offices for the purpose of prior art searches.[59]

Besides this defensive approach, a more trade-related approach also exists, which places registration in the context of the access and benefit sharing regimes and identifies the goals for promoting the utilization of the stored information through bioprospecting and for ensuring the sharing of the resulting benefits with the local communities.

The position of custodians of the knowledge. The holders and custodians of traditional knowledge are displaying increasing concern about the documentation of their knowledge. The fear is that by documenting the information in databases, secret knowledge and traditions could be put into the public domain or more easily accessed by interested third parties, thus promoting biopiracy instead of preventing it.

For instance, in Venezuela indigenous people are claiming that before collection or storage of TK in the Venezuelan databank on indigenous knowledge on flora and fauna (Biozulua) is continued, a decision about its future use is to be taken with their (equal) participation. They protest against the patenting of the collected knowledge *by*

---

[55] See below the discussion of the Indian National Innovation Foundation.

[56] See WIPO/GRTKF/IC/3/6, p. 16 and Eugui D.V. (without year).

[57] See, for instance, the Honey Bee Network as described by Gupta (undated a, b).

[58] See, for example, the case of the field bean cultivar, WIPO/GRTKF/IC/5/6, pp. 6–8.

[59] See, for example, the initiatives by the Intergovernmental Committee on Intellectual Property, and Genetic Resources, Traditional Knowledge and Folklore (IGC) to study the feasibility of electronic exchange of public domain TK documentation data through the establishment of international online TK databases; WIPO/GRTKF/IC/3/6 and http://www.wipo.int/globalissues/databases/index.html. As an example at the national level, see the Peruvian Law No. 27811, Articles 5(f) and 23.

the state (Dalton, 2002). In India similar attitudes and fears have emerged (Jayaraman, 2002).

In sum, the following arguments are brought forward: first, that it is up to the holders of TK to decide what information is integrated in the database; secondly, that prior to the integration of the knowledge into the database the PIC of the knowledge holders is necessary; and thirdly, that it is the holders of TK who are the ones to decide how the stored information is used.[60]

The position of industry. For industry, according to the results of a stakeholder dialogue process on the issue of Intellectual Property Rights in Biotechnology (WBCSD/WZB, 2002), for the purchasers of biological resources and TK, in particular for industrial R&D, accessibility to the information and legal security plays a pre-eminent role[61] (see also Section 7.3).

So, for instance, the chemical industry (European Chemical Industry Council, CEFIC, 2002) believes that it is necessary to create inventories of traditional knowledge for the following reasons:

- Stocktaking of TK, for its conservation and as 'a background on which further innovations may be documented'.
- The determination of public domain knowledge.
- The determination of the possible co-proprietorship between different groups.

CEFIC points to the necessity of identifying the inventor or creator of an information or its entitled owner, including possible collective ownership. It further proposes that rights to TK be created and formally registered, registration being the starting point for the duration of the protection.[62]

In sum, from the users' perspective, registration of TK is welcomed to secure transparency regarding the entitled holders or custodians of the information, to facilitate the ABS procedure and to further legal security.

ANALYSIS: MULTIFUNCTIONALITY OF DOCUMENTATION OR REGISTRATION (MULTIPLE-PURPOSE DATABASES)

The examples mentioned above illustrate that a variety of (to some extent contrasting) functions and goals of databases exists. One basic distinction, as elaborated upon by WIPO, differentiates between defensive and positive protection.[63]

Defensive protection consists of measures aimed at preventing so-called

---

[60] Mgbeoji (2001), who advocates the creation of *sui generis* rights for the protection of TK, takes the critique a step further. He points out that: (i) the negative protection by publication in the sense of prior art does not go far enough; (ii) the registry perpetuates the unfair economic paradigm that conceptualizes indigenous peoples as mere producers of raw materials and importers of finished products; (iii) he perceives registration as 'reducing the claims of indigenous peoples and non-western local communities to begging for funds'; and (iv) cautions that contracts based on the registry of uses model only provide bilateral protection, neglect the inequality of bargaining power and produce 'paternalistic bureaucracy overseeing the resulting licensing agreements'.

[61] These statements mainly refer to TK in connection with PGR that are of interest for use in pharmaceutical (and related) R&D processes. They do not take account of the situation as to traditional varieties of PGRFA.

[62] CEFIC (2002). Compare also 'Pharmaceutical Executive', September 2002, which takes up the Venezuelan conflict and in principle welcomes the collection of TK in a database, acknowledging that an essential part of the scheme is that the ethnic groups retain the intellectual property rights associated with the plants' healing properties. The databases are considered to be advantageous to the pharmaceutical industry as it makes the 'task of pinpointing potentially useful plants much simpler. ... And if the databases become a tool to ensure that the indigenous people who have used the plants receive suitable payments for their knowledge, that will minimise the industry's potential for charges of biopiracy'.

[63] WIPO/GRTKF/IC/3/6, 5–8. WPO/GRTKF/IC/5/6.

biopiracy, i.e. the acquisition of IPR over TK (or products of TK) or genetic resources by parties other than the customary custodians of the knowledge or resources. In order to fulfil this defensive function, the databases need to ensure that the information is available as prior art to search authorities in patent procedures and patent examiners. According to WIPO this would encompass measures to improve the availability of the information, the searchability of the database for instance, by indexing or classification systems, and the exchangeability of TK between databases, in order to facilitate proof of prior art. Likewise it is important to ensure that the information is stored in such a way as to meet the legal criteria to be counted as prior art in the jurisdiction concerned.[64]

Even if defensive protection prevents third parties from gaining IPR over the information, it does not prevent others from using it. So the problem is that by documenting TK, in particular by making possible public access to TK that is otherwise undisclosed, secret or inaccessible, the unauthorized use of it by others may actually be facilitated.

To prevent unauthorized use by third parties, it is necessary to assert *positive* rights to the stored information. According to WIPO, such positive legal protection could consist either in the use of existing IPR, in the development of new *sui generis* rights to TK, or in the use of contractual rights. Such a solution would encompass the rights of the holders to restrict the way TK is used by others or at least to claim compensation for its use.[65]

For these reasons it is proposed to create multiple-purpose databases that serve both the defensive and positive protection of TK and genetic resources.[66]

The third option, which consists of the utilization of databases and registries as marketing instruments to further trade in TK and genetic resources in an *offensive* way, i.e. by displaying or advertising the marketable information, seems not to be considered in this concept.

Different emphasis: examples

PERU

The Peruvian law on the protection regime for the collective knowledge of indigenous peoples derived from biological resources, which was adopted in 2002,[67] establishes a *sui generis* protection regime for the collective knowledge of indigenous peoples connected with biological resources (Article 3). A registration system is part of this regime. It consists of three types of register: the Public National Register, the Confidential National Register and the Local Registers (Article 15).

The registers have a twofold purpose: on the one hand they are meant as instruments to preserve the collective knowledge of indigenous peoples; on the other they are intended to serve as a tool to defend the interests of indigenous peoples in their TK (Article 16).

The registration does not constitute rights over the traditional knowledge. All knowledge which is collective in nature, developed by indigenous peoples and which is not in the public domain, irrespective of whether it is registered or not, is protected against disclosure, acquisition or use without the consent of the indigenous peoples who possess it, and against unauthorized disclosure involving a breach of a duty of reserve (WIPO, 2003).

---

[64] WIPO/GRTKF/IC/5/6, No 7.

[65] WIPO/GRTKF/IC/3/6 No 6.

[66] WIPO/GRTKF/IC/4/14, p. 2.

[67] Law No. 27811: Ley que establece el régimen de protecctión de los concoimientos colectivos de los puelbos indígenas vinculados a los recursos biológicos, adopted on 10 August 2002; reprinted in English in WIPO/GRTKF/IC/INF/2 Annex III. See also Chapter 2. See WIPO, Advance Copy (2003).

So, there is a certain degree of protection but no exclusive rights are conferred (Article 42).[68]

The Public National Register encompasses the knowledge that is in the public domain (Article 17). Therefore, this register is public. This knowledge is either compiled by the responsible national agency (INDECOPI[69]) that is in charge of it, or contributed to by indigenous peoples (Article 19). The Public National Register has clearly been created for defensive purposes: INDECOPI is obliged to send the information stored in the Public Register to the main patent offices of the world (Article 23).

The second register, the National Confidential Register, is also under the responsibility of INDECOPI. This register relates to collective knowledge that is not in the public domain. Indigenous peoples may apply for the integration of their knowledge in this register. Its main goal is to preserve and safeguard the collective knowledge of indigenous peoples and their rights therein. Accordingly, no third party has access to this register (Article 18).

Finally, the indigenous peoples may organize Local Registers of collective knowledge in accordance with their collective practices and customs (Article 24).

INDIA: THE NATIONAL INNOVATION FOUNDATION

The Indian National Innovation Foundation (NIF) is an autonomous society, which was set up in 2000. Its goal is to foster and support grassroots inventorship and the maintenance of TK by its documentation and dissemination on a commercial as well as a non-commercial basis. Its structure and objectives are explicitly meant to replicate the philosophy of the Honey Bee Network, which was founded in 1989 as a private initiative (Gupta, undated (a), p. 4). Its main objective is to serve as a knowledge network that pools solutions developed by people on the grassroots level to solve their problems in different sectors.[70]

In the framework of NIF and the Honey Bee Network, documentation is part of a broader strategy that has the goal to serve as a clearing house mechanism to link innovation, enterprises and investments (the golden triangle of creativity (Gupta, undated (a), p. 4)). The Network is backed by the Society for Research and Initiatives for Sustainable Technologies and Institutions (SRISTI), a voluntary organization to provide institutional support, which was later supplemented by GIAN (Grassroots Innovation Augmentation Network, Gujarat). GIAN is a private organization that provides support for the innovators and holders of knowledge in development, marketing and legal protection. NIF is about to set up four more GIANs in different regions of the country.

The activities of NIF encompass: the scouting and documentation of innovations and outstanding examples of TK; verification of the claims; obtaining PIC from the providers of the knowledge; and sharing the innovations permitted in the public domain.

So NIF initiates value addition, helps to develop product development plans and to generate funding, and facilitates the development of the product by development teams on a contractual basis. It further assists in concluding licensing agreements and in matters concerning IPR, and helps the promotion and marketing of the product.[71]

---

[68] Collective knowledge is considered to be in the public domain when it has been made accessible to persons other than the indigenous peoples by mass communication media, or when its properties, uses or characteristics have become known extensively outside the confines of the indigenous peoples and communities (Law No. 27811, Article 13). That means, for example, that the publication of the information in a scientific paper does not put it in the public domain.

[69] Instituto Nacional de Defensa de la Competencia y de la Protección de la Propriedad Intelectual.

[70] See NIF homepage http://www.nifindia.org/activity1.htm#sd

[71] See NIF homepage http://www.nifindia.org/activity1.htm#sd

Accordingly, the resulting database (the National Register of Grassroots Innovations and Traditional Knowledge), which is the centrepiece of this system, is meant to serve as a tool to: (i) acknowledge individual and collective creativity; (ii) grant entitlements to grassroots innovators [and holders of TK] to receive a share of returns that may arise from commercial application of their knowledge; (iii) link investments, enterprise and innovations, in particular to help small-scale investors at the local, regional or national level to explore opportunities for investment; and (iv) build linkages between excellence in formal scientific systems and informal knowledge systems.[72]

ANALYSIS

Considering the various functions of documentation and registration of TK, which might be: (i) its preservation for the sake of the custodians and their descendants; (ii) defensive protection against illicit appropriation by IPR; (iii) positive protection, which also protects against unauthorized utilization of the information; and (iv) the offensive function, fostering marketing, the following lessons can be drawn from the two given examples.

The *Peruvian system* takes account of the different degrees of protective needs in creating different types of register, the Public National Register being the only one that is generally accessible. Thus the emphasis lies rather on the defensive function, and on the aspect of the preservation of TK. This last function is highlighted by the option given to the communities to create their own local, community registers. The autonomy of the communities to decide whether they want to make use of the facilities made available by INDECOPI is of great importance. This is underlined by the fact that there is no need for registration in order to get the information protected legally, as the law grants *sui generis* protection for all collective TK that is not in the public domain (Article 42).

On the other hand, the option to foster the wider application of TK and the sharing of the resulting benefits is not implemented very prominently in the Peruvian law. Only the National Public Register, in which public domain knowledge is documented, is openly accessible. However, its use only sets off a benefit-sharing mechanism if the knowledge has been in the public domain[73] no longer than 20 years (Article 13). Whether the national register might be of use for the marketing of the stored information also depends on its accessibility in the sense of the searchability of the database, for instance, by way of indexing or classification systems.

An element in the Peruvian legislation facilitating ABS and trade might be the obligation that the application for registration has to be made through the representative organizations of the respective community (Article 20) and that the law provides for a dispute settlement mechanism against third parties, including between indigenous peoples (Articles 45 and 46).[74] Thus the holders of the registered information and their representatives are identified to a certain degree, and it is also probable that the decision-making procedures regarding collective TK within the community have been

[72] See also the recent initiative by Asia–Pacific nations to create a network for sharing expertise in communicating and marketing indigenous knowledge; SciDevNet, 29 September 2003 (http//www.scidev.net/News/index.cfm?useaction=readNews&itemid=1026&language=1).

[73] See note 69 above.

[74] Here it might be a problem that the information of other communities possessing the same knowledge is only necessary at the moment of engaging in negotiations. There is only a duty to inform and to take due account of their interests and concerns, but no regulation as to common negotiations, benefit sharing or approval of the other communities; Article 6 (see also Tobin and Swiderska (2001) p. 44).

established.[75] However, a legal obligation to inform communities holding identical knowledge is only set off at the moment negotiations are opened (Article 6). This might complicate and draw out the negotiations.

The system of documentation found in the Indian NIF has a different emphasis. The basis is acknowledgement of the creativity and skill of grassroots inventors and custodians of TK, and of the diffusion and sharing of knowledge for fostering communication and learning, to create synergies in the solution of problems, and thus to improve the livelihoods of 'knowledge rich economically poor' (Gupta ?) people. This strong communicative element is also inherent in the second element characterizing the network, which is the empowerment of holders of the information by fostering the development and commercial use of the information stored in the database. The goal is to apply information technologies to 'democratise knowledge, reduce transaction costs of innovators, potential investors, and entrepreneurs' and to build 'bridges between the excellence in formal and informal science' (Gupta, undated (a)).[76]

So, the system of the NIF has a strong communicative character. The main goal is to create networks between holders of TK and inventors at the grassroots level, but also to connect informal and formal science, technology and invention and to disseminate the information to potential buyers or investors as well. The conflict between the dissemination and protection of the documented information is obvious.

This system is specific in two ways. First, the pre-eminent paradigm is the idea of communication and networking, which also applies to the goal of fostering trade of the information stored in the documentation. Secondly, it perceives the mechanism of access and benefit sharing as a pre-eminently intra-national opportunity and responsibility (Gupta, undated (a), p. 2), creating business opportunities especially for small investors and entrepreneurs, thus deviating from the 'classic' picture which involves a North–South and a (multinational) industrial company–local community scenario.

The objective of preservation of TK for the sake of traditional uses within the communities, and the defensive function, seem to be less prominent. The example of the village of Pattuvam in Kerala (see Chapter 3), which took the initiative by itself to document all of its biological resources, but then did not make the information available to outsiders other than local communities, may be an indicator of the necessity and usefulness of this type of register.

According to these premises, the defensive protection against illicit appropriation by IPR and the positive protection against illicit use are more difficult to implement.

For this reason NIF has developed an elaborated procedure of PIC, which consists of a Prior Informed Consent form and an accompanying explanatory note. The goal is to 'balance the twin goals, partly in conflict, of dissemination and promotion of [the] ... innovation/traditional knowledge so that other communities and individuals can benefit from it, vis-à-vis the protection and potential commercialisation of the same through contractual arrangements'. It is explained that this procedure is meant to fill the gap of the lack of intellectual property protection for grassroots innovation and traditional knowledge (National Innovation Foundation, undated (a and b)).

The PIC form serves to obtain written consent and authorization from the knowledge providers to disclose and add value to the information submitted for inclusion in

[75] As to the importance of this element for the ABS process, see Chapter 7 and the case study by Baruffol (2003).

[76] The work of Honey Bee and NIF is motivated by a strong background of development and social change. Gupta argues 'that we need a new paradigm of envisioning social change and development built around overcoming information asymmetries. Knowledge can indeed become a means of power if coalition/networks of relevant actors evolve ... networks which connect information, institutions, incentives with innovations and enterprises' (Gupta, (a) p. 7).

the register. It contains questions as to the degree of publicity; the competence of NIF to negotiate or mediate for the development, marketing and legal protection of the innovation or knowledge; identification of the type of beneficiaries of benefit sharing; and negotiation on the interface between community and individually developed knowledge. The advantages and risks of partial or full disclosure are clearly described in the explanatory note. So, for example, it is clearly stated that even a partial disclosure or a disclosure only in summary form may bear the risk that the idea may be used by third parties, if this is possible on the basis of the disclosed information alone.

Registration as a marketing tool?

From both of the above examples, on the utilization of documentation and registration devices for fostering marketing and trade, certain lessons can be learnt.

The harnessing of documentation and registers for utilization as marketing tools has two different facets: one concerning the internal relationship to the providers of the information, and the other in view of the accessibility of the stored information for prospective interested investors, entrepreneurs or buyers.

If documentation and registration is evaluated according to this perspective, it becomes apparent that the central issues to be resolved are the following:

- It must be possible to clearly allocate the information to its holders: to this end, the legitimate holder(s), which can be an individual, a community or several communities, are identified and the question of their representation in the negotiation process resolved.
- It is necessary that the holders approve of the marketing of their knowledge and agree with the utilization of the registers as a marketing tool.
- It is necessary to find ways and means to resolve the conflict or dichotomy between the defensive and the offensive functions of the databases, i.e. between the prevention of the unauthorized utilization of the information by third parties and the dissemination of the existing information to foster its use by outsiders.

It is submitted that one of the basic prerequisites to transforming TK databases into positive marketing tools on the 'internal' side is to secure the trust of the holders or custodians of the information. This implies various facets.

The autonomy of the holders/custodians of the knowledge in deciding about its use must be secured. In particular the option to keep the knowledge secret and/or to document it only for the sake of preservation within the community must be granted. The measures envisaged must be transparent, the prior informed consent for the further steps in the value-adding chain be secured, and the holders be involved in or informed about further steps as they wish. Control over the stored information must be assured. The possibility of retiring information from the database must be given.

These criteria imply a bottom-up approach. In order to facilitate the active participation of the holders and custodians, capacity-building initiatives on the side of the holders/custodians of the information, as well as for the designers of the registers and databases, are necessary.

In any case, the dichotomy between the twin goals of dissemination and protection of the information must be balanced and the solution made transparent for the custodians/holders of the knowledge. A clear legal basis which defines the ownership of the registries, including the right to define access and the rights to the information stored in the registries, would provide transparency and legal security, and thus contribute to this end. The option of *sui generis* intellectual property rights is to be evaluated under this viewpoint too. The rather complicated PIC procedure in the Indian example amply demonstrates the advantage of such a system for information that is meant to be traded.

Considered under the viewpoint of prospective investors, the accessibility of the information and the ease of the negotiations may play an important role. Accordingly the following elements are deemed to be of importance: (i) the clear identification of the representatives or entities competent to negotiate upon the utilization of the information. If there are several individuals, a community or several communities involved, it may be helpful if the procedures to reach decisions are clear from the outset. (ii) The database is easily accessible; the stored information is classified according to market-relevant criteria.

For the sake of equity, in order to create a level playing field and to become operational in a broader trade context, registries ought to be embedded in devices supporting the holders of the information in negotiating and enforcing the (licence) contracts, and facilitating the marketing of the information. The two model devices presented and discussed below are to be seen in this context.

6.3.2 Collecting societies[77]

General considerations

This section analyses the most efficient methods for implementing intellectual property rules to achieve the policy goals underlying the granting of exclusive rights to traditional knowledge holders relating to plant genetic resources. These goals include equity- and biodiversity-related considerations as well as the public interest objective of maintaining, promoting and disseminating traditional knowledge to the benefit of society at large. So far we have explored some of the complex issues related to the allocation of intellectual property titles to holders of traditional knowledge relating to plant genetic resources. From the perspective of local communities, the process of obtaining exclusive rights requires a corresponding knowledge of the functioning of the intellectual property system. Without such knowledge, concerned individuals and communities will not be able to take full advantage of the system for the purposes of obtaining the rights, managing, commercially exploiting and, as the case may be, enforcing them (Leestli and Pengelly, 2002).

COMMERCIALIZATION

There are two main approaches for the commercial exploitation of intellectual property titles: the right holders can perform this task by themselves, or can assign it to a third party, such as a publisher, who performs this activity on a professional basis. In both cases, commercial exploitation means that all the necessary tasks in bringing the protected work to the user in exchange for remuneration are fulfilled. In addition to entrepreneurial skills, this also requires legal knowledge with respect to contract negotiation and enforcement. The broader the scale of the commercial exploitation of the intellectual achievements, the more international their exploitation is likely to be. This cross-border feature, in turn, requires that the entrepreneurial and legal know-how of rights holders be accordingly sophisticated. One must become familiar with foreign business and legal environments, or work with local partners who have the necessary prerequisites in order to appropriately access new markets. Even if rights holders do not want to become active outside of their neighbouring environment, they may want to keep track of and control over the uses performed by third parties with respect to their intellectual property. They will probably at the very least want to avoid third parties registering their rights, which, in the worst-case scenario, could eventually even impede their own further use of them (in order to reduce such risks so-called 'defensive rights' should be provided for, see point (b) below).

ECONOMIC VALUE OF IPR

Obviously, intellectual property rights can be expensive to manage, especially on a large scale: the cost of negotiating licences,

[77] Author: Christophe Germann.

collecting royalties, litigating infringements and fighting against piracy on a worldwide level are high. Private players have an interest in assuming these costs if they are either able to recoup them, or if they need to protect their exclusive rights in order to bargain them for third parties' rights that are necessary for their own business purposes.

As a matter of fact, copyright and patent protection is not only a competitive advantage granted in exchange for a qualified creative or innovative effort, but also a kind of commodity in corporate practice, i.e. a value that is subject to trade. A product or a service typically requires a multitude of tangible and intellectual components. The producer or service supplier may not own all of such components and therefore must acquire them from third parties. Subject to competition laws, if these third parties are competitors they may refuse to enter into an agreement authorizing the use of their intellectual property. For example, if compulsory licensing is not an available option, production could be jeopardized, unless one can bargain with one's own exclusive rights, which are necessary to the competitors for their products or services. In the latter case, competitors proceed to an exchange of titles commonly called 'cross-licensing': where competitor A is authorized to use competitor B's patented invention for their product and vice versa.[78]

Coming back to the classic value of exclusive rights manifested as a competitive advantage limited in time, we face two main scenarios: either the intellectual property owners recoup their investments into creativity and innovation (research and development) by marketing the protected goods or services themselves, or they sell the rights to third parties. To summarize, in practice exclusive rights appear as tools, to allow their owners: (i) to negotiate a better price for their innovative or creative achievements; (ii) to bargain for third parties' intellectual property titles; (iii) to obtain competitive advantages *vis-à-vis* competitors; and (iv) to exclude free riders. These four benefits or opportunities designed for innovators and creators acting within the framework of an industrialized society arguably have only a limited relevance for holders of traditional knowledge related to plant genetic resources. If TK holders have no intention of commercializing their traditional knowledge on a scale that exceeds the immediate environment, then all four benefits and opportunities are lost to a large extent. In this situation, the only incentive for TK holders to seek protection is to enable them to have a title, which they can enforce as a so-called 'defensive right' against a third party who has usurped their rights (for instance, against a corporation that has been granted a patent for an invention that was based on the holder's traditional knowledge). The TK holder is thus obliged to participate in the intellectual property system in order to simply defend the *status quo*.

'MARKET ACTIVATION'

To achieve results that go beyond a merely defensive position from this system, TK holders should adopt a proactive approach. Exclusive rights can generate the above-listed benefits and opportunities (a) to (d). However, these rights have no value if the creative or innovative achievements are not marketed. Intellectual property rights can provide an incentive for commercialization. However, in the absence of commercialization, such rights have only a potential value. In order to reach the desirable policy goals underlying the granting of intellectual property rights, i.e. in our specific case including an increase in equity between the North and the South, the maintenance of traditional knowledge as a global public good, and indirectly, beneficial effects on bio-diversity, the so-called 'market-activation' of exclusive rights becomes necessary. If holders of traditional knowledge refuse to play this game, exclusive rights will not procure any of the contemplated benefits and opportunities listed above. Here we are

[78] See Section 3.3.6, this volume.

faced with an issue that is cultural in addition to being commercial. In this context we can refer to the Suva Statement of April 1995 and the Coica Statement of September 1994, which consider the intellectual property system as a colonialist instrument.[79] On the other hand, appropriate institutions are required to implement a level playing field when TK holders agree to act according to the rules of the intellectual property system, as adapted to their specific needs.[80] Such a fair environment for the exchange of traditional knowledge in consideration of other values such as money, technological transfer or innovative goods and services, requires an institutional interface between TK donors and receivers.

Collecting societies as a possible model

INDIVIDUAL AND COLLECTIVE RIGHTS MANAGEMENT

A distinction must be made between the individual and collective management of rights. In the former case, the right holders or their assignees or licensees grant authorizations to use the intellectual property on a case-by-case basis. They conclude the agreement and collect the royalties personally. In the latter case, a separate entity such as a collecting society grants these authorizations on behalf of the right holders on a general basis, most often in the context of large-scale commercial exploitation of the work. As opposed to the situation that occurs in the field of technological innovation, we currently find the institution of collecting societies in the area of creative achievements that enjoy copyright protection (works of literature and music, films, software, etc.). This institution provides indispensable services to right holders as well as users, especially where mass reproduction and representation of the works are concerned.

THE FUNCTIONS OF A COLLECTING SOCIETY

We suggest that inspiration be taken from collecting societies for the purpose of designing this institutional interface. The functions of collecting societies include:

- The negotiation of tariffs for the use of specific intellectual property rights.
- The gathering of royalties from licensees and the distribution of royalties to licensors.
- The enforcement of exclusive rights and the building of capacity amongst owners and users of rights.

In the majority of jurisdictions, collecting societies act as independent private bodies, which are subject to public scrutiny through various administrative and judicial forms of control, including competition law mechanisms.[81] New technologies may contribute towards lowering the cost of rights management and thus reducing the bureaucracy that is generally inherent in larger collecting societies.

THE FUNCTIONS OF PUBLISHERS AND COLLECTING SOCIETIES

As mentioned, the commercial exploitation of intellectual property in the field of authors' rights or copyright may be assumed either by the original rights holder, or by a publisher who assumes the exploitation of the rights based on a corresponding assignment or licence agreement. In the former case, the original creator grants the authorization to use the protected work directly in exchange for money or other benefits. In the latter case, the original author assigns this task to a publisher. The publisher typically invests in the reproduction and, as the case may be, in the distribution and marketing costs. The publisher will pay the author either a flat fee

[79] Final statement from the UNDP Consultation on Indigenous Peoples' Knowledge and Intellectual Property Rights of April 1995 (Suva Statement) and Coica Statement of September 1994; these statements are quoted in Nordmann (2001).

[80] The Mataatua Declaration on Cultural and Intellectual Property Rights of Indigenous Peoples of 1993, for example, adopts a favourable approach to IPRs; see Nordmann (2001).

[81] See, for example, the Swiss copyright act in the CLEA database http://clea.wipo.int/clea/lpext.dll?f=templates&fn=main-h.htm&2.0

and/or, in case of a profit participation arrangement, a share of the revenues generated by the sale of the protected work. Publishers usually deal either directly with wholesalers or retailers, or via distributors. In addition, for certain types of exploitation, collecting societies may act on behalf of original rights owners and their successors or assignees in a legally binding way.

In the area of authors' rights or copyright under continental law, collecting societies have a long-standing tradition. In France, the first collecting society, Société des auteurs et compositeurs dramatiques (SACD), was founded in 1776. In fact, the management of exclusive rights for the purposes of representing a copyrighted work requires a collective approach in order to be economically rational. Further, in cases of representation, a production such as a musical or theatre play may be shown simultaneously in several places. On the other hand, the individual management of exclusive rights pertaining to the reproduction did not cause difficulties until recently and could therefore be assumed on an individual basis. For this reason, the collective management of rights appertaining to certain forms of mass exploitation became compulsory by law, for example, music dissemination through television and radio broadcasting and a levy on blank cassettes.[85] Nowadays, with the advent of new digital reproduction and dissemination techniques, the management of these rights is substantially even more complex. In many areas of commercial exploitation it is therefore neither in the interests of the right holders, nor of the rights users, to manage certain rights of reproduction on an individual basis.

Modern collecting societies grant the authorization to use copyrighted works for purposes that have been specifically determined, i.e. to gather royalties from users and distribute them to the entitled rights holders, to provide legal advice and to defend the common interests of their members in court and during the political and legislative decision-making process. In addition to these classical functions of collecting societies, today they also contribute to capacity-building amongst rights owners and users as well as assuming certain functions that promote social and cultural concerns. In this way, these collecting societies act as an interface between authors and users, and, in a broader sense, society at large. In many jurisdictions, their management is subject to supervision through the public administration or the courts.

The collecting societies' internal distribution modalities of the revenues among the right holders are in most jurisdictions subject to state control. These measures ensure transparent and fair procedures. This feature arguably meets the specific needs of TK holders (communities and individuals) who often have neither the means nor the will to enforce their rights among each other. We recommend that policy makers further explore whether adapted forms of collective society could work as an interface between local holders of traditional knowledge related to plant genetic resources and third party users. These users may be individuals or corporations having a commercial or academic interest in traditional knowledge. In this context, one may envisage different roles for collecting societies to be set up, specifically in order to facilitate the interaction between holders of traditional knowledge and users. Such institutions must be designed to fulfil several goals. First, these collecting societies could be instrumental in raising the awareness of holders of traditional knowledge on the intellectual property system and the benefits that this system may procure them. Secondly, these collecting societies could enter into agreements with TK holders in order to manage their rights within a clearly defined framework. This management would provide authorization for interested third parties who wish to use the rights, collecting royalties for this and distributing them to the right holders. Thirdly, collecting societies could enforce the rights of TK holders against infringing parties. Fourthly, collecting societies may act as know-how centres, advising their

[82] See Swiss copyright act, note 81 above.

members with respect to individual management of rights, contract negotiations or litigation. Eventually, they can contribute to the elaboration and articulation of their members' common interests, and promote these interests as policies in the political and legislative context. In this position they would act as liaison between TK holders and society at large.[83]

Domaine Public Payant

In addition to the forms of intellectual property protection for traditional knowledge related to plant genetic resources that we have outlined so far in Chapter 3,[84] policy- and law-makers may also envisage the implementation of the concept of *Domain Public Payant*. This mechanism subjects certain transactions to a levy even if the intellectual values at stake are already in the so-called public domain. This concept applies to systems such as the levy on blank cassettes, known in many jurisdictions in the field of copyright. The blank cassette levy, for example, overcomes the impossibility of exactly assessing mass private copying of copyrighted content.[85] This system imposes a fee on the sale of blank cassettes to consumers. The revenues coming from this levy are distributed through collecting societies to the right holders of protected content based on a distribution key that takes into account the popularity of the content measured in other contexts.[86] This system is arguably much less cumbersome to implement than the patent system, and relies on agreed-upon average figures for the distribution of royalties. It nevertheless requires the definition of traditional knowledge that falls under the scope of protection (for instance, all neem farmers) and a consensus on how to distribute the revenues among the right holders, based upon the value of their respective contributions.

Further considerations

Appropriate collective rights management institutions should be conceived and set up in order to implement intellectual property laws and policies that are adapted to protect traditional knowledge related to plant genetic resources. These organizations could be modelled upon collecting societies in order to act as interfaces between various legal cultures to promote the understanding of different systems and collaboration during the pursuit of public policy goals. In concrete terms, they should advise their members, represent their interests, manage rights (collection and distribution of royalties), and coordinate and monitor international prior art search mechanisms ('defensive rights' advocacy).

A system of collecting societies that is similar to the one in place in the area of copyright may contribute to better implementation of the various kinds of intellectual property laws and policies contemplated in the context of the management of rights on traditional knowledge related to plant genetic resources. Collecting societies could negotiate individual licence agreements as well as tariffs for mass use. In addition to the classical tasks assumed by collecting societies, this infrastructure could also work as a registration office for the rights at stake. Users would have the possibility of seeking rights clearance from a single entity that has both the capacity and the legitimacy to give genuinely prior informed consent (PIC) on behalf of the rights holders. This type of function implies that the collecting society should be liable *vis-à-vis* the rights holders

[83] See, for example, the various services provided by the Swiss collecting society for audiovisual works Suissimage: http://www.suissimage.ch

[84] Compare also World Intellectual Property Organization (WIPO), 2001.

[85] Another form of levy concerns the sale of photocopying machines, and the number of photocopies made by owners of such machines. The revenues generated by this levy are typically distributed amongst writers.

[86] For example, the composer of more popular songs, as expressed in market shares of compact discs, will receive a bigger share from the blank cassette levy since one may imply that his songs will be more often privately copied by the users.

with respect to the PIC that they grant to users. In practical terms, this means that rights holders have a legal entity that they can sue in case of non-compliance with PIC-modalities, based on predictable rules of law. It arguably would represent a substantial improvement to a claimant's situation in the case of litigation, so long as the defendant is located in the same jurisdiction. The defendant, in turn, would have recourse against the users. In this situation, the collecting society is likely to have a better financial standing than the original rights owners when fighting for relief against infringement by users who may be powerful corporations located in a foreign jurisdiction. Finally, the more efficient enforcement of rights over traditional knowledge relating to plant genetic resources will, in turn, improve compliance with such rights.

Classical collecting societies are designed primarily to facilitate the management of mass exploitation of rights, i.e. where the mass of users makes it no longer rational to commercially exploit the rights on an individual basis (for instance, broadcasting or blank cassettes). This type of commercialization may not always be relevant for most TK holders. However, the other features of collecting societies, the publicly controlled distribution of monies, could be of great benefit to TK holders for obvious transparency and fairness reasons. As opposed to arrangements inspired by publishing agreements, the collective gathering and distribution of royalties is likely to fulfil both an efficient management of rights and an equitable distribution of the revenues among TK holders.

6.3.3 Clearing house mechanisms[87]

As concluded above, the mere documentation and registration of TK and PGRFA does not necessarily lead to an increasing demand from the purchaser side,[88] and the presently existing national regimes on Access and Benefit Sharing seem to hinder rather than promote trade relationships (see Chapter 6).

It was concluded that an essential element to fostering trade in TK and related biological resources is the minimization of transaction costs in time, manpower and investment. Therefore, transparent and swift procedures, based on clear rules in order to promote legal security and minimize risks, are asked for. In addition, the fact has to be taken account of that trade in biological resources and associated TK frequently takes place at the interface of different cultures, including of course different legal cultures. Negotiations, in order to be successful, need to be built on a basis of trust and mutual understanding. This corresponds to the emerging insights into the value of the 'social capital', in this case the value of human networks that facilitate transactions (Krattiger, 2004). According to Krattiger, 'in order to get something used by as many people or institutions as possible, one must sell or license it. This requires transactions, and these happen between people who know, trust, and value each other' (Krattiger, 2004, p. 9).[89]

In turn, the discussion of the documentation and registration systems has shown the importance of a proactive approach to furthering trade. The mere documentation of the information, or its protection by an IPR, does at most serve a defensive end, but is commercially quite useless if nobody is interested in buying. Accordingly, it was concluded that, if the idea of market-based incentives for sustainable use of biodiversity was to bear fruit, it is necessary to think of additional strategies and devices to support communication and exchange.

Various institutions and instruments have been proposed under different head-

[87] Authors: Susette Biber-Klemm and Jonathan Curci.

[88] For the experience of the countries of the Andean Pact, see Ruiz (2003, pp. 2, 5–7).

[89] One of the above-cited examples, the Indian NIF (National Innovation Foundation (www.nifindia.org, accessed 5 February 2006)), has its emphasis on this communication-oriented networking between the stakeholders involved in the innovation processes regarding/integrating traditional knowledge and genetic resources.

ings to this end. Krattiger and Lesser (1995) put forward the function of a 'facilitator' to strengthen the equitable and sustainable use of biodiversity. Drahos (2000) recommends the creation of a 'global bio-collecting society'; and Krattiger (2004), in his recent discussion on means to facilitate biotechnology transfer, analyses a broad range of instruments, from different types of clearing houses and technology transfer agencies to brokers and other types of facilitators.

In spite of the varying designs, the instruments are meant to support the same underlying goal, which is the promotion of the networking and matchmaking processes between prospective sellers and buyers of goods and the fostering of equitable transactions.

This will be discussed in more detail, based on the model of the 'clearing house mechanism' (CHM) as a starting point. First an overview over the different types of CHMs in the field of biodiversity and TK will be given. On this basis, secondly, two proposed models will be described and the additional elements that would have to be integrated into a CHM specifically aimed at fostering trade in TK and related biological resources analysed and evaluated.

What is a clearing house mechanism?

The term 'clearing house' is originally linked to bank jargon referring to financial establishments where cheques and bills are exchanged among member banks so that only the net balances need to be settled in cash. This term has been extended to identify any agency that brings together seekers and providers of goods, services or information, thus matching demand with supply. In the wake of the electronic revolution, the advances made in the development of the Internet and in the fields of information management technology and computer networking, the virtual communication system is made use of to create online clearing house mechanisms (CHMs), creating worldwide information networks. The concept promotes the advertising, discovery, access, dissemination and use of information and data held by numerous organizations, using the decentralized capabilities of the Internet.

A CHM typically consists of different 'nodes', i.e. participating sites, which usually are coordinated through a central node. The function of the central node is to facilitate and coordinate the decentralized nodes, by for example creating common protocols, linking the different nodes, designing and providing structured queries and searches to member sites and translation services. The responsibility for the decentralized nodes remains with the initial providers, and the central node typically is to remain independent, its operators having no interest in controlling or selecting the information.

CHMs are also being created within the framework of Multinational Environmental Agreements[90] to foster technology transfer or as a tool for capacity-building, and within WIPO to facilitate research in IPR-related matters.

CHMs in the realm of TK management: the CBD CHM and the WIPO Platform

THE CBD CHM

The CBD's clearing house mechanism was established to promote and facilitate technical and scientific cooperation within the scope of the Convention,[91] namely to further the Convention's three objectives of conservation, sustainable use of biodiversity and fair and equitable sharing of the benefits resulting from its use. This will foster the development of a global mechanism for exchanging and integrating information on biodiversity and of the

[90] See, for instance, the CHM of the CBD (http://www.biodiv.org/chm/default.aspx), and of the Biosafety Protocol (http://bch.biodiv.org/Pilot/Home.aspx) and the CHM of the GPA (Global Programme of Action for the Protection of the Marine Environment from Land-based Activities (http://www.gpa.unep.org/links/default.htm). See also the overview in Krattiger (2004, pp. 3, 18–31).

[91] CBD Article 18(3) and Conference of the Parties to the Convention on Biological Diversity, Decision I/3.

necessary human and technological network.[92]

So the CBD CHM has a twofold objective: first to provide improved access to information, and secondly to foster technology transfer, i.e. to promote scientific and technical cooperation.[93] According to the Strategic Plan,[94] this involves use of the CHM for identifying, developing and promoting opportunities for collaboration.[95] This is to be reached by, for example, providing a collaboration promotion mechanism that institutions, experts and service and technology providers can use to introduce themselves, and to identify areas of potential collaboration which they are interested in pursuing.

Thus the CBD CHM, which originally was based on Article 18.3, promoting mainly technical and scientific cooperation, also includes the objectives of Article 17 on exchange of information.[96] According to Article 17, this exchange is to encompass the results of technical, scientific and socio-economic research, information on training and surveying programmes and specialized knowledge, including indigenous and traditional knowledge as such and in combination with biotechnologies. This exchange is limited to information from publicly available resources (Article 17.1).

Decision VI/18 has further spelled out the aspect of exchange of TK. It asks for the development of specific communication networks, in the sense of a thematic focal point, for the use of indigenous and local communities. These networks are explicitly not to be used as repositories or for the public exchange of traditional knowledge (Dec. VI/18; UNEP/CBD/AHTEG/TK-CHM/1/3, Annex I). According to the *ad hoc* technical expert group on traditional knowledge and the clearing-house mechanism, its objective would be to support initiatives of indigenous and local communities in the use of communication technologies and networks to enable information sharing, mainly between the indigenous groups themselves, and to establish links between the many existing networks.[97]

Hence, the CBD CHM has a broad approach, going beyond the 'classic' concept of the CHM. It aims at supporting the implementation of the entire range of goals of the CBD and at fostering communication processes within the political debates. It not only promotes information exchange but offers additional services: it initiates capacity-building and supports the implementation of the CHM at the regional and national level. In its objective to connect people in order to promote collaboration, it comes close to the model of the NIF register described above.

THE WIPO ONLINE PORTAL

The WIPO Intergovernmental Committee on Intellectual Property and Genetic Resources, Traditional Knowledge and Folklore (IGC)[98] at its third session established online inventories and databases, containing TK databases and compilations of other information relevant for the inter-

[92] Meeting of the Informal Advisory Committee, Clearing-House Mechanism of the Convention on Biological Diversity, Canada 11 November 2001, UNEP/CBD/CHM, available at http://www.biodiv.org/doc/meetings/chm/chmiac-2001/official/chmiac-2001–11–11–ann-agenda-en.pdf Compare also the CHM strategic plan (UNEP/CBD/COP/5/INF/3 and doc. UNEP/COP/7/Inf/12, p. 1).

[93] This second aspect has been particularly emphasized by COP VII (Decision VII/23).

[94] A.a.o. note 95.

[95] Decision-making tools and processes, training and capacity-building, research, funding, access to and transfer of technology, repatriation of information.

[96] For the chronology of the CHM establishment and development, see UNEP-WCMC, CHM Review Project, Final Report, UNEP/CBD/COP/7/INF/12.

[97] See the compilation in doc. UNEP/CBD/AHTEG/TK-CHM/1/3.

[98] The WIPO Intergovernmental Committee on Intellectual Property and Genetic Resources, Traditional Knowledge and Folklore (the 'IGC') was established by the WIPO General Assembly in October 2000 (doc. WO/GA/26/6) as an international forum for debate and dialogue concerning the interplay between intellectual property (IP), and traditional knowledge, genetic resources and traditional cultural expressions (folklore).

face between TK and IPRs. This portal encompasses the following databases:

- A non-exhaustive inventory of TK-related periodicals.
- A non-exhaustive inventory of TK-related databases.
- Online Databases and Registries of Traditional Knowledge and Genetic Resources which include hyperlinks to databases on TK and PGR compiled by China and India, to the TK database of the World Bank and the CGIAR's (Consultative Group on International Agricultural Research) genetic resources database.

The portal is a pilot project, meant to facilitate the study of intellectual property issues resulting from the establishment, management and use of such databases. In particular, the objective of the portal is to provide a trial product through which users can test in practice the potential of traditional knowledge databases for improving the availability of traditional knowledge as prior art, in particular the efficiency of online searches for prior art investigations by patent examiners (WIPO/GRTKF/IC/3/6). So the rationale behind the project is a purely defensive one, the portal encompassing only databases and references to databases which document TK in the public domain. This reflects the problem of the disclosure of publicly inaccessible TK by registration, as long as no positive protection exists, a topic which was prominent throughout the debates on the online databases in the IGC (WIPO/GRTKF/IC/3/17).

A capacity-building tool on the IGC website is the online, searchable database of biodiversity-related Access and Benefit-Sharing Agreements – a selection of model agreements and MTAs and a sample of actual ABS and licensing agreements – all with a particular emphasis on the intellectual property aspects of such agreements.

Thus the WIPO site corresponds more closely to the classic pattern of a CHM.

CHMs to facilitate trade in TK: examples

The goals of a TK CHM, which are to foster technology transfer in a broad sense, and the marketing and trade of PGR and TK, are broader than the above-described CHMs, even if some elements are rather similar. The specifics and objectives of a CHM for trade in traditional knowledge and associated PGR originate in the complexities and intricacies of the bioprospecting process. Bioprospecting is a long-term venture, taking place between different systems of innovation, which might also be geographically wide apart and are based on different cultures, including different legal, business and negotiation cultures, with a lack of know-how and opportunity to procure the relevant information for both providers and purchasers.

Krattiger (1996, p. 9) understands the transfer of TK and genetic resources as 'reverse' technology transfer (technology transfer also defined as geographic movement of productive capacity), but with significant differences: the transfer is predominantly South–North and secondarily South–South; the materials are natural products and the related know-how creates technical and institutional complexities (see also Krattiger, 2004, p. 30) as they follow patterns differing from the access to resources such as, for example, minerals.[99]

According to Drahos (2000), the relationship between provider and purchaser of information is characterized by a great amount of uncertainty and imperfect information on both sides. He mentions the uncertainty of a specific compound achieving marketing maturity, or the difficulty of indigenous groups in knowing the value of their knowledge and in judging whether the use made of it will be consistent with their cultures. The information necessary to reduce uncertainty and risk includes a broad range of elements, such as what information is traded, who is the owner of the information, information on the IPR-systems and on the means of tracking patent procedures and of controlling infringements.

---

[99] One of the main differences being the physical impossibility of controlling access (see Ruiz, 2003).

Both start out from the insight that the specifics of the trade relationships in matters of resources and traditional knowledge held by indigenous and local communities require specific support mechanisms. Drahos points out the 'intriguing and potentially unstable combination: some of the world's most globalised and hypermodern companies seeking deals with some of the world's most local and traditional people'. Accordingly they identify as basic goals of such a mechanism the promotion of equitable relationships between providers and buyers. In this context, Krattiger and Lesser (1995) point out the necessity of valuing the contribution of the providing countries,[100] i.e. to improve the market operations so as to reflect the legitimate contributions of genetic resources and value-added activities by the source countries and to assist in the determination of the value of TK for the groups that choose to share it (p. 212).

A second important point is to minimize risks and transfer costs, inherent for both parties in these transactions, given by the specific character of the technology transfer (Krattiger and Lesser, 1995, p. 212)[101] and the relative novelty characterizing the legal side of these transactions. Drahos (2000) perceives the risks as resulting from imperfect information on the value of the information on the side of the providers, and the uncertainty of the development of a marketable product on the side of the purchasers (p. 3).

Accordingly, to make supportive mechanisms operational for both TK providers and buyers, a series of specific prerequisites and objectives must be fulfilled.

- The basic objective would be to create a level playing field between traditional innovators and/or holders of biological resources and TK, and the bioprospecting companies, entrepreneurs or investors. To this end, capacity-building for both sides is needed – to provide negotiating skills and to bridge the cultural gap between the parties. This function might also be taken over by an intermediary, an 'honest broker' (see also Krattiger, 2004, p. 23), who preferably is familiar with both cultures.[102]
- There is a need to base the social networks on a foundation of mutual trust, which can be fostered by transparency of the information and negotiation processes, easy access to the essential information and at the same time respect for its possible confidentiality. Generally accepted codes of conduct or of good business practices might be means to this end.
- Another important objective would be the reduction of transaction costs and risks by creating a certainty regarding the partners, the tradable goods and the general conditions, e.g. by creating model contracts.
- A necessary function would consist of the often complex monitoring of the contract implementation and the settlement of disputes, in particular in cases where intellectual property rights are involved.

As mentioned above, Krattiger and Lesser (1995), Krattiger (1996) and Drahos (2000) each propose a model to overcome the problems and tasks outlined above: Drahos (2000) submits the idea of a global biocollecting society,[103] whereas Krattiger and Lesser propose to create 'facilitator'

[100] Drahos perceives the question of the valuation problem as a problem of imperfect information, adding to the risks of the transaction (p. 3).

[101] The transfer of genetic material (and TK?) perceived as productive capacity in a raw form. The differences in this kind of technology transfer consisting of its predominant South–North transfer, and the technical and institutional capacities created by the fact that the technology transferred is a natural product (or closely related to it).

[102] See, for example, the case of the UZACHI-Novartis negotiation process (Baruffol, 2003), published on CBD http://www.biodiv.org/programmes/socio-eco/benefit/cs.aspx

[103] The model of the global biocollecting society as suggested by Drahos in its proposed functions – to overcome the information problems inherent in/ingrained in the transcultural exchange of information – rather corresponds to the model of the CHM and accordingly is subsumed under this part of the chapter.

(1995) or 'honest broker' services (Krattiger, 2004).

Drahos' Biocollecting Society

Drahos developed his model of a global biocollecting society in order to enhance the transfer of information, to promote the goals of the CBD and to stimulate a process of private ordering among companies and indigenous groups.

Drahos suggests the creation of a global biocollecting society rather than a great number of national collecting societies. His main arguments are first that it would be more easy to monitor an only and global society; therefore there would be more transparency; and secondly that an international organization may serve the interests of the often marginalized holders of TK and related PGR better than state organizations.

He proposes that a biocollecting society could be best established as a private organization outside the context of any inter-state treaty negotiation. He argues that the 'politicised waters of treaty negotiation would make it difficult for any initiative to reach the shoreline' (Drahos, 2000, p. 248). Membership of the society would be open for both companies and indigenous groups, and entirely optional. He proposes that the membership could be considered as an implicit acceptance of the principle of national indigenous property rights. Thus, a sort of respect for the use of indigenous knowledge could be secured by this simple acceptance, even though the state to which the indigenous group belonged had failed to provide legislative protection.

Drahos proposes that a collecting society offer the following functions and services:

- To act as a repository of and to assume custody over community registers of indigenous knowledge which is either in the public domain or is meant to be traded by its holders, the latter under strict obligations of confidentiality.
- To provide assistance with any contractual negotiations, possibly by maintaining a register of independent legal experts.
- To set up a monitoring service for the use of TK, which might also involve a regular check of patent applications around the world.
- To create a dispute resolution function, exercised by recommendations of a committee constituted by people of 'impeccable independence'.
- To function as a standard-setting body by developing an authoritative code of conduct, containing for instance standards including a pricing scheme, terms of contract and a royalty disbursement accounting system agreed upon by representatives of industry, indigenous groups and states.

To fulfil these functions, Drahos deems it necessary to have supportive, external funding. He proposes the World Bank as sponsor of the system.

Krattiger and Lessers' Facilitator[104]

According to Krattiger and Lesser, the purpose of a facilitator is to:

> enable, on a voluntary basis, equitable and sustainable deals to be made between sources and users of genetic resources; to promote cooperation in the transfer of technological, human and information resources and skills, by providing information and training that are directed at making the market more efficiently and at rendering the negotiators more nearly equal in skills. (p. 213)

Krattiger insists that the facilitator is to be an independent entity, with no vested interests in the brokered arrangements and operating at the interface of genetic technology providers and users, development agencies and information.

He proposes the following functions:

- To provide 'honest broker' services, encompassing assistance in the germplasm marketing, including the

[104] The following is based on Krattiger and Lesser (1995, pp. 211–212).

identification and execution of relative arrangements, supporting the parties in recognizing the implications of the sales conditions and raising the necessary funds to underwrite the initial brokered agreements.
- To identify agreements which will provide for the necessary technical training of national marketing specialists, scientists and policy makers, providing training in technology marketing and contract negotiation; encourage sharing of knowledge and experience by the users of the resource, including training and access to information.
- To assisting governments, on their request, in the identification and implementation of legislation suited to the country's role as a technology seller.

Krattiger proposes start-up funding and organization by a non-profit entity, while the access and payment issues are worked out.

Analysis

The models of CHMs proposed for the transfer of TK-related technology start out from a broad concept. They advocate a proactive approach, encompassing various elements to secure a level playing field and to provide transparent relationships for both providers and purchasers.

The two concepts concentrate on slightly different aspects of the marketing of TK and biological resources: Drahos' focus is on the bioprospecting process in its entirety, thus including also instruments for monitoring, control and dispute settlement. Krattiger and Lesser include the aspect of technology transfer. Their model encompasses the (non-monetary) elements of the benefit sharing as criteria for the selection of prospective partners and consultancies to support governments in their role as technology sellers.

The link to the discussion of registration and documentation mechanisms is obvious; SRISTI and the described NIF database could be named as practical examples of a CHM.

To some extent the CHM corresponds to the model of the collecting society as proposed above (p. 266): both support a proactive approach to the marketing of TK and related biological resources and aim at creating equitable relationships between sellers and buyers. Accordingly, additional services and facilities are proposed, which are not included in the respective models *stricto sensu*, such as capacity-building, brokering and representation of right holders in the case of infringements.

However, there are also important differences: the collecting society is a partisan institution, representing the interests of the providers of the information. It is created bottom-up and remains under the control of its members. The CHM in turn has the function of an independent, neutral intermediary between the involved parties, facilitating communication and negotiation in taking the role of 'honest broker'.

THE CHM: ELEMENTS AND DISCUSSION

The described models of CHM are a combination of various existing technology access and transfer systems,[105] such as:

- Royalty collection agencies.
- Pure information clearing houses.
- Technology clearing houses.
- Honest brokers/facilitators.

Krattiger (2004) has analysed the pros and cons of the various systems. His conclusions might be important for assessing the cost–benefit ratio of the propositions.

He considers royalty collection agencies as advantageous as they are easy to set up, working at low cost with minimal overheads. They can be created by the users themselves and in principle need little support by governmental agencies. However, this applies only if many players participate.

Information clearing houses he considers relatively easy to set up. They can provide easy global access to the stored

[105] From Krattiger (2004).

information. However, they need considerable input as they are useful only if comprehensive, with user-friendly structures, e.g. by offering analytical tools. That means that they are to be constantly maintained and updated. Krattiger judges that they are useful for exchanging information on specific issues, but not by themselves useful for technology transfer. Accordingly he opts for broad models for clearing houses, identifying sellers and buyers and assisting in negotiations.

Honest brokers and other forms of facilitators as described above typically fulfil a range of integrated functions. They are appropriate for charting new territory and bringing (public and private) actors closer together. They are effective in setting new models of collaboration which are specific to technologies, industry types and needs. However it must be noted that they demand complex institutional arrangements and significant funding. Moreover, according to Krattiger, they are limited to serving non-profit, directly humanitarian activities.

INSTITUTIONALIZATION AND STRUCTURE

With a view to the institutionalization and structure of a CHM for promoting trade in TK and biological resources, two essential questions need to be answered: the first relating its basic structure. The question is whether the CHM is created and controlled bottom-up in a decentralized manner, or rather initiated top-down, e.g. by an internationally acting society from a global level. The second question is whether a CHM ought to be privately organized (Drahos' point), integrating state members at most in their function as providers of biological resources, or whether it would be advantageous to base such an institution on some sort of international state consensus.

The bottom-up structure we find in the various local and national TK registers and in the collecting-society model. The advantage is that such locally managed institutions are closer to their users, remaining under their control. This might be useful in the case of a collecting society. However, for marketing purposes, i.e. for connecting sellers and buyers internationally, the creation of one address might be advantageous. It can be imagined that the nodes of the structure involve many national CHM focal points, e.g. the local or national registers of tradable TK, which remain under the control of its providers, but are linked by common protocols, systems and search criteria. Such a CHM might also integrate the international registers of protected TK (Chapter 5), which could be connected electronically to form a global one. In this case the main remaining question is whether it is (yet) possible to provide the additional functions and services (brokering, capacity-building) electronically, by a global institution, a solution which would be advantageous from the financial point of view.

With regard to the participation of state actors, it can be argued that sidestepping international negotiations would be detrimental to the activity of the CHM itself, especially if new types of IPRs are at stake for the protection of TK, and if the CHM aspires to a high level of credibility. The states' legal recognition of these eventual new IPRs or of the legitimacy of the entitlement of this traditional know-how might lead to a higher degree of success in terms of ample participation and compliance on the part of the private sector.

On the other hand it can be maintained (e.g. Drahos, 2000) that to reach a consensus in this matter, in particular if the question of rights to traditional knowledge is involved, is a long-term venture. None of the negotiations in the realm of the CBD, WIPO and TRIPS up to now has reached a conclusive level. So it could be advantageous to set up one or several pilot projects, inspired by the mechanisms and experiences of the CHMs existing in the field.[106]

Whether such CHMs, in the long run, could be managed as self-supporting entities can only be judged by experience. For the start-up financing and support during a

[106] As for example proposed by Krattiger and Lesser (1995, pp. 213–214).

pilot phase, external sponsoring would be necessary.

Conclusions

The CHM mechanisms might be applicable and prove fruitful and advantageous for both types of access: access governed by the contractual ABS regime only (Chapter 7) or by seeking a licence for TK protected by a *sui generis* IPR (above, Chapter 5).

However, as has been pointed out above, the communicative aspect of such a CHM, in particular the exchange of information related to TK which is not in the public domain, is encumbered if such information can not be legally protected.[107]

A CHM may provide freedom to operate with substantial savings for traditional innovators and bioprospecting companies. A CHM on the Internet may inspire a host of entrepreneurs to communicate speedily and eventually to enter into contractual relations with the rightholders. By entering this marketplace entrepreneurs would have instant access to a cornucopia of content. Clearly, one-stop electronic shopping of this sort would make it much easier to create and market traditional grassroots innovations and technologies.

In sum, a CHM could enhance: (i) contacts between TK stakeholders and bioprospecting companies interested in the exploitation of the registered TK; and (ii) international contracts on the sharing of trans-boundary commercial benefits in a real, informed manner.

Dealing directly with TK stakeholders through a global CHM rather than with many national bureaucracies would lower industry's transaction costs. One or more electronic CHMs would consolidate all these transactions by bringing together indigenous groups and members of the life sciences and agricultural industry. This will lower the search of costs for both and the additional services granted by a CHM can reduce the risks which are otherwise implied in bioprospecting contracts for both sellers and buyers.

Since normal channels of communication do not enhance sources of creativity and immediate exchange of communication and contacts, an on-line CHM may reduce the very high transaction costs for innovators around the world in learning from each other and thereby improve the livelihood options.[108]

This series of proposals on the nature, structure and functions of a CHM should build the outline of what could be an international electronic infrastructure for the management of rights, whether in the form of know-how on the use of plants, existing IPRs related to TK or even a new generation of TIP-Rights. Although, as an end result, we rather opt for a CHM structure with global outreach, possibly embedded in the CBD and/or WIPO structure, the aforementioned essential services could be also adapted to various private law-based CHMs and a global CHM should be the result of many national efforts in an organic manner built upon the synergies of relevant international organizations.

---

[107] See above (on registration) and the same in Drahos (2000, p. 2).

[108] In the field of agricultural biotechnology, no single online exchange provides access to all of the listings of relevant patents that – on the contrary – turn up scattered or empty results on even the most developed online exchanges. Therefore the costs of laboratory researchers or technology managers seeking access to a technology become significant. Graff and Zilberman describe the types of search costs: 'Those in search of a specific kind of technology have to go site to site, registering numerous times for web site memberships, remembering passwords, and in some cases paying significant fees for membership or pay-per-view for patent listings in which they are not yet sure they are interested. Two things would help to alleviate this problem, at least for a given industry such as agriculture: 1) a drastic consolidation of the online patent exchanges into a unified marketplace; or 2) a universal cross listing of current offerings across all of the online patent exchanges' (Graff and Zilberman, 2001, p. 8).

Bibliography

Baruffol, U. (2003) Contractual regulation of access to information on biodiversity for scientific and commercial use – the Novartis-UZACHI Biolead Project. *Forest Science Contributions* 30; Swiss Federal Institute of Technology, Zurich. http://www.biodiv.org/programmes/socio-eco/benefit/cs.aspx

Bhandarai, S. (2002) Need of legislation for plant registration and trade promotion: interface with TRIPS. In Ministry of Forests and Soil Conservation, Department of Plant Resources, Nepal Flora Implementation Project: *Proceedings of the Orientation Training Workshop on Flora Writing*, 11–19 June 2002.

Blakeney, M. (2001a) Geographical indications and TRIPS. Occasional Paper No. 8. Quaker United Nations Office, Geneva.

Blakeney, M. (2001b) Proposals for the international regulation of geographical indications. *Journal of World Intellectual Property* 4, 629.

Commission on Intellectual Property Rights (2002) *Integrating Intellectual Property Rights and Development Policy*. CIPR, London.

Dalton, R. (2002) Tribes query motives of knowledge databases. *Nature* 419, 866.

de Carvalho, N.P. (2000) Requiring disclosure of the origin of genetic resources and prior informed consent in patent applications without infringing the TRIPS Agreement: the problem and the solution. *Washington University Journal of Law and Policy* 2, 371.

de Carvalho, N.P. (2005) From the shaman's hut to the patent office – in search of effective protection for traditional knowledge. *Washington University Journal of Law & Policy* 17, pp. 111–186.

Downes, D.R. (2000) How intellectual property could be a tool to protect traditional knowledge. *Columbia Journal of Environmental Law* 25, 253.

Downes, D. and Laird, S.A. (1999) *Innovative Mechanisms for Sharing Benefits of Biodiversity and Related Knowledge: Case Studies on Geographical Indications and Trademarks*. Biotrade Initiative. UNCTAD, Geneva.

Drahos, P. (2000) Indigenous knowledge, intellectual property and biopiracy: is a global bio-collecting society the answer? *European Intellectual Property Review* 22, 248; 251–261.

Dutfield, G. (2000) *Intellectual Property Rights, Trade and Biodiversity: Seeds and Plant Varieties*. Earthscan, London.

Dutfield, G. (2001) TRIPS-related aspects of traditional knowledge. *Case Western Reserve Journal of International Law* 33, 233.

Eugui, D.V. (w.y.) El regimen legal y la experiencia Venezolana en materia de acceso a los rescursos geneticos, los conocimientos tradicionales y la propriedad intelectual. See http://www.icstd.org/dlogue/2001-02-22/Eugui.doc (accessed 4 January 2006).

European Chemical Industry Council (CEFIC) (2002) The chemical industry comments on the legal protection of traditional knowledge & access to genetic resources – patenting. Available at: http://www.cefic.be/Files/Publications/NM2002IP41bis.pdf

Gervais, D.J. (2002) The internationalization of intellectual property: new challenges from the very old and the very new. *Fordham Intellectual Property, Media and Entertainment Law Journal* 12, 929.

Graff, G. and Zilberman, D. (2001) Towards an intellectual property clearinghouse for agricultural biotechnology. *IP Strategy Today*, 3, http://www.nature.com/cgi-taf/dynapage.taf?file=/nbt/journal/v19/n12/index.html

Gupta, A.K. (undated, a) Securing traditional knowledge and contemporary innovations: can global trade links help grassroots innovations? Honey bee perspective. Available at: http://www.sristi.org/papers/A3.htm

Gupta, A.K. (undated, b) Portfolio of incentives for conservation of biodiversity: honey bee perspective. Available at: http://www.sristi.org/papers/A48.htm

Huft, M.J. (1995) Indigenous peoples and drug discovery research: a question of intellectual property rights. *Northwestern University Law Review* 89, 1678.

Janis, M.D. (1999) Second tier patent protection. *Harvard International Law Journal* 40, 151.

Jayaraman, K.S. (2002) Biopiracy fears cloud Indian database. SciDev Net, 5 December 2002. Available at: http://www.scidev.net/News

Krattiger, A.F. (1996) Building novel North–South partnerships through biodiversity: challenges, opportunities and a vision. Friedrich-Ebert-Stiftung (ed.)/Nord-Süd-Biodiversität (Electronic edn) Bonn. http://www.fes.de/fulltext/iez/00089001.html (accessed 20 December 2004).

Krattiger, A.F. (2004) Financing the bioindustry and facilitating biotechnology transfer. *IP Strategy Today*, August.

Krattiger, A.F. and Lesser, W.H. (1995) The 'facilitator': proposing a new mechanism to strengthen the equitable and sustainable use of biodiversity. *Environmental Conservation* 22(3), 211–215.

Lebot, V., Merlin, M. and Lindstron, L. (1992) *Kava the Pacific Drug*. Yale University Press, New Haven, Connecticut.

Leestli, M. and Pengelly, T. (2002) Study Paper 9, Institutional Issues for Developing Countries in Intellectual Property Policymaking, Administration & Enforcement. Available at http://www.iprcommission.org

Mgbeoji, I. (2001) Patents and traditional knowledge of the uses of plants: is a communal patent regime part of the solution to the scourge of bio piracy? *Indiana Journal of Global Legal Studies* 9, 163.

National Innovation Foundation (undated, a) Explanatory note for prior informed consent. Available at: http://www.nifindia.org/picnote.htm

National Innovation Foundation (undated, b) Prior Informed Consent Form for Traditional Knowledge. Available at: http://www.nifindia.org/pic.htm#trad

Nordmann, M. (2001) *Rechtsschutz von Folkloreformen*. Baden-Baden, Germany.

O'Connor and Company (2003) *Geographical Indications in National and International Law*. O'Connor and Company, Brussels, No. 6, p. 157.

Ragavan, S. (2001) Protection of traditional knowledge. *Minnesota Intellectual Property Review* 2, 1.

Rangnekar, D. (2002) *Geographical Indications: A Review of Proposals at the TRIPS Council*. UNCTAD/ICTSD Capacity Building Project on Intellectual Property Rights and Sustainable Development, Geneva.

Reichman, J.H. (1994) Legal hybrids between the patent and copyright paradigms. *Columbia Law Review* 94, 2432.

Reichman, J.H. (2000a) Of green tulips and legal kudzu: repackaging rights in subpatentable innovation. *Vanderbilt Law Review* 53, 1743.

Reichman, J. (2000b) The TRIPS Agreement comes of age: conflict or cooperation with the developing countries? *Case Western Reserve Journal of International Law* 32, 441.

Ruiz, M. (2003) ¿Es necesario un nuevo marco jurídico para la bioprospección en la región Andina? Breve revisión crítica de la Decisión 391. *Sociedad Peruana de Derecho Ambiental* 14, Febrero, 1–8.

Stern, R.H. (1994) Solving the algorithm conundrum: after 1994 in the Federal circuit patent law needs a radical algorithmectomy. *AIPLA Quarterly Journal* 22, 167.

Suthersanen, U. (2001) Incremental inventions in Europe: a legal and economic appraisal of second tier patents. *Journal of Business Law*, 319.

Tobin, B. and Swiderska, K. (2001) Speaking in tongues: Indigenous participation in the development of a sui generis regime to protect traditional knowledge in Peru. Available at: http://www.iied.org/blg/pubs/partpolicy.html

WIPO (2001) *Intellectual Property Needs and Expectations of Traditional Knowledge Holders. Report on Fact-finding Missions on Intellectual Property and Traditional Knowledge (1998–1999)*. WIPO, Geneva.

WIPO (2003) *Descriptions of National and Regional Experiences with Existing Sui Generis Measures and Laws for the Protection of Traditional Knowledge*. WIPO, Geneva.

WIPO/UNEP (2002) Case Study One – Mali. In: *The Role of Intellectual Property Rights in the Sharing of Benefits Arising from the Use of Biological Resources and Associated Traditional Knowledge – Selected Case Studies*. WIPO/UNEP, Geneva.

WTO (2001) World Trade Organization; Review under Article 24.2 of the application of the provisions of the section of the TRIPS Agreement on geographical indications. WTO doc. IP/C/W/253.

Part III

Options for Collective and Trade Policy Measures

7 New Collective Policies

Susette Biber-Klemm,[a] Philippe Cullet[b] and Katharina Kummer Peiry[c]
[a]*World Trade Institute, University of Berne, and Faculty of Law, University of Basel, Maiengasse 51, CH 4056 Basel, Switzerland;* [b]*School of Law, School of Oriental and African Studies, University of London, Thornhaugh Street, London WC1H 0XG, UK;* [c]*Kummer EcoConsult, Chemin de la Forêt 4, CH 1752 Villars-sur-Glane, Switzerland*

7.1 Introduction

This and the following chapter move on to consider a number of other aspects of the links between PGR and IPR. This chapter is divided in three sections that examine different areas which are linked to IPR but have not necessarily been directly integrated in discussions on TK and IPR. It considers issues concerning PGR and TK that are not assignable to a specific individual or group of persons. The first section examines the concept of farmers' rights, its development in international law and its relevance in the context of the protection of TK. The second section analyses issues related to access to PGR and examines both the legal framework provided by international law instruments – such as the Biodiversity Convention – and issues concerning contractual arrangements for access between private parties. Finally, the third section considers the extent to which financial mechanisms, mooted in particular in the context of international environmental treaties, could be used as models for compensating TK holders.

7.2 Farmers' Rights[1]

The rights that farmers have under international law – or absence thereof – with regard to diverse elements such as seeds and knowledge related to PGRFA, have been and remain contentious. First, in the context of the development of a protection regime for commercial plant breeders, what became known as the farmer's privilege is a simple recognition that plant breeders' rights are not absolute and do not stop farmers from using the protected variety that they have grown for further development.[2] Secondly, the rights of farmers can be con-

[1] Author: Philippe Cullet. Thanks to Dr Martin Girsberger for his detailed and specific comments on a previous version of this text.

[2] For further developments on the UPOV Convention, see Chapter 2, Section 2.3.4, this volume. The farmer's privilege is not included in this review of farmers' rights.

ceived as fully fledged rights over traditional knowledge (TK). This option has not been recognized in any international legal framework, but some developing countries have attempted to develop legal frameworks that provide for the rights of farmers over their TK.[3] Thirdly, farmers' rights can be conceived as a form of compensation for services rendered by the community by all farmers to the conservation and enhancement of PGRFA over time. This is the option that has been developed over time in the context of the two main international legal instruments recognizing farmers' rights, the International Undertaking (IUPGR) and the International Treaty on Plant Genetic Resources for Food and Agriculture (ITPGRFA).

At present, the ITPGRFA provides the only existing recognition of farmers' rights in a binding instrument. The ITPGRFA does not, however, provide a substantive definition of farmers' rights and in particular does not provide any form of property rights for farmers over their knowledge. As a result, farmers' rights at the international level are currently little more than a policy tool to foster the recognition of the contribution of farmers and farming communities in the overall conservation and development of plant genetic resources. Further elaboration of farmers' rights can only be found in specific domestic legal regimes. This section analyses the notion of farmers' rights as it has evolved in international law over the past couple of decades and examines the extent to which the existing notion of farmers' rights can contribute to the protection of TK.

7.2.1 Farmers' rights under the International Undertaking

Farmers' rights were first enshrined in a legal instrument whose primary intent was to promote the free exchange of plant genetic resources. As noted in Chapter 2, the IUPGR, which promoted the management of plant genetic resources, on the basis of the principle of common heritage of humankind, proved unacceptable to some countries in its original form. The revised version of the IUPGR maintained a reference to common heritage as a basis for regulating plant genetic resources, but at the same time recognized the validity of claims over these resources. As a result, it accepts both plant breeders' rights and farmers' rights as valid claims over plant genetic resources and drastically narrows down the relevance of the concept of common heritage by recognizing states' sovereign rights over their plant genetic resources.

The IUPGR is rather unclear in its formulation of farmers' rights. It first specifically mentions that the basis for the concept of farmers' rights is the 'enormous contribution that farmers of all regions have made to the conservation and development of plant genetic resources, which constitute the basis of plant production throughout the world'.[4] It further emphasizes that the context for the adoption of farmers' rights is the dichotomy between the role of farmers in developing countries in developing and maintaining plant genetic diversity and the need to access these resources for the purposes of conservation and plant breeding in the commercial sector.[5] The specific definition of farmers' rights under the IUPGR reads as follows:

> Farmers' Rights mean rights arising from the past, present and future contributions of farmers in conserving, improving, and making available plant genetic resources, particularly those in the centres of origin/diversity. These rights are vested in the International Community, as trustee for present and future generations of farmers, for the purpose of ensuring full benefits to farmers, and supporting the continuation of

[3] See also Chapter 2, Section 2.3.3, this volume.

[4] Res. 4/89, Agreed Interpretation of the International Undertaking, 29 Nov. 1989, Report of the Conference of FAO, 25th Session, Rome, 11–29 November 1989, Doc. C89/REP.

[5] Res. 5/89, Farmers' Rights, 29 November 1989, Report of the Conference of FAO, 25th Session, Rome, 11–29 November 1989, Doc. C89/REP.

> their contributions, as well as the attainment of the overall purposes of the International Undertaking.

This definition does not include the rights of farmers over their TK. In fact, what is provided in the IUPGR is little more than the recognition of the collective contribution of farming communities to the development and maintenance of agricultural biodiversity. This recognition is materialized through the provision of financial resources to ensure the conservation of plant genetic resources and the need for farmers to participate in the benefits from the use of their varieties in genetic engineering.

Insofar as the implementation of the recognition of farmers' contribution to plant genetic resource management is concerned, the rationale provided is that 'the best way to implement the concept of Farmers' Rights is to ensure the conservation, management and use of plant genetic resources, for the benefit of present and future generations of farmers'.[6] The only specific implementation mechanism provided is the International Fund for Plant Genetic Resources. In keeping with the recognition that farmers' rights are vested in the international community and not in farmers themselves, the International Fund is conceived as a financial mechanism that rewards countries rather than individual farmers or farming communities. In fact, the Fund is generally conceived more along the lines of a development aid programme to developing countries for capacity building in the field of agricultural biotechnology than as a tool to reward individual farmers or farming communities for their contribution to the development or improvement of plant varieties.

On the whole, the system of farmers' rights provided in the IUPGR seeks to counterbalance the existence of plant breeders' rights and other IPR in the field of genetic engineering with some form of incentive and compensation for farmers. However, it stops short of defining any type of individual or collective property rights of farmers over their TK. It only provides recognition of the contribution of farmers to the conservation and sustainable use of plant genetic resources. A general form of compensation is offered under the aegis of the International Fund, but even the material benefits from the International Fund are targeted in a very general manner and are not meant to directly reach specific farmers or farming communities that have made a specific contribution to the development and conservation of plant genetic resources. In practice, the concept of farmers' rights provided in the IUPGR has never had much impact, since the International Fund never became a reality because funds were not made available by donor countries. However, developments in the context of the undertaking have had an important influence on subsequent developments leading to the adoption of the ITPGRFA.

7.2.2 Farmers' rights under the International Treaty on PGRFA

The ITPGRFA addresses a number of important issues concerning the conservation and use of PGRFA. Farmers' rights are only one of many issues considered, but it is noteworthy that negotiations concerning the definition of farmers' rights constituted one of the important issues on which the successful completion of the treaty depended. In fact, one of the main goals of the revision of the IUPGR was the further development and concretization of farmers' rights. In the end, farmers' rights were retained as one of the important elements of the treaty, but negotiators stopped short of recognizing and comprehensively defining farmers' rights at the international level.

The concept of farmers' rights under the ITPGRFA has evolved, as compared to the first formulation in the context of the IUPGR, but a clear affiliation can still be seen. Most importantly, the ITPGRFA does not go beyond the concept of farmers' rights as a form of compensation for farmers' contribution to the development and mainte-

[6] Res. 4/89, note 4 above, p. 284.

nance of agricultural biodiversity. It does not include property rights but provides, however, that member states are free to develop their own forms of protection.

The basis for the recognition of farmers' rights is again the

> enormous contribution that the local and indigenous communities and farmers of all regions of the world, particularly those in the centres of origin and crop diversity, have made and will continue to make for the conservation and development of plant genetic resources which constitute the basis of food and agriculture production throughout the world.[7]

The Treaty does not provide any other definition of farmers' rights insofar as their protection at the international level is concerned. It only makes explicit the minimum level of protection of farmers' interests that should be protected. In effect, only existing rights of farmers to save, use, exchange and sell farm-saved seed/propagating material are protected under the Treaty.[8] Further, even these residual rights are protected only to the extent that national law does not differ from the provisions of the Treaty.

Insofar as domestic law is concerned, the ITPGRFA provides that member states should in principle undertake to protect and promote farmers' rights. The Treaty highlights some of the areas that should be addressed in priority. This includes the protection of TK in the context of agricultural biodiversity, benefit sharing and participation in decision-making with regard to the management of agrobiological resources.[9]

7.2.3 Farmers' rights as a means to protect traditional knowledge

The preceding sections indicate that the concept of farmers' rights in international law has a specific meaning in the context of the IUPGR and the ITPGRFA. In the context of these two instruments, farmers' rights constitute in effect a policy decision at the international level to formally acknowledge the contribution of farmers and farming communities to the development and conservation of agricultural diversity. Apart from the call to member states to realize farmers' rights at the national level, the only practical consequence of this acknowledgment at the international level has been the progressive development of the notion of benefit-sharing, which focuses on ways to provide financial and other compensation for the farming community's contribution to fulfilling humankind's food needs. As such, the concept of benefit sharing is not specific to the case of farmers and has been developed more generally to compensate holders of TK for the use of this knowledge in different contexts.[10]

Despite the relative underdevelopment of farmers' rights in the two FAO instruments, it is noteworthy that a treaty that does not specifically focus on agriculture, like the Desertification Convention, indirectly considers farmers' rights through the lens of TK. The Desertification Convention is a typical treaty of the law of sustainable development. This implies that it takes a broad view of the environmental challenges linked to land degradation and includes agriculture among the important fields that must be taken into account to successfully address desertification. In this context, the recognition that state parties must not only promote and use TK, but also protect it, is significant.[11] In this sense, the Desertification Convention constitutes another instrument indirectly promoting the development of farmers' rights at the national level as part of the more general protection of TK. The importance of the Desertification Con-

---

[7] Article 9.1 of the International Treaty on Plant Genetic Resources for Food and Agriculture, Rome, 3 November 2001 [hereafter ITPGRFA].

[8] Article 9.3 of the ITPGRFA.

[9] Article 9.2 of the ITPGRFA.

[10] Further examination of benefit-sharing is undertaken in Section 7.3 of this chapter.

[11] See Article 18 of the Convention to Combat Desertification in those Countries Experiencing Serious Drought and/or Desertification, Particularly in Africa, Paris, 17 June 1994, 33 *ILM* 1328 (1994).

vention in this area stems from its recognition of a need to *protect* TK rather than the more limited aim of compensation for services rendered, as is the case under the ITPGRFA.

The limits of the concept of farmers' rights at the international level are no bar to their further development at the national level. In fact, this is what the ITPGRFA calls for. At the national level, farmers' rights have been conceived as a tool for implementing benefit sharing in accordance with developments at the international level. Some countries have, however, attempted to go beyond this limited notion of farmers' rights, while a number of others are still developing legal frameworks in this area. It is therefore useful to review some of the options that countries have chosen, or may choose, to implement the notion of farmers' rights in their legislation.

In some countries, like India, the development of farmers' rights has taken a route that focuses at least in part on the individual rights of farmers to their varieties. Since this individual right aspect was covered in Chapter 5, this section does not consider it again. In fact, one of the important contributions of the concept of farmers' rights is the recognition that innovation and incremental improvements in the field of agricultural biodiversity cannot always be assigned to a particular individual or group. In such cases, traditional notions of IPR are not particularly well suited to offer protection for this TK. Farmers' rights thus constitute an interesting departure point to developing rights that could be collective in scope. There have been a number of proposals seeking to grant property rights to local communities or other groups.[12] The challenge has already been taken up by some countries like Venezuela, which has instituted a system that distinguishes the rights of TK holders and individual IPR in cases where knowledge is developed incrementally.[13] In other words, the Venezuelan law goes further than merely acknowledging the importance of biodiversity-related TK with the usual compensation for conservation of biological resources, and specifically recognizes the associated rights of indigenous and local communities.[14] This tends to indicate that where knowledge acquisition is either collective or incremental, the concept of farmers' rights may offer opportunities to devise an alternative form of protection to the dominant intellectual property model.

Where collective rights are allocated, it is necessary to make sure that the benefits reach all the rights holders. Some countries have focused on exploring the possibility of assigning collective rights to local democratically elected bodies rather than to legally amorphous local communities. In countries with strong traditions of local democratic governance, as in India, which has a highly developed panchayat system, using them has obvious advantages since it guarantees to a certain extent that benefits arising from the assignment of property rights will at least in principle be shared with a measure of public accountability. This is in fact what the Indian Biodiversity Act attempts by requiring the setting up of biodiversity management committees in each panchayat whose functions include the promotion of conservation, sustainable use and documentation of TK.[15]

Beyond the collective rights dimension, farmers' rights may provide an interesting departure point for examining alternative forms of knowledge protection. Registration may, for instance, constitute a ground for differentiating farmer varieties from commercial breeders' varieties. This can be achieved by providing that protection for a farmer variety can only be obtained for varieties which are particular to a given locality. The rights that can be conferred include the right to develop,

[12] For instance, Biodiversity (Rights and Protection) Bill, 1998 (New Delhi: Research Foundation for Science, Technology and Ecology & Lawyers Collective, 1998).
[13] Article 85, Venezuela, Ley de diversidad biológica, 27 October 1999.
[14] Articles 13 and 84, Venezuela: Ley de diversidad biológica, 27 October 1999.
[15] India: Biological Diversity Act, 2002.

produce, sell and export the protected variety.[16] The distinguishing element is the non-exclusivity of the protection which is offered. This is due to the fact that exclusivity may not be practical in the fields of farmers' varieties since they may exist in more than one location and possibly more than one country. Another argument against exclusivity is the strong link between farmer varieties, agro-biodiversity management and food security. These links seem to preclude a focus on the commercial dimension of property rights at the expense of the broader goals of environmental conservation and food security. As a result, non-exclusive farmer rights could provide an alternative scheme whereby all rights holders are entitled to separately produce and commercialize their own products without infringing on the rights of other similarly placed rights holders.

Another interesting dimension of collective farmers' rights is that they can be used as a tool to take into account not only the commercialization of knowledge, but other goals as well. These include, for instance, agro-biodiversity management, biosafety and food security. The management of agro-biodiversity is particularly interesting insofar as diversity has historically been conserved and enhanced by farmers. This contribution of farmers is likely to remain important in the future.[17] One of the central points with regard to agro-biodiversity and farmers' rights is that while farmers directly benefit from agro-biodiversity conservation, national governments and the international community also benefit in direct and indirect ways. This may require the sharing of conservation obligations on an equitable basis between all actors benefiting from the exploitation of agro-biodiversity, from farmers and local firms marketing seeds to research institutions, private seed companies and states. In terms of property rights, this tends to indicate the need for farmers' rights that are appropriate to giving incentives to farmers not only to conserve existing agro-biodiversity and associated TK, but also incentives to further develop this knowledge.

Overall, the notion of farmers' rights is interesting because it fosters a broader understanding of the links between innovation, rights over knowledge, biodiversity conservation and the sustainable use of agro-biodiversity. While patents and plant breeders' rights are on the whole completely unlinked from concerns over conservation and sustainable use, farmers' rights are quite different and much more amenable to a broader perspective. This is now enshrined in the ITPGRFA, which clearly recognizes the links between conservation of plant genetic resources and the use that can be made, including the commercialization of products derived from plant genetic resources-related knowledge.

7.3 Reviewing Access Legislation[18]

7.3.1 Introduction

Both the CBD and the ITPGRFA enunciate the sovereignty of states over their genetic resources. Yet the exercise of sovereign rights is balanced by the obligation of donor states to facilitate access to their genetic resources. Both conventions provide frameworks to this end: the CBD establishes a basic regime on Access and Benefit Sharing (ABS) (Articles 1, 15, 16, 19; see Chapter 2, this volume); the ITPGRFA concretizes this ABS regime for a defined part of the resources by the Multilateral System (MLS) (see below). The concept of the sharing of

[16] Note that Thailand has, for instance, adopted a farmers' rights regime which entitles the local legal entity to 'have the exclusive right to develop, study, conduct an experiment or research in, produce, sell, export or distribute by any means the propagating material [of the registered local domestic plant variety]'. See Section 47 of the Plant Varieties Protection Act, B.E. 2542 (1999).

[17] See, for example, Article 9 of the ITPGRFA.

[18] Author: Susette Biber-Klemm. I thank Alwin Kopse, Philippe Cullet and Michael Halewood for their thoughtful and inspiring comments and Danuta Szymura Berglas for her careful language editing.

benefits resulting from the use of PGR and TK is a response to the experiences in the exchange of this information and its appropriation due to different levels of protection of informational values in different countries (see Chapter 3, this volume).

The focus in the following discussion will be access to TK in the framework of the two instruments. TK can be associated either with wild or domesticated genetic resources, or – in the case of traditional PGRFA – be expressed in a product resulting from the use of TK.[19] This means that the regulation of both the CBD and the ITPGRFA will be examined regarding the provisions for: (i) access to TK associated with PGR in general (CBD); (ii) TK associated with PGRFA (ITPGRFA); and (iii) access to the PGRFA themselves (ITPGRFA). The goal is to describe and evaluate ABS regimes as instruments for promoting trade in the informational values of TK and traditional PGRFA. The criteria for the evaluation of the ABS system are the goals discussed in Chapter 1, namely the goal of creation of incentives and economic means to support and maintain TK and traditional PGRFA. In this context it is important to take account of the findings of Young and Gunningham (1997) and of Swanson and Göschl (2000), who argue that incentives to maintain TK and PGRFA are best created at the level where the costs occur (see Chapter 4, this volume). In the case of the conservation and maintenance of TK and traditional PGRFA, these costs occur not only in the *ex situ* facilities but, as has been shown (Chapters 1 and 4), to a great degree also at the level of *in situ* conservation by traditional on-farm breeding. Therefore one of the key questions is to what degree the stakeholders at the level of local farming communities and indigenous peoples can be involved in the respective ABS regimes in order to benefit from the values they have created. In respect of PGRFA, the focus will therefore be put on access to *in situ*-maintained traditional PGRFA.

The second point will then be to evaluate the ABS system in view of its capacity to create benefits in a trade context. The thesis is that at present rather cumbersome procedures[20] ought to be rethought in order to be more market and trade supportive. Accordingly, the emphasis will not primarily be put onto the technicalities of the ABS system but rather the questions arising from its implementation at the interface between national and international trade and local stakeholders.

In the present discussion the trade aspects of genetic resources and TK are predominantly perceived as a North–South relationship, in that the biodiversity-rich countries of the South are the providers, and the industrialized countries of the North the recipients and users of the information. However, in evaluating the ABS system, it is to be kept in mind that the relationships of access to genetic resources and TK show a much more varied pattern. First, not only international, but also domestic trade relations, might be of importance, and secondly, exchange of genetic information and TK may also take place in a South–South or North–North relationship. The global interdependence of all players is in particular true for PGRFA, where all countries may have the role of both purchaser and provider.[21] Finally, one should avoid the over-simplification that 'less-developed' necessarily equates with 'biodiversity-rich', and 'industrialized' with 'biodiversity-poor': an obvious example being Australia, which is both a developed country and one containing 'mega biodiversity' (see Chapter 2, this volume).

7.3.2 Legal background

In general

From the factual, legal and political point of view, access to (wild) genetic resources is

[19] In the following, the term 'information' is meant to encompass both elements.

[20] See the case study on Peru in Chapter 2.

[21] See Fowler *et al.* (2000).

different from access to TK and domesticated varieties of PGRFA, respectively. Whereas the primary right to regulate ownership of wild genetic resources is attributed to the sovereign state, in the case of TK associated with PGR there seems to be a common legal conviction that the right and competence to deal with the product of creativeness lies with its creators or holders.[22] Thus it is considered inequitable and politically unfeasible for states to attempt to expropriate individual and collective rights of communities to their knowledge (Ruiz 1997, cited in Glowka, 1998, p. 37).

The question is whether the same can be said for varieties of PGRFA based on the understanding – confirmed by the last generation of scholarship – that maintenance and breeding of farmers' varieties involves technical know-how and creative skills (see Chapter 4, this volume). In the case of PGRFA, we find ourselves with the problem of creative information being embodied in a physical, self-propagating entity, which, moreover, can also be traded as a commodity. Plant Breeders' Rights are the solution to protect the informational value contained in varieties bred in formal, industrial processes. However, no such protection exists for the protection of informational value incorporated in farmers' varieties. The question, therefore, is whether compensation for the utilization of the informational value is possible on the basis of the ABS system.

The regulation of the CBD[23]

The subject matter of the CBD is in principle all biological resources: wild and domesticated, and related TK. However, the Conference concluding the convention delegated the issues regarding PGRFA to the negotiations in the framework of the revision process of the FAO IU on PGRFA, now the IT on PGRFA.[24] So the latter is primarily relevant for PGRFA, including wild relatives, and related TK, while the CBD covers the remaining wild biological resources and TK.

The regime of access to genetic resources as laid down by the CBD is based on a contractual approach. It encompasses the three elements of 'prior informed consent', 'mutually agreed terms' and the 'fair and equitable sharing of benefits' (Article 15). It is important to note that the authority for determining access to genetic resources is vested in the state and subject to national legislation. Thus the convention only entitles the providing state, and does not confer any rights to the individual holders of the PGR.

The same is true for the regulation of issues regarding access to TK associated with PGR. The implementation of the general principles that are stated in Article 8(j) of the CBD is referred to the national level. Yet the CBD does give some interpretative guidelines. States are – as far as possible and appropriate – to involve the holders of the TK in its wider application, and to encourage the equitable sharing of the benefits arising from its use.

In the follow-up process of the CBD, its ABS regime has been further developed. The 'Guidelines on Access to Genetic Resources and Fair and Equitable Sharing of the Benefits Arising out of their Utilisation' (Bonn Guidelines)[25] provide a more detailed framework for the development of

---

[22] See, for example, Article 27(2) of the Universal Declaration of Human Rights which states that 'Everyone has the right to the protection of the moral and material interests resulting from any scientific, literary or artistic production of which he is the author…', and Article 15 of the International Covenant on Economic, Social and Cultural Rights, which states that 'The States Parties to the present Covenant recognize the right of everyone … to benefit from the protection of the moral and material interests resulting from any scientific, literary or artistic production of which he is the author.'

[23] For details see Chapter 2.

[24] International Treaty on Plant Genetic Resources for Food and Agriculture, Rome, 3 November 2001 [hereafter ITPGRFA].

[25] Accepted as Decision VI/24 by the VI meeting of the Conference of the Parties, The Hague, The Netherlands (7–19 April 2002).

ABS regimes, including with particular reference to Article 8(j) on TK (Nos 1 and 9). One of the explicit goals is to 'contribute to the development ... of mechanisms and access and benefit-sharing regimes that recognise the protection of traditional knowledge ... in accordance with domestic laws and relevant international instruments'. The Guidelines assert the importance of the involvement of all relevant stakeholders in the ABS process, explicitly including indigenous and local communities. In particular, stakeholders should be integrated in the negotiation and implementation of ABS arrangements on the one hand, and in the development of national strategies, policies or regimes on ABS on the other. The provision of pertinent information and capacity-building is identified as an important measure in supporting negotiations (Nos 17–21).[26]

Thus, the framework of the CBD presently evolves in the direction of strengthening the position of the (individual or community) holders, which in turn leads to the marketing problems discussed below.

The regulation of ABS by the ITPGRFA[27]

The development of a regime regulating access to PGRFA is directly linked to the principle of state sovereignty over genetic resources. The reiteration of this principle in the CBD has prompted the revision of the IU (see Chapter 2, this volume). Accordingly the two Conventions are closely interrelated. The definition of the objectives of the treaty – conservation and sustainable use of PGRFA and the fair and equitable sharing of the benefits arising out of their use – corresponds to the CBD goals. The goals of the IT are to be attained 'in harmony' with the CBD, and by closely linking the IT to the CBD (Article 1.2).

The ITPGRFA adopts the concept of the CBD that the authority for determining access to PGR rests with national governments and is subject to national legislation (Article 10.2), and also limits national sovereignty by the obligation to facilitate access for contracting parties.

Given this, and the recognition that PGRFA are a common concern of humankind, the treaty establishes a specific system of facilitated access to a selection of varieties specified in an Annex (Annex 1) to the Treaty, the so-called Multilateral System of Access and Benefit-Sharing (MLS) (Article 10).

Thus, within the ITPGRFA, two basic regimes exist for access to PGRFA and the sharing of the benefits resulting from their use: (i) the MLS for the Annex 1 material and, as the Treaty is silent on non-Annex 1 material, (ii) by default the more general principles of the basic ABS regime covering the remaining varieties.

THE MULTILATERAL SYSTEM

In the background of the MLS lies the insight into the global interdependence regarding PGRFA, or worded differently, in their common resource nature (Preamble para. 2 ITPGRFA). Accordingly, the MLS in some ways takes up the concept of PGR as being the common heritage of mankind, which at its outset ruled the IU (see Chapter 2, this volume), and adapts it to the system of state sovereignty. Consequently the goal of the MLS is to support the free flow of germplasm, and to bypass the complica-

[26] Some of the Contracting Parties are concerned about the voluntary character of the Guidelines. It is feared that due to their voluntary nature, they remain ineffective (see, for example, Brazil in IP/C/W/228, No. 34). Accordingly, the COP explicitly decided to keep the Guidelines under review, considering that they are but a first step of an evolutionary process (Dec. VI/24). The Plan of Implementation adopted by the World Summit on Sustainable Development in 2002 calls for action to negotiate an international regime to promote and safeguard the fair and equitable sharing of benefits (Report of the World Summit on Sustainable Development, Johannesburg, South Africa, 26 August–4 September 2002; A/CONF.199/20, chapter I, resolution 2, annex; see also Doc. UNEP/CBD/MYPOW/6, Nos. 1, 15–20). A regime is a 'set of principles, norms, rules and decision-making procedures' and not *per definition* a legally binding instrument, but might include binding norms (UNEP/CBD/MYPOW/6 No. 19).

[27] For an overview of the entire Treaty see Chapter 2, this volume.

tions, which might occur by the CBD ABS regime, between those that are parties to the treaty.

The system is based on the idea of solidarity and mutual benefit. The listed varieties are selected according to the criteria of food security and interdependence (Article 11.1); access is to be provided for research, breeding and training for food and agriculture; no IPR of any kind limiting facilitated access might be claimed on the PGRFA in the form received; and any fees imposed may cover only the minimal cost incurred in providing access. Access is to be granted expeditiously, and – of importance for our question – without the need to track individual accessions back to their origin. Together with the accession, all available passport data and other associated descriptive information are to be made available 'subject to applicable law'. This information may well encompass TK elements. From this argument it follows that, if TK elements are included in the passport data of varieties integrated in the MLS and stored in an *ex situ* facility, there is no necessity for PIC to be given by the original holders of the knowledge for the transfer of the information.[28]

Transfer of the PGRFA and associated information integrated in the MLS is facilitated by the utilization of standardized contracts, i.e. the standard material transfer agreement (MTA), which is also to apply to all subsequent transfers of the PGRFA it covers. The MTA is to be adopted by the Governing Body of the treaty.

As to benefit sharing, facilitated access to PGRFA included in the MLS is in itself considered to be a major benefit. Further types of benefit sharing include the exchange of information; access to and transfer of technology; capacity-building; and the sharing of benefits arising from commercialization (Article 13.2).[29] There is a monetary benefit sharing clause that is triggered when a recipient of material from the MLS incorporates that material into a new product, and subsequently restricts access to the product to disallow its use for research purposes. In such cases, the party would be obliged to make a contribution to an international fund, at a rate yet to be established by the Governing Body, to be used to support conservation efforts (Article 13.3).

If the MLS is examined in view of the creation of direct incentives for local *in situ* conservators of traditional PGRFA, the following aspects must be considered.

The Multilateral System includes only PGRFAs that are under the management and control of the Contracting Parties and in the public domain. Natural and legal persons who are holders of listed varieties are thus not automatically part of the system, but are invited to include their varieties in order to participate in the system and have access to the stored resources. However, during a 2-year pilot period, access to the resources included in the MLS is to be granted to all legal and natural persons who are under the jurisdiction of a contracting party. The decision as to whether this regime will continue in the future is to be taken by the Governing Body at the end of the 2-year period.

The system seems to be primarily meant to provide facilitated access to the public *ex situ* collections of the listed varieties. Access to listed PGRFA in *in situ* conditions is to be provided according to national legislation, otherwise according to

[28] But PIC would be needed at the time of collecting the accessions *in situ*; PIC would have to encompass the fact that the knowledge is handed on without getting back to the original holder once the sample and the associated knowledge is integrated in the MLS (see also below). However, according to the research by IPGRI, at present there is very little ethnobotanical information stored in the passport data (Chapter 4, this volume).

[29] The Governing Body is mandated to determine the level of the payments. It may differentiate the levels of the payments for various categories of recipients who commercialize products resulting from the use of the PGRFA integrated in the System. In particular it may exempt small farmers in developing countries from payments (Article 13(d)(ii)).

standards set by the Governing Body of the Treaty (Article 12.3(h)).[30]

The position of the small-scale farmers within this system of access and benefit-sharing at present is not clear. On the one hand, the access to *in situ* resources, either for collection for an *ex situ* facility or in the sense of direct bioprospecting, is not (yet) specified. On the other hand, at present it is not quite clear how these farmers will be able to benefit directly from the system.

As the treaty has only recently been concluded and come into force,[31] the details are still in a state of flux. With regard to the strengthening of the position of the individual small-scale farmers, and farming communities or corporations within the MLS, the following points must be taken account of.

In view of access to PGRFA in *in situ* conditions,[32] two goals are to be kept in mind. First, it must remain possible to collect *in situ* accessions and related additional information for integration into the MLS; and secondly, incentives must be created for the maintenance and further development of traditional PGRFA. This is bound to remain difficult as long as the legal position and benefit for the *in situ* breeders is not clear.[33]

Two types of benefits can be imagined in this context. First, the benefits of the MLS itself; secondly, the benefits of marketing or selling traditional varieties which have been improved, such as in semi-formal on-farm breeding (see Chapter 4, this volume), using a variety from an *ex situ* facility.

An initial question, therefore, is how to make the benefits of the MLS, namely, facilitated access in itself, the exchange of information, access to technology, capacity-building and the sharing of benefits of commercialization (Articles 31.1 and 13.2) *directly* beneficial for small-scale farmers, thus creating an incentive for them to contribute their varieties to the MLS.[34] This is part of the more general issue of how the challenge of extending the MLS to small-scale farmers and farming communities, and of making it directly operational and beneficial, can best be met at the legal, political and also practical levels.

Secondly, there is a need to clarify under which conditions varieties bred on the basis of varieties integrated in the system, or making use of them, can be commercialized by their breeders, such as on local or regional markets. Two regulations governing the MLS might be relevant: Article 12.3(b) prescribes that within the system varieties are to be exchanged 'free of charge'. Article 12.3(d) prescribes that no IPR of any kind limiting facilitated access might be claimed on the PGRFA in the form received. Thus, it can be concluded that only varieties 'in the form received' fall under the system and therefore must be exchanged free of charge. Accordingly, whether a variety can be commercialized

---

30 It is submitted that in the case of absence of national legislation or standards set by the Governing Body, the CBD ABS regime applies.

31 In accordance with Article 28, the Treaty entered into force on the 90th day after the deposit of the 40th instrument of ratification, acceptance, approval or accession, provided that at least 20 instruments of ratification, acceptance, approval or accession have been deposited by Members of FAO. On 31 March 2004, 13 instruments (including the European Community) were deposited with the Director-General of FAO. Accordingly the date of entry into force was 29 June 2004.

32 As long as national legislation or subsidiary regulation by the Governing Body are lacking, access to these resources depends on the CBD ABS regime. See below.

33 It seems that the present insecure legal situation, and the hope of gaining benefits from the ABS procedures, prevents farmers donating their varieties.

34 There is at present no indication as to how the facilitated access of holders of *in situ* resources will be handled after the expiry of the 2-year trial period of facilitated access. If access to the system is being restricted, without including the *in situ* preserved PGRFA, access by local farmers or farming communities would be limited to communities maintaining *ex situ* collections of their seeds and integrating these collections into the system. It is submitted that, in any case, facilitated access should be granted for small-scale farmers, their communities and organizations.

depends on the term 'in the form received from the system', which is highly in need of interpretation.

It seems that this wording allows for the IPR or other protection of 'new forms' of varieties that are developed on the basis of materials received from the MLS. The question in this case is what would be the criteria to define a 'new form'? Do they correspond to the PBR or patenting criteria? In this case traditionally bred landraces would in any case remain under the prescriptions of the system and under its obligation to facilitate exchange. It is submitted that for reasons of equity, means ought to be found to recognize 'new forms' (contra Article 12.3(d))[35] of varieties developed by farmers traditionally breeding on the basis of a MLS variety received from an *ex situ* facility. This would provide a legitimization to limit the facilitated access and allow the breeder to sell his variety to interested formal breeders or in regional markets, thus rewarding his breeding efforts.

VARIETIES AND TK OUTSIDE THE MULTILATERAL SYSTEM

Given this, a rather complex pattern of access situations arises, as by far not all PGRFA are yet included in the MS.

For instance, the MLS is not applicable to:

- All varieties in countries which are not parties to the Treaty (unless they are held in public international institutions which volunteer that they are in the MLS).
- Varieties which are not listed in Annex 1 and associated knowledge.
- Varieties in private *ex situ* facilities which are not integrated in the MS.
- Varieties *in situ* and associated knowledge.
- Varieties protected by intellectual property rights.
- Materials under development.[36]

It is submitted that in these cases if the country is a signatory to the CBD, then the rules of the CBD regime on ABS apply. The CBD provisions on ABS cover access to PGR of all kinds, so they can thus be considered as a general and subsidiary rule. This solution is supported by Article 1 of the ITPGRFA that points out the need to closely link the IT to the CBD.

With a view to *in situ* accessions, it can therefore be argued that in the absence of national legislation or specification by the Governing Body of the ITPGRFA, access to and collection of accessions *in situ* is submitted to the ABS regime of the CBD. That means that for each accession the conditions of this system are to be fulfilled. As for the varieties which are included in the MLS and collected to be stored in an *ex situ* facility, this also encompasses the informed consent to future facilitated access, which means, first, that neither the accession nor associated TK is to be tracked back to its original holders; and, secondly, that the sharing of benefits follows the rules of the MLS.

In this context, the question arises as to how the collection from *in situ* resources for *ex situ* facilities and other uses can be made operational in the future. First, whether it is possible to find the means and consensus to simplify the PIC requirements for those PGRFA and, secondly, what would be the benefit to the holders of *in situ* resources for contributing their varieties to *ex situ* facilities? It is submitted that in order to create an incentive for granting access, it is necessary that some sort of benefit accrues directly to the donors of the germplasm.

Conclusions and discussion

The contractual ABS system is highly complex and diverse, covering various subject matters, a great variety of stakeholders and different systems for ruling ABS.

The CBD system, which is applicable

[35] From the rather obscure formulation of Article 12.3(d) it seems to follow that the interdiction to limit the facilitated access covers only the varieties as received by the *ex situ* facility and not changed by any breeding efforts (cases of 'pure' biopiracy).
[36] For further exemptions see Articles 11–13 ITPGRFA.

for PGR (including PGRFA and TK), can be used as the basis upon which to establish national frameworks to facilitate bilateral contractual agreements negotiated between the involved stakeholders. The corresponding provisions regulating ABS are directed to the Contracting Parties, and their implementation is explicitly subject to national legislation. Thus, the details for the operation of the system – such as the determination of the stakeholders to be involved in the negotiations, the procedures and the sharing of benefits between the stakeholders of the providing countries – mostly need to be defined on the national level, the Bonn Guidelines giving some interpretative assistance to this end. This is also true for the holders of *in situ* traditional PGRFA and associated TK.

Notwithstanding the provisions of the Bonn Guidelines and the greater degree of differentiation provided by the ITPGRFA, many questions remain, in particular regarding the involvement of grassroots stakeholders. The main question is, of course, how to implement the system of ABS on both the national and international level, and also to make it rewarding for small-scale farmers. With regard to the MLS it is to be asked how to make the system work for local stakeholders. In order to be supportive of the *in situ* maintenance of diversity of PGRFA it is essential that the advantages of the system, be they the access to *ex situ* collections or the other benefits provided by the system, have a low threshold of accessibility for farmers, their communities and organizations.

With regard to operationalizing the ABS system for PGRFA not otherwise integrated into the system, the challenge will be to create standardized instruments (MTAs) which allow for transparent and swift negotiations.[37] Supporting institutions, such as clearing house mechanisms or private collecting societies (Chapter 6) could prove useful to facilitating information in support of the marketing and the conclusion of contracts.

7.3.3 The ABS system at the interface with markets: issues to consider

Introduction

Bilateral contractual instruments for the management of ABS to TK associated with PGR and traditional PGRFA have the advantage of adaptive flexibility, allowing for solutions to be determined on a case-by-case basis according to the needs of the parties.

However, up to now, few success stories have been reported relating to both the integration of local people in the ABS process and their sharing of benefits. The operation of the principles proves to be difficult in practice.[38]

The following problem areas can be identified in the context of international trade:

- The marketability of the goods and the information *per se*.
- The identification of the relevant stakeholders and their integration into the process and the possible inequality of the negotiating partners.

---

[37] It has been suggested that the FAO code of conduct should be referred to for the basic elements of such agreements.

[38] See, for example, the Case Studies on Access to Genetic Resources and Benefit Sharing of the CBD (http://www.biodiv.org/programmes/socia-eco/benefit/case-studies). The major part of the studies treats institutional matters or describes projects. The same is true for WIPO's contracts database: http://www.wipo.int/globalissues/databases/contracts/summaries/ind One (or maximum two) model agreements out of 16 are about the cooperation between a research institute and a holder of TK (Model Agreement between the National Institute for Pharmaceutical Research and Development, Nigeria and a consultant Herbalist, 1997); the actual agreements published on the website all involve governmental agencies, research institutes and companies. But see the initiatives in India (Case study, Chapter 3 and the description of the initiatives of SRISTI/Honeybee and the National Innovation Foundation in Chapter 7 (Registration)). Compare also Ruiz Muller (2000).

- The internationality of the trade relations and the limited reach of national legislation, which could bring about problems of control, implementation and enforcement.
- The dichotomy between the rationale to protect PGR and associated TK from illegitimate use on the one hand and the wish to promote and facilitate trade on the other.

It is submitted that further measures are essential to meet the goal of successful ABS arrangements, such as the development of markets, information and capacity-building, and also the establishment of supportive measures on the side of the recipients. These will be discussed in the following section.

Further, the (contractual) ABS system itself seems not to be without controversy. The respective positions of the USA and Brazil, as expressed in their statements to the Council for TRIPS in regard to review of the provisions of Article 27.3(b), may serve as an example for two opposite positions:[39]

> The US is in favour of the instruments of bilateral contracts on ABS backed by national or local legislation. It perceives the contracts as means to make clear from the start (all of) the conditions for access, including the rights and duties of parties involved as well as additional issues such as questions of court jurisdiction, and conditions to be met by contracts concluded with third parties.[40]
>
> Brazil claims that a contractual approach is clearly insufficient to protect transfer of TK. It points out that bilateral contracts are not easily enforceable, in particular in international trade relationships.[41] Furthermore, it argues that there is no way to ensure that the consent received from the communities is an informed one, neither that the benefit sharing will be fair and equitable, as they may be the result of negotiations between unequal parties. It is submitted that only a proprietary protection will ensure that market forces will operate to generate fairness and equity.[42]

An important step in the future evolution of the ABS regime is the recommendation of the Johannesburg World Summit to negotiate a binding instrument or binding instruments for its concretization and implementation. It will also be crucial that in this process the aspects of the market and the marketing of PGRs and TK are taken account of.

Value of PGRFA and TK and the creation of markets

The value of PGR, PGRFA and associated TK for research and development in industrial production processes is controversial (see Chapter 1, this volume).

In the pharmaceutical sector, industries show some interest in TK as a 'potentially valuable source of creativity and invention outside the communities from which the knowledge originates' (WBCSD/WZB, 2003). However, some firms like Novartis point out that 'the potential contribution of bio-prospecting to the biotechnology industry and the communities involved may not be as substantial ... as previously believed' (1999, p. 5).[43] Accordingly, Baruffol (2003) concludes that, even if natural compounds and their derivatives may continue to be of interest for the pharmaceutical industry, 'bioprospecting, at least within the next years, will not have the potential to create ... conservation incentives to a larger extent' (p. 112).

---

[39] WTO, Council for TRIPS. The Protection of Traditional Knowledge and Folklore: Summary of Issues raised and Points made. IP/C/W/370, 8 August 2002, No. 23.

[40] WTO Doc. IP/C/W/209, Review of the Provisions of Article 27.3(b): Further views of the United States, p. 6.

[41] WTO Doc. IP/C/W/228, No. 23. Compare also the statement by India: 'the time, effort and money involved in getting individual patents examined and revoked in foreign patent offices is prohibitive' (WT/CTE/W/156, No. 10).

[42] WTO Doc. IP/C/W/228, Review of Article 27.3(b), Communication from Brazil, No. 34.

[43] See also the statements in Baruffol (2002, p. 112).

For PGRFA, Correa (2000) draws a similar conclusion. He maintains that the market value produced by the gene flow from farmers' varieties to privately marketed cultivars is very modest,[44] and concludes that 'though it is expected that the demand for primitive materials may increase in the future ... it would be unrealistic to think that substantial value may be derived from current gene flow of farmers' varieties held in *in-situ* conditions' (p. 10). The International Association of Plant Breeders (ASSINSEL) points out that genetic diversity in landraces and wild species represents resources with limited present value for breeding purposes for most species. It is argued that they require a great deal of time and effort to explore, investments which – as a rule – commercial breeders cannot afford (1998, p. 1). From this it can be concluded that the benefits resulting from industrial companies prospecting PGR, PGRFA and additional TK will be less substantial than formerly expected or hoped for.

Therefore, it is submitted that the generation of benefits first requires the development of respective markets. In this context, two aspects are perceived as being essential: first, it seems to be crucial not only to market the genetic, biochemical or informational resource as such, but also to identify and develop markets for primary, semi-processed and processed products, produced on the basis of the biological resources and their associated TK. Accordingly, the markets for the 'raw' products need not only to be situated in an international ('South–North') context, but production chains and markets for processed products may also be created and found on the local, regional and national level.[45] This line of thought indicates that the issue of access to genetic resources and associated TK might not only be an issue in the international North–South context, but also on the local or national level. From this follows, secondly, that 'benefits' may not exclusively consist of direct, economic remuneration and compensation, but may also be generated by the sharing of the outcomes of (cooperative) research, which allows for the generation of a value-added product in the provider country itself.

The creation of markets and successful marketing, including the exploitation of instruments to support the marketing of particularly high-value or high-quality products, such as geographical indications, trade marks and labelling, all need to be evaluated and the option to create supporting measures in the framework of the world trade order are to be assessed (Chapters 6 and 8, this volume).

Marketing and access legislation

Important factors influencing the market for genetic resources and associated TK are the transaction costs and legal security in view of the legitimacy of the transaction.[46] Novartis, for example, cautions against too complicated access procedures and points out the connection between access legislation and marketability of the information: 'if access to biodiversity becomes too bureaucratic, time consuming and expensive, then the importance of biodiversity may become limited to research and academic arenas, instead of market-oriented industrial applications' (Novartis, 1999).[47]

---

[44] According to one study, materials from *ex situ* gene banks contributed 3% of the germplasm used by industrial breeders, and materials from *in situ* conservation areas a further 1% (Swanson and Luxmoore, cited in Correa, 2000, p. 10).

[45] See Gupta (1999/2000) Securing traditional knowledge and contemporary innovations: can global trade links help grassroots innovation? See also as examples the initiatives of community-based commercial cultivation of indigenous medicinal plants by the International Centre of Insect Physiology and Ecology (ICIPE) and the University of Nairobi (http://www.icipe.org).

[46] No pharmaceutical company finds favourable, after having carefully negotiated a contract with a community, acquired the information, and developed a product, to be then 'in the last minute' accused of biopiracy.

[47] See also WBCSD-WZB (2002, pp. 18–23, in particular Nos 76 and 77).

In this context, Cabrera Medaglia (2003) points out that – instead of promoting access – the national regulations of some countries have focused more on defensive measures, introducing strict controls to prevent 'biopiracy'. This generates high transaction costs and bureaucratic procedures and the absence of applications for access. He argues that 'as long as the idea persists that access represents a way of colonialism instead of a mechanism for the generation of appropriate joint initiatives for all participating parties, the possibilities of generating desirable results will be much more limited'. That means, first, that in order to promote the marketing of TK and PGRFA, access procedures must be transparent, flexible and as simple as possible.[48] They must assure legal security in providing a clear legal basis for the transaction, and designate the parties to be integrated on the provider side in the negotiation and conclusion of the contract.

The question is whether ways can be found to make these criteria compatible with the option to integrate the grassroots providers of the information into the process. It is suggested that such integration might even be advantageous: the assent of all stakeholders could be secured, mutual trust created and the risk of (political) opposition minimized.

However, specific conditions must be fulfilled in order not to complicate the procedure: the process and its participants, including the holders or owners of the resources, must be defined clearly. From the perspective of the communities or individual holders of the information, it would be advantageous to clearly identify the marketable information; this might in particular be the case where the information is integrated in a sacred, spiritual context. Documentation and registration might be a suitable means to this end (see (d), below and Chapter 6). Further, the competencies and decision processes within the communities must be established.[49]

Control on the side of the users of the information

According to Cabrera Medaglia (2003), there exists a connection between the defensive access legislation of donor states and the lack of control mechanisms on the side of the users of the information. He maintains that implementation problems are the reason behind the defensive, 'draconian' and prohibitive access regulations adopted by the provider countries to prevent biopiracy.

The problem is that bilateral contractual agreements on ABS are effective only between the parties to the contract: third parties are not bound by their terms. Therefore it is difficult for the providing party to follow the R&D process, to be aware of possible resulting IPR and to control the legitimate use of the information by its recipients. This is true for state agencies of the providing countries,[50] but even more so for providers at the community level.

The same is also true for the control of benefit-sharing agreements. Apart from any up-front payments, a fair sharing of the benefits requires the disclosure of acquired benefits and transparency with a view to net gains, which is all based on mutual trust.

Decision 391 of the Andean Pact countries provides for sanctions of infractions against its provisions outside the Andean Community. According to Articles 46 and 47 and the second complementary provision, illicitly attributed property rights must not be acknowledged within the Andean Community, and further access to PGR and TK by the violator can be denied. This solution, which applies only to the

[48] This is in particular true for access for academic research; the ITPGRFA might provide valuable inspiration to this end.

[49] See, for example, the Novartis-Uzachi case, where it was an important prerequisite for success that the communities had established decision processes in the management of their natural resources. Baruffol (2003, p. 114).

[50] See Communication from Brazil to the Review of Article 27.3 TRIPS (IP/C/W/228) No. 24.

Member States of the Cartagena Agreement, illustrates the imperfection of the system in view of the international or global dimension of the interests involved and the international character of the markets.

In sum: the implementation problems brought about by the internationality of the issue and the details of the contractual solution encompass not only the control of the legitimate use of the information, but also the question of bringing about fairness in the sharing of benefits.

The proposed remedy is the adoption of control measures on the side of the recipients and users of the information. It is argued that this would enable the providers to streamline their legislation in a more user-friendly way, supporting the building of trust and the generation of joint initiatives for all participating parties.[51]

This conclusion is furthermore backed by an equity argument. It is submitted that the principle of equity demands that the burden of regulating ABS be borne by both providers of genetic resources and TK and the recipients. This is a strong argument for the obligation to create supportive measures, such as the control of the legitimate access to genetic resources and related TK, to be applied on the side of the recipients.

The designation and integration of holders of TK, including capacity-building

Another problem of the contractual approach is that we have to deal with a complex situation involving a variety of stakeholders on different political levels and in different countries. As mentioned above, in order to determine participation in access negotiations, it is crucial to determine the holders of TK and/or PGRFA. The absence or uncertainty regarding ownership implies difficulties in securing PIC.

This issue has two aspects: on the one hand, the original owners or holders of the information must be determined; and on the other hand, transparency is to be secured throughout the process, which might imply various 'layers' of stakeholders.

THE NECESSITY OF IDENTIFYOMG THE HOLDERS OF THE INFORMATION

The approach for identifying or designating the holders or owners of PGRFA seems to differ from that applied to associated TK.

The issue is complex because of the great variety of possible situations. The nature of TK, access to it and competence over it all depend on the social, environmental and cultural context in which it is found. It might be in some cases that farmers' varieties can be clearly allocated to specific farming communities (see Chapter 4, this volume), whereas TK associated with medicinal plants is considered to be common to an entire biogeographical region. On the other hand, medicinal TK can be entirely in the hands of one individual shaman, whereas PGRFA may not even be specifically considered as the result of a creative process.

Whereas it seems to be generally acknowledged that the negotiation of an ABS contract has to include the holders of the knowledge, and that contracts must be based on their prior informed consent and on terms mutually agreed with them,[52] this is not always the case for the on-farm-bred PGRFA. For instance, in the Andean Common Regime on Access to Genetic Resources[53], the rights over all genetic resources are considered to belong to the patrimony of the Nation of each Member Country. They are – as 'goods of the Nation' – assigned to the state, independent of the legal regime applicable to the biological resources, meaning the physical entity that

[51] See CBD COP Dec. V/26, No. 4, reiterated in COP CBD Dec. VI/24, No. 8 (c).

[52] See, for example, Guidelines Nos. 17 ff.; Peru, Law No. 27811 on the Protection of Collective Knowledge of Indigenous Peoples derived from Biological Resources, Chapter 2, this volume; Costa Rica, Law on Biological Diversity, described in Cabrera Medaglia (2003).

[53] Decision 391 'Common Regime on Access to Genetic Resources', Commission of the Cartagena Agreement, July 1996.

contains them or the associated knowledge.[54] Similarly, in Costa Rica, genetic and biochemical wild and domesticated resources apparently belong to the state (Cabrera Medaglia, 2003).

However, it is also maintained that in the case of PGRFA[55] the farmers or traditional on-farm breeders ought to be designated as the holders of their varieties wherever possible, particularly in the case of semi-formal on-farm breeding of traditional varieties. As discussed earlier, this would allow their participation in the contractual ABS system or in the MS.

The creation of registration systems, possibly combined with a licensing system, is proposed to resolve the problem of the identification and allocation of associated TK to its holders (Chapter 6).

THE NECESSITY OF TRANSPARENCY IN TRADE

In respect of companies and research institutes seeking access, the body responsible for negotiations is the agency designated by the state. The problem frequently is that both within purchasing and providing countries several 'layers' of stakeholder exist, and the information is passed on by several intermediary institutions or agencies.[56] In order to be able to verify the legitimacy of the access it is necessary to make such procedures transparent. Proposed instruments to do this include certificates of origin, or licensing systems[57] to prove the legitimacy of acquisition, and thus guarantee legal security. In all these systems, however, the tracing of processes and the proof of legitimacy are bound to be complicated. In any case, a great amount of capacity-building is necessary, not only on the side of the holders of TK, but also on the side of intermediary agencies and purchasers.

7.3.4 Conclusions

The application of the system of ABS to TK reveals a highly complex situation. Despite interpretative assistance provided by the Bonn Guidelines and the more exact differentiation found in the ITPGRFA, many questions remain unanswered, especially regarding the involvement of grassroots stakeholders. The management of ABS of TK associated with PGR and traditional PGRFA in bilateral contractual instruments has the advantage of flexibility, which allows its adaptation to different situations, on a case-by-case basis. However, until the present time few success stories have been reported relating to either the sharing of benefits or to the integration of local people, and, as described above, the operation of the principles proves to be difficult in practice and it has the negative impact of actually discouraging exchanges of materials that are essential to breeding, research and conservation.

Approaches for the solution of these problems might lie in the further evolution of the Bonn Guidelines and, within the ITPGRFA, in the implementation of farmers' rights and the further development of the elements of the Multilateral System (Material Transfer Agreements and Funding System).

---

[54] See in detail Rosell (1997). Rosell takes a critical approach to this solution, stating that the property rights regime envisaged by the Decision might be less effective than one based on private property rights. 'From a policy point of view, Decision 391 could have gone further towards recognizing the individual as the owner of the genetic resource, so as to benefit directly the national who has conserved and made available the resource. This in turn might have enhanced the control and enforcement mechanisms which constitute its weak aspects' (p. 282).

[55] As to the problems of allocation of traditionally 'on farm'-bred varieties see Chapter 4, this volume.

[56] Example of Zimbabwe: Local healers transmitted information and plant samples to the local university. The local university transmitted the samples to a Swiss university. The Swiss university sold the results of research to an American company.

[57] Tobin (1997, p. 337); Peruvian Law No. 27811 on the protection of the collective traditional knowledge of indigenous peoples, Articles 25–33. Certificates of origin to prove the legitimate acquisition of a good are a common instrument to control, for example, trade in endangered species.

It is submitted that a broad variety of supporting measures are needed to allow the providers of TK and traditional PGRFA to increase the benefits received from their resources and wisdom. The following points are considered to be essential:

- Instruments and institutions for the clear identification of local stakeholders are to be created, such as documentation, registration and *sui generis erga omnes* rights.[58] It is necessary to support a bottom-up approach, where the entitled stakeholders are clearly designated and integrated in the instruments.
- Capacity-building measures are necessary for both the providers and purchasers of the biological resources.
- The obligation by the providers and providing countries to facilitate access must be balanced against the obligation of recipients to create control instruments. Control instruments (for legitimate access) are to be created either on the national or international level (or both).
- Technology transfer is to be furthered, in particular to support the development of processed products and to encourage trade on the local, regional and/or international level. Instruments to support trade in high-level products should be evaluated and the options within WTO assessed.
- It might be advisable to find means to assess the cost-benefit ratio of measures, e.g. the potential market value of the resources in question in particular countries, so as not to go down the route of expensive regulation for no return benefits.

7.4 Financial Mechanisms for Compensation of Non-assignable Traditional Plant Genetic Resources and Traditional Knowledge[59]

7.4.1 The role of a future financial mechanism in the context of access to genetic resources

Aims and functions of a future financial mechanism

This section examines the possibility of establishing an international mechanism for providing financial resources for the valuation of TK associated with PGR and traditional PGRFA, where neither IPR nor *sui generis* rights can be allocated.[60] In the absence of allocated rights, the challenge will be to design a financial mechanism that can substitute the compensation of TK holders through the exercise of the relevant rights by these holders. Ways and means must be found to generate financial means and to direct them to the holders of the knowledge in a way that supports the preservation and maintenance of TK, the promotion of its wider application with the approval and involvement of the holders, and the equitable sharing of the benefits, as set out in Article 8(j) of the Convention on Biological Diversity (CBD). Ultimately, the financial mechanism should contribute to furthering the aims of the CBD as they relate to plant genetic resources, and those of the ITPGRFA, namely the conservation and sustainable use of the genetic resource base and the equitable sharing of benefits derived from it.

In order to achieve this overarching objective, the mechanism should fulfil the specific functions outlined below.

First, a central function of a future financial mechanism will be to promote the equitable sharing of benefits derived from plant genetic resources, and to compensate

[58] See WTO IP/C/W/370, 24–28.

[59] Author: Katharina Kummer Peiry. I wish to thank Juliette Voïnov, Martin Girsberger and Susette Biber-Klemm for their input and comments on earlier drafts of this section.

[60] The problem of allocation of intellectual property rights and *sui generis* rights is discussed in Chapter 4, this volume.

holders of TK associated with PGR and traditional PGRFA for the service they provide to users of the resources and to the international community at large. As far as the financial burden of contributing to the mechanism is concerned, this should be shared among interested persons and entities in a manner that is perceived as fair and equitable by all. This will enhance the legitimacy of the mechanism and the willingness of actors to contribute to it. Criteria to be considered in equitable burden sharing include the financial capacities of the actors on the one hand and the extent of their use of the resources on the other.

Secondly, sufficient financial means, the availability of which is assured on a long-term basis, are indispensable for a functioning financial mechanism. Without such a basis, a mechanism is not able to operate in accordance with an ongoing work plan and budget. In addition, projects aimed at sustainable resource management often have a long life span. An important element of the predictability of funding is quantification of the incoming funds. The creation of a broad financial basis also calls for the inclusion of as many sources of funding as possible. In building a financial mechanism, a central question thus concerns the means by which the most funding can be generated in the most reliable and predictable manner.

Thirdly, a future financial mechanism should ideally influence the behaviour of relevant actors by creating incentives for preserving and maintaining TK associated with PGR and traditional PGRFA. This includes the valuation and support of efforts at conservation and sustainable use of plant genetic resources, and the empowerment of TK holders. In accordance with the nature of the mechanism, these incentives will be of a financial nature. The mechanism should provide the appropriate incentives both for users of PGR and PGRFA, who are the prospective contributors to the mechanism, and for holders of TK associated with PGR and traditional PGRFA, who are the prospective recipients. As regards users, the creation of incentives for the private sector (in particular, relevant industry depending on access to specific resources) to contribute to the mechanism is a particularly important aim.

Elements of a future financial mechanism

If an international financial mechanism is to be established for the purpose of achieving the above aims and functions, the following elements of the mechanism will need to be defined.

First, as the financial mechanism is to operate at the international level, it should be established in an international framework, either as part of an existing international body or as an independent international institution. The mechanism should be established by an international legal instrument, which would constitute the legal basis of its operation. As a minimum, it needs to feature an executive body with the competence of generating financial means, in general by collecting contributions, and of disbursing funds to recipients. It should also have a supreme organ to which the executive body is accountable. Rules must be elaborated for the generation, administration and disbursement of funds. This will include a procedure for submission of claims and for decisions regarding the disbursement of funds. Depending on the size of the operation, the body will need to be supported by a secretariat. The future mechanism must have a legal personality recognized by all parties involved in its operation. As it will be established at the international level, this must include recognition by the legal systems of participating states.

Secondly, the sources of funding and the methods of generating funds must be defined. There are various possible avenues. In view of achieving a stable and predictable financial basis, the most obvious is the establishment of a system by which defined contributors (states and/or private entities) must make regular financial contributions to the financial mechanism. This can be done, for example, by using a scale of assessment setting out criteria for the level of contribution, in accordance with the financial position of the contributor.[61] Another method to be considered is a levy on a certain type of activity or on the use or consumption of a certain type of good. In the present case, this would mean any direct or indirect use of TK associated with PGR and traditional PGRFA. This approach has the advantage of placing the financial burden on the persons benefiting directly from the pertinent activity, and making the size of the contribution directly dependent on the extent of economic benefit derived from the TK and resulting genetic resources. This would constitute a contribution to equitable burden sharing.

Thirdly, the contributing entities must be defined. Bearing in mind the function of compensating holders of TK associated with PGR and traditional PGRFA, for their service to those using the resulting resources, contributors to the financial mechanism should be entities that reasonably can be assigned financial responsibility in this context. An obvious motivation for assigning such responsibility is the fact that the contributor directly or indirectly benefits from TK associated with PGR and traditional PGRFA. This is the case for any 'user' of such knowledge, notably private enterprises (for instance, seed companies) and research institutions. The motivation for assuming financial obligations could also be the more general role of the state in ensuring the protection of fundamental interests of its nationals as well as the conservation and sustainable management of its genetic resources base. Thus it would be conceivable also to devise a system by which contributions are made by states (Girsberger, 1998). In view of achieving a broad financial basis, the establishment of a financial mechanism involving contributions both by states and by private entities may be the most efficient approach.

Fourthly, the claimants need to be defined. As the mechanism has the function of providing financial compensation to holders of TK associated with PGR and traditional PGRFA, in cases where the allocation of traditional or *sui generis* IPR is not possible, the definition of the claimants should not be too restrictive. At the substantive level, it should be possible to define as claimant any entity that is either a holder of TK, or working to promote the interests of TK holders, independently of whether or not the entity in question assumes a recognized legal form, in order to accommodate the often informal character of TK holders. At the procedural level, however, it will be necessary to define the holder as a natural or legal person, in particular for the purposes of the procedure to be established for the submission of claims and the disbursement of funds. In order to reconcile these conflicting needs, it will be necessary to find a way of designating persons or organizations as formal representatives of relevant informal communities, and to require appropriate legitimization. Hence claimants could be defined as organizations formed by holders of TK associated with PGR and traditional PGRFA (for instance, village council or similar), or NGOs with a recognized curriculum in the pertinent field.[62] In the case of utilization of generally known information in the context of PGR (for instance, neem), it is conceivable that entire countries, provinces or communities be defined as recipients. The verification of a claimant's legitimacy will be one of the tasks of the body operating the fund.

Fifthly, it will be necessary to define the criteria for disbursement of funds. The flow of financial resources must be designed so as to support the aims as outlined above. In view of the often informal nature of holders of TK associated with PGR and PGRFA, and traditional PGRFA, which may make representation by a natural or legal person necessary, it is important that such representatives can only obtain funding for activities that demonstrably support the aims. An approach to be investigated is to

[61] This is the method of defining contributions to certain UN institutions or conventions by member states. The UN scale of assessment is based on the GNP of the states.

[62] Girsberger (1998, p. 308) mentions the possibility of claimants being NGOs representing holders of Farmers' Rights.

disburse funds for the implementation of programmes or projects that promote one or more of the aims outlined above. Such projects would have to be submitted to the body administering the fund, which would determine whether or not the project met the criteria.[63]

7.4.2 Analysis of existing mechanisms as possible models

This paragraph analyses a number of existing multilateral financial mechanisms in international law, with a view to establishing their potential usefulness as models for a future financial mechanism for TK associated with PGR and traditional PGRFA, taking into account the aims discussed above. Emphasis is placed on financial mechanisms that feature the elements discussed above. Where it is considered to be of particular interest, instruments will also be discussed that are not financial mechanisms but include one or more of the above elements. The following paragraphs discuss two different types of financial mechanisms used in international environmental law, namely assistance funds and compensation funds, as well as the Flexible Mechanisms of the Kyoto Protocol, which may provide some interesting insights. It then goes on to provide an overview of pertinent ongoing work within the framework of FAO.

There are a variety of existing and prospective financial mechanisms in international environmental law. Their structure and functioning are conceptually different in accordance with the aims for which they have been established. The following fundamentally different purposes of a mechanism of this type can be distinguished.

Financial mechanisms aiming at the provision of assistance to developing and transitional countries ('assistance funds')

This type of financial mechanism is used in the field of environmental protection. Its aim is generally to assist recipient states in meeting their obligations under international law, and to develop or strengthen their capacities and infrastructure for adopting environmental protection measures at the national level. Where the mechanism is established in the framework of an MEA, its function is to assist party states in the implementation of that MEA at the national level. Assistance funds are established by an international legal instrument, and operated by a multilateral organization. The establishing instrument determines their management and administration, as well as the means of generating funds and the criteria for distribution. Both donors and recipients can only be states: this type of fund does not provide for contributions or claims by non-state entities. The establishing instrument defines donor and recipient states, and the scale of assessment for contributions of donor states and the criteria for financing activities in recipient states. As a general rule, donor states are industrialized countries, whereas recipient states are developing countries and countries with economies in transition. Where the financial mechanism is linked to an MEA, only contracting parties can be donors and recipients.

Special types of assistance funds are trust funds for particular activities to be carried out in the framework of an MEA. These funds also aim at supporting the implementation of the MEA, but their focus is narrower: they support activities related to the operation of the MEA and the participation of a particular category of countries in these activities. Examples of MEAs establishing trust funds are the conventions for which UNEP acts as Secretariat, including the CBD, the Basel Convention on hazardous wastes, CITES, and the Vienna Convention and its Montreal Protocol on ozone-depleting substances. A trust fund has also been established for the Climate Convention and its Kyoto Protocol. Under the Kyoto Protocol, a number of funds to support climate-related activities are to be established after its entry into force.[64] This type of trust fund

[63] For a similar line of argument see Girsberger (1998, p. 308).

[64] COP Decision 5/CP.6 (2001), Section I.

is established and operated as a part of the institutional framework of the relevant agreement; no special body is established for this purpose.

In some cases, the entire budget of a convention is financed through a trust fund, to which all contracting parties contribute in accordance with an agreed scale of assessment (for instance, Basel Convention, CBD, CITES, ozone treaties). In these cases, the budget is determined by the Conference of the Parties (COP). By contrast, contributions to trust funds established for special purposes are usually voluntary. Such trust funds support a fairly narrow and clearly defined range of activities. Beneficiaries are usually developing countries, in some cases including countries with economies in transition. Purposes for which special trust funds are set up include the participation of developing and transition countries in negotiating meetings within the framework of the convention (Climate Convention, CBD, Basel Convention), technical assistance to developing countries (Basel Convention, Ramsar Convention on wetlands), and financing of special Secretariat support to Parties, for instance, workshops, seminars and websites (Climate Convention).[65]

THE GLOBAL ENVIRONMENTAL FACILITY[66]

The Global Environmental Facility (GEF) was established in 1991 and restructured in 1994 through the adoption of the Instrument for the Establishment of the Restructured Global Environment Facility ('the Instrument'). The Instrument lays down the fundamental principles of the operation of the GEF, including, *inter alia*, governance and structure, principles of decision-making, beneficiaries, as well as contributions of participating countries during the first replenishment period (from 1994 to 1997).

The objective of the restructured GEF is to serve as a mechanism for international cooperation for the purpose of providing new and additional grant and concessional funding to meet the agreed global environmental needs in the following focal areas: (i) global warming/climate change; (ii) pollution of international waters; (iii) loss of biological diversity; (iv) depletion of the stratospheric ozone layer; (v) persistent organic pollutants; and (vi) land degradation. The GEF supports activities in the above areas through projects on a grant or concessional basis. The projects and other activities are generally carried out in cooperation with the institutional mechanism of the conventions addressing the issues in question, where such a convention exists.

The GEF is jointly operated by the World Bank, the United Nations Development Programme (UNDP) and the United Nations Environment Programme (UNEP). Its principal mechanism is the GEF Trust Fund. Other features are the independent Scientific and Technical Advisory Panel to assist in the development of criteria for project selection and to review and comment on project proposals, and the Small Grants Programme for NGOs. There are also additional co-financing agreements. The World Bank carries out the secretariat functions for the GEF and is Trustee of the GEF Trust Fund. It is also responsible for the GEF-financed investment projects. UNDP provides technical assistance, identifies projects and runs the Small Grants Programme for NGOs. UNEP provides the secretariat for the Scientific and Technical Advisory Panel and contributes environmental experience. Any member state of the UN or of any of its specialized agencies may become a participant in the GEF by depositing an instrument of participation in accordance with the Instrument. With 174 participants (as of April 2003), membership of the GEF is nearly universal.

---

[65] This is discussed by Ten Kate and Lasén Diaz (1997, p. 289); in the *Yearbook of International Cooperation on Environment and Development 1999/2000*, p. 92 (Climate), p. 108 (Basel Convention), p. 169 (CBD) and p. 174 (CITES); and by Bragdon (2001, p. 2).

[66] See Glowka *et al.* (1994, p. 107); *Yearbook of International Cooperation on Environment and Development 1999/2000*, p. 220; GEF website: http://www.gefweb.org

The governing bodies of the GEF are the Assembly and the Council. The Assembly, in which all participating states are represented, reviews the general policies of the GEF, and evaluates its operation on the basis of reports submitted by the Council. The Council is the main governing body, responsible for developing, adopting and evaluating the operational policies and programmes for GEF-financed activities. It is composed of 32 members (16 from developing countries, 14 from developed countries and two from Central and Eastern European countries).

In accordance with the Instrument, beneficiaries of the GEF are the countries eligible to borrow from the Word Bank, or eligible for technical assistance from UNDP, i.e. countries with a per capita income of less than US$4000 per year. The following principal criteria for project selection (grants) are applied: (i) the project must benefit the global environment, and (ii) the project must be innovative.

Any country (developed, developing or transitional) can pledge contributions to the GEF in accordance with the criteria laid down in the Instrument. Contributions by developed countries as laid down in Annex C to the Instrument are roughly in line with a formula based on their shares in the World Bank's International Development Association. For the second replenishment, 28 countries announced pledges to the GEF Trust Fund, including ten developing countries.

In accordance with Article I para. 6 of the Instrument, the GEF operates the multilateral financial mechanisms of several environmental conventions. In operating the financial mechanism of a convention, the GEF carries out projects and other activities related to the aims, priority areas and work programmes of the convention in question. These activities are determined in close cooperation between the COP to the agreements in question and the Council of the GEF. The COP to the multilateral agreements, in accordance with the applicable provisions of the agreements, regularly provides guidance to the GEF with respect to the operation of the financial mechanism, and the Council of the GEF regularly reports to the COP of the agreements on activities carried out and planned in accordance with guidance received. Under the UN Framework Convention on Climate Change (Climate Convention), the COP has adopted a number of decisions providing guidance to the GEF in fulfilling its function as the financial mechanism of the Convention, in accordance with its Articles 11 and 21 and with Article I para. 6 of the GEF Instrument.[67] At its second session in 1996, the COP adopted a Decision that brought into force a Memorandum of Understanding between the COP and the GEF Council concerning operation of the financial mechanism by the GEF on an interim basis.[68] At its fourth session in 1998, the COP designated the GEF as the entity operating the financial mechanism on a permanent basis, subject to review every four years.[69] The GEF is also available to meet the agreed full costs of activities under Article 12 of the Climate Convention. Investment projects financed by the GEF under the Climate Convention focus on the reduction of greenhouse gas emissions by increasing energy efficiency and the use of renewable energies, as well as support to developing countries to implement the Convention and to prepare national communications to the COP. The GEF also promotes bilateral and multilateral co-financing and the leveraging of private sector participation and resources. In its role as the operating entity of the financial mechanism of the Climate Convention, the GEF has to date provided about US$1.125 billion in the form of grants from the GEF Trust Fund for climate change projects in

[67] Cf. COP Decisions 10/CP.1, 11/CP.1, 12/1 (1995), 10/CP.2, 11/CP.2 (1996), 2/CP.4, 12/CP.4 (1998), and 8/CP.5 (1999).

[68] COP Decision 12/CP.2 (1996).

[69] COP Decision 3/CP.4 (1998).

non-Annex I countries.[70] In accordance with Article 11 of the Kyoto Protocol, financial assistance to developing countries adopting measures under the Protocol is to be provided, *inter alia*, through the GEF as the financial mechanism of the Climate Convention.[71] This task of the GEF will become operational after the entry into force of the Protocol.

On an interim basis, the GEF also operates the financial mechanism of the CBD and its Cartagena Protocol on Biosafety, and that of the Stockholm Convention on Persistent Organic Pollutants (POPs), as provided in these agreements. The modalities are largely the same as under the Climate Convention.

THE MULTILATERAL FUND OF THE MONTREAL PROTOCOL ON SUBSTANCES THAT DEPLETE THE OZONE LAYER[72]

The Multilateral Fund of the Montreal Protocol is the best-known example of a financial mechanism set up within the framework of an MEA to assist developing country parties in complying with their obligations under the agreement in question. Its establishment in the early 1990s was considered a major breakthrough in favour of developing countries, which generally prefer an independent financial mechanism to a mechanism under the responsibility of the Bretton Woods Institutions, including the GEF (Lang, 1992). Indeed, the Multilateral Fund of the Montreal Protocol remains the only independent financial mechanism in a multilateral environmental agreement to date, later treaties having entrusted the operation of their financial mechanism to the GEF (see above). In accordance with Article 10 of the Montreal Protocol, which forms part of the so-called London Amendment of 1990, the Fund was established initially on an interim basis to meet agreed incremental costs to developing countries of implementing the control measures of the Protocol. In 1993, the Multilateral Fund was established on a permanent basis. The implementing agencies are UNEP, UNDP, the World Bank and – as of 1992 – UNIDO, with UNEP also serving as treasurer. Contrary to the treaties discussed in the previous section, the Montreal Protocol does not entrust the governing of the financial mechanism to an outside UN institution, but assigns this function to the Executive Committee, which is composed of Parties to the Protocol. The Meeting of the Parties (MOP) to the Montreal Protocol periodically approves a three-year budget of roughly US$500 million.

Beneficiaries of the Multilateral Fund are Parties that benefit from a more favourable schedule under Article 5 of the Montreal Protocol (i.e. Parties that are developing countries and have a yearly per capita consumption of less than 0.3 kg of CFCs and halons). Contributions are made by Parties that do not operate under Article 5, in accordance with the UN scale of assessment. Other Parties may contribute on a voluntary basis. The Fund has been replenished four times to date: US$240 million (1991–1993), US$455 million (1994–1996), US$466 million (1997–1999) and US$440 million (2000–2002). As at 28 February 2001, the contributions made to the Multilateral Fund by some 32 non-Article 5 countries amounted to US$1.22 billion.

The Executive Committee of the Multilateral Fund, in which 14 Parties are represented (seven Article 5 and seven non-Article 5 countries), governs the operation of the Fund. It is elected by the MOP for one calendar year. The positions of

[70] The countries not listed in Annex I of the Climate Convention comprise all developing countries parties to the Convention.

[71] See *Yearbook of International Cooperation on Environment and Development 1999/2000*, p. 90 ff.; UNFCCC website: http://www.unfccc.de

[72] On the Multilateral Fund of the Montreal Protocol, see generally *Yearbook of International Cooperation on Environment and Development 1999/2000*, p. 98; website of the Ozone Fund: http://www.unep.org/ozone/finances.shtml

Chairperson and Vice-Chairperson alternate between the two groups of countries. The Executive Committee is responsible for the development of operational policies of the Fund, criteria for project eligibility as well as other guidelines and administrative arrangements, monitoring of the implementation of these policies, approval of implementing agencies' business plans and work programmes, approval of expenditures for investment projects and other activities, allocation and disbursement of resources, and the monitoring and evaluation of performance.

The Fund Secretariat was established in 1991 to assist the Executive Committee in the discharge of its functions. Its activities include the development of a 3-year plan and budget as well as a system for disbursement, the management of the business planning cycle of the Fund, the monitoring of expenditures and activities of the implementing agencies, the preparation of policy papers and other documents, the review and assessment of investment projects, country programmes and the business plans and work programmes of the implementing agencies, liaising between the Committee, governments and implementing agencies, and servicing meetings of the Executive Committee. The Secretariat also carries out the function of monitoring and evaluating the effectiveness of the Fund, which was introduced by the Executive Committee in May 1997. The Secretariat thus has important substantive functions in the operation of the Fund, which go far beyond the administrative and supporting functions generally entrusted to secretariats of international agreements and institutions.

OTHER ASSISTANCE FUNDS IN THE ENVIRONMENTAL FIELD

Additional assistance funds include notably the World Heritage Fund of the UNESCO Convention Concerning the Protection of the World Cultural and Natural Heritage, the Small Grants Fund of the Ramsar Convention on wetlands, and the Global Mechanism of the UN Convention to Combat Desertification (Bragdon, 2001). The objective of these mechanisms is to fund projects in developing countries to support the implementation of the agreement in question. The institutional structure and management of funds are similar to the Montreal Protocol Fund. Some funds provide for assessed obligatory contributions from industrialized states parties to the MEA, which are generally rather modest. However, the larger part of the contributions is of a voluntary nature. Accordingly, the financial basis is generally more modest than that of the Montreal Protocol Fund.[73]

ASSISTANCE FUNDS AS MODELS FOR PLANT GENETIC RESOURCES AND TRADITIONAL KNOWLEDGE

Funds of this type feature all the elements discussed above. As far as the institutional infrastructure governing the mechanism is concerned, the existing assistance funds may well have a certain model function for equivalent efforts in the field of PGR. In terms of generating a stable and predictable financial basis, the system of compulsory contributions based on a scale of assessment could also provide a useful model.

The function of assistance funds, however, is a different one from that underlying a future mechanism for PGR, namely capacity-building for states to undertake environmental protection activities at the national level. Consequently, states are the only actors in this type of financial mechanism, and the mechanisms depend entirely on contributions from states, whether on a compulsory or on a voluntary basis. In order to fulfil the function of compensating TK holders who often have an informal character and are nearly always non-state actors, appropriate changes would therefore need to be made if the concept of an assis-

[73] The 1997 Budget of the UNESCO Fund was US$3.5 million: the yearly budget of the Global Mechanism of the Desertification Convention is similar, and that of the Ramsar Small Grants Fund is a few hundred thousand Swiss francs. See Bradgon (2001).

tance fund were to be used as a model.

Another constraint is the absence of incentives for contributors. As discussed above, an important factor in devising a financial mechanism for plant genetic resources is the creation of incentives for preserving the genetic resource base. Also, a financial mechanism is likely to be more successful if there is a direct or indirect benefit flowing from contribution. By contrast, the assistance funds offer no incentives for contributors. Donor states derive no concrete benefit from their contribution, but contribute on the basis of the concept of a historical and economic responsibility of the richer towards the poorer states, and of the shared interest in the conservation of the natural resources of the globe. The absence of direct incentives makes this mechanism inherently less attractive for donors. As far as recipients are concerned, an assistance fund does provide certain incentives for acting in accordance with the aims of an MEA, as only pertinent activities are financially supported in the recipient state.

Precedents show that the establishment of an assistance fund is generally a difficult issue in international negotiations, and one that creates a fundamental North–South divide. Developing countries usually favour an independent financial mechanism, as in the Montreal Protocol, an attitude in part due to these countries' inherent mistrust of the GEF. Developing countries also generally want to endow the fund with far-reaching responsibilities. Arguing that they lack capacities for implementing MEAs, they strongly support making the extent of their implementation of a given treaty dependent on the level of funding they receive from the mechanism.

Developed countries, on the other hand, are usually opposed to the establishment of an assistance fund. Since such mechanisms are generally administered by a body with equal representation of recipient and donor countries, as in the case of the Multilateral Fund of the Montreal Protocol, or with a majority of recipient countries, as in case of the GEF, developed countries feel they would be contributing to a mechanism over which they do not have control. They are also reluctant towards the creation of new institutional infrastructure with the resulting additional costs and bureaucracy. Developed countries therefore generally prefer existing bilateral channels, namely overseas development cooperation, to be used for financial support of implementation of an MEA by developing and transition countries. In addition, budget cuts in many government administrations in recent years have led to a reduction of the amount of money available for international financial mechanisms, which is an additional reason for prevailing reluctance on the part of most developed countries. This political difficulty should also be taken into account if this type of financial mechanism were to be considered as a model.

Financial mechanisms forming part of a treaty system on civil liability for environmental damage caused by potentially hazardous activities ('compensation funds')[74]

The financial mechanisms discussed in this paragraph, often referred to as 'compensation funds', are designed to supplement an international civil liability regime through provision of compensation to persons having sustained damage as a result of a potentially hazardous transaction, in cases where compensation is unavailable or only partly available under the civil liability provisions. They are established within the framework of an international legal regime on civil liability for damage caused by potentially hazardous activities, either by the treaty addressing civil liability, or by a separate treaty that has a link to the civil liability treaty. The objectives of a compensation fund are to ensure adequate compensation of victims of environmental damage, and to spread the financial burden of compensation among the potential perpetrators of damage (Doeker and Gehring, 1992; Kummer, 1995/1999). In this respect, it fulfils a similar function to insurance.

There is an inherent close relationship

[74] For an analysis see Kummer (1995/1999, p. 252 ff.).

between the civil liability agreement and the compensation fund. The civil liability agreement sets out the obligation of the perpetrator of damage, generally an operator carrying out a potentially hazardous activity, to pay compensation to persons having suffered damage as a result of that activity, if his liability is established. This approach constitutes a unification of laws, designed to overcome the procedural obstacles inherent in the perpetrator and the victim being subject to the jurisdiction of different countries with different legal systems. An international agreement on civil liability is self-executing, i.e. creates enforceable rights and obligations for private persons under the jurisdiction of contracting parties to the treaty. The liable persons, as well as the victims, are generally private persons. The state as such does not have a role except where it is the perpetrator (for instance, as operator of a hazardous installation) or the victim of damage (for instance, as owner of contaminated land).

As a general rule, civil liability treaties establish a financial limit of liability of the operator. In addition, there is usually an exclusion clause, for instance, where damage is the result of *force majeure*, armed conflict or acts of terrorism. Also, practical reasons may make it impossible to obtain full compensation under the liability regime: for instance, where the identity of the perpetrator is not known, the perpetrator cannot be held liable in accordance with the liability treaty, the damage exceeds the financial ceiling set by the regime and/or compulsory insurance, or the liable operator is unable to meet his financial obligations. Thus, in some instances, full compensation for damage sustained is not available under a civil liability treaty. In these cases, the role of the fund is to replace or supplement the compensation received under the liability treaty.

Due to its aims and nature, compensation funds directly involve private persons, in contrast to the assistance funds described previously. In accordance with the nature of a civil liability agreement, the contributors to a compensation fund, as well as the claimants, are natural or legal persons under the jurisdiction of a contracting party. The state can be a contributor or a claimant if it is an operator, or if it has suffered damage. Otherwise, the state may have the role of facilitator, i.e. by collecting funds from the contributors and forwarding them to the compensation fund. However, it is also possible for the international institution operating the fund to collect the contributions without the involvement of national authorities.

Compensation funds have a legal personality that is recognized by the legal systems of the contracting parties to the constituting international legal instrument. The constituting instrument determines the criteria for contributions, conditions for disbursement of funds and the claims procedure. Contributions are made by operators that are potentially liable under the corresponding regime, often levied on the goods or services they provide. Claimants are persons having suffered damage as a result of a hazardous activity who do not receive full compensation under the corresponding liability agreement. Most compensation funds place a limit on the amount to be paid in a single incident.

The funds have an institutional infrastructure established by the constituting treaty. This normally consists of an executive body responsible for making decisions regarding collection and disbursement of financial means; a supervisory body consisting of all parties to the establishing agreement, which determines the policy and management criteria, and gives guidance to the executive body; and a secretariat responsible for administrative matters.

In practice, both international agreements on civil liability for environmental damage and compensation funds established within their framework have proved so far to be difficult to negotiate.[75] Only the

[75] For an overview of ongoing work in international fora and difficulties encountered, see the Synthesis of developments in the field of liability and redress prepared by the Secretariat of the Convention on Biological Diversity (Document UNEP/CBD/COP/5/16, 1 March 2000).

International Oil Pollution Compensation Funds discussed below are based on a legal instrument that has entered into force; they are thus the only ones that have been in operation for a number of years. The other instruments are not operational, and there is hence no possibility of judging their effectiveness in practice.

THE INTERNATIONAL OIL POLLUTION COMPENSATION FUNDS (IOPC FUNDS)[76]

The first IOPC Fund was set up under the 1971 Convention on the Establishment of an International Fund for Compensation of Oil Pollution Damage (1971 Fund Convention), which supplements the 1969 Convention on Civil Liability for Oil Pollution Damage. Both conventions were adopted under the auspices of the IMO. Prior to entry into force of these instruments, the oil industry established its own funding schemes, namely the Tanker Owners' Voluntary Agreement concerning Liability for Oil Pollution (TOVALOP, 1968) and the Contract Regarding an Interim Supplement to Tanker Liability for Oil Pollution (CRISTAL, 1971), both financed by cargo interests. The industry schemes covered nearly 90% of the world tanker fleet within a short time, which shows the degree of support from the relevant industry for this type of scheme.

In 1992, both the 1969 and the 1971 Conventions were amended by respective Protocols. The amended instruments are known as the 1992 Civil Liability Convention (CLC) and the 1992 Fund Convention (FC), which set up the 1992 IOPC Fund. The amended Conventions entered into force in 1996. The industry schemes TOVALOP and CRISTAL, having become obsolete with the entry into force of the amended Conventions, were terminated in 1997. After a transitional period, during which both the 1971 Fund and the 1992 Fund were operational concurrently, the 1992 regime replaced the 'old' 1969/71 regime. On 24 May 2002, the 1971 Fund Convention ceased to be in force. The fundamental aims, structure and mode of operation basically remain the same. This discussion focuses on the 1992 IOPC Fund.

The 1992 IOPC Fund covers damage occurring in connection with the bulk transportation of oil by sea. It has a dual aim: first, to provide a compensation system supplementary to the system established by the 1992 CLC, in order to ensure full compensation to victims of damage caused by persistent oil spilled from laden tankers; and secondly, to distribute the economic burden among the shipping industry and cargo interests. Compensation from the Fund can be claimed in cases where full compensation is not available under the CLC. As the CLC permits ship owners to limit their liability under certain conditions, this may be the case where the actual damage sustained goes beyond the limit established. It may also be the case if the tanker owner cannot be identified, or is insolvent and uninsured, or is exonerated from liability under the provisions of the CLC.

The IOPC Fund is an international organization with legal personality, independent of IMO or other UN organizations. Every Party to the Convention automatically becomes a member of the Fund.

The Fund is financed by levies on certain types of oil carried by sea. These are collected by the Fund directly from the entities that receive oil after sea transport, which can be private or state-owned companies, or a state itself. Annual contributions are levied on entities receiving more than 150,000 tons of crude oil and/or heavy fuel oil in a party state, after sea transport, during a calendar year. The contributions are determined in proportion to the quantity received, and on the basis of anticipated payments of compensation and estimated administrative expenses during the forthcoming year. Each party must communicate annually to the Fund Secretariat a

[76] For a further discussion see Doeker and Gehring (1992, p. 418); Rengifo (1997); Report in the *Yearbook of International Cooperation on Environment and Development 1999/2000*, pp. 128–129; Gold (1999, p. 31); and White (2001).

list of oil-receiving entities under its jurisdiction, and the amount of oil received by each. These lists are confidential, and are closely monitored by the Secretariat. The Fund Convention does not provide for contributions by states except where they are oil-receiving entities.

Persons having suffered pollution damage in a state that is a party to the Convention may make a claim against the IOPC Fund for compensation. Under the 1992 Fund Convention, the maximum amount of compensation payable from the Fund for a single incident, including the amount paid by the ship owner or his insurer under the 1992 CLC, is 135 million SDR[77] (about US$174 million). Where at least three party states have received at least 600 million tons of oil in the previous year, the limit may be increased to 200 million SDR (about US$257). Under a 2000 Amendment that entered into force in 2003, the maximum was increased to 203 million SDR (about US$260 million) for a single incident.

The institutional infrastructure consists of the Assembly as the supreme governing body of the Fund, composed of all parties to the Fund Convention; the Executive Committee, composed of 15 members, with the main function of approving settlement of claims, to the extent that the Director is not authorized to do so; and the Secretariat headed by the Director, responsible for the conduct of business, including collection of contributions and settlement of claims up to a certain amount.

The CLC Convention has been ratified by 91 countries, representing 91% of the world tonnage, and the 1992 Fund Convention has been ratified by 85 countries, representing 87% of the world tonnage.[78] In this respect, the system can thus be considered successful (White, 1999).

COMPENSATION FUNDS TO BE ESTABLISHED UNDER THE HNS AND BASEL CONVENTIONS

Under the International Convention on Liability and Compensation in Connection with the Carriage of Hazardous and Noxious Substances by Sea (HNS) of 1996, a compensation fund is to be established to supplement the liability provisions under the Convention.[79] The HNS Convention covers potentially hazardous chemical substances transported by sea. The objectives, mode of operation and structure are similar to the IOPC Funds. As only two states have ratified the Convention,[80] it has not yet entered into force, and it is increasingly unlikely that the HNS Fund will ever become operational.

Under the Basel Convention on hazardous wastes, two types of funds with the purpose of compensating damage caused by hazardous wastes have been under consideration for many years, although neither is likely to be established in the foreseeable future. The first fund under discussion is a compensation fund to be set up in the framework of the 1999 Protocol on civil liability to the Basel Convention, with a structure, mode of operation and objectives similar to the IOPC and HNS Funds. However, when adopting the Protocol, states were unable to agree on an explicit legal basis for a compensation fund. The second prospective mechanism is a so-called revolving fund, for which Article 14 para. 2 of the Convention provides the legal basis. This fund would provide financial resources for emergency measures in the event of damage caused by hazardous wastes. The parties to the Convention would contribute on the basis of a scale of assessment. After settlement of the liability claim, the revolving fund would have the right of recourse to the liable person

[77] The 1992 Fund Convention uses the Special Drawing Right (SDR), defined by the International Monetary Fund as a unit of account. As of October 2001, the applicable exchange rate was 1 SDR = US$1.287.

[78] Data as of February 2003. See list of ratifications published on the IMO website: http://www.imo.org

[79] On the HNS Convention and Compensation Fund, see generally Rengifo (1997) and the *Yearbook of International Cooperation on Environment and Development 1999/2000*, p. 130. The information provided here has been obtained from the IMO website: http://www.imo.org

[80] See status of Conventions published on the IMO website: http://www.imo.org

or the compensation fund, as the case may be.[81]

COMPENSATION FUNDS AS MODELS FOR PLANT GENETIC RESOURCES AND TRADITIONAL KNOWLEDGE

As is the case with the assistance funds analysed above, the compensation funds feature all the necessary institutional elements outlined previously (pp. 302–303). The fundamental difference between this type of mechanism and a potential mechanism in the context of PGR is that this type of fund is tied to the concept of liability and compensation: funds are disbursed in the event of damage caused by acts entailing civil liability. The element of providing an incentive for recipients of funds to adhere to a certain manner of behaviour is thus absent in this context. As concerns incentives for contributors, these are directly linked to the related legal regime establishing civil liability. Within this framework, the funds serve as insurance, and there is thus an incentive for industry to participate, as the example of the IOPC Fund demonstrates. This can, however, not be easily replicated in the context of PGR, and incentives would therefore need to be created by other means.

In addition, two features of the compensation funds appear interesting in the context of a future mechanism for PGR, namely the involvement of private entities both as contributors and as claimants, and the generation of funds through levies on specified activities carried out by the relevant industry. In this sense, the compensation funds appear to be better placed to serve as potential models of a future mechanism for PGR than the assistance funds. It should be noted, however, that the disbursement of funds requires the status of a legal or natural person, on the one hand, and a clearly defined claim, established under the corresponding civil liability regime, on the other. In this sense, adjustments would need to be made if this type of fund were to serve as a model for the area of TK associated with PGR and traditional PGRFA. As in the case of an assistance fund, the political difficulties inherent in adopting such an instrument should not be underestimated. As stated above, the IOPC Fund, though obviously successful, remains the only financial mechanism of this type that is actually operational. As the analysis shows, this is in large part due to the interest of the oil industry itself, which set up its own scheme even before the entry into force of the intergovernmental scheme.

Mechanisms to provide incentives for implementation under the Kyoto Protocol[82]

The Kyoto Protocol to the Climate Convention provides three market-based instruments, the so-called Flexible Mechanisms or 'Flex Mex', to promote the aim of reducing the emissions of greenhouse gases by providing incentives to states and to private entities to act in conformity with this aim. In contrast to the mechanisms discussed above, these instruments are not financial mechanisms. They do not feature a multilateral fund, and adopt a fundamentally different approach to the problem of providing incentives. However, the underlying concepts may be interesting *vis-à-vis* a future financial mechanism in the context of PGR. In particular, the approach taken by the Kyoto mechanisms to the problem of providing incentives to the private sector to contribute to the solution of an environmental problem is worthy of investigation, as this has been identified as one of the functions of a future financial mechanism for PGR.

The Flex Mex comprise Joint Implementation (JI) (Article 6), the Clean Development Mechanism (CDM) (Article 12) and Emissions Trading (Article 17). In accordance with the relevant provisions of the Protocol, the modalities and operation of JI and the CDM will be elaborated by the COP to the Convention serving as the Meeting of the Parties to the Protocol (COP/MOP), and those of Emissions Trading by the COP to the Convention. The Kyoto Protocol has not

[81] For a more detailed discussion see among others Kummer (1995/1999, p. 253).
[82] See Arquit Niederberger (1998, p. 9) and Yamin (1999, p. 265).

yet entered into force,[83] but the COP has been working on the elaboration of the modalities of all three Flex Mex for the last few years,[84] with the aim of developing the mechanisms prior to entry into force of the Protocol[85] and laying down the concrete rights and obligations of countries in this context in a sufficiently clear way to provide a basis for ratification. At the resumed sixth session of the COP, held in Bonn in July 2001 (COP 6), and at the seventh session, held in Marrakesh in November 2001 (COP 7), agreement was reached on the modalities of the mechanisms (generally referred to as the Bonn Agreement and the Marrakesh Accords, respectively). The COP has prepared relevant decisions for adoption by the COP/MOP after entry into force of the Protocol.[86]

In view of this situation, the Flex Mex can at the present time be discussed only in a theoretical way. It is not possible to assess their functioning in practice, as they are not as yet operational.

Under the Kyoto Protocol, only a specific category of countries, namely developed countries and countries with economies in transition, have quantified obligations to reduce greenhouse gas emissions. These countries are listed in Annex I to the Climate Convention. The reduction obligations allocated to each Annex I country are specified in Annex B to the Protocol, and can be measured in quantified units. In other words, every Annex I country must achieve a certain number of quantified units within a given time period. The Flex Mex provide the possibility for countries to exchange these units. The terms designating the units and their definitions are distinct for each of the three mechanisms: they are denominated as Emission Reduction Units (ERUs) in the context of JI, as Certified Emission Reductions (CERs) in the context of the CDM and as Assigned Amount Units (AAUs) in the context of Emissions Trading. In addition, COP 7 agreed to identify as Removal Units (RMUs) net greenhouse gas removals resulting from sinks activities.

The concept underlying this system is to couch the reduction units in a tradable form. This allows a Party to fulfil a part of its reduction obligations by assisting another Party in meeting its own obligations, or to exchange ERUs against a financial contribution.[87] As Parties may impose corresponding obligations on their nationals, including industry, the possibility of trade-offs or financial compensation may provide an incentive for the private sector to comply with these obligations.

JOINT IMPLEMENTATION

Joint Implementation (JI), regulated in Article 6 of the Kyoto Protocol, is a mechanism by which an Annex I country, through a pertinent project, supports activities in another Annex I country, by which a certain number of ERUs are achieved, either by reducing greenhouse gas emissions or by enhancing sinks. Through provision of this support, the donor country may fulfil part

[83] In accordance with Article 25, the Kyoto Protocol will enter into force upon ratification by 55 countries, including industrialized countries accounting for at least 55% of the total carbon dioxide emissions for 1990 from this group. As of July 2003, 111 countries, including industrialized countries accounting for 44.2% of the emissions, had ratified the Protocol (see UNFCC website: http://www.unfccc.int).

[84] At COP 4 in 1998, Parties agreed on a work programme to elaborate principles, modalities, rules and guidelines on all three mechanisms, with priority to be given to the Clean Development Mechanism, for submission to COP 6 (COP Decision 7/CP.4 (1998), the so-called Buenos Aires Plan of Action). COP 6 was held in two parts in November 2000 and July 2001, respectively. The so-called Bonn Agreement, adopted in July 2001, provided the basis for finalizing the modalities of the mechanisms at COP 7 in November 2001 in the framework of the so-called Marrakesh Accords (COP Decision 5/CP.6 (2001)).

[85] See, for example, UNFCCC Press Release 'Bonn Decisions Promise to Speed Action on Climate Change' (27 July 2001); Yamin (1999, p. 268).

[86] See the text of the Marrakesh Accords, available on the UNFCCC website: http://www.unfccc.int

[87] For a full analysis of the quantified units defined under the Kyoto Protocol, and of the nature of countries' rights to them, see Yamin (1999, p. 268).

of its own reduction commitment, i.e. the number of ERUs achieved through the project is in part allocated to the donor country. The project can be financed from state or private sources in the donor country. This mechanism is likely to be applied primarily between more developed and less developed countries of the Annex I category, namely between Western industrialized countries and Central and Eastern European countries with economies in transition. A project must meet the following conditions to qualify for JI:

- It must be approved by the countries involved.
- It must be proved that the reduction in emissions by sources, or the enhancement of removals by sinks, is additional to what would have occurred without the project.
- The countries involved must have met their obligations to set up a national system for estimating emissions (Article 5) and to have submitted inventories and national communications (Article 7) under the Kyoto Protocol.
- Acquisition of ERUs by the donor country must be supplemental to reductions achieved by domestic action, i.e. domestic action shall constitute a 'significant element' of the effort made by each Party to reduce its emissions.

The incentive for donor countries and their industries lies in the fact that in a recipient country with a less developed economy and infrastructure, a given financial investment will achieve a more substantive reduction of greenhouse gas emissions than in the donor country itself. The concept of JI has initially been subject to criticism, based on the argument that this possibility detracts from the more developed countries' domestic obligations at the expense of less developed countries (Gupta, 1999/2000).

THE CLEAN DEVELOPMENT MECHANISM

The establishment of the CDM (Article 12) will enable Annex I Parties to implement projects that reduce greenhouse gas emissions or enhance sinks in non-Annex I Parties, which do not have reduction obligations under the Protocol, and to credit the reductions achieved in this manner to the achievement of their own reduction targets. The CDM thus allows developed countries to achieve a part of their reduction commitments through projects in developing countries, with the additional goal of assisting non-Annex I Parties in achieving sustainable development and contributing to the ultimate objective of the Climate Convention. Under the CDM, emission reductions generated by project activities in non-Annex I Parties will be certified by operational entities designated by the COP/MOP, on the basis of measurable criteria. These reductions are denoted as Certified Emission Reductions (CERs). The CDM will be supervised by the Executive Board of the mechanism. A 'share of the proceeds', i.e. a part of the CERs generated under a CDM project, will be used to assist particularly vulnerable developing countries in meeting the costs of adaptation.

The CDM can be seen as the counterpart of JI: both are an extension and further development of the AIJ pilot phase, but JI takes place among Annex I Parties, whereas the CDM allows for joint projects between Annex I and non-Annex I Parties. The incentives for donor states and their industry is essentially the same as in the case of JI. Like JI, the concept of a CDM was initially criticized by some developing countries, with the argument that it constitutes a way for developed countries to solve a problem for which they are primarily responsible in the developing world instead of seeking domestic solutions.[88]

EMISSIONS TRADING

Emissions Trading, as set out in Article 17, permits an Annex I Party to transfer AAUs,

[88] See, for example, Centre for Science and Environment India: The Kyoto Protocol – What It Says, New Delhi (1998, p. 10).

i.e. a part of the quantified units of reduction that Party is in a position to achieve during the commitment period, to another Annex I Party. This permits the second Party to add the so acquired AAUs to its own emission reduction. In general, the first Party will receive payment for this service.

Like the other mechanisms, the concept of Emissions Trading is not undisputed, since it basically provides the possibility for a country to transfer its obligations under the Protocol to another country. Emissions Trading is likely to be of interest to countries that – for reasons of economic decline – achieve an emission reduction greater than their commitment under the Protocol, the so-called 'hot air' (which is the case for a number of Central and Eastern European countries following the collapse of the former Soviet Union), and to countries that have difficulties in achieving their reduction targets without resorting to expensive and politically difficult domestic measures, and hence prefer to acquire AAUs at a lower cost (which is the case for some developed Western states).

THE ROLE OF NON-STATE ACTORS

If the Flex Mex are to become truly market-based instruments, concrete ways will need to be designed to involve non-state actors, in particular private companies and other investors, in their application. In other words, it must be possible for non-state actors to exercise the rights and duties pertaining to the allocation of the quantified units under the three mechanisms. In order to create a true incentive, private entities will have to be able to obtain the financial benefits to be gained by the application of the mechanisms.

The obvious way of involving private actors is through the domestic legislation of the Parties to the Protocol. This is expressly provided by the Kyoto Protocol in Articles 6 para. 3 (for JI) and 12 para. 9 (for the CDM), as well as in the respective guidelines for all three mechanisms. The alternative option of giving non-state actors standing under international law, their actions to be subject to control by the COP/MOP, has a weak basis both in theory and in practice (Yamin, 1999). Accordingly, states will need to enact domestic legislation, under which private actors under their jurisdiction can apply the mechanisms of the Kyoto Protocol. In order to comply with its emission reduction obligations, an Annex I state will impose corresponding obligations on its subjects by domestic law. By the same token, the Kyoto mechanisms can be transposed into national law, and non-state actors can be assigned the corresponding rights. Thus, a company in an Annex I Party initiating or supporting a project in another state Party to the Protocol, which contributes to emission reductions in that state, will be able to deduct the reductions achieved from its own national reduction obligation in accordance with JI (if the project is carried out in an Annex I Party) or the CDM (if it is carried out in a non-Annex I Party). Likewise, non-state actors will be able to acquire or sell AAUs under the Emissions Trading system. Involvement of national actors requires a national emissions inventory that includes reductions achieved by non-state actors under the Kyoto mechanisms.[89] In Switzerland, the legal basis for the involvement of private actors has been created with the new CO_2 Law, which entered into force in May 2000.

The functioning of this concept in practice can only be fully assessed once the Kyoto Protocol is in force and all Annex I Parties have enacted and implemented relevant national legislation.

THE KYOTO MECHANISMS AS MODELS FOR PLANT GENETIC RESOURCES AND TRADITIONAL KNOWLEDGE

As stated above, the main interest in the Kyoto mechanisms in the context of a future financial mechanism for plant genetic resources is the way in which financial incentives are created for the imple-

[89] For a more detailed discussion of the role of the private sector using the Swiss example, see Arquit Niederberger (1998, p. 9).

mentation of an international agreement both for states and for private entities. This is the element that may possibly have a model function in that context. As is the case with the depletion of the genetic resource base and related TK, climate change is an area where it is not possible to clearly identify actors that are entitled to financial compensation. This is a fundamental difference for issues such as oil pollution or contamination with hazardous materials, where an approach based on civil liability and compensation can be taken. With genetic resources as with the global climate, the interest in protection, for the individual, is thus a fairly abstract and long-term one, and methods for ensuring protection must be adjusted to this reality. The approach of providing economic incentives instead of assigning liability for damage caused is thus a concept to be considered. A prerequisite for the use of this concept as a model is a substantive obligation to private entities that can be fulfilled through a financial contribution, as is the case in the Kyoto mechanisms. This type of obligation has yet to be created in the context of plant genetic resources.

More specifically, the concept of providing an economic incentive in the shape of the CDM, and imposing a levy on the proceeds from this mechanism as a contribution to a fund which is to be used for the support of adaptation measures, may be a very interesting approach that is as well adapted to the nature of the problem of genetic resources as it is for climate change. The nature and functioning of the fund will need to be further developed, possibly through reliance on models elaborated elsewhere in this study.

Relevant work in the framework of the FAO

The issue of a financial mechanism for PGR has thus far been addressed only in a very preliminary fashion within the framework of the FAO. Under the International Undertaking on Plant Genetic Resources (IU), a non-binding instrument adopted by the FAO Conference in 1983, the FAO Fund for Plant Genetic Resources was established on an interim basis in 1988. Donors (governments, NGOs and individuals) were to contribute to the Fund to support plant genetic resource conservation and use. By Resolution 4/89 adopted in 1989, the FAO Conference agreed that the conservation, management and use of plant genetic resources could be achieved through financial mechanisms, in particular the FAO Fund for Plant Genetic Resources. However, no contributions have been made to date, and the Fund has therefore never become operational. In 1991, the FAO Conference by Resolution 3/91 approved Annex III to the IU, which endorsed the concept of implementing Farmers' Rights through an international fund on plant genetic resources to support conservation and utilization programmes, particularly in developing countries. The priorities of the fund are to be overseen by the FAO CGRFA.[90] The Secretariat of the FAO CGRFA considers that the FAO Fund for Plant Genetic Resources should assume the role of this mechanism. However, no further work has been undertaken to render it operational, and it therefore remains 'dead letter'.[91]

In November 2001, the FAO Conference adopted the ITPGRFA, a binding legal instrument elaborated by the CGRFA.[92] The Treaty constitutes a revision of the IU. One of the core elements of the ITPGRFA is the establishment of the Multilateral System for Facilitated ABS, in which the resources listed in Annex I shall be included. One of the aims of the System is to share monetary benefits of these resources through a financial mechanism, which is to be established in accordance with Article 19.3(f).

Unlike the CBD, which set up a financial mechanism from the beginning (see

[90] Girsberger (1998, p. 304), *Yearbook of International Cooperation on Environment and Development 1999/2000*, p. 181.

[91] Information received from the Secretariat of the FAO/CGRFA, December 2001.

[92] For a detailed discussion of the ITPGRFA see Chapter 2, this volume.

point 1 above), the ITPGRFA addresses the issue of funding in a much more preliminary way. Articles 18 and 19 constitute enabling provisions for a funding strategy and a financial mechanism. Article 18 sets out a number of elements of a financial strategy to be elaborated by the Parties. The purpose of the strategy is to promote the support by developed countries of efforts to implement the ITPGRFA undertaken by developing and transition countries. Article 19 para. 3(f) mandates the Governing Body of the Treaty to 'establish, as needed, an appropriate mechanism, such as a Trust Account, for receiving and utilizing financial resources that will accrue to it for purposes of implementing this Treaty'. Finally, Article 13.2(d) states that sharing of monetary benefits resulting from the use of plant genetic resources included in the Multilateral System should be subject to payment of an equitable share of the benefits into a future mechanism, to be established under Article 19.3(f).

Article 19.3(f) is an enabling provision of the more open variety: the wording 'as needed' and 'appropriate' leaves it to the Governing Body to decide what form of mechanism would be appropriate. The provision does not provide much guidance concerning elements of such a mechanism. The wording of Article 19.3(f) ('for the purposes of implementing this Treaty') does give an indication that the type of fund to be established may be what has been termed here an assistance fund. Article 13.2(d) points in the direction of a similar method to the one to be used in the context of the CDM under the Kyoto Protocol, namely the generation of funds through levies on benefits derived from related economic transactions.

As in equivalent discussions in the context of most MEAs, there is fundamental disagreement between industrialized and developing countries as to whether or not a financial mechanism should be established. However, the issue did not figure among the prominent topics in the debates leading to the adoption of the ITPGRFA.

By a Resolution on interim arrangements,[93] the FAO Conference mandated the CGRFA to act as Interim Committee pending the entry into force of the Treaty, at which time the Governing Body will be established. The Resolution outlines a number of priority actions to be undertaken by the Interim Committee, concluding with the general clause 'perform such other functions as may be necessary for the effective implementation' of the Treaty. As work on a future financial mechanism does not figure on the list of priority actions, but comes within the purview of the general clause, one may conclude that the Conference did not consider this a priority. As the financial strategy is to be adopted by the Governing Body of the Treaty at its first meeting, the issue will be addressed at the second meeting of the Interim Committee.[94]

7.4.3 Options for a financial mechanism in the context of access to genetic resources

Based on the above discussion, we may derive a number of central elements and options for a future financial mechanism from existing models.

Structure, organization and legal personality

As is the case of all the existing financial mechanisms that have been discussed, an international legal instrument should establish the future mechanism for plant genetic resources. It should be endowed with a legal personality recognized by all contracting parties to the instrument in question, and all parties should be members of the mechanism. Its infrastructure should comprise an assembly as supreme governing body, consisting of all members, an executive body to direct the generation and attri-

[93] Conference Resolution 3/2001, 3 November 2001, Section B. Interim Arrangements, para. 3 (published on the FAO website: http://www.fao.org).

[94] Provisional work programme of the Interim Committee at its second meeting (Doc. CGRFA/MIC/-1/02/9, October 2002).

bution of financial resources, and a secretariat. All assistance funds and compensation funds discussed could serve as models in this respect.

The question arises as to whether a new legal instrument should be negotiated for this purpose and a new mechanism established, or whether the functions of the financial mechanism could be attributed to an existing mechanism. The latter would be more efficient in terms of avoiding duplication of efforts and resources. It would, however, presuppose the existence of a financial mechanism with a similar scope, structure, aims and functions. None of the mechanisms established in environmental law that are described above would meet this requirement: the assistance funds do not provide for a role of private entities as contributors and recipients of funding, and the compensation funds are conceptually tied to a legal instrument on civil liability and compensation. The only feasible option seems to be the attribution of the pertinent functions to the financial mechanism to be established within the framework of the ITPGRFA. As substantive work on this mechanism has yet to begin, there may be sufficient leeway to accommodate the functions discussed here. This would of course be subject to decision by the Governing Body of the ITPGRFA. It must also be noted that the ITPGRFA – and hence also its financial mechanism – applies to plant genetic resources for food and agriculture only, whereas the scope of the mechanism discussed here should cover all plant genetic resources. Should the difference in scope between the ITPRGFA and the future financial mechanism be considered too significant an obstacle, the option of establishing separate funds in the framework of different related treaties could be investigated. Thus, a mechanism for plant genetic resources food and agriculture could be set up in the framework of the ITPGRFA, a mechanism for other types of plant genetic resources in the framework of the CBD.

Generation of funds: contributors

In order to achieve stability and predictability, a compulsory contribution to the mechanism on a regular basis should be instituted. In view of the creation of a broad financial basis, the funds should be derived from as many sources as possible. However, the concepts of fair and equitable distribution of the financial burden and creation of incentives call for collecting funds specifically from persons and entities benefiting from the resources. Equally, the contributions should be based on criteria such as the financial situation of the contributor (in this context, exemptions could be provided for specific actors such as small farmers in developing countries) and the extent of the use of resources. In this respect, the assistance funds cannot provide guidance, being based on the concept of assistance to states lacking financial capacity by more affluent states, and excluding private entities as contributors and as recipients of funding. The compensation funds, on the other hand, could serve as models in this respect. In particular, the system by which contributions are levied on specific activities is a more useful model than the scale of assessment for contributions used by the assistance funds, which takes into account only the financial status of the contributor. An option to be considered would thus be to levy contributions on plant genetic resources accessed or used by a given entity, which could be a private enterprise or a government institution. If this were envisaged at the international level, the difficulty would be that this would require monitoring of the access to and use of the resources. An institution to undertake this task, as well as a relevant procedure, would have to be created. The same international legal instrument that establishes the financial mechanism could establish this. If the functions of the mechanism were assigned to the financial mechanism of the ITPGRFA, this problem could largely be avoided, as the Treaty's Multilateral System for Facilitated ABS creates the framework for tracking resources, and the future financial mechanism of the ITPGRFA will administer funds derived from resources included in the Multilateral System. Should this solution be pursued, issues of political acceptability and scope would arise, as mentioned above. Alternatively, levies could be considered as a method for states to generate resources to pay their contributions to the financial mechanism. While this would need to be addressed at the national level and is thus beyond the scope of this study, it would be an interesting option for states to investigate.

As concerns creating incentives for financial efforts that go beyond paying contributions for the access to certain resources, the concept underlying the mechanisms of the Kyoto Protocol could be of interest as a model. The Kyoto Protocol establishes concrete obligations that can be imposed on private entities, coupled with the possibility of fulfilling these obligations through the provision of financial resources – for instance, funding of a project in another country. In the case of JI and the CDM, the incentive lies especially in the fact that it is often more cost-effective to implement the obligations by financing projects through these mechanisms than by adopting direct emission reduction measures. In sum, the incentive lies in the fact that contributing to the mechanism is a simpler or more cost-effective way for the contributors to meet their obligations. This approach presupposes an international legal instrument imposing obligations on states and providing for states in turn to impose obligations on their nationals, which the nationals may then implement through contributing to, or participating in, a pertinent mechanism. An equivalent obligation remains to be established in a future regime on PGR.

Disbursement of funds: claimants

The absence of clearly defined claimants or recipients of financial means, due to the fact that traditional or *sui generis* rights cannot be allocated, is not accommodated by the concept of the compensation funds, which function on the basis of claims submitted by persons having suffered damage. The assistance funds could serve as models in that they provide funding for projects in a certain field that are submitted by national authorities, rather than disbursing funds to individuals. This concept could be developed in the present context. In this respect, the Adaptation Fund to be established under the CDM of the Kyoto Protocol, which will use financial means generated by levies on proceeds of the CDM for supporting concrete adaptation projects and programmes in developing countries, could be a useful precedent. The system used by the Montreal Protocol Fund and the GEF to assess and approve pertinent projects would also be interesting. Accordingly, a system could be established under which potential recipient countries would be defined on the basis of criteria that ensured equitable access to the funds, such as their financial capacity, the amount and nature of genetic resources held under their national jurisdiction, and the presence of local or indigenous communities that are TK holders in their territory. The national authorities of these countries would be eligible to submit project proposals to the executive body of the mechanism for possible funding. The projects would have to be proven to benefit TK holders who are producers of genetic resources, and to support the aims outlined in the first section of this chapter; this would be verified by the executive body as part of the decision-making process. Projects approved by the mechanism for funding would be carried out by competent national authorities, or under their responsibility. The authorities would be accountable to the financial mechanism. This approach would support the aim of providing incentives for the valuation of TK and conservation and sustainable use of genetic resources, since only projects that further these aims would be financed by the mechanism.

7.4.4 Conclusions and recommendations

As a first conclusion, it should be noted that a future financial mechanism that could substitute financial compensation of TK holders in cases where traditional or *sui generis* IPR cannot be allocated, would need to be established by an international legal instrument. The negotiation of such an instrument and the establishment of the requisite institutional infrastructure would require the relevant political will of the international community, the leadership of an international organization and financial resources that may be considerable. The general wariness of the international community in relation to the establishment of new international mechanisms would be a factor to be taken into account. The required infrastructure would comprise a supreme body made up of all member states, an executive body and a secretariat. In this respect, there are a number of precedents in international environmental law that could serve as models.

An alternative would be to assign the functions of such a mechanism to an existing institution established in the framework of an existing international legal instrument. The most obvious choice would be the ITPGRFA, which establishes a legal basis for a financial mechanism. The necessary adjustments would have to be made, in particular as concerns the scope of the Treaty. A pertinent decision would need to be taken by the Governing Body of the ITPGRFA. Alternatively, the establishment of separate mechanisms within more than one international legal instrument in accordance with their respective scope could be considered. Apart from the ITPGRFA, an obvious candidate is the CBD.

In order to create a broad, predictable and stable financial basis, indispensable to the success of the mechanism, the legal instrument establishing the fund should set out an obligation to users of genetic resources – whether private entities or government institutions – to make contributions to the fund. The levels of contribution could be detailed on the basis of criteria that take into account the aim of equitable sharing of the financial burden, such as the amount and type of resources used, and possibly the financial situation of the contributor. In order to create an incentive to users of the resources to contribute, contributions could be levied on the amount and nature of resources used. Existing compensation funds could serve as models in this respect. This could also be envisaged at the national level, as a means for states to generate their contribution to the fund. Looking to the future, a further incentive to contribute to the fund could be created if substantive obligations imposed on private entities by an international legal instrument could be met by making financial contributions, as in the Flexible Mechanisms of the Kyoto Protocol.

As regards disbursement of funds, the legal instrument should designate a category of countries that could submit relevant projects to the mechanism for funding. Criteria for designating countries as recipients should include financial capacity below a measurable level, the amount and nature of resources held or generated under their national jurisdiction, and the presence of indigenous or local communities that are TK holders. The projects submitted would have to be proven to benefit holders of TK and support the fundamental aims of the mechanism. The decision on funding would be taken by the executive body based on these criteria. The projects would be carried out under the responsibility of the competent government authorities, which would be accountable to the financial mechanism. The assistance funds and the Adaptation Fund of the Kyoto Protocol could serve as models in this respect.

Bibliography

Arquit Niederberger, A. (1998) Marktwirtschaftliche Anreize für den Klimaschutz: Fallbeispiel 'Joint Implementation'. *Bankverein/Der Monat Nr. 4.*

ASSINSEL (1998) Position on access to plant genetic resources for food and agriculture and the equitable sharing of benefits arising from their use. http://www.worldseed.org./multie.htm

Baruffol, U. (2003) Contractual regulation of access to information on biodiversity for scientific and commercial use – the Novartis-UZACHI Biolead Project. Contribution 30: Forest Science, Swiss Federal Institute of Technology, Zürich (on file with author).

Blakeney, M. (2002) Protection of plant varieties and farmers' rights. *European Intellectual Property Review* 24, 9.

Bragdon, S.H. (2001) Financing treaty operations and implementation: a survey of mechanisms. FAO Background Study Paper No. 13.

Cabrera Medaglia, J.A. (2003) Access to genetic resources, protection of traditional knowledge and intellectual property rights: lessons learned from the Costa Rican experience. WIPO Intergovernmental Committee on Intellectual Property and Genetic Resources, Traditional Knowledge and Folklore, Fifth Session, Geneva, 7–15 July 2003, advance copy (on file with author).

Correa, C. (2000) Options for the implementation of farmers' rights at the national level. South Centre, Geneva, T.R.A.D.E. Working Paper No. 8.

Doeker, G. and Gehring, T. (1992) Liability for environmental damage. In: Sand, P.H. (ed.) *The Effectiveness of International Environmental Agreements.* Grotius, Cambridge, UK.

Dutfield, G. (2001) TRIPS-related aspects of traditional knowledge. *Case Western Reserve Journal of International Law* 33, 233.

Esquinas-Alcázar, J. (1996) The realisation of farmer's rights. In: Swaminathan, M.S. (ed.) *Agrobiodiversity and Farmers' Rights.* Konark Publishers, New Delhi, India, p. 2.

Fowler, C., Smale, M. and Gaiji, S. (2000) Germplasm flows between developing countries and the CGIAR: an initial assessment. Paper prepared for the Global Forum on Agricultural Research, FAO and IPGRI, Rome. http://www.sgrp.cgiar.org/GermplasmFLowsgfar0065.pdf

Girsberger, M.A. (1998) The Protection of Traditional Plant Genetic Resources for Food and Agriculture and Related Know-How by Intellectual Property Rights in International Law. Berne, Switzerland.

Girsberger, M.A. (1999) *Biodiversity and the Concept of Farmers' Rights in International Law – Factual Background and Legal Analysis.* Peter Lang, Berne, Switzerland.

Glowka, L. (1998) *A Guide to Designing Legal Frameworks to Determine Access to Genetic Resources.* IUCN, Gland, Switzerland.

Glowka, L. *et al.* (1994) *A Guide to the Convention on Biological Diversity.* IUCN, Gland, Switzerland.

Gold, E. (1999) Liability and compensation for ship-source marine pollution. *Yearbook of International Cooperation on Environment and Development 1999/2000.* Earthscan, London, pp. 31–38.

Gupta, J. (1999/2000) Evaluation of the climate change regime and related developments. *Yearbook of International Cooperation on Environment and Development 1999/2000.* Earthscan, London, pp. 19–30.

Kummer, K. (1995/1999) *International Management of Hazardous Wastes: The Basel Convention and Related Legal Rules.* Oxford University Press, Oxford, UK.

Lang, W. (1992) Ozone layer. *Yearbook of International Environmental Law* 3, 225.

Novartis (1999) Bio-prospecting and benefit sharing. Report of a UNED-UK/Novartis Workshop hosted by the Rockefeller Foundation New York. Towards Earth Summit III-2002.

Rengifo, A. (1997) The International Convention on Liability and Compensation for Damage in Connection with the Carriage of Hazardous and Noxious Substances by Sea, 1996. *Review of European Community & International Environmental Law* 6, 191–197.

Rosell, M. (1997) Access to genetic resources: a critical approach to Decision 391 'Common Regime on Access to Genetic Resources' of the Commission of the Cartagena Agreement. *Review of European Community & International Environmental Law* 6, 274–283.

Ruiz Muller, M. (2000) ¿Es necesario un nuevo marco jurídico para la bioprospección en la región andina? Breve revision crítica de la Decisión 391. In: *Serie de Politica y Derecho Ambiental* 14 (February), 1–8. Sociedad Peruana de Derecho Ambiental, Lima, Peru.

Seiler, A., van den Daele, W. and Döbert, R. (2003) Protection of Traditional Knowledge. Deliberations from a Transnational Stakeholder Dialogue Between Pharmaceutical Companies and Civil Society Organizations. Wissenschaftszentrum Berlin (WZB) Discussion Paper No. SPIV 2003–102.

Swaminathan, M.S. (ed.) (1995) *Farmers' Rights and Plant Genetic Resources – Recognition & Reward.* Macmillan, Madras, India.

Swanson, T. and Göschl, T. (2000) Property rights issues involving plant genetic resources: implications of ownership for economic efficiency. *Ecological Economics* 32, 75–92.

Ten Kate, K. and Lasén Diaz, C. (1997) The undertaking revisited – A commentary on the revision of the International Undertaking on Plant Genetic Resources for Food and Agriculture. *Review of European Community & International Environmental Law* 6, 284–292.

Tobin, B. (1997) Certificates of origin: a role for IPR regimes in securing prior informed consent. In: Mugabe, J. *et al.* (eds) *Access to Genetic Resources. Strategies for Sharing Benefits.* IUCN, Bonn, Germany.
WBCSD/WZB Stakeholder Dialogue Process on the Issues of Intellectual Property Rights in Biotechnology: Summary Report (2003) Available at: http://www.wz-berlin.de/ipr-dialogue/wbcsd_wzb_final_report.pdf
White, I.C. (1999) Oil spill compensation. Available at: http://www.itopf.com
White, I.C. (2001) The International Oil Pollution Compensation Fund: Explanatory Note prepared by the 1992 Fund Secretariat. London, available at http://www.imo.org
Yamin, F. (1999) Equity, entitlements and property rights under the Kyoto Protocol: the shape of 'things' to come. *Review of European Community & International Environmental Law* 8, 265–274.
Yearbook of International Cooperation on Environment and Development 1999/2000. Earthscan, London.
Young, M. and Gunningham (1997) Mixing instruments and institutional arrangements for optimal biodiversity conservation. In: *Investing in Biological Diversity.* OECD, Paris, pp. 141–165.

8 International Trade Regulation for Plant Genetic Resources and Related Products

Part 1: The Impact of Agricultural Subsidies and Food Aid on Agrobiodiversity

Susette Biber-Klemm[a] and Michael Burkard[b]

[a]*World Trade Institute, University of Berne, and Faculty of Law, University of Basel, Maiengasse 51, CH 4056 Basel, Switzerland;* [b]*World Trade Institute, University of Berne, Hallerstrasse 6, CH-3012, Berne, Switzerland*

8.1 Introduction[1]

8.1.1 General question

In the former chapters, the means to foster trade in the informational value of TK and traditional PGRFA were discussed. However, the value of traditional knowledge and traditional PGRFA not only consists of information, but other essential components as well, such as the biological resource itself, a specific preparation or process, and end product. Such TK products can be of agricultural or non-agricultural origin, and include, besides foodstuffs, such things as pharmaceutical remedies, textiles, materials for construction or other specific uses, handicrafts and so on.

Accordingly, if ways and means to create incentives for the maintenance of TK and traditional PGRFA are sought, trade in TK products ('traditional products') has to be taken into consideration. This aspect is of importance at all levels, local, regional and global. Regarding the latter, customer preferences in industrialized countries indicate an increasing market for products from developing countries which are supported by labels. Consumers are willing to pay higher prices for such products, due to environmental concerns, and sustainability and equity considerations.

The question asked in this chapter is, therefore, which measures in the framework of the world trade order could be used for the promotion of TK-based products, and how they would have to be adapted to better serve the goals of equity and food security.

In the following, measures in the world trade order/GATT for the promotion of trade in TK-based products based on biological resources will be examined. Two types of measures are taken into consideration: (i) positive measures, such as the direct support of production of TK products, or the facilitation of their market access; and (ii) negative measures such as removing competition-distorting mechanisms such as export subsidies.

[1] Author: Susette Biber-Klemm. I thank Marion Panizzon and Elisabeth Bürgin for their comments.

  Rights to Plant Genetic Resources and Traditional Knowledge (eds S. Biber-Klemm and T. Cottier, assoc. ed. D. Szymura Berglas)

Relevant measures and instruments can be found in all GATT thematic areas that deal with trade in natural products, such as the Agreements on Agriculture (AoA), Sanitary and Phytosanitary Measures (SPS) and Technical Barriers to Trade (TBT). Besides the cross-cutting issues of special and differential treatment for developing countries, issues of trade and environment are highly relevant in this context.

8.1.2 Marrakesh Agreement: cross-cutting issues and recent developments

Relevance of topics: environment and development

Issues of environment and development within the WTO are of direct concern in the context of agriculture and TK (see Chapters 1 and 4, and below). The Marrakesh Agreement integrates these concerns in its preamble, thereby defining the objectives of the entire package of agreements as a whole.

With regard to environment and development, the preamble states the objective of optimal use of the world's resources 'in accordance with the objective of sustainable development, seeking both to protect and preserve the environment'. It further recognizes the need for specific efforts to ensure that developing countries, especially the least-developed countries, receive a share in the international trade corresponding to the needs of their economic development (Preamble, paras. 1 and 2).

Conversely, the basic WTO instruments (GATT 1947 and Marrakesh Agreement) integrate only in a marginal way the interface of trade and environment in the substantial law. The exception of Article XX b GATT, authorizing measures necessary to protect human, animal or plant life or health if they are not applied in a discriminatory manner, allows exceptions only under narrow conditions. Yet, at the time of the signature of the Marrakesh Agreement, the Ministers also adopted a Decision on Trade and Environment, which mandates the first meeting of the General Council to establish a Committee on Trade and Environment and defines its responsibilities. Its basic duty is to identify the relationship between trade and environmental measures, in order to promote sustainable development, and to make recommendations about changes required in the trade agreements. Such changes must be compatible with the principles of the trading system.[2]

As to the interface of trade and development, Article IV of the Marrakesh Agreement establishes a Committee on Trade and Development. Its task is to review the special provisions in the multilateral trade agreements in favour of the least developed country members and to report to the General Council for appropriate action (Article IV.7). The duties of this Committee are spelled out in a Ministerial Decision that also specifies modalities for special and differential measures taken in favour of the least-developed countries.[3]

The objectives of trade and environment and development were integrated and more concretely formulated in the overall reform process of the WTO and now play an important role in the 'Millennium negotiation round' (see below, Doha Declaration and Doha Work Programme). They are also framed in the specific contracts such as the Agreement on Agriculture (Preamble and Article 6), the Agreement on Technical Barriers to Trade (Article 2) and the Agreement on Subsidies and Countervailing Measures (Article 8.2).

Doha Declaration and Doha Work Programme

At the conclusion of the Uruguay Round, the WTO members were conscious that in various areas the consensus reached was but a first step, and that further negotiations were to follow. Accordingly, a remarkable series of agreements contain a built-in

[2] Decision on Trade and Environment (LT/UR/D-6/2).

[3] Decision on Measures in Favour of Least Developing Countries (LT/UR/D-1/3), 15 April 1994.

agenda for their reviews,[4] and the WTO itself was perceived as a forum for further negotiations (Article III.2 Marrakesh Agreement).

At their second meeting in Geneva in 1998, the Conference of Ministers mandated the Governing Council to start a process to initiate and to coach the in-built reform process and to submit recommendations in view of further trade liberalization.[5] This process led – after the failure of the third Ministerial Conference in Seattle – to the Ministerial Declaration of Doha, Qatar, in 2001, which launched the so-called Doha negotiation round.

The DOHA Declaration[6] initiated a comprehensive round of trade negotiations and defined its scope, goals and work programme. It puts weight on the needs and interests of the developing countries in placing them 'at the heart of the Work Programme' (2), and in particular to take account of the vulnerability of the least-developed countries (3). It contains a strong commitment to the objective of sustainable development. The basis is the conviction that it is possible and necessary to meet the goals of an open and non-discriminatory multilateral trading system, acting for the protection of the environment and the promotion of sustainable development in a mutually supportive way (6).

It sets goals for a broad range of sectors – such as agriculture, market access for non-agricultural products, intellectual property rights, trade and investment, trade and competition policy, trade and environment, least developed countries and special and differential treatment, and defines the organization, management and timetable[7] of the work programme.

Further developments: the July package

At the fifth Ministerial Conference in Cancun in 2003, an impasse over the so-called Singapore issues (investment, competition policy, transparency in government procurement and trade facilitation, the most controversial being the first two), led to a deadlock in the negotiations, which came to a halt. Issues of interest in our context such as special and differential treatment, non-agricultural market access, environmental issues and cotton, were not negotiated.

Subsequent intense informal and formal preparatory work and negotiations by the General Council led to the adoption of the 'Doha Work Programme'[8] on 31 July 2004, hailed as a breakthrough in the negotiations. The programme contains a general part confirming the Doha Decisions, and frameworks and other agreements designed to streamline the negotiations, such as the frameworks for establishing modalities in agriculture, market access for non-agricultural products and for negotiations on trade facilitation.[9] These frameworks have reached a varying degree of specification, the Framework for Establishing Modalities in Agriculture being more detailed than the

[4] Agriculture, Services, Anti-Dumping, Customs Valuation, Dispute Settlement Understanding, Import Licensing, Preshipment Inspection, Rules of Origin, Sanitary and Phytosanitary Measures, Safeguards, Subsidies and Countervailing Measures, Technical Barriers to Trade, Textiles and Clothing, Trade Policy Review Mechanism, Trade-Related Aspects of Intellectual Property Rights and Trade-Related Investment Measures (Ministerial Declaration Singapore).

[5] 'We decide that a process will be established under the direction of the General Council to ensure full and faithful implementation of existing agreements, and to prepare for the Third Session of the Ministerial Conference. This process shall enable the General Council to submit recommendations regarding the WTO's work programme, including further liberalization sufficiently broad-based to respond to the range of interests and concerns of all Members, within the WTO framework, that will enable us to take decisions at the Third Session of the Ministerial Conference.' (Ministerial Declaration, 20 May 1998.)

[6] Ministerial Declaration adopted at the fourth Ministerial Conference, Doha, 14 November 2001, Doc. WT/MIN(01)/DEC/1.

[7] The negotiations were to be concluded not later than 1 January 2005.

[8] WT/L/579, 2 August 2004.

[9] Annexes A, B, D.

Framework for Establishing Modalities in Market Access for Non-Agricultural Products. The work programme depicts the stage of the consensus reached, leaving open a wide range of issues for further negotiations. For these negotiations it provides a roadmap and a timetable, delaying the conclusion of the Round at least until December 2005.

8.2 The Agreement of Agriculture and the Maintenance of TK and PGRFA[10]

8.2.1 Introduction

The questions discussed here are whether the WTO legal framework on agriculture in its present form offers instruments for achieving the objectives to preserve and sustainably use TK and traditional PGRFA, or what would be the openings to integrate these goals into the ongoing reform and negotiation process.

The question needs to be differentiated. According to the reflections in Chapter 1, it must be asked whether there are options for the creation of positive incentives supporting the maintenance of TK and traditional products, or for the elimination of negative incentives, hindering the successful commercialization of the traditionally produced goods. The question is of particular relevance with regard to the new round of trade negotiations, which have made agriculture and the interests of developing countries key topics.[11] The issue will be discussed against the background of a characterization and analysis of the types of agriculture producing traditional products. Next, it has to be examined under which 'headings' of the AoA the goals to maintain TK and traditional PGRFA can be subsumed. To lay the ground for these reflections, an overview of the principles reigning over the AoA, its goals and substantive obligations, and the (post-Cancun) state of play (as of December 2004) are given.

8.2.2 Legal background

The AoA in general

The Agreement on Agriculture (AoA), concluded in the Uruguay Round of the international trade negotiations, found a large number of countries in agreement over a set of multilateral rules, principles and disciplines with the long-term objective to 'establish a fair and market oriented agricultural trading system' (Preamble Para. 2).

The AoA has two main objectives. It establishes the framework for a process to reform trade and domestic policies in agriculture. This was at the time of its conclusion already perceived as a necessarily long-term development. It also lays the ground for negotiations to reach substantive commitments from the member states.

The overall objective is to provide for an increased market orientation of trade in agricultural products, and to correct and prevent restrictions and distortions in world agricultural markets (Preamble, para. 3). This goal is to be reached by specific commitments to reducing support and protection in the areas of market access, domestic support, export competition, and through the establishment of strengthened and more operationally effective GATT rules and disciplines (Preamble, para. 2).

On the operational level, the member states have agreed to provide for substantial progressive reductions in agricultural support and protection, addressing three main (sensitive) areas: (i) market access, that is, protective measures confronting imports; (ii) domestic, internal support of agricultural products and production, in particular subsidies bound to farming products; and (iii) export subsidies and other methods used to make exports artificially competitive. First steps to start reductions in protection and trade-distorting support were undertaken in the Uruguay Round.

---

[10] Author: Susette Biber-Klemm. I thank Elisabeth Bürgin for her comments on the AoA part of this chapter.

[11] See the DOHA Ministerial Declaration, agreed at the Fourth WTO Ministerial Conference in November 2001 (WT/MIN(01)/DEC/1, 20 November 2001) para. 2.

Member states have bound themselves to maximum tariffs, which are to be reduced according to an agreed timetable: they are committed to reducing agricultural export subsidies and trade-distorting domestic support.

The programme to reform agriculture is to create equity among all members and to take account of the so-called 'non-trade concerns'. Non-trade concerns include, in particular, food security and the need to protect the environment. The needs and conditions of developing country members are to be taken account of. Special and differential treatment for developing countries is to be an integral element of the negotiations; further, developed country members are to provide for a greater improvement of opportunities and terms of access for agricultural products of particular interest to the developing country members, including the fullest liberalization of trade in tropical agricultural products (Preamble, paras 5 and 6).

Coverage

According to its Annex 1, the Agreement covers a large range of agricultural products.[12] Included are the 'classical' agricultural products such as cereals, fruit and vegetables; coffee and tea; live animals and animal products (including meat from cows, pigs, poultry, horses and dairy produce); and honey. Additional animal products such as hair, bones, bristles, ivory, horns and tortoiseshell are also covered. Excluded are all fish and crustaceans, molluscs and other aquatic invertebrates.

Yet the agreement covers additional products which could be of interest in the context of TK, such as:

- Animal meat and other products from non-domesticated, possibly non-farmed animals such as primates, whales, dolphins, and reptiles including snakes and turtles.
- Hides and skins, and raw fur skins.
- Raw fibres for textile utilization, such as silk, wool and other animal hair, and cotton.
- Plaiting materials (e.g. rattan, reeds, including bamboo), stuffing materials, materials used in dyeing or tanning.
- Spices, and vegetable saps and extracts.
- Essential oils, including resinoids and concentrates; aqueous distillates and solutions (but not mixtures of a kind used as raw material in industry).
- Live trees and other plants, bulbs, roots and cut flowers and ornamental foliage.
- Industrial plants, e.g. for the production of insecticides, fungicides.
- Plants and parts of plants used directly in perfumery and in pharmacy (excluding processed products such as pharmaceutical products, perfumery, cosmetic and toilet preparations).
- Animal or vegetable fats and oils and waxes.
- Prepared foodstuffs, beverages, spirits and vinegar, tobacco, and manufactured tobacco substitutes, with the exception of preparations of fish or crustaceans, molluscs or other aquatic invertebrates.

Hence the AoA defines a wide range of products as 'agricultural', and covers products resulting from a great variety of agricultural systems, comprising for instance agroforestry, bush meat production and crafted dyes, but also industrialized agricultural production.

In this context, it can be asked whether these products can be subsumed under the category of the so-called environmental goods and services which are included in the Work Programme of the Doha Declaration (WT/MIN(01)/DEC/1). According to this programme, specific

[12] Annex I is based on the World Custom Organization's 'Harmonised System Nomenclature' (Harmonized Commodity Coding System, HS).

conditions such as the reduction or elimination of tariff and non-tariff barriers are to be negotiated for these products (31 (iii)).

The definition and coverage of environmental goods and services are presently discussed in the Committee on Trade and Environment. The discussion is based on lists of environmental goods provided by member states. The focus is on products contributing to pollution control and the concept of predominant environmental end use (TN/TE/10). So products resulting from TK are not subsumed under this category.

TK products were, however, discussed in a presentation by UNCTAD, under the heading of the so-called Environmentally Preferable Products (PPEs), which include, for example, organics, tropical produce, sustainable fisheries and forestry products and natural fibres (see TN/TE/R/10 Nos. 68–74 and TN/TE/INF/6 and TN/TE/INF/7). UNCTAD concluded that 'given the interest of developing countries in EPPs, the difficulties in capturing some of these in the HS,[13] and the low tariffs that prevailed with respect to these products, it might prove easier and more productive to focus the negotiations concerning EPPs on NTBs' (non-trade barriers) (No. 72).[14]

8.2.3 Reform programme: mandates and current stage

The negotiation mandates

Negotiations on trade in agricultural products have a long history in the international trade debates. The controversial issues tabled in the Doha Round – product and export subsidies, and market access, involving import restrictions in order to protect one's own agricultural producers – appear as a 'leitmotiv' since the establishment of the GATT, and continue to be at the centre of the debates.

Agriculture has always been perceived as a special sector of the economy for various reasons, such as ensuring national food security and independence in the supply of foodstuffs, and consequently had a unique political status.[15] Therefore, already by the stage of GATT (1947), a few important exceptions to the general regulations of trade in goods were gradually created for agricultural goods (Article XVI.3 and 4). As a consequence, there was no one true agricultural discipline: the exceptions led to important problems and distortions in world agriculture.

Thus, the conclusion of the Agreement of Agriculture in the Uruguay Round was a real breakthrough.[16] For the first time, agriculture had a firm, defined place within the world trade order, and disciplines and commitments for member states were introduced. Yet the AoA, at the time of its conclusion, was considered only a first step in the process of reforming agricultural trade. As a result, the agreement itself contains a mandate and a built-in agenda for the continuation of the reform process and for negotiations on further trade liberalization: according to Article 20, member states were committed to start negotiations on continuing the reform process in 2000.

The general goal of the reform process is clearly set out in the agreement. The long-term objective is substantial progressive reduction in support and protection (Article 20). The above-mentioned concerns and objectives, which are of particular importance for the developing countries, are to be taken specifically into account in

[13] Harmonized System, see note 12.

[14] See also the discussions in Chapter 8 Part 2.

[15] For a succinct historical overview see Sharma (2000). See also Desta (2002, pp. 5–7), in particular the citation from Smith (1776, p. 6).

[16] But see Desta's 'caveat' on this analysis regarding the regime of Green Box direct payments (2002, p. 421).

the negotiations for further liberalization of the agricultural markets: Article 20 of the AoA mandates the members – among other things – to have regard to the non-trade concerns, special and differential treatment to developing country members and the principles of equity and fairness.

Doha Declaration and Doha Work Programme: agriculture

THE NEGOTIATIONS

The negotiations on the reform process in agriculture started in early 2000. The first phase, until the Doha Ministerial Conference in November 2001, was characterized by a stocktaking of the negotiation proposals of the member states.

The Doha Declarations Work Programme set the task of establishing 'modalities'[17] for further commitments, including provisions for special and differential treatment until March 2003, and planned the submission of participants' comprehensive draft schedules for the fifth Ministerial Conference in September 2003. In the work programme agreed upon by the members in March 2002, it was decided that negotiations would focus on all three pillars of the agriculture agreement: export subsidies, market access and domestic support. Non-trade concerns, including food security and 'special and differential treatment' were to be an integral part of all of these topics. The goal was to reach a common position on draft modalities as a basis for discussion at the Cancun Ministerial. However, diverging interests and opposing views stalled the process of reaching a common position.[18] The last-minute solution was the submission of a 'framework', first put forward by the USA and EU, and then submitted to the Conference by the Chair of the General Council, Carlos Pérez del Castillo, under his own authority, as there was no consensus to submit this or any other draft.

So, ministers at the Cancun Conference were not then expected to agree on the modalities for negotiating the substantive issues of tariff and subsidy cuts, but rather, to set a new deadline for agreeing on modalities for reaching agreement after further negotiations in Geneva (*Bridges*, 2003). As the ministerial talks collapsed over the Singapore issues before the members had a chance to turn to the agenda item 'agriculture', no discussion and agreement on a text as a further working basis was possible.

However, the discussions on the framework were continued, first in informal meetings and discussions, then in negotiation meetings in March, April and June, the most difficult subject being market access. At the July 2004 meeting of the General Council, arduous round-the-clock negotiations finally led to a 'historical breakthrough': the consensual adaptation of the elaborated package, the 'Framework for Establishing Modalities in Agriculture' by the General Council.

THE DOHA DECLARATION

The mandate of the AoA was reformulated in the Doha Ministerial Declaration[19] and concretized in the Doha Work Programme, which was adopted after intense negotiations by the General Council on 1 August 2004.[20]

The Declaration built upon the work carried out under Article 20 of the AoA in the first phase of the negotiations that took place from 2000 to March 2001, reflected in the negotiating proposals already submitted.[21] It made the objectives more explicit and set deadlines for the negotiations.

It defined the scope, goals and work

---

[17] 'The term "modalities" refers to the numerical targets for further commitments in the different areas of disciplines established in the AoA during the Uruguay Round and new rule elements that define the framework within which trade under the new commitments will take place. The modalities agreed will determine the depth, scope and speed of further trade liberalisation in agricultural trade...' (Bernal, 2003, p. 7).

[18] For an overview over the main positions see Bernal (2003, pp. 8–10); WTO (2004).

[19] WTO Doc. WT/MIN(01)/DEC/1.

[20] WTO Doc. WT/L/579.

[21] Listed in WTO (2004).

programme of the negotiations. It put weight on the needs and interests of the developing countries in placing them 'at the heart of the Work Programme' (No. 2), and in particular on taking account of the vulnerability of the least-developed countries (No. 3). It contains a strong commitment to the objective of sustainable development. The basis is the conviction that it is possible and necessary to meet the goals of an open and non-discriminatory multilateral trading system, acting for the protection of the environment and the promotion of sustainable development in a mutually supportive way (No. 6).

The interests of the developing countries are taken account of in the commitment to integrate special and differential treatment (SDT) for developing countries in all elements of the negotiations. Developing countries are to be enabled to effectively take account of their development needs. These needs explicitly include food security and rural development. It is further confirmed that non-trade concerns are to be taken account of.[22]

With regard to the AoA, the Declaration reaffirms the general goal to 'establish a fair and market-oriented trading system'. The negotiations are to be aimed at: (i) a substantial improvement in market access; (ii) reductions with a view to phasing out all forms of export subsidies; and (iii) substantial reductions in trade-distorting domestic support.

THE DOHA WORK PROGRAMME AND THE 'FRAMEWORK FOR ESTABLISHING MODALITIES IN AGRICULTURE'

As mentioned above, the Doha Work Programme differentiates the subject matter of the Doha Declaration and provides a roadmap for further negotiations. Yet, with regard to the negotiations in agriculture,[23] the detailed modalities still have to be negotiated, the framework defining guidelines for the debates to come. In particular, the debates on formulae for reductions of support levels and schedules for their entire elimination are still open.[24] So whether the end result will live up to the goals set by the Doha Declaration depends on the next stage of the negotiations.

PARTICULAR SPECIAL AND DIFFERENTIAL TREATMENT IN CONNECTION WITH AGRICULTURE

As mentioned above, the specific consideration of the needs and conditions of developing country members forms in principle an integral element of the AoA (see Preamble para. 6). It is contained in the following general commitments: (i) the aspect of special and differential treatment is to be integrated into the negotiations; (ii) developed countries are to provide greater market access for agricultural products of particular interest for developing countries; and (iii) negative effects of the implementation of the reform programme on least developing countries are to be considered. Specific provisions for SDT provide for longer implementation periods and for lower reduction rates for tariffs, domestic support and export subsidies. As to domestic support, besides the higher *de minimis* threshold and the lower reduction rate for Total Aggregate Measure of Support (AMS),[25] an important SDT is the exemption from reduction commitments of specific rural development measures according to Article 6.2 AoA.[26]

The Doha Declaration commits the member states to integrating special and differential treatment for developing countries in all parts of all elements of the negotiations 'so as to be operationally effective

[22] See also the Decision on Implementation-Related Issues and Concerns, complementing the Declaration, adopted at the Fourth WTO Ministerial Conference (WT/MIN(01)/DEC/1).

[23] And even more so in the Framework to Establish Modalities in Market Access for Non-Agricultural Products.

[24] For details see, for instance, *Bridges* Year 8 No. 7 July–August 2004 (http://www.icstd.org).

[25] See details below.

[26] See details below; see also the concise overview by Mamaty (2000).

and to enable developing countries to effectively take account of their development needs, including food security and rural development' (No. 13). This obligation is entirely taken up and reinforced by the Doha Work Programme. All SDT provisions are to be reviewed with a view to their strengthening and making more precise, effective and operational (WT/L/579, p. 2).

In the stocktaking phase of the negotiations, propositions were tabled as to the creation of a specific 'development box' or 'food security box' that would provide greater flexibility for developing countries' agricultural support measures, in analogy with the boxes system of domestic support. The propositions encompassed a set of SDT provisions to allow developing countries, or parts of them, to protect and subsidize their agricultural production.[27]

While issues of food security and rural employment are still on the table, 'development box'-like concepts have become somewhat muted. Instead, developing countries have started focusing on instruments such as strategic products and a new special safeguard mechanism (International Institute for Sustainable Development, 2003b, p. 2).[28]

The Doha Work Programme incorporates the concept of 'special products' which are eligible for more flexible treatment. Developing country members can designate a number of products that are of importance for food security, livelihood security and rural development needs. Criteria and treatment of these products are to be negotiated (WT/L/579 No. 41). Further, a Special Safeguard Mechanism (SSM) for use by developing countries is to be established and the 'fullest liberalization' of trade in tropical agricultural products. The diversification of production away from the growing of illicit narcotic crops is considered to be overdue and is to be addressed effectively (Nos 42 and 43).

These measures are presently being discussed in the AoA negotiation round. In December 2004 the G33, which is a group of over 40 developing countries, initiated the discussion on the Special Safeguard Mechanism, which is to protect developing country members from sudden import surges that threaten to disrupt domestic production and employment. The paper argues that the current measures are inadequate for developing countries, as they frequently lack the administrative capabilities and infrastructure to apply the existing measures. It puts forward a set of criteria that should guide the negotiations on the SSM: the mechanism is to be simple, effective and easy to implement; it is to be automatically triggered and available to all agricultural products; both price and volume-triggered safeguards are to be contemplated; and both additional duties and quantitative restrictions are to be envisaged as response measures.[29] The proposition was supported by Kenya, the Philippines, China and Brazil, but encountered objections by developed country members, including the EC, the USA and New Zealand.

'NON-TRADE CONCERNS'

The Doha Declaration (No. 13) as well as the Doha Work Programme (Annex A, 2) 'take note' of the non-trade concerns reflected in the negotiation proposals submitted by members and confirms that they will be taken into account in the negotiations.

This wording was promoted by the supporters of the concept of multifunctionality of agriculture (the EC, Switzerland, Norway, Japan, Korea and Mauritius). The basic argument is that the public good output of agriculture such as food security,

[27] See Diaz-Bonilla *et al.* (2003, p. 2). For more details on the proposals for a development and/or food security box, see ibid. p. 29.

[28] For details on SPs and SSMs, see Bernal (2003); Committee on Agriculture, Special Session: Negotiations on Agriculture, Report by the Chairman, Mr Stuart Harbinson, to the TNC, Doc. TS/AG/10; iiss, ICTSD, Doha Round Briefing Series, August 2003.

[29] See *Bridges Weekly Trade News Digest* Vol. 8 No 44, 22 December 2004, and Third World Network (TWN) Info Service on WTO and Trade Issues (Dec04/5) (www.twnside.org.sg/title2/twinfo176.htm) December 05.

rural development and environmental services is to be taken account of and flexibility is needed to address these issues through agricultural production (International Institute for Sustainable Development, February 2003a, p. 3). This is mainly related to the discussions on the decoupled payments of industrialized countries under the 'green box measures' (see below).

8.3 The Structure of Agriculture in Developing Countries[30]

Generally speaking, one can divide agriculture into an industrialized, highly productive and intensive type, carried out by large commercial farms and mainly found in OECD countries, and a low-productive type, found in developing countries. In the following, the focus is on the structure of peasant agriculture in Latin America, Asia and Africa. There are similarities and differences. One similarity is contained in the sometimes misunderstood term of subsistence farming.

8.3.1 Terminology

In order to make a clear definition within an economic perspective, we use the term 'traditional agricultural system' for an economic entity covering most of its needs by self-sufficient production, and intra-community exchange transactions (goods for goods) instead of inter-community trade (goods for money). Self-sufficiency and intra-community exchange transactions can be regarded as economic features of specific indigenous communities. Traditional agricultural systems of this kind, managed by indigenous communities, can be described as economically and ecologically integrated entities with minimal need to interchange with external markets. Territories and people of indigenous communities of this kind are self-determining and their extent remains limited over time. In contrast, widespread subsistence farming (about 70% of the world's poorest people are estimated to rely on subsistence farming) is family-oriented. In order to get access to basic goods and services apart from pure nourishment, for example, health and education, the family needs money, earned by trade or family members working in urban areas. According to this definition, subsistence farmers are in a way part of local markets, while traditional agricultural systems form in themselves integrated agroeconomic systems. In the Todaro (2000) development model, subsistence farming is the first step in a straight line of economic development through the transitional stage of diversified farming up to specialized farming. In contrast, the economic principle of self-sufficiency reflects the sovereignty of indigenous communities to self-determine the development of their society. With regard to biological diversity, the 'ecologic value' of traditional agricultural systems is inversely proportional to their market power.[31]

8.3.2 Economics of subsistence farming

In the past, subsistence farming was often characterized as technologically backward, irrationally producing, not responsive to economic incentives and resistant to technological innovation.

In contrast, modern studies have established that 'peasant farmers do act rationally and are responsive to economic

[30] Author: Michael Burkard.

[31] In fact, the distinction between subsistence farming and traditional indigenous agriculture is tenuous. For example, the community of farmers in the Cotahuasi Valley in Arequipa, Peru, are descendants of the Inca and Wari. They export organic quinoa as a nutragenic product to Europe. A strategic alliance of their Association of Organic Crop Growers (Asociación de Productores de Cultivos Orgánicos APCO) with the specialized non-governmental development organization AEDES allows them to process their nutragenic products in accordance with the norms of regulation 2092/91 of the European Union and to get the corresponding certification.

incentives' (Todaro, 2000, p. 388). As being essential for the understanding of peasant farming, Todaro has pointed to two key factors of the economics of subsistence farming: risk and uncertainty. Upon a closer look, subsistence agriculture is a highly risky business. If it fails, there is not only money at stake, but the whole family engaged at the farm will be exposed to the real and concrete danger of starvation. There are many risks and uncertainties, like variable rainfall, animal pests and crop failure, diseases diminishing the human workforce, and so on.[32] In such circumstances, the main motivation of the farming family is not maximization of income but minimization of risk, thus optimizing the family's chances of survival.[33] If risk and uncertainty are high, a peasant farming family will stick to the traditional technologies and crops that have proven to be successful rather than engage in costly and unknown ones, embodying a greater risk of crop failure. When survival is at stake, it is more important to minimize the risk even in a bad year (total crop failure) than to maximize the output and the income in better years. In terms of economics, a rational peasant farmer prefers traditional technologies and crops that combine a low mean per-hectare yield with low variance (less fluctuation around the average), rather than commercial technologies and crops, which may promise a higher mean yield on the one hand but also the risk of a greater variance on the other. Thus, the major motivation of subsistence farming is the economic rationale of risk aversion.

Thinking in terms of economic development, the transitional stage from subsistence farming to specialized farming is described as diversified or mixed farming (Todaro, 2000, p. 389). In this transitional stage, the farmers cultivate cash crops like fruits, vegetables, coffee, tea and pyrethrum beside the staple crop. Studies have established through evidence from Colombia, Mexico, Nigeria, Ghana, Kenya, India, Pakistan, Thailand and the Philippines that farmers are responsive to market signals like price incentives and business opportunities.[34] But such price incentives must be based on fair market prices. Diversified farming can be observed in Latin America, Africa and Asia. Thus, transition from subsistence farming to diversified farming is a second similarity of agrarian structure in developing countries, as soon as price signals and fair market conditions are identified and key elements like infrastructure[35] are assured.

The third step of farming is known as specialized commercial farming (Todaro, 2000, pp. 390–391). Here, maximum per-hectare yields are derived from capital-intensive and labour-saving techniques. The emphasis is on the cultivation of one particular crop or product. There is no difference in the concept of commercial farming than in other large industrial enterprises. Some of the largest commercial

[32] Even more severe than natural risks might be social and political uncertainties, e.g. private businesses like landlords and moneylenders capturing all gains and profits, policy failures like prices 'guaranteed' by government which are never paid, and additional inputs like development credits, fertilizers, pesticides, 'assured' water supplies never delivered (Todaro, 2000, p. 387).

[33] The optimization of food security in a subsistence farming community framework is described in Sterkenburg (1987, pp. 183–184).

[34] In terms of development, an improvement in public transport in order to reach local markets to supply a pile of cash crops, a new road or loans under fair market conditions, especially when granted to women, are some key factors for the transition from subsistence farming to diversified farming. According to this experience, efforts for rural development should directly address subsistence farming families and improve the basic conditions for a transition from subsistence farming to diversified farming.

[35] Infrastructure and transport are crucial in order to improve rural output in developing countries. For example, farmers in Rukwa, a productive region in Tanzania, cannot get their crops to Dar es Salaam or other domestic markets because of poor roads. Although Tanzania is threatened by food shortage, the government had to lift the ban to export cereals from Rukwa to external markets in the neighbouring Zambia and the Democratic Republic of Congo (*The East African*, 4 August 2003).

farms in developing and especially in least developed countries are owned and managed by multinational agribusiness companies.[36] In order to maximize profit, those agroenterprises are often specialized in high-yielding production of cash crops like fruit, vegetables and cut flowers for export to OECD countries. Thus, the existence of commercial farms is not only the third similarity of agrarian structure in developing countries, but can be observed in developed as well as in developing countries.[37]

Today, subsistence economies are not only challenged by natural, but also by economic and political interference.

According to Todaro (2000), the operating entity of subsistence farming in Asia, Africa and Latin America is the family structure. But family structures are not able to fulfil all the needs of their members. In order to make available manufactured goods or services like health or education, the farming family must have sources of income. Besides labour migration of family members to urban areas, a source of income is to supply to local or urban markets any agricultural products surplus to daily family needs or any additionally planted cash crops.[38] In urban areas in Africa, Latin America and Asia a hierarchy of markets for agricultural products can be observed. The official market is in the hands of large-scale farmers and commercial companies, whereas dozens of individuals sell small piles or even handfuls of vegetables and other agricultural products (and processed foods) on improvised stands or spread out on cloths in the streets.[39] Thus, the link between subsistence farming and local or urban markets for agricultural products is established as part of the so-called informal sector.

8.3.3 Rural–urban migration

Socioeconomic research has established that the informal sector is not only linked with the formal urban sector but with the rural sector as well. Migrants from rural areas are seeking jobs in the formal and informal sectors of urban areas, thus causing one of the major problems of developing countries, rural–urban migration.[40] Until recently and following the historical model, rural–urban migration was regarded as favourable for developing countries in the

[36] The biggest multinationals in the agrobusiness, according to their turnover, are: (i) Nestlé SA (CH), (ii) Cargill Inc. (USA), (iii) Unilever (GB), (iv) Procter&Gamble, Co. (USA), (v) Diageo PLC (GB), (vi) Kraft Foods Inc. (USA), (vii) Conagra Inc. (USA), (viii) Supervalu Inc. (USA), Novartis AG (CH), PepsiCo Inc. (USA) (http://www.alliancesud.ch/deutsch/filesD_PnDf2.pdf, accessed December 2005).

[37] I observed the transition of a diversified farm to a specialized one in Uganda, East Africa. In the beginning, the staple products were milk and dairy products for local markets and the nearby urban area. Besides this, there was vanilla production as a cash crop for direct marketing and export. As soon as there was a business opportunity and loans under fair market conditions were available, the farm specialized in chicken breeding. Today, this farm is the biggest chicken producer in Uganda and independent from foreign investment (see http://www.ugachick.com).

[38] Theissen and Voigt (1986, p. 34) underline these findings.

[39] I noticed this striking market structure, e.g. in Uganda, East Africa. According to my observations, the majority of the individuals selling small amounts of agricultural products were women. This observation corresponds with the important role of women established in the agricultural sectors of developing countries. Women provide additional work outside the household in an estimated 60–80% of agricultural labour in Africa and Asia and about 40% in Latin America (Todaro, 2000, pp. 380–383). Taking into account the gender gap in education (the female literacy rate in developing countries is about 30% lower than male literacy), it has to be stressed that projects targeting the preservation of traditional knowledge in agriculture must carefully address women and their specific needs in rural areas. The huge share of agricultural production in developing countries carried out by women and their specific knowledge of women's crops like garden vegetables call for a specific gender approach in projects targeting traditional knowledge in agriculture.

[40] The concrete interchange between job seeking and rural–urban migration is highlighted by Bope, presenting a case study of the *kuba* people in Kongo (Bope, 1991, p. 105).

same way as it was for the economic development of Western Europe and the USA. But in developing countries, the paradox has occurred that rural–urban migration has not effected an accelerated development in urban areas but caused rising urban unemployment. Todaro (2000) has identified the imbalance of economic opportunities between urban and rural areas as a main source of the high rates of urban unemployment in most economically slow-developing countries.[41] He stresses that policies designed to reduce urban unemployment may even cause higher levels of urban unemployment, due to induced migration. Thus, policies operating on the demand side of urban employment like wage subsidies, scarcity factor pricing, direct government hiring and employers' tax incentives can be counterproductive in the sense of urban employment (Todaro, 2000, p. 310–311).[42] By accelerating rural–urban migration, such policies also lead to lower levels of agricultural domestic output. As an additional external factor, the structural adjustment programmes (SAPs) of the World Bank have to be mentioned. In recent years, SAPs have promoted the production of cash crops for export.[43] But the primary beneficiaries of past agricultural programmes have been large, high-yielding commercial farms (Todaro, 2000, p. 573). Thus, domestic agricultural policies were in conflict with export orientation due to balance of payments deficit and the necessity of rural development programmes designed for subsistence and small-scale farmers.[44]

It has to be added that policies other than those designed to reduce urban unemployment may have inducing effects for rural–urban migration. If relative costs and benefits are key factors in stimulating migration, all policies designed to reduce living costs for the increasing urban population in order to prevent poverty have to be taken into account. Among those, urban development and poverty reduction programmes designed to cheapen basic goods like fuel and food are of specific interest. If basic agricultural products like rice,

[41] Todaro has evolved a new model of rural–urban migration, known as the Todaro migration model (Todaro, 2000, pp. 305–310). Some features of the Todaro migration model are: (i) migration is rational, motivated by economic considerations of relative benefits and costs; (ii) migration is based on expectations; the decision to migrate depends on expected urban–rural wage differentials; (iii) migration remains attractive as long as there is a reasonable expectation for rural–urban wage differentials in the long term. Thus, rural migration remains attractive even if there is a high rate of urban unemployment; (iv) in sum, rural–urban migration is stimulated by the imbalance of economic opportunities between rural and urban areas.

[42] Accordingly, Todaro (2000) advocates policies designed to regulate the supply of labour to urban areas by creating attractive economic opportunities in rural areas. However, Todaro does not discuss the adverse effects of export subsidies.

[43] Export-oriented development strategies rely upon the theory of comparative advantage, developed by David Ricardo in 1817. Using a historical approach, Professor Mathias Binswanger examined the examples of English clothes and Portuguese wine. He found that the application of the theory of comparative advantage led to a rapid decline in the Portuguese clothing industry without a significant improvement in wine exports to England. As reasons for these findings, Binswanger identified that the premises for the theory were not there in the cited example. England would not be able to build up significant wine production due to climate conditions, and workers in the clothing industry in Portugal could not shift massively into wine production due to limited land resources and limited demand for Portuguese wine. Binswanger concluded that the situation of developing countries might be similar to the historical example of Ricardo insofar as premises are missing from the beginning (Mathias Binswanger, 'Wein gegen Tuch: Doch nur einer gewinnt,' in: *Die Weltwoche* No. 16, 19 April 2001). Recently, Ha-Joon Chang (2002, p. 52) evaluated development strategies of now-industrialized countries within a historical perspective. He found that these strategies were seldom based on the theory of comparative advantage of David Ricardo. On the contrary, he established that the now-industrialized countries used to rely on protectionist instruments and strategic trade policies, which are now restricted by the WTO, both for already-industrialized countries as well as for developing countries.

[44] A historical analysis of the export-orientation of many African countries can be found in Theissen and Voigt (1986, p. 35).

wheat[45] and maize are cheapened, then rural wages decline. Additionally, in many developing countries policies tend to tax agriculture and support the consumption of an increasing urban population (Hoekman and Kostecki, 2001, p. 210).[46] In sum, a vicious circle of rural–urban migration, urban unemployment, urban employment programmes, and induced and accelerated rural–urban migration occurs.

For developing countries, there are several instruments for the funding of policies designed for urban employment and poverty reduction, for example internal financing or department financing through loans from the Word Bank and International Monetary Fund (IMF). But due to tariff reductions committed in the Uruguay Round, some developing countries – and especially least developed countries – suffer from budgetary imbalances, because tariffs formed a major source of public income. In this situation and due to financial constraints, governments may be willing to bring down the prices for essential commodities like food by an accelerated liberalization of import tariffs, instead of providing subsidies for the increasing urban population. This result corresponds with the theoretical findings of the economic analyses evolved below (see Section 8.4.2).

8.3.4 Structures of agriculture

Having identified subsistence farming as a point of similarity between the structures of agriculture in developing countries, the differences will now be considered.

The agrarian structure in Latin America is dominated by the so-called latifundio–minifundio dualism. Latifundios are defined as large-scale farms providing employment for more than 12 people. Latifundios comprise less than 7% of all farms, but occupy up to 82% of the agricultural land. In contrast, minifundios are defined as very small farms providing employment for only two people, not even enough for a family. Minifundios comprise up to 90% of the farms, but occupy not more than 17% of the total agricultural land in most Latin American countries.[47] Studies have established that this process of land concentration in Latin America is still going on. But a third category of farms called family farms, medium-sized farms or multifamily farms, has to be introduced. They employ between 4 and 12 workers. These intermediate farm organizations are highly efficient and account for up to 50% of total agricultural output in Venezuela, Brazil and Uruguay, for example. Many of those medium-size farms can be subsumed under the mentioned term of diversified or mixed farming.

In Asia, fragmentation and monetization due to scarce arable land in relation to population density can be identified as key elements of the agrarian structure. Population growth leads to an accelerated process of fragmentation until production falls below the subsistence level. At the same time, the development of commercial farming leads to cash needs of peasant farmers, driving them into the hands of moneylenders. The result is a shift to cash crops such as tea, rubber and jute, often carried out by commercial farms after the peasant farmers are driven off their land.

[45] The OECD reports that, among those policies, the subsidization of wheat flour or bread to urban consumers is the most widely used instrument. The average level of support is estimated to be more than 50% of the total costs of providing flour to bakers' consumers in traditional wheat-consuming developing countries, and above 20–30% in the mixed cereal-consuming developing countries in South and East Asia. Furthermore, because wheat is considered as an industrial output, customs tariffs have generally been maintained at low levels. Thus, the real costs of imported wheat are significantly lowered relative to domestically produced substitutes (OECD, 1993, p. 45).

[46] In OECD countries, the opposite policies are imposed due to reasons explained by political economists and the theory of public choice: the urban population is taxed to support the agrarian sector.

[47] In the context of this research it may be of interest to note that in countries with a high quota of indigenous populations minifundios are particularly widespread, for example in Ecuador, Guatemala and Peru (Todaro, 2000, p. 373).

The structure of agriculture in Africa is the result of elements, both natural and historical, which have restricted an increase in productivity. Shifting cultivation and a relatively low productivity were the most economic solutions within the given conditions, as long as population size remained relatively stable. But as soon as population density increased, shifting cultivation was replaced by sedentary small-scale farming and subsistence practices. As the name suggests, subsistence farming is unable to keep up with the population growth in urban areas. At the same time, the neglect of rural development, land taxes, soil erosion and deforestation have led to a dramatic decline in agricultural productivity. As a result, dependence on food imports increases.

8.3.5 Agriculture and genetic erosion in PGRFA

In the literature, many reasons for the loss of agrobiodiversity can be found. The *World Atlas of Biodiversity* (Groombridge and Jenkins, 2002, p. 40) points out that industrial-scale agriculture and commercial breeding have replaced local varieties, of both crops and livestock, by genetically uniform types such as HYVs.[48] Other authors (e.g. Carrol *et al.*, 1990, p. 173) point to the conversion of diverse natural ecosystems into monocultural systems of production and to the narrowing of the genetic diversity of crops as being greatly responsible for ongoing genetic erosion.

The *Report on the State of the World's Plant Genetic Resources for Food and Agriculture* (FAO, 1996) has identified the spread of commercial agriculture as the main cause of the loss of genetic diversity. As a consequence of the introduction of HYVs, the report mentions the replacement and loss of traditional, highly variable farmer varieties. In summing up, a shift can be established from highly variable varieties (HVVs) to HYVs.

This summary corresponds with the conclusion made by Swingland (2001, p. 385). He admits that it is difficult to demonstrate erosion of diversity in crop gene pools quantitatively. But Swingland argues that the genetic erosion can be 'indirectly assessed in terms of the increasing proportion of world crop land planted in high-yielding, but genetically uniform, varieties'.

[48] The danger of a narrow genetic base recently became apparent when diseases like the so-called Panama disease and black sigatoka widely attacked the Cavendish banana, which is mostly exported to Western countries. 'Dessert-type' bananas like the Cavendish banana are cultivated mostly by the large-scale banana companies for international trade. Their vulnerability is regarded as a consequence of growing one genotype on a large scale, leading to uniformity. Cavendish bananas are mostly sterile mutants that cannot breed and the fruit cannot evolve to develop resistance to pests and diseases. The scale of the problem can be estimated when it is recalled that in the 1950s the Gros Michel banana, the predecessor of the Cavendish banana, was wiped out by Panama disease. Therefore, the FAO called for the development of more diversity in the banana, especially for export bananas. The FAO mentioned that, 'fortunately, small-scale farmers around the world have maintained a broad genetic pool which can be used for future banana crop improvement'. Yet, though important for world trade, the Cavendish banana accounts for only 10% of bananas produced and consumed globally (FAO, in: http://www.fao.org/english/newsroom/news/2003/13120–en.html, accessed December 2005). In contrast to this FAO statement, the BBC reported that edible bananas might disappear within a decade (http://news.bbc.co.uk/2/hi/science/nature/2664373.stm, accessed December 2005). The source of this report was Dr Emile Frison, a plant pathologist who heads the International Network for the Improvement of Banana and Plantain (INIBAP). Dr Frison assembled scientists from 11 countries in the newly founded Global Musa (Banana) Genomics Consortium. He suggests sequencing the banana genome for 'improving the varieties on which Africans depend for their survival'. Following rice and the small weed *Arabidopsis thaliana*, the banana will become the third plant to be sequenced (http://news.bbc.co.uk/2/hi/science/nature/1445357.stm, accessed December 2005). Regarding the source of the alarming BBC news on the one hand and the de-escalating FAO information on the other hand, the conflict regarding genetically modified organisms (GMOs) becomes apparent. While driving forces around science and agrobusiness boost the acceptance of GMOs, the FAO and other institutions rely on the maintenance of biological diversity in PGRFA by virtue of adequate policies.

Going into more detail, Rangnekar (2003, p. 22) has identified the separation of breeding from farming and its commercialization as the origin of breeding genetically uniform varieties. The separation of breeding from farming and its commercialization have led to two reinforcing consequences. First, the farmer becomes dependent on the breeder for the supply of varieties, because genetically homogenous varieties reduce and eliminate the farmer's option of practising on-farm varietal selection. Secondly, the farmer is dependent on the commercial breeder to supply fresh seeds on a regular basis, because he is forced to maintain the purity of the variety in order to uphold the productivity gains.

Research seems to be more advanced in explaining losses in animal breed diversity than losses in crop diversity. The Swanson dominance effect explains the tendency of breeds selected in higher-income countries to replace those in less developed countries. According to Swanson, diversity declines as 'a result of the uniformity of the development process across heterogeneous states' (Tisdell, 2003, pp. 365–376). In his analysis, Tisdell (2003) has worked out that a driving force of the ongoing loss of genetic diversity is the process of extending markets. The principle of comparative advantage and the economies of scale encourage regional and international specialization[49] in production, thus leading to genetic uniformity and the extinction of less productive landraces and breeds. Going into further detail, Tisdell (2003) determined that the decoupling of breeds and animal husbandry from local natural environments under highly productive industrial farming systems is a key element of genetic erosion. Because environmentally tolerant breeds and those breeds not ideal for intensive husbandry suffer a demise, industrial farming accelerates genetic erosion.

Summarizing the findings from both Tisdell (2003) and Rangnekar (2003), a double alienation qualifies a farming system as narrowing genetic diversity: the separation of breeding from farming, and the decoupling of farming from the local natural environment.

Considering the contrary position, scientists (e.g. Schulze *et al.*, 2002, p. 676) have established evidence that small-scale and subsistence farming systems correspond to high biological diversity. This finding is based on the fact that small-scale and subsistence farming systems are less disturbed by agrochemicals and high yielding farming techniques than are industrial farming systems ('medium disturbance hypothesis').

8.4 The Impact of Agricultural Export Subsidies and Food Aid on Agrobiodiversity[50]

8.4.1 Introduction

In the relevant international legal instruments, subsidies and food aid are mentioned in different ways. On the one hand, Article 11 of the Convention on Biological Diversity (CBD) characterizes the removal of domestic subsidies in agriculture as an economic incentive in favour of local biological diversity. On the other hand, restrictions on agricultural export subsidies and food aid are regarded as favourable for market-oriented international trade in agricultural products by the Agreement on Agriculture (AoA) of the WTO.

In this section, the two approaches are integrated, and it is evaluated whether the removal of agricultural export subsidies can be considered as an incentive measure in the light of the CBD at a global level. The findings indicate that the removal of agricultural export subsidies and a strengthening of food aid disciplines favour

[49] This viewpoint is underlined by Pearson, who looked at the relationship between trade and environment as follows: 'Trade and sustainability are linked in yet another way. Trade increases specialization of production within countries and regions of countries. Specialization promotes monocultures (...). Monocultures are the antithesis of ecological sustainability – the diversity and resiliency of ecosystems' (2000, p. 483).

[50] Author: Michael Burkard.

biodiverse agriculture in developing countries (DCs).

8.4.2 Economics of agricultural export subsidies and food aid

Economics of agricultural export subsidies

DEFINITION

In an economic sense, an export subsidy is defined as 'a payment to an exporter of either a fixed sum of money per unit exported or a proportion of the value of the goods exported' (Habte, 2001).

At first glance, subsidizing exports of agricultural products seems to be a paradox: why should domestic consumers and taxpayers subsidize consumers abroad?[51] The answer may appear on closer analysis of the economic effects of export subsidies at the domestic level.

ECONOMIC EFFECTS OF AGRICULTURAL EXPORT SUBSIDIES AT THE DOMESTIC LEVEL

According to Rieder and Anwander Phan-Huy (1994, pp. 260–262), export subsidization has a double effect: exports increase and domestic prices rise.

This raises the question of how the costs and benefits of export subsidization are allocated. The costs of export subsidization are first burdened on domestic taxpayers. Their payments are beneficial, first, to consumers abroad, and secondly to producers at home. In turn, domestic consumers are confronted with a price rise. As a consequence, demand at home declines.

So, export subsidization has the effect of higher prices for the subsidized goods at home, and lower prices for the subsidized goods abroad. The costs are borne by taxpayers and consumers at home, and the benefits are transferred to consumers abroad and producers at home.

In political terms, it is important to note that the beneficiaries of export subsidization are not only consumers abroad. In the first place, the domestic producers profit from this kind of financial transfer.[52]

In order to explain export subsidization in the field of agriculture and to investigate whether it may affect biological diversity, the scope of the study will be widened to include the effects of export subsidies on the international market.

ECONOMIC EFFECTS OF AGRICULTURAL EXPORT SUBSIDIES ON THE INTERNATIONAL MARKET

According to Egger *et al.* (1992, pp. 8–9) and Rieder and Anwander Phan-Huy (1994, pp. 348–351), export subsidies lower the price for export goods at the border in order to increase the sale of goods abroad. In other words, export subsidies influence the demand abroad. Export subsidies are often used if the goods concerned cannot fetch a reasonable price in the producing country.

Subsidization of agricultural goods affects the market mechanisms of the global agricultural market. With the increase of price support measures combined with export subsidies, the EU, for example, has turned from being an importer to an exporter of agricultural products. The product surpluses in the EU have led to a rise in supply on the world market for the subsidized agricultural products. At the same time the role of the EU as a demander for those products has decreased, so demand on the world market has dropped. From

[51] A wide range of answers can be found in the literature. See, *inter alia,* Myers and Kent (2001). In the scope of this paper, discussion will be limited to questions relevant for agriculture. For this reason, no discussion will be undertaken of the so-called strategic trade policy (Brander-Spencer model), which may be relevant for industry policies (Siebert, 1997, pp. 169–170; Siebert, 2000, pp. 123–127).

[52] More detailed studies carried out by the OECD, however, do differentiate between farmers and landowners. In fact, the main beneficiaries of support are found to be the landowners, regardless of the measure providing it. Thus, farm households are among the biggest recipients of support measures in agriculture only if they own the land they farm (OECD, 2001a, p. 8). See also the rent-seeking model of the theory of public choice. According to Mueller, the iron law of rent seeking is that 'whenever a rent is to be found, a rent seeker will be there trying to get it' (1989, p. 241).

this a lower world price for agricultural products subsidized by the EU results.

In view of DCs and LDCs (least developed countries) with little or no means to protect their agricultural sector, imports increase, and domestic agricultural production of the subsidized products decreases. On the one hand, consumers in urban areas profit from lower prices for imported agricultural products. On the other hand, incentives for local rural production diminish. Incomes in rural areas fall, and tendencies for rural exodus arise. This leads to an increasing dependence on imported agricultural products.

Economic effects of food aid

From an economic point of view, according to Rieder and Anwander Phan-Huy (1994, p. 263, pp. 330–332), analogous to export subsidies, food aid takes economic pressure away from domestic supply of the producer country, and thereby upholds the price at home. Like export subsidies, food aid reduces surplus domestic production. And, favourably, expenditure for food aid is suitable for inclusion in development budgets.

An example of such economically motivated food aid is the rice delivered by Japan to Burkina Faso. Following international trade rules, Japan has to allow some imports of rice into its territory. But after being counted by the trade statistics, this rice will not appear on the Japanese market but is further delivered to Burkina Faso according to a food aid cooperation programme between the two countries. According to this food aid cooperation programme, the government of Burkina Faso gets financial support from Japan. But on the other hand, Burkina Faso is obliged to buy rice from Japan with this money. This food aid cooperation programme allows Japan to be in line with international trade obligations without affecting its domestic rice producers. And furthermore, the rice delivery can be statistically counted as food aid and be included in the development budget. In Burkina Faso, the dumped rice is sold at a lower price than the domestic one, thus competing with local rice producers.[53]

Economic effects of a removal of agricultural export subsidies

On the basis of data established by Tokarick *et al.* (2002) in an essay for the International Monetary Fund (IMF), the welfare effects of a subsidy removal by industrial countries (as percentage of the GDP[54]) on both industrial countries and DCs can be outlined as follows.[55]

Looking at the industrial countries, two groups are likely to profit from a subsidy removal: first the major agricultural producers like Australia, Canada and New Zealand, organized in the Cairns Group.[56] They would profit from higher world prices. Secondly, groups with huge expenditures for export subsidies such as the EU. The total welfare gains for industrial countries are estimated at US$14.1 billion.

In contrast to industrial countries, the negative welfare effects of the removal of subsidies by industrial countries on DCs are established. Only Brazil, a member of the Cairns Group, would benefit substantially. The total welfare losses for DCs are estimated at US$4.7 billion.

The explanation given by the IMF

[53] See http://www.abcburkina.net/vu_vu/vu_18.htm (accessed December 2005).

[54] GDP (Gross Domestic Product): the total value of goods and services produced by a country in a year. To assess the static effects of agricultural liberalization, the economists of the IMF used the general equilibrium model of the world economy (GTAP).

[55] It has to be taken into account that the IMF research examines the welfare effects of all kinds of subsidies spent for agriculture, in comparison with the welfare effects of a tariff removal. The IMF research does not quantify the welfare effects of a removal of export subsidies only.

[56] Named after the Australian town of Cairns, the group emerged in the GATT-Uruguay Round and consists of 18 (originally 14) highly competitive exporters of agricultural products. These are Argentina, Australia, Bolivia, Brazil, Canada, Chile, Colombia, Costa Rica, Fiji, Guatemala, Indonesia, Malaysia, New Zealand, Paraguay, Philippines, South Africa, Thailand and Uruguay (Desta, 2002, p. 208).

experts for the negative welfare effects on DCs corresponds with the findings of the theoretical approach of this paper. As outlined above, export subsidization of agricultural products from industrialized countries leads to a decrease in domestic production in DCs and to an increase in imports of agricultural products. While consumers mainly in urban areas profit from lower prices for imported agricultural products, producers in rural areas are confronted with falling incomes due to lower prices for their products and competition by dumped goods, respectively.

In contrast, the removal of export subsidies by industrialized countries would burden the established welfare losses of US$4.7 billion mainly upon the increasing urban population in DCs. Thus, a removal of agricultural export subsidies would especially affect Net Food-Importing DCs (NFIDCs).[57]

8.4.3 Impact of agricultural export subsidies and food aid on agrobiodiversity: the PLA model

A review of the economic effects of export subsidies may indicate the impact on agrobiodiversity. The starting point of the impact analysis is the economic effects of agricultural export subsidies on the international market. It was summarized that incentives for local production decline, as, at the same time, imports increase. The sum of additional imports reflects the substitution of local agricultural goods by imported ones. If, in a theoretical approach, the sum of additional imports caused by export subsidies reflects the substitution of local agricultural products by imported ones, the substitution rate thus indicates a potential loss of agrobiodiversity (PLA) on a local market. This presumption, of course, requires further underlining. In order to better illustrate the effects of subsidized agricultural products exported to a local market, the following case study, reported by Shiva (2002, pp. 305–326) is presented.

In August 1999 there was a scandal in India because of adulterated mustard oil. The government prohibited Indian mustard oil and abolished all import restrictions for cooking oils. The import of soybeans and soya oil was liberalized. The prices for oil seeds fell and Indian farmers were no longer able to cover the costs of production. Within only one cultivation period the liberalization of soya imports destroyed all the production and manufacturing processes for oil seeds in India. Besides the effects on employment and agrarian income, there were ongoing changes in the market structure of oil seeds and oil products in India. Mustard, sesame, linseed, gingelli plant and groundnut were disappearing from the fields and the Indian market was flooded with imported soybeans. The import of soybeans increased by 60%, whereas the price for groundnuts, for example, dropped by 23%. One reason for these changes in the market structure for food oil in India is the fact that US soybean is highly subsidized. The low price for one ton of soybean of only US$155 can only be obtained by subsidizing each ton of soybean by US$193.

From an economic point of view, the OECD corroborates the findings mentioned above by giving the following data:

> In 1992–93 Indian imports [of soybeans] were barely 100 000 tonnes. Imports following economic growth and trade policy reforms (removal in 1995 of import restrictions and subsequent tariff reduction to 16.5%, though recently increased to 27.7%) rose to an estimated 4.6 million tonnes in 1999–2000. (2000, p. 57)

The OECD does not stress the issue of export subsidies in the context of Indian trade policy; however, it does question whether US oilseed policies do affect world

[57] In terms of political economy, the argument may arise that one reason for the increase of urban population is precisely the rural exodus in developing countries, which is caused by export dumping by industrialized countries. This line of argument refers to the analysis of the structure of agriculture in developing countries.

prices. The OECD has made a comparison of projected market outcomes with those which would have prevailed if US soybean producers had not received additional support under the loan programmes, but only from market returns. On the one hand, the OECD found that marketing loans for soybeans have a price-depressing impact on world prices of soybean. On the other hand, the OECD states that this effect 'dissipates over time'. To estimate the impact of US support policies for soybean producers on local markets, such as in India, it has to be taken into account that loan programmes for US soybean producers cover up to 88% of US soybean production (1998), and that the USA produces nearly half the world's soybean crop (OECD, 2000, pp. 58–59).

From this case study we can draw the following provisional conclusions:

- Substitution of local crops: the US soybean was highly subsidized in order to be competitive on the world market. On the local level abroad, the subsidized soybean overtook traditional agricultural products on the Indian market. The traditional oil plants were no longer competitive on the Indian market and their cultivation was abandoned. Finally, imported and highly subsidized US soybean products substituted traditional oil plants in India. In the long term, local landraces may become extinct.
- Cultivation shift and changes in local peasant structure: as demonstrated, subsidized agricultural exports from industrialized countries lead to a decline in local rural production in DCs and to an increase of imports. Local subsistence or diversified farmers are already in danger of becoming indebted under normal circumstances. When local prices fall, their income diminishes even more, and moneylenders may take control over the land of the indebted farming families. Thus, this process brings the land formerly used for subsistence or diversified family farmland into the hands of commercially oriented owners. The commercially oriented landowners can take more risk and shift to a specialized farming method by replacing traditional staple food crops with cash crops like cotton. As presented by Shiva (2002), an example of such a shift in cultivation methods and its effects on agrobiodiversity can be found in the changes cotton production has undergone in Warangal, Andhra Pradesh. The region of Warangal is semi-arid and the soil was traditionally used for rice, pulses, millet, oilseed and vegetables. Because cotton was praised as 'white gold', many farmers tripled the land they used for cotton production and correspondingly reduced the cultivation of traditional cereals such as jawar and bajra.[58]
- Distorting effect on local markets and distribution channels: the case study demonstrates that imported products may substitute for traditional agricultural products as soon as subsidized products flood a local market. Thus, in the light of agrobiodiversity, the criterion is not whether supporting measures have a distorting effect on the world market (meaning, according to Article XVI of GATT 1994: 'more than an equitable share of world export trade in that product'), but if they have a distorting effect on a concrete local market. From this point of view, the distinction between domestic support and export subsidies loses its validity. Domestic support, like loan programmes, may have the same distorting effects as export subsidies, as soon as the resulting products appear on a local market abroad. In order to protect agrobiodiversity, the 'tariff-light' approach, with its three boxes and the criterion of a distorting effect on the world market, needs to be revised.

[58] Rangnekar has indirectly underlined this case study (2003). He stated that the adoption of 'green revolution' rice in Andhra Pradesh led to the loss of 95% of the traditional varieties without their collection and/or documentation.

8.4.4 Impact of food aid on agrobiodiversity

First, like export subsidies, food aid substitutes local agricultural products with imported ones. Analogous to subsidized imports, the sum of additional agricultural goods brought in by food aid reflects the amount of this substitution. Thus, the substitution rate indicates the potential loss of traditional agricultural products supplied on local markets.

Because food aid is cheap or even free of charge, local producers of agricultural products cannot compete and are put out of business: the result is rural exodus and urbanization. If not limited in time, dependence on food aid may become of a structural nature. As food aid programmes are usually implemented in moments of surplus production, changes in the world market situation may bring the programme to a sudden end, thus intensifying fluctuations of food supply.

Secondly, food aid changes consumer preference directly by replacing traditional food, and indirectly by increasing tendencies of rural exodus and urbanization due to falling rural income. Based on collected data, the OECD has estimated that food aid may be responsible for long-enduring structural changes in cereal consumption habits in recipient DCs in favour of wheat and white bread. The OECD (1993, p. 48) mentions that these structural changes can persist long after such food aid is reduced. In conclusion, food aid challenges traditional crops on both sides, supply and demand.

Thirdly, genetically modified organisms (GMOs), such as GMO-based cereals introduced as food aid and not milled before distribution, may be used as seeds and replace traditional crops.

In sum, food aid may have the following effects on agrobiodiversity:

- Like export subsidies, food aid may substitute for local agricultural products.
- Food aid may be responsible for structural changes in cereal consumption habits in recipient DCs, for example in favour of wheat and white bread.
- Food aid challenges traditional local crops on both sides, supply and demand.

8.4.5 Conclusions

Having examined the economics of agricultural export subsidies on the one hand and the structure of agriculture in DCs and the reasons for genetic erosion in PGRFA on the other, the conclusions can be drawn that the impact of a removal of export subsidies has to be examined at the level of a concrete local market for food and agricultural products. Furthermore, the local agrarian structure and the economic and political circumstances have to be taken into account, for example the tendency of rural–urban migration. To estimate the impact of a removal of export subsidies on agrobiodiversity, the following indicators may be used:

- A significant substitution of traditional agricultural products by imported ones on a concrete local market since the start of identified export-dumping measures indicates that the substitution could be imposed by distorting effects.
- Significant changes in peasant agrarian structures since the start of identified export-dumping measures indicates that the peasant farmers are already being driven off their land by moneylenders or commercial farmers, or have migrated from their land to urban areas. In those cases, a subsidy removal would not reverse this occurrence but would replace the dumped products by domestic, but now commercially produced, ones.

According to these indicators, the conclusion is twofold. Generally, the removal of export subsidies would privilege farmers who were able to enhance the input of land and/or financial resources in order to increase productivity. Thus, the crucial question is whether in a given area there are subsistence farmers ready for transition into diversified farming due to development policies like investment in infrastructure and public transport, and to loan programmes targeting subsistence farmers.

Or whether, conversely, there are just some already big commercial farms which would become even greater and more specialized due to new business opportunities.

In Latin America this might be the large-scale latifundios and the efficient medium-size farms, the prevailing agrarian structure for example in Venezuela, Brazil and Uruguay. If commercial latifundios prevail in an area, it can be assumed that these entities would profit most from better business opportunities due to a removal of subsidies. Thus, it can be assumed that commercial farms in member countries of the Cairns group[59] would profit from a removal of agricultural export subsidies. But the effect of a removal of subsidies may be different in countries where the agrarian structure is largely dominated by minifundios, as in Peru and Ecuador. For example, the Cotahuasi farmers in the Cotahuasi Valley in Arequipa, Peru, use ecological and organic cultivating methods from their Wari and Inca ancestors. Their products, like amaranth, quinoa, beans and broad beans, maize, Ilacón and anise, are marketed by the Association of Organic Crop Growers (APCO) in La Union. An increase in demand due to better price signals after a subsidy removal would not change their successful production methods, but would be an incentive for further development.

In conclusion, for Latin America the question must be answered individually for each region and each agricultural product.

In Asia, dependency on moneylenders and the falling into indebtedness could be decelerated due to higher rural income. Because of higher prices for imported agricultural products, the decision between cash crops and traditional food crops could be made in favour of the latter. And already widespread diversification farmers could be stimulated to make the step to specialized farming. As in certain Latin American countries, it can be assumed that commercial high-yielding farms in Asia would profit most from a removal of agricultural export subsidies by expanding their production. As already mentioned, it can be assumed that commercial farms in member countries of the Cairns group[60] would be able to expand their agricultural output.

For Africa, the removal of subsidies seems to be most beneficial. Due to the lack of additional arable land and/or financial resources in many parts of the continent, the removal of subsidies would be beneficial to the diversified or mixed farmers as well as to the subsistence farmers. Subsistence farmers could bring quantities of cash crops to local markets, enabling them to later make the step to diversified farming. And already diversified farms could be enabled to make the step to specialized farming, thus reducing dependence on imports. But in this case, as already mentioned, the latter transition would be beneficial in terms of development, but questionable in terms of agrobiodiversity. Although specialized farming in general is not favourable in terms of agrobiodiversity, this statement needs some clarification. If the farm specialized in the production of a regional variety[61] of one specific crop which forms an independent centre of diversity, commercial production of this one crop could be beneficial for agrobiodiversity, because otherwise this crop might become extinct in the long term. A second reservation has to be enunciated. As mentioned above, commercial farms, especially in least developed countries, are often owned and managed by multinational agrobusiness corporations. In order to maximize profit, they usually specialize in cash crops for

[59] Latin American members of the Cairns group include Argentina, Bolivia, Brazil, Chile, Colombia, Costa Rica, Guatemala, Paraguay and Uruguay.

[60] Member states of the Cairns group located in Asia are Indonesia, Malaysia, Philippines and Thailand.

[61] For example, quinoa in southern Peru and Bolivia, fonio in West Africa, tare in China and Pacific islands, tannia in West Africa (used for the preparation of fufu), pigeonpea in agroforestry systems in India and chickpea in southeast Turkey (Groombridge and Jenkins, 2002, Appendix 2).

export purposes.[62] It can be assumed that a slight increase in prices for domestic staple food crops would not be a commercial incentive for those entities to diversify away from cash crops for export to traditional crops for domestic food supply.

8.4.6 Summary and outlook

Agricultural export subsidies

The findings indicate that a removal of export subsidies for agricultural products by OECD countries under certain conditions is beneficial to agrobiodiversity and therefore could be subsumed under the term 'incentive measure' used by the CBD. These conditions are twofold.

First, in OECD countries, agricultural export subsidies have to be removed and not replaced by domestic support measures with similar inducing effects on production and the same distorting effects on international trade or local markets in DCs.

Secondly, in DCs, it has to be assured that the increase in demand for agricultural products could mainly be responded to by small-scale farming structures and does not cause a significant concentration process in favour of commercial high-yielding farms.

In terms of political economy, the findings indicate that a removal of agricultural export subsidies would be most beneficial to the highly competitive net exporters of agricultural products, namely Australia, New Zealand, Canada and Brazil, formerly organized in the Cairns group. In contrast, a removal of agricultural export subsidies would burden welfare losses to DCs and LDCs by an estimated US$4.7 billion, namely to NFIDCs.

Food aid

The findings indicate that under certain circumstances food aid might also have negative effects on agrobiodiversity as soon as the delivered agricultural products result from production surplus and the delivery becomes permanent, thus jeopardizing local markets in developing countries. Food aid should be granted according to international conventions like the Food Aid Convention (FAC, 1999) only. In order to protect local seeds from GMOs, it must be assured that grain is not contaminated with GMOs or that grain given as food aid is milled before distribution.

Outlook: an integrated approach towards global multifunctionality

In this perspective, the agrobiodiversity issue may be an additional argument in favour of the acknowledgment of non-trade concerns (NTCs) and the concept of multifunctionality. As long as multifunctionality is regarded as a shield to protect rich farmers in OECD countries,[63] acceptance will be low, as the small G-10 grouping gives evidence. But if 'multifunctionalism' is widened to a concept appropriate not only for OEDC countries with means to support their farmers, but also for DCs and LDCs, an integrated approach towards trade in agricultural products may be achieved. The challenge of such an integrated approach consists of integrating the various NTCs and different 'multifunctionalist policies' both from developing and industrialized countries with vulnerable agricultural structures in an appropriate manner.

An integrated approach, for example, would balance the NTC about food

[62] For example, the area for banana production for export purposes in the Philippines increased from 61,000 hectares in 1989 to 370,000 hectares in 2001, and the production area for cut flowers increased from 1008 hectares in 1990 to 1427 hectares in 1996. Over the same time period, production of rice and maize on Mindanao Island decreased by about 63% and 46%, respectively. In regions of Latin America and Africa there was also a dramatic increase in the production of cut flowers, vegetables and fruits for export to OECD countries (http://www.alliancesud.ch/deutsch/files/D_PnDf2.pdf, accessed December 2005).

[63] Developing countries characterize multifunctionality as 'a form of special and differential treatment for rich countries' (Desta, 2002, p. 8).

safety,[64] stressed by the EU, with the NTC about food sovereignty,[65] addressed by some DCs. And the NTC about rural–urban migration in DCs[66] deserves the same respect as the goal of a decentralized inhabitation of the country, as expressed, for example, in Article 104 of the Federal Constitution of the Swiss Confederation. The system of tariff peaks, which protects the production of sensible agricultural products in Switzerland, is addressed with a similar seriousness as the request from DCs for equivalent tariff barriers on any subsidy embodied in exported products, for example in the Development Box Initiative.[67]

But an integrated approach in favour of agrobiodiversity reaches beyond WTO policies. For instance, it includes flanking policies addressing the welfare losses of a removal of agricultural export subsidies. For this purpose, flanking policies must include development policies, especially of the other Bretton Woods Institutions, the World Bank and IMF, respectively. If, for example, structural adjustment programmes (SAPs) focus on the production of cash crops for export, they may conflict with agrobiodiversity concerns.

8.5 Direct Payments[68]

8.5.1 Introduction

In this part, the question is whether domestic support to further the maintenance of TK in agriculture, provided by both national and international sources, is compatible with WTO law. The question is of relevance not only with a view to the national agricultural politics of the members, but also in the context of the harmonization of WTO law with the various agriculture-relevant International Environmental Agreements, which institutionalize funding mechanisms to support their implementation in the member countries, as for example the Biodiversity Convention and the International Treaty on Plant Genetic Resources for Food and Agriculture. If international funding to support the maintenance of traditional farming methods can be subsumed under the green box payments or the development measures according to Article 6, these would be exempt from the integration in the calculation of the Current Total Aggregated Measures of Support (AMS).

In the following, an overview is given of the present regulations regarding domestic support in general and, in particular, direct decoupled payments (the 'green box' and the development measures). Next, it will be examined how the protection of TK could be subsumed under the present regulations. Thirdly, it will be discussed if the issue of the conservation and maintenance of farmers' TK ought to be made an integral part of the ongoing negotiations on non-trade concerns, and if so, based on which arguments.

[64] Food safety issues are addressed in the Agreement on the Application of Sanitary and Phytosanitary Measures (SPS). To ensure that such measures are not misused for protectionist purposes, the SPS agreement requires that such measures must be based on scientific evidence. The EU tries to strengthen the precautionary principle in order to extend the scope of SPS measures (Desta, 2002, p. 438).

[65] Food sovereignty is a concept introduced by international farmers' associations in the 1990s. It declares the right of each country to have food and agriculture policies designed for its self-defined needs and in cooperation with domestic producers and consumer associations (see plate-forme 'Souveraineté alimentaire', cited from 'Recht auf Freihandel oder Recht auf Nahrung? Positionspapier zur internationalen Agrarpolitik,' in: http://www.alliancesud.ch (accessed December 2005).

In a seminar held at the World Trade Institute (WTI) on 12–13 June 2003, in Berne, the relationship and conflict between trade, agriculture, food and human rights were discussed. From a human rights perspective, the question may rise whether the right to food is adequately reflected in the AoA.

[66] The interplay between internal rural–urban migration and international migration is outlined in Knox and Agnew (1998, p. 40) and Gerster (2001, p. 60).

[67] Pearson (2000, p. 483) evaluates the idea of implementing trade restrictions in order to enforce multinational environmental protection agreements (MEAs).

[68] Author: Susette Biber-Klemm.

8.5.2 WTO regulation of domestic support: overview

Domestic price support measures can be provided in two different forms: either through administered prices, involving transfer from customers, or through direct payments from governments.

The Agreement on Agriculture (AoA) has introduced disciplines to control such domestic support measures as follows: members that during the base period had non-exempt domestic support have to reduce this support according to their agreed schedules (20% over 6 years for developed country members, and 13% over 10 years for developing country members), the maximum level per year being specified in the schedule. Members that did not have any domestic support measures in place during the base period have to maintain any domestic support that does not correspond to the exception categories within the so-called *de minimis* levels.[69,70]

The basis for the calculation of the domestic support is the 'Aggregate Measure of Support' (AMS) and the 'Total Aggregate Measure of Support' (Total AMS). The AMS is defined as 'the annual level of support, expressed in monetary terms, provided for an agricultural product in favour of the producers of the basic agricultural product or non-product-specific support provided in favour of agricultural producers in general' (Article 2 a)). According to this definition the AMS is the monetary expression of the annual transfers made for the benefit of agricultural producers, either in the form of product-specific support to the producers or the non-product-specific support to all agricultural producers in general[71] (Desta, 2002, p. 396). The Total AMS is the sum of all AMS within a year of the implementation period (Currrent Total AMS).[72]

In principle, all domestic support measures are to be submitted under the agreed discipline, if they are not explicitly exempted. There are two types of domestic support measures in favour of agricultural producers that are not to be included in the reduction commitments: the SDT measures (development programmes) under Article 6.2, and the 'green box measures' under Article 6.1 and Annex 2 of the AoA.

8.5.3 The 'green box' measures

The AoA puts domestic support measures in three categories: the so-called 'amber box' measures, which are considered highly trade-distortive and thus subject to reduction commitments; the 'blue box' measures, the goal of which is to limit production and which are exempt from reduction commitments; and the 'green box' measures, which are considered to have no or at most minimal trade-distorting effects and are subject to no reduction at all.

To be subsumed under the 'allowed' green box payments, they have to fulfil two types of criteria: (i) the basic criteria as specified in Annex 2, paras 1(a) and (b); and (ii) the policy-specific criteria as specified under the enumerated practices.

The basic criteria all measures have to meet are: (i) that – among others, for transparency reasons (Desta, 2002, p. 414) – the support is to be provided through a publicly funded programme and not to involve

[69] For developed country members, the *de minimis* level for product-specific domestic support is fixed at 5% of the total value of production of a basic agricultural product during the relevant year; and non-product-specific support at 5% of the value of the total agricultural production. For the developing country members, this *de minimis* percentage is 10%.

[70] This mechanism leads to the situation that developing countries with no direct support during the base period are now bound by the *de minimis* level, whereas countries that provide non-exempt domestic support are legitimized to continue to do so, albeit reduced according to the agreed schedule. For a critical commentary upon this, see: UNCTAD, Doc. TD/B/COM.1/EM.11/2, June 2000, paras 44 and 45.

[71] This is calculated by multiplying the gap between the applied price and the 'world market price' by the quantity of production. See details in Annexes 3 and 4 of the AoA.

[72] See Article 1 (h) AoA.

transfers from consumers; and (ii) they are not to provide price support to producers.

Annex 2 to the AoA provides for a non-exhaustive (but, according to Desta (2002, p. 421), a virtually exhaustive) list of practices for which governments may claim exemptions for their reduction commitments.

With regard to the creation of incentives to maintain TK and traditional PGRFA, the category of direct payments to producers is interesting. The AoA gives a series of options, but, in principle, the list for direct payments is open (para. 5). Basically, existing or new types of direct payments have to correspond to the basic requirements mentioned above; additionally, such payments are to be completely decoupled from production (para. 5 in connection with paras 6 (b) to (e)).

The specified categories of direct payments that could be of interest in our context encompass decoupled income support (para. 6); government financial participation in income insurance and income safety-net programmes (para. 7); payments for relief from natural disasters (para. 8); payments under environmental programmes (para. 12); and payments under regional assistance programmes (para. 13).

These measures (except the income insurance and safety net programmes) allow for an unlimited amount of direct income support so long as the payments are made in a manner that is decoupled from production decisions and trade. Payments which take the form of income insurance, structural adjustment assistance or environmental and regional development programmes have to comply with the requirement that they may not be given in excess of the actual losses suffered or extra costs incurred to implement the government programme (Desta, 2002, p. 421).

The eligibility for such programmes is further to be determined by clearly defined criteria. The basis for payments under environmental programmes, for instance, is to be laid down by a government environment or conservation programme, and the payments must be dependent on the fulfilment of specific conditions under this programme. Payments under regional assistance programmes are limited to producers in disadvantaged regions; these regions need to be clearly designated and have a 'definable economic and administrative identity, considered as disadvantaged on the basis of neutral and objective criteria, clearly spelt out in law or regulation and indicating that the region's difficulties arise out of more than temporary circumstances' (para. 13 (a)).

It is presumed that if the measures meet the criteria and conditions set out in Annex 2, they have no, or at most minimal, trade-distorting effects or effects on production. If they conform fully to the provisions of Annex 2, they are immune from challenge under any of the provisions of the GATT (Desta, 2002, p. 421).

8.5.4 The development programmes

Under Article 6.2 – which is the answer to the claim for special and differential treatment for developing countries – specific measures within a governmental development programme designed to encourage agricultural and rural development are exempt from domestic support reduction commitments and must not be included in the calculation of the Total AMS.

Such measures are: (i) investment subsidies, which are generally available to agriculture in developing country members; (ii) agricultural input subsidies, which are generally available to low-income or resource-poor producers in developing country members; and (iii) domestic support to encourage diversification from growing illicit narcotic crops (Article 6.2).

In comparison with the definition of the direct payments falling under the so-called 'green box' in Annex 2 of the Agreement, the formulation of these exceptions is remarkably vague. With regard to the second criterion, in particular, which seems the most probable candidate for our purpose, the terms 'low-income' and 'resource poor' producers are far from clear. There is no indication that this instrument could specifically be applied to supporting diversified farming in marginal and perhaps remote areas, or specifically to create incentives for maintaining TK and traditional PGRFA, particularly since by 'agricultural inputs' technical support such as fertilizers, pesticides, irrigation systems and power seems to be implied. So, if applied in a generalized way, these support measures could even be detrimental to diversified farming.

The development programmes are exempt from the reduction commitments, but can, under certain circumstances, be challenged under the so-called 'peace clause' of Article 13 (b) of the AoA (see Desta, 2002, p. 413 and below).

8.5.5 Assessment in view of support of agrobiodiversity

The question at the outset of this section is whether incentives to further and sustain diversity in (traditional) PGRFA and TK could be subsumed under the rules exempting domestic support from the reduction commitments, meaning the green box measures, and measures to encourage agricultural and rural development.

It is submitted that, according to the arguments laid out above, programmes to support agricultural TK could, under the heading of environmental or perhaps regional assistance programmes (Annex 2, paras 12 and 13), in principle be subsumed under the green box measures, but not under the development measures under Article 6.2 of the AoA.

However, there are also various problems in the application of the green box measures to supporting agricultural TK. As will be discussed in more detail in the following paragraph, the green box measures have been designed against the background of the situation in the developed countries. So the concept does not entirely fit the situation of smallholder agriculture in developing countries.

For instance, with regard to payments under environmental programmes, one problem is the calculation of the actual losses suffered. A condition set by the AoA is that the amount of payment, in order not to be trade-distorting, is limited to the extra costs or the loss of income foregone by compliance with the government programme. However, programmes to support TK and traditional farmers' know-how are aimed at improving the situation of the small-scale subsistence farmers in non-optimal conditions who, in joining the programme, hardly forgo higher benefits. So, at first sight, it is not clear how in this case the losses incurred could be defined. In this context, in referring to the general criteria green box measures have to meet, it could be stipulated that decoupled domestic support in smallholders' agronomies a priori cannot be trade-distorting as the small-scale farmers produce for their own needs or, at most, for local markets.

As to the payments under regional assistance programmes, neither do they meet exactly the prerequisites described. Their objective is apparently to reduce migration from marginalized areas; therefore their payments must be available to all producers in areas eligible for this type of support, without further conditions.

The question remains whether programmes to support the maintenance of crop diversity and TK could be subsumed under the general rule on decoupled income support. In principle, the AoA permits an unlimited number of different forms of direct payments, under the condition that they are not trade distorting and meet the general requirements and the specific conditions as set in para. 6. However, one condition is that the amount of the payments given is not to be related to the *type* or volume of production (b) and *no production* shall be required in order to receive such payments (e). So again, the rules do not clearly fit the described situation, even if as stated above, small-scale farming is by no means distorting (international) trade.

From this follows that the subsuming of measures to support the maintenance of agrobiodiversity under the green box or development measures is rather questionable and would require a generous interpretation of the relevant rules. However, the foremost problem in this context lies not primarily in the state of legislation, but in the fact that the concept of direct payments demands not only financial capacity, but also resources in organizational and institutional infrastructure. So, the basic problem is that even if measures to support traditional farming know-how can be subsumed under the 'green box' measures, the options of direct payments and other similar measures do not help if the necessary financial means are lacking.

8.5.6 The debates

Critique of the present state in general

The inclusion of the controversial issue of agricultural domestic subsidies in the multilateral trading system has been praised as 'the most innovative' and 'the most important' element of the Uruguay Round (Desta, 2002, p. 425). However, from the beginning, and increasingly with growing experience, there was also substantial critique directed either at the entire concept as enshrined in the AoA or through its expression in the regulation of the direct payments.

At first, in comparing the green box and development provisions, a marked difference is apparent. Whereas the green box measures are laid out in some detail, and (apparently) based on clear-cut concepts of the possible design of non-trade-distorting domestic support, the terms of Article 6.2 specifically targeted at agricultural and rural development in developing countries seem rather poor and unimaginative, not taking account of the varying realities in developing countries. In particular, there are two critiques in this context. First, that the provisions comprise all DCs, with the exception of the least developed countries, in one group, whereas there are clear differences in view of the economic position of their agricultural production (FAO, 2002).[73] And secondly that the provisions are not flexible enough to take the institutional constraints affecting developing country agriculture into account.

A further critique is that the development measures are not integrated in the 'green box'. That means first that, in contrast to the green box measures, they are in principle considered to be 'trade-distorting' And secondly, that they do not to the same extent benefit from the so-called peace clause. Whereas the green box measures under Article 13 were until a short time ago exempt from challenge under any of the provisions of the General Agreement (Desta, 2002, p. 421), the development measures referred to in Article 6.2 are in principle actionable to the extent defined in Article 13(b). Even if it is not probable that such an action will in fact be filed, the principle mirrors an inequity of the rules.[74]

Desta submits that nearly all the green box policies may undercut the effect of any tariff or non-tariff concessions achieved during the negotiations, and the effect of the peace clause was a significant weakening of the Uruguay Rounds results on domestic support (p. 421).[75]

This seems still to be true, even if the peace clause, which was limited to the

[73] The FAO (2002) differentiates between: (i) the Cairns Group as major exporters of agricultural commodities; (ii) the net food-importing developing countries (NFIDCs); (iii) countries with significant agricultural sectors, exporting various agricultural products, which also import food (e.g. India); (iv) countries with small, non-diversified agricultural sectors (because of climatic conditions and land constraints); and (v) higher-income developing countries, which may attach priority to the different functions agriculture plays in their society.

[74] Desta (2002, p. 413).

[75] The peace clause was limited to the implementation period and therefore has expired. The consequences seem to be highly unclear. Different propositions have been submitted (see WTO (2004) Agriculture negotiations, where we are now p. 73). In the 2004 framework it does not appear. According to Steinberg and Josling (2003) the expiry of the peace clause would lead to the application of Articles 6.3(a)–(c) of the Agreement on Subsidies and Countervailing Measures. This position is discussed controversially (see reference to Chambovey, 2002, in Steinberg and Josling, 2003, footnote 1). Up to now, there have been no challenges of, for example, the USA or the EU.

implementation period, by now has expired. The consequences of this fact seem to be highly unclear. Different propositions have been submitted.[76] In the 2004 Framework for Establishing Modalities in Agriculture (in the 'July package') the peace clause is not mentioned. According to Steinberg and Josling (2003) the expiry of the peace clause would lead to the application of Articles 6.3(a)–(c) of the Agreement on Subsidies and Countervailing Measures under which the green box measures could be challenged. This position is discussed controversially.[77] Up to now, no challenges by, for instance, the USA or EU green box measures have been filed.

In spite of the obligation to reduce (trade-distorting) subsidies, the amount of subsidies spent in particular in industrialized countries has in fact grown (Desta, 2002, pp. 7, 307), or remains high[78] with a clear shift from the amber box to the green box outlays between 1988 and 1996 (FAO date). The large amount of domestic support under the green box leads to the question of whether this type of subsidy is in fact not or only minimally trade distorting as legally defined by the AoA. Developing countries increasingly argue that subsidies are in any case trade distorting, including the decoupled green box subsidies, as income supports cut farmers' costs, reduce risks and sustain supply (WTO, 2004, p. 55).

A further imbalance exists in the way support limits were defined. The caps for developed countries have their base period in high-support years with 'a historic peak for many commodities and countries' (OECD, cited in Desta, 2002, p. 426). The reduction commitments were fixed accordingly and, in combination with the shift to the green box measures, relatively easy to attain, whereas countries with no declared support in the base period are bound by their much lower *de minimis* levels.

Against this background, some developing countries have argued that, in effect, the AoA amounts to a SDT for rich countries. Desta suggests that the domestic support provisions of the AoA, innovative and important as they definitely are, could not become of much practical significance, at least in the short term (Desta, 2002, p. 426 with further references). In this critique he nevertheless leaves the option that on the basis of the improvements achieved in the Uruguay Round, the revision process could lead to further progress in favour of equitable solutions.

The multifunctionality vs *food security debate*

With regard to agricultural TK, the themes of food security, rural development, maintenance of biodiversity, support of small-scale farmers and poverty alleviation are all relevant. The ongoing debates in these topics focus on: (i) the 'green box' measures; (ii) the non-trade concerns; and (iii) the argument of SDT, formerly under the heading of a 'Development and/or Food Security' Box, which in the pre-Cancun negotiations was integrated into the concepts of Special Products and a new Special Safeguard Mechanism.

These debates have taken place within two controversial discussion groups, which can be characterized by the argument of multifunctionality of agriculture on the one hand, and of flexibility for developing countries in connection with livelihood concerns for the population on the other.

THE ARGUMENT OF MULTIFUNCTIONALITY

The principle of multifunctionality acknowledges that agriculture also renders public services and produces public goods, neither of which is supported by markets.[79] The term does not appear in the documents mandating the negotiations. It seems at present to be subsumed under the discussion of 'non-trade concerns'.

---

[76] See WTO (2004, p. 73).

[77] See reference to Chambovey (2002), in Steinberg and Josling (2003, footnote 1).

[78] See members' usage of domestic support categories, export subsidies and export credits. WTO Doc. TN/AG/S/1.

[79] OECD (2001b).

The term 'multifunctionality' originates from Agenda 21, where it was particularly meant to enclose the issues of food security and sustainable development. It signals there a holistic approach to agricultural policy encompassing food security, in the sense of the production of adequate and culturally appropriate food supplies. It stipulates that environmental considerations are to be integrated into all economic activities, including agriculture. This line of argument was taken up in a variety of follow-up processes, such as in the FAO's Leipzig Declaration and Plan of Action (June 1996);[80] the Rome Declaration on World Food Security and World Food Summit Plan of Action (1996) and then focused on the 'Multifunctionality' of Agriculture by The FAO/Netherlands Conference on the Multifunctional Character of Agriculture and Land (Maastricht, 1999). The FAO/Netherlands Conference on the Multifunctional Character of Agriculture and Land (Maastricht, 1999) concluded that: (i) agriculture has the capacity to contribute in several ways to welfare (direct impact on nature and environment, primary food material); (ii) it is necessary to balance social, environmental and economic objectives (in particular in intensive and specialized forms of agriculture); (iii) the growing attention to non-food functions of agriculture makes it necessary to develop policies which should be targeted, transparent, cost-effective, do not distort production and trade and contribute to food security.[81]

In its submission to the Conference on Non-Trade Concerns in Agriculture (Norway, 1–4 July 2000),[82] Switzerland pointed out that agriculture, in comparison with the other sectors, has specific characteristics for creating positive externalities and public goods, like the protection of the environment and of finite resources, the contribution to food security or to socio-economic development of rural areas (No. 8). Therefore,

> for countries with an agriculture responding to non-trade concerns as 'food security and the need to protect the environment' [Uruguay Agreement on Agriculture], the market orientation alone will not be sufficient to respond to multiple societal objectives which need to be met in order to enhance national and global welfare, as well as to the criteria 'fair' of the preamble and Article 20 of the Agreement. (No. 7)

In particular, the obligation to increase market access for basic commodities may reduce incentives for local production. However, given the necessity to safeguard a certain level of domestic production for the fulfilment of societal objectives, 'local conditions may require specific instruments to allow farmers to continue production in less favourable areas...' (27).

It is concluded that market failure for public goods may call for government action to secure optimal supply. 'In order to do this, governments must be able to implement targeted, effective, efficient and least trade distorting instruments' (24).[83]

THE ARGUMENT OF FOOD SECURITY AND LIVELIHOOD CONCERNS

Developing countries argue that their needs and policies are quite different from those of the developed or industrialized countries. They have large agrarian populations, whose farming is different from the scales and methods in the developing countries and whose livelihoods they have to

[80] Adopted by the Fourth International Technical Conference for Plant Genetic Resources, Leipzig, Germany, 17–23 June 1996.

[81] Chairman's Report (approved during closing session).

[82] Specific characteristics of agriculture and the need to treat agriculture separately within WTO. International Conference on Non-trade Concerns in Agriculture, Norway, 1–4 July 2000. Discussion Paper One, Presented by Switzerland. Integrated in Doc. G/AG/NG/W/36, pp. 2–10.

[83] See also: Twenty-Second FAO Regional Conference for Europe, Porto, Portugal, 24–28 July 2000; OECD (2001b); Committee on Agriculture (2000); Studies on the Implementation and Impact of the Agreement on Agriculture; G/AG/NG/S/16; WTO, Committee on Agriculture (2000) Operation of the Green Box; Issues Raised by Members in AIE Papers and Pre-Seattle Submissions G/AG/NG/S/18.

ensure.[84] They argue that, according to the analysis of FAO,[85] the promotion of economic growth, reduction of poverty and the enhancement of food security are closely linked with the necessity to develop the potential capacity of agriculture. Therefore subsidies and protection are needed to ensure food security, to support small-scale faming, to make up for a lack of capital, or to prevent the rural poor from migrating into cities.[86]

As the subsistence farmers participate only marginally in international trade, India argues that WTO arrangements should be more flexible to allow the support and protection of agricultural and rural development within the framework of the green box. According to the line of argument of the Indian proposals, the high vulnerability of agriculture in developing countries justifies the extension of special provisions to the developing country members for ensuring their food and livelihood security concerns. The proposal is, for example, to include provisions for the general development of agriculture, including its diversification in developing countries into the green box provisions, and/or to exempt product-specific support given to low-income and resource-poor farmers from AMS calculations.[87]

Some developing countries make a clear distinction between their needs and what they consider to be the desire of much richer countries to spend large amounts subsidizing agriculture at the expense of poorer countries.[88] India, for example, points out that

> in this context it should also be noted that the 'Food Security and Livelihood Concerns' of developing countries are on a totally different plane and should not be confused or equated with the non-trade concerns advocated under 'Multi-Functionality of Agriculture' by a few developed countries with a view to provide legitimacy to and thereby perpetuate their trade distortion subsidies.

The green box subsidies are deemed a special and differential treatment in favour of the developed countries and considered as *de facto* trade distorting by their sheer volume (Cairns Group and the Like-Minded Group of Developing Countries). Therefore, it is maintained that the green box is to be fundamentally renegotiated.[89]

8.5.7 Discussion and conclusions

The existence of non-trade concerns of agriculture (being food security, structural adjustment, rural development and poverty alleviation) seems to be accepted to a large degree. The question debated is, rather, what instruments are needed, and to what extent, to support these non-trade concerns both in industrialized and developing

[84] WTO Agriculture Negotiations: The Issues, and where we are now (updated 21 October 2002), p. 21.

[85] FAO, Commodity Policy and Projections Service, Commodities and Trade Division, 'Issues at Stake Relating to Agricultural Development, Trade and Food Security'. FAO Symposium on Agriculture, Trade and Food Security: Issues and Options in the forthcoming WTO Negotiations from the Perspective of Developing Countries, Geneva, 23–24 September 1999, Paper No. 4.

[86] WTO (2004). See also the proposals by India (G/AG/NG/W/102) and Nigeria (G/AG/NG/W/130).

[87] For further information see: FAO Symposium on Agriculture, Trade and Food Security: Issues and Options in the forthcoming WTO Negotiations from the Perspective of Developing Countries, Geneva, 23–24 September 1999, Paper No. 5; Aldington, T. J., Food Security and the Forthcoming Trade Negotiations: Key Issues Raised by the World Food Summit; Paper No. 6, Measures to enhance agricultural development, trade and food security in the context of the forthcoming WTO negotiations; FAO (1999) Agriculture, Trade and Food Security: Issues and Options in the Forthcoming WTO Negotiations from the Perspective of Developing Countries; Report of an FAO Symposium held on 23–24 September 1999 at the Palais des Nations, Geneva, Switzerland.

[88] WTO (2004, p. 21).

[89] See International Institute for Sustainable Development (August 2003); contrary to the position of the EC and Japan, which maintain that Green Box subsidies are explicitly exempted from the Doha mandate, as support measures under this category must – by definition – have 'no, or at most minimal trade distorting effects'.

countries.[90] The general context is the need to create a coherent policy framework for Sustainable Development in the agricultural sector, a goal that is captured in the formula of sustainable agriculture and rural development (SARD).

The controversy mirrors the differing problems of developing and industrialized countries in view of the sustainability of their agriculture: the first concern of the poorer developing countries is to feed their rural and urban populations. They need to enhance production and at the same time they have to strive for an environmentally sound and sustainable agriculture, which also succeeds in supporting the population in remote areas. In turn, for the industrial countries, agriculture is an instrument for environmentally sound, sustainable land-management practices to balance the detrimental influences of economic development.

The problem seems in both cases to be the integration of (positive) services rendered by agriculture, the present policy favouring the cheapest and not the most sustainable products. It is submitted that this topic – i.e. the question how public good services can be integrated into and acknowledged by the system of domestic support – has to be asked in an approach taking both positions into account: on the one hand the services to balance detrimental influences of industrialization, and on the other, the contributions to maintain agricultural biodiversity by small-scale traditional farming. The main problem will be to define legal criteria for the definition of 'sustainable agriculture', which are generally applicable.

So, it is put forward that measures to support small-scale farmers in order to maintain agrobiodiversity and TK can be subsumed under the non-trade concerns according to Article 20 and are therefore also under the combined headings of 'environment and development' to be integrated in the ongoing debates. It is recommended to do so explicitly, especially in the context of the harmonization of the WTO instruments with the MEAs, in particular the ITPGRFA.

According to the Framework for Establishing Modalities attached to the July package (WT/L/579) the debate on the green box is not to be opened up entirely. But the negotiations are to include a review of the criteria and their clarification with a view to ensure that green box measures have no, or at most minimal, trade-distorting effects or effects on production. The review is to ensure that the basic concepts, principles and effectiveness of the green box remain and take due account of non-trade concerns (No. 16). So, in principle, there is no obstacle to taking account of the measures discussed above in the negotiations.

The question is how a consensus on this issue between the two groups can be achieved. To this end it makes sense to analyse clearly where the real problem in the debate between the 'multifunctionalists' and 'food securists' lies. The thesis is that it is primarily an economic problem. One argument is that the policy to shift the economic support of farmers from product subsidies to decoupled payments may lessen the trade-distorting effect. In turn, the counter-argument that direct payments do have a distorting effect by, for example, indirectly reducing market prices, reducing farmers' risks and possibly increasing food quality standards is valid. And secondly, it must be stated clearly that the option of domestic support and direct payments is not viable, if the financial and institutional means to put it into effect are lacking.

In our opinion, both lines of argument are legitimate: the need for the industrialized countries to support positive, public good externalities should be recognized. However, if the policy of direct support under the argument of multifunctionality is to be maintained to the present extent, the principle of equity and fairness asks for measures to balance the situation between developing and developed countries. Therefore, if the multifunctionality line of argument is to be successful, appropriate

[90] WTO (2004).

flanking measures and concessions or trade-offs are necessary. A coordinated approach in and outside the WTO negotiations is asked for, taking the goals of the Doha Round seriously.

Two ways of thinking are possible. Either direct payments in industrialized countries are reduced, or support and flanking measures in the interest of developing countries are increased, as, for example, by providing additional funding from developed to developing countries to support the global public good agrobiodiversity, which is in the global interest. This funding is, according to set criteria, to be exempt from the calculation of the *de minimis* and the total AMS calculations. Another option is the support of marketability and trade of products out of traditional small-scale farming by facilitating the importation of such products (see Part 2 of Chapter 8).

So, the discussion of both the removal of subsidies and the question of green box direct payments leads to the same conclusion: that an integrated and possibly cross-cutting and coordinated approach is necessary to support the maintenance and sustainable use of agrobiodiversity.

Part 2: Enhancing Market Access[91]

Thomas Cottier and Marion Panizzon
World Trade Institute, University of Berne, Hallerstrasse 6/8, 3012 Berne, Switzerland

8.7 Introduction

The relevance of the WTO multilateral trading system for the protection and promotion of traditional knowledge, understood as the intellectual value expressed in a discovery of circumstance (see Chapter 5), may be expanded to include the products based upon TK as well as traditional process and production methods (PPMs) of these. This approach situates itself in the context of the emerging consideration for environmentally friendly products and services (EGS), including environmentally preferable products (EPPs; see Araya, 2003; Chaytor, 2003; Hamway *et al.*, 2003). Insofar as an analogy to EGS may be drawn for TK-based products, one realizes that the potential to protect and advance TK is not limited to issues of intellectual property and the problem of agricultural subsidies (see Hoekman *et al.*, 2003, pp. 17–20; Cottier and Panizzon, 2004, pp. 371ff.).[92] In analogy to EGS and EPPs, for the purposes of distinguishing a traditionally produced from an industrially produced good, TK may be discussed as a PPM, which should be, but is not yet, part of the established definition of likeness of a product under GATT 1994.

Of equal importance is the regulation of market access in terms of tariffs and technical restrictions, as well as sanitary and phytosanitary restrictions to trade in TK-based products. Market access for TK products originating in developing countries will be further defined by the Special and Differential Treatment (SDT) as provided for in the Preamble of the WTO (Marrakesh) Agreement, Parts IV GATT and GATS, TRIPS Articles 66 and 67, the Enabling Clause and the Waivers pursuant to Article IX:3 WTO (Marrakesh) Agreement, namely the 1999 General Council Decision on Waiver regarding Preferential Tariff Treatment for Least-Developed

[91] The authors wish to thank Dr Gloria Pasadilla, M.I.L.E. and the Philippines Institute for Development Studies, Manila, Philippines for assistance, Ambassador Matthias Meier, State Secretariat for Economic Affairs (SECO), Didier Chambovey, Deputy Permanent Representative of Switzerland to the WTO, Lothar Ehring, Directorate-General for Trade, European Commission, Dr Susette Biber-Klemm, WTI and University of Basel and Helen Morris, Research Fellow, WTI for their comments on a previous draft.

[92] The TRIPS Agreement is said to 'affect the supply of international collective goods in several areas', even if it is 'weak in reflecting developing country interests' (Hoekman *et al.*, 2003, p. 20); or, as put by Cottier and Panizzon, 'in its present form and scope primarily responds to the needs of industrialized countries' (2004, p. 381). Both authors propose amongst other measures that the TRIPS should be redesigned to better take into consideration developing country concerns, so that one of the goals should be to 'provide [...] for the establishment of new forms of IPRs over collective and traditional knowledge' (Hoekman *et al.*, 2003, p. 20), because the '[l]east developed countries would [...] benefit from such an arrangement, and it would thus help to rebalance the social costs of the TRIPS Agreement' (Cottier and Panizzon, 2004, p. 399).

Countries[93] (see below, Section 8.9). Special trade preferences have proved to be inefficient, non-competitive, incomplete and overly conditional for developing country products. For such reasons, this chapter will be critical of SDT and argue in favour of regulating market access for TK-based products from developing countries through liberalization of the classical market access border measures, such as reciprocal tariffs, and define appropriate technical sanitary and phytosanitary domestic, regional and international standards for TK-based products and process and production methods (PPMs) (see Perkins, 2003, p. 3; Hoekman *et al.*, 2004b, p. 175).

High levels of tariff protection still impede the importation of agricultural products from developing and least developed countries; existing schemes of the General System of Preferences (GSP) are not specifically aimed at promoting production in a way that takes into account equitable goals (including recognition of TK), both in terms of ecological and distributive concerns. Current technical standards, in particular those relating to food, fail to privilege TK-based products. In many cases, they are likely to amount to important obstacles for the importation of products, which are based upon traditional, artisan methods of production as opposed to industrial ones.

The emphasis on IPR-related issues, as discussed in the Committee on Trade and Environment (CTE) and the Council for TRIPS, discussed in Chapter 5 (this volume), should not distract from the fact that the abovementioned non-IPR areas of trade policy are of equal relevance to the promotion and protection of TK-based products. Non-IPR trade regulations often directly address the special concerns of developing countries (DCs) and least-developed countries (LDCs). Trade instruments are complementary to IPR protection of TK insofar as these attach to the product benefiting from the IPR-protected information and foster access for the TK-based product into developed country markets, independently of whether or not the information behind the TK *per se* is IP protected or not. Thus, many developing countries, which are still in the process of building effective IPRs, will first rely on and crusade for open markets in industrialized countries and for a reduction in agricultural tariffs relating to such products, before extending to IPRs.[94]

First, we shall focus on tariffs under the system of GSP and under the World Customs Organizations' (WCO) Harmonized System (HS). Our focus will be the different categories that exist within the HS for identical products originating from different sources, such as synthetic vs artificial fibres or identical products processed differently (e.g. textiles with different sorts of weaving machines; cheese with fresh or pasteurized milk). In addition to introducing a category for TK-based as opposed to industrial products, we ask to what extent differentiating otherwise 'like' products based on their production method can be put to use for the promotion of TK products, which are 'like' an industrially produced one, but for their production method. Since discussing a tariff-based differentiation for different types of PPMs was introduced for environmentally friendly goods and services (EGS), an analogy may be drawn with TK-based PPMs. Linking trade with development concerns, perhaps even more so than linking trade and environment, will create a win–win situation for developing countries, namely strengthening trade in traditionally produced goods and building trading opportunities for

[93] WTO, Development: Trade and Development Committee, Work on Special and Differential Provisions, available at: http://www.wto.org/english/tratop_e/devel_e/dev_special_differential_provisions_e.htm (last accessed 8 February 2005).

[94] For an overview of the results of the Cancun Ministerial Conference, including members' statements, http://www.wto.org/english/thewto_e/minist_e/min03_e/min03_e.htm (accessed December 2005).

developing countries, which may thus capture the benefits of liberalization (see Araya, 2003). In this context, this section will also briefly explore Canada's system which certifies TK-produced goods to distinguish these from industrially produced ones in view of granting TK-produced goods a lower tariff treatment. Canada's tariff differentiation could be a model applied multilaterally.

Secondly, this section explores the impact of two other agreements relevant to trade regulation for TK-based products and PPMs: the Agreement on the Application of Sanitary and Phytosanitary Measures (SPS) and the Agreement on Technical Barriers to Trade (TBT). The first is of prime importance in assessing food quality standards, which may be detrimental to the importation of TK-based products, since developing countries will often lack the technical know-how to make their exports of both foods and pharmaceuticals 'safe' pursuant to industrialized standards. From a more general point of view, the question to be considered will be whether the WTO legal system is equipped to address questions of distributive justice, which is the main issue relating to TK. Distributive justice entails going beyond regulating against protectionist tendencies through the obligation not to discriminate. It imposes a positive obligation upon developed countries to recognize (offensive measure) or not apply (defensive measure) certain trade tools in order to give developing country products the chance to increase their market share (see Howse and Türk, 2001, pp. 284–285).

Thirdly, the question is whether preferential rules could be contemplated for TK-based products in terms of taxation and market regulations. This issue will be addressed under the heading of MFN and national treatment and the related analysis of like products. As TK-based products are often defined by their method and means of production rather than by particular features of the product, the issue of so-called product and process methods (PPMs) plays a key role in this analysis. The WTO AB and panels have identified four criteria, and left it to the discretion of future panels and the AB to define more criteria for a finding of 'likeness'. We explore to what extent distinctions could be based upon existing case law.

Among the constitutive criteria for likeness as defined for the GATT 1994 in the *Japan–Alcohol* case, figure physical characteristics, consumer end-use and tariff treatment.[95] Not among these are production methods (see Araya, 2003). Thus today, different production methods will not make a product unlike, and consequently will not allow for discrimination between production methods. The mainstays of WTO, jurisprudence and doctrine, concur that production methods have no place in the interpretation of Article III GATT which requires equal, i.e. non-discriminatory, treatment to be accorded to 'like products'. But there are also opinions to the contrary which need to be explored, insofar as they could assist in supporting future negotiations. As some authors have pointed out, the fixation of WTO members on the rather narrow criteria of end use no longer reflects the market reality today. Trends in consumer behaviour and public opinion suggest the opposite, that 'fair-trade', consumer rejection of genetically modified foods and boycotts of unsustainable corporate practices, are given at least as much attention as to how a product is made, as opposed to what a product is like (see Araya, 2003).

All these problems, as of today, have hardly been addressed and properly localized within the WTO. Tariffs, TBT, SPS and non-discrimination were not touched upon in negotiations in relation to TK. Trade instruments have one advantage over IPR-protection and one disadvantage. IPRs require the identification of the knowledge holder, while trade instruments apply even

[95] *Japan-Taxes on Alcoholic Beverages*, Appellate Body Report, 4 October 1996, WT/DS8/AB7R, WT/DS10/AB/R, WT/DS11/AB/R, p. 25.

if the plant genetic resources for food and agriculture (PGRFA) as well as plant genetic material relating to pharmaceuticals (PGM) remain unassignable (Cottier, 1998, pp. 561–562, 582, 584). The disadvantage of trade instruments is that they do not compensate the indigenous communities when non-original holders (usually industry) appropriate their knowledge. Trade instruments relate to the product *based* upon TK and only incidentally validate the TK by increasing the market share for the TK-based *product*. Both areas therefore are complementary. In the light of the Uruguay Round single package approach, the protection of TK should be approached comprehensively throughout all WTO Agreements. We stress this point since, up to this point in time, emphasis has been mainly given to TK in relation to IPRs, while both the regulatory potentials relating to subsidies and to market access remain to be developed.

Discussion of these issues will not be able to draw substantially from explicit WTO regulations relating to TK, since there is neither specific protection for TK under the current state of play of the WTO Agreements, nor implicitly from case law, the latter being virtually non-existent. In fact, regulations are currently absent with three minor exceptions. The WTO Uruguay Round Agreement Texts mention indigenous knowledge in two unexpected places: in Article 12.4 of the TBT Agreement, and in the Annex to the Agreement on Textiles and Clothing (ATC) Article 3(a) Annex ATC. However, the ATC was phased out by year-end 2004, and therefore is no longer applicable law. Thirdly, Article 27(3)(b) TRIPS implicitly refers to traditional knowledge. So far none of the three provisions has been tested in adjudication. The only attention paid was to Article 27(3)(b) TRIPS, when the Doha Ministerial Declaration of 14 November 2001 instructed the Council for TRIPS to include TK as a subject of a review under Article 71.1 TRIPS in para. 19 of the Doha Ministerial Declaration with regards to the development dimension.[96] So far, WTO members have essentially considered TK as an IPR-related issue, which they are currently discussing in the TRIPS Council and, as it relates to biological diversity, also in the Committee on Trade and Environment (CTE). Both the Council for TRIPS and the CTE have addressed the protection of TK in terms of IP instruments.

In contrast, the meaning of indigenous knowledge as a technical standard in Article 12.4 of the TBT agreement and as the rationale behind the non-application of the textiles safeguard measures under Article 3(a) Annex ATC have not been discussed in a WTO Committee, nor has any member brought up the subject as a term of reference for future negotiations.

At the WTO Cancún Ministerial, TK was mentioned only by certain African delegations in the context of reviewing TRIPS.[97] The issue was not discussed in the context of SDT, tariff cuts as part of non-agricultural market access (NAMA), or in tariff cuts for agricultural products. Outside the WTO framework, only the Convention on Trade in Endangered Species of Wild Flora and Fauna (CITES) has discussed – for pharmaceutical products – the inverse issue of authorizing developing countries to use safeguards for protecting their resources from biopiracy by industrialized countries.[98] Finally, the subject was dropped completely by the July 2004 decisions from the agenda of the TRIPS Council.[99]

[96] Doha Ministerial Declaration, adopted 14 November 2001, WTO Doc. WT/MIN(01)/DEC/1.

[97] See Mozambique, Statement by H.E. Mr. Carlos Alberto Sampaio Morgado, WTO Doc. WT/MIN(03)/ST/74, 12 September 2003, p. 3; see also ACP Declaration on the Fifth Ministerial Conference of the WTO, Communication from Botswana, WTO Doc. WT/MIN(03)/4, 21 August 2003.

[98] See CITES, Conf. 10.19, available at: http://www.cites.org (last accessed 16 September 2003).

[99] See WTO Doha Work Programme, Decision adopted by the General Council on 1 August 2004, WTO Doc. WT/L/579 of 2 August 2004 [hereinafter WTO, Doha Work Programme], the Doha Work Programme is also called the 'General Council's post-Cancún decision', p. 2; however, a progress report on GIs was due to be submitted to the Trade Negotiations Committee (TNC) and the General Council no later than May 2005.

Despite the lack of current regulations and negotiations, we are optimistic that new and appropriate solutions can be found within the multilateral trading system, which support and take into account the goals of biodiversity and equity enshrined in related agreements such as the CBD. We see a potential for dealing with the matter in the context of facilitating market access for environmentally friendly goods. The shared interests of both North and South in preserving long-term biodiversity and a broad nutritional base for global food security, emphasized throughout this book, are key to this effort. Moreover, it will be seen that DCs and LDCs can find common interests and create common ground for making a strong impact in the negotiations, more than in other areas where such a common element is missing. As has been stated: 'despite their diversity of circumstances and commercial interests, they have it in their power to make common cause for their collective good' (Short, 2001, p. 59). Yet, special efforts need to be made if enhanced market access of TK-based products is to be achieved. If incentives are to be created for the conservation of a large variety of plant genetic resources through use and commercialization, they cannot be exempted from general tariff reductions and liberalization of trade in agriculture.

8.8 Market Access for Developing Countries in PGR Products

8.8.1 Tariffication of trade barriers and the prospects of tariff reductions

The multilateral trading system of the WTO is founded in theory on the comparative advantage paradigms of classical economists – that increased trade leads to increased prosperity and that open markets, with low tariffs, the absence of subsidies and non-tariff barriers, bring both growth and poverty alleviation (Short, 2001, p. 60). The WTO is based upon the idea of progressive liberalization of international markets. It does not amount to a free trade agreement, but recognizes the legitimacy of tariffs, in the pursuit of legitimate policy goals, and a proper balance between the interest of market access and those of domestic production. It also allows for other trade restrictions, but clearly favours tariff over other measures (Hoekman *et al.*, 2003, pp. 17–20). The rationale for preferring tariff protection over all other types of trade regulations at the border is well known: tariffs are the trade instrument most closely related to the price of a product and thus do not distort the international allocation of resources and the mechanisms of comparative advantage (Perkins, 2003, p. 3). Most recently, the Consultative Board to the Director-General Supachai Panitchpakdi of 2005, on the future of the WTO, called for substituting Special and Differential Treatment (SDT) by an overall tariff reduction to zero tariffs (Sutherland *et al.*, 2005).[100]

Tariffs are subject to certain obligations. Reduced tariffs are mostly bound, i.e. they cannot be increased without paying

[100] *The Economist*, 'Free Trade's Best Friend, the WTO is Worth Boosting' of 22 January 2005, referring to the report's section on non-discrimination, which suggests that to counter the trend that discrimination has become the rule and non-discrimination the exception, rich countries ought to set a date by which all their tariffs would be cut to zero.

compensation under Article XXVIII GATT to other parties mainly affected by the increase.

GATT law thus essentially provides a framework for gradually reducing tariffs and preventing or remedying circumvention of tariff cuts by other means of trade policy. Conceptually, measures other than tariffs are exceptional. To this effect, the GATT establishes a general prohibition of quantitative restrictions, subject to narrowly defined exceptions. In reality, however, these exceptions were the rule in the field of agriculture.[101] Quantitative import restrictions, combined with sanitary and phytosanitary measures, export and domestic subsidies, formed the mainstay of post-World War II agriculture policies of Western industrialized countries and constituted, with the Common Agricultural Policy (CAP), the very heart of the European Economic Community. Efforts in agricultural negotiations thus primarily focused on bringing about enhanced disciplines in relation to domestic and export subsidies as well as quantitative restrictions, rather than on tariff reductions.

Tariffication of trade barriers in agriculture was seen to be one of the main achievements of the Uruguay Round negotiations. Its main merit consists of abolishing quantitative restrictions, which had often been applied in a trade distorting way prior to the entry into force of the AoA. Existing measures, however, were not abolished in practical terms, but translated into tariff equivalent protection. Newly defined, increased tariff levels were subsequently subject to defined tariff reductions amounting to 36% on average (with a minimum rate of 15% per tariff line for sensitive products) over a period of 6 years. It nevertheless left most industrialized countries and developing countries with high import tariffs bound under Article II of GATT.

Developing countries, and thus more than 100 of the 149 members of the WTO, remain particularly affected. The Uruguay Round brought about a new system, but not substantially enhanced market access. As described in the previous chapters of this book, a substantive part of TK relating to PGRFA, as well as plant genetic material relating to pharmaceuticals (PGM), lies in the developing world. Products derived therefrom thus generally remain under high tariff protection.

About 15% of the tariff peaks in agricultural products (tariffs over 100%) are concentrated in products that are of interest to developing countries. In 1999, 15.7% of dutiable imports originated in LDCs (Hoekman, 2002, p. 25). In addition, industrialized nations have countered the effect of lowering tariff barriers for agriculture by subsidizing their own production, discussed in Chapter XVIII A (see also Rosenberg, 2003). Moreover, as tariffs are lowered, industrialized countries are tempted to use sanitary and phytosanitary measures in a protectionist way to stop imports from developing countries. Despite much rhetoric to the contrary in the public debate, many developing countries genuinely crusade for trade liberalization and against protectionism in agriculture, in line with the precepts and goals of the WTO (see also Short, 2001, p. 61). The Doha Development Agenda is essentially dedicated to this goal and overall results will depend on progress in relation to agricultural trade.

Developing countries pursue a two-track liberalization. First, they seek general reductions of tariffs. Alternatively, they ask for privileged market access, reduction of tariffs or the non-application of otherwise permitted trade remedies. Such privileges, in principle, run against the best allocation of production factors, which the WTO wants to achieve through 'substantial reduction of tariffs' and 'the elimination of discriminatory treatment' (Preamble, Mar-

[101] 'The Erosion of Non-Discrimination', in: The Future of the WTO: Addressing Institutional Challenges in the New Millennium, Report by the Consultative Board to the Director-General Supachai Panitchpakdi, pp. 19–27, available at: http://www.wto.org/english/thewto_e/10anniv_e/future_wto_e.pdf (last accessed 27 February 2005).

rakesh Agreement). However, SDT may be justified by arguments of equity, which the WTO equally subscribes to by recognizing the 'need for positive efforts designed to secure that developing countries can secure a share in the growth in international trade...' (Preamble, Marrakesh Agreement).

The request for enhanced and privileged market access for developing country exports has been countered by the argument of developed countries that they need to protect their production by means of domestic subsidies or higher import tariffs from cheaper foreign products that have been produced with damage to the environment and/or in disregard for labour standards (Meinheit, 1995, p. 1). Indeed, preserving the environment is, in addition to freer and fairer trade, another of the WTO's goals (Preamble, Marrakesh Agreement) and has to be taken into account with a view to developing structures suitable for sustainable agriculture. A number of countries also rely upon additional non-economic functions of agriculture, in particular the need to support decentralized habitation. Besides protecting their own production, the refusal of developed countries to lower import tariffs also stems from a valid concern for the global environment and reflects the recognition that environmental depletion and the loss of biodiversity is a problem common to the entire world (Meinheit, 1995, p. 2).

It is thus a matter of finding a new balance, and there are many obstacles to achieving this goal in due course. Agricultural structures, all over the world, are essentially conservative, to the effect that they can only be changed gradually. Despite recognition 'to put developing country concerns at the centre of the WTO' in the 2001 'Doha Development Agenda', the real outcome in terms of market access to be expected is relatively minor (Hart and Dymond, 2003, pp. 395–415). In addition to structural resistance in agriculture, there is currently little pressure for change in industrialized countries due to the 'absence of strong support by OECD manufacturing industries for the [Doha] agenda, which reflect the fact that market access barriers to many industries are already quite low as a result of the past zero-for-zero tariff reductions and the spread of free trade arrangements' (Hoekman, 2002, p. 24). Another argument for the rather relaxed consideration of the Doha Development Agenda by governments is 'that sector-specific industries seek to maintain the high rates of protection – in particular the textile industry in Canada and the US and the agricultural lobbies in the EU and Japan' (Hoekman, 2002, p. 24).

While the commitments to reduce agricultural tariffs remained, as discussed before, relatively modest, tariffication sets the stage for long-term tariff reduction in coming decades. We should not forget that the reduction of industrial tariffs from an average of 47% to around 4% took some 50 years since the founding days of the GATT in 1947, and no less than eight rounds of multilateral negotiations. With firmer disciplines on agriculture, the potential to see substantial tariff reductions over coming decades is realistic. It will no longer be possible to develop WTO law further in non-agricultural goods and services without substantial commitments in agriculture.

Efforts to enhance market access in the services sector and intellectual property protection in developing countries cannot be achieved without a new deal in agriculture. The developed countries' refusal to substantially reduce tariffs for agricultural products and thus to offer increased market access to developing countries was an important factor leading to the Cancún Ministerial's failure.[102] It must not be repeated. Existing non-economic functions of agriculture call for a long-term approach towards structural adjustment, in compliance with the goals of global sustainable development. The process of general tariff reduction and improved market access is

[102] See The Cancun Failure, Editorial, *The New York Times*, 16 September 2003, available at: http://www.nytimes.com (accessed 17 September 2003).

bound to remain slow for developing countries, where the constituencies need to be convinced of adaptation towards a multifunctional and sustainable agriculture (Brenton, 2003, pp. 623, 642–643). A fast track may be achieved with the EC initiative of EBA (everything but arms) for least developed countries.[103] It will not extend to developing countries in general.

8.8.2 Impact on traditional knowledge-based products

Products based upon PGRFA, including food and textiles, are likely to remain under relatively high tariff protection in coming decades. There are currently no policies in place that privilege sensitive products from the point of view of conservation through use and preservation of biodiversity and sustainable agriculture. Tariffs are applied across the board for different like products, independently of their way of production and of the knowledge and information involved. Without special efforts, it is evident that TK-based products will continue to suffer from high import barriers for decades to come. Efforts to alleviate the situation, so far, have not materialized. Pre-Cancún agricultural negotiations had opted to create a special products category with reduced tariff-cutting commitments for developing countries. A proposal by the Dominican Republic and others would have contained the possibility for developing countries to self-designate the products falling under such a category. Under a self-designation regime, a developing country could have opted to set up the special products category with no tariff cuts to encompass all industrially produced goods it imports from outside that are equivalent to TK-processed goods it is able to produce itself, in order to provide the incentive for consumers to buy the TK-produced home products.[104] Yet, the Annex A to the Draft Cancún Declaration omits the self-designation by developing countries of such products. The findings confirm that the relationship of market access, conservation through use and biodiversity, as well as sustained livelihood for traditional farming communities and the relevance of SDT in this context, have not been sufficiently explored.

8.9 The Prospects of Differential Tariffs for Traditional Knowledge-based Products

8.9.1 Special and differential treatment

Dispersed among the various agreements under the WTO umbrella are rules providing for more favourable treatment to small and low-income countries, which make up the concept known as special and differential treatment (SDT). Originally a result of the Tokyo Round of multilateral trade negotiations, SDT rules today are 'part of the WTO's legal *acquis*' (Sutherland *et al.*, 2005, para. 89). They are intended to support and further the DC's integration into the multilateral system. These privileges, however, often remain without real practical impact and are even counterproductive.

SDT has been criticized for 'sheltering developing economies from the full application of trade liberalization rules', as 'wholly political' and 'bereft of any economic underpinning' (Hart and Dymond, 2003, p. 395). In addition, they undermine the negotiating position of DCs and LDCs, which demand that developed countries reduce agricultural tariffs. The SDT concept distinguishes three types of rules: first, preferential access for developing countries to developed country markets; secondly, promises by developed countries to provide

---

[103] See Council Regulation (EC) No. 2820/98 as amended by Regulation (EC) 416/2001, O.J. L 60, 1.3.2001; the USA knows a similar, Africa-specific initiative, the US African Growth and Opportunity Act (AGOA).

[104] See Agriculture: A Blank Check Bitterly Contested, *Bridges*, July–August 2003, available at: http://www.ictsd.org (last accessed 17 September 2003).

technical assistance to lower-income economies; and thirdly, exemptions from WTO rules, which can be distinguished into transitory ones, such as relating to customs valuation, abolition of trade-related investment measures under the TRIMs and the implementation of stronger IPR protection under the TRIPS, and permanent ones, such as the waiver and the enabling clause's limited reciprocity (Hoekman *et al.*, 2003, p. 5). Another differentiation among the exceptions from WTO rules may be made between explicit 'preferences' on the one hand, and the non-application of trade remedies against developing countries on the other, the latter called 'instruments of contingent protection' by Hoekman *et al.* (2003, p. 22).

Overall, SDT has weakened DCs within the WTO and also prevented these countries from adjusting their structures and liberalizing their trade in due course. SDT should not prevent developing countries from liberalizing market structures, stepping up production capacities and increasing value by adding production processes to agricultural goods. SDT should be reformed as to being product-specific, with the effect that it lowers the prices on the world market by increasing the quality and variety of consumption goods (as opposed to non-reciprocal tariff cuts), and 'fostering the specialization of economic activity into areas where countries have a comparative advantage' (Hoekman *et al.*, 2003, pp. 1, 18).

Unfortunately, the fact that developing countries' exporting economies consist to date only of agriculture and/or textiles is an additional reason why SDT will not remedy the basic imbalance between developed and developing countries, because while the mainstay of agricultural rules are tilted in favour of rich countries, SDT is only an exception. Audley *et al.* (2003, pp. 5–6) maintains that even if the WTO provides for fair play rules (in the sense of SDT), there is no guarantee of fair competition because the 'teams are mismatched'. In addition, the opening of the textiles quota in 2005 has not provided the developing countries with the trade opportunities they had hoped for, since China's accession to the WTO distorted the global textile market, specifically the trading opportunities of developing countries with their comparatively much smaller production capacity (Audley *et al.*, 2003, pp. 5–6).

Current efforts seek to reinstate effective differential treatment provisions into the agreements extending beyond capacity building and transfer of technology. The Draft Cancun Ministerial stresses '[b]uilding upon our commitment in the Doha Declaration we shall continue to expeditiously pursue the objective of duty-free and quota-free market access for products originating from LDCs'.[105] As Hart and Dymond say, '[s]pecial and differential treatment in favour of developing countries lies at the heart of the agenda of the new "Development Round" of World Trade Organization negotiations launched by ministers at Doha...' (2003, p. 395). It will be a matter of finding appropriate solutions and effective rules, such as to encompass both 'rule development' and 'expectations of reciprocity', as well as an aggrandized catalogue of plurilateral agreements which are not mandatory, such as the present regime on government procurement in order to avoid the downside of past SDT policies. To the extent that SDT offers a real possibility to support TK-based products, this avenue should thus be further explored.

8.9.2 The case of special and differential treatment for TK-based products

We submit that preferential treatment for TK-based products, supporting functions of equity and sustainability, offers an effective means in the present context. The high rates of agricultural tariff protection (EC:

[105] Full text available at: http://www.wto.org/english/thewto_e/minist_e/min03_e/draft_decl_e.htm (last accessed 17 September 2003).

average 16.1% with many tariff peaks, e.g. Switzerland: average 34.3% with peaks up to 678%) offer a unique opportunity to seek special and differential treatment for TK-based products. Based on our assumptions and findings that TK-based products are important from both an ecological as well as a social perspective, ways and means should be sought to lower tariffs for such products faster than for conventional like or similar products. The elimination of agricultural tariffs mostly by industrialized countries would encourage efforts to market TK-based products abroad.

This offers the potential to seek specific reductions in support of products that are essential for the conservation of biodiversity as a public global good. Efforts therefore are possible which particularly focus on relevant products and link tariff policy and market to efforts at conservation through use. It is apparent that TK-based products, given their importance, should be a prime candidate for such efforts. They could make an important contribution to fostering sustainable agriculture and alleviating problems of equity caused by generally high tariff protection. To the extent that privileged access is possible, it could be granted to include, and even focus on, products based upon the use of TK. Industrialized countries would thus make a contribution towards compensation for what often amounts to more cumbersome, more costly and inefficient production, whilst traditional PGRFAs and methods of production are employed in developing countries. Rural communities in these countries could enhance their livelihood. Products sold on markets would generate a more adequate level of income, and work would be validated way beyond current levels. It goes without saying that the same approach could also be applied to other types of TK products, such as TK-based textiles and handicrafts of importance to the livelihood of rural communities.

Discussions should build upon the concept of environmentally preferable products (EPPs), which, so far, has been limited to industrial goods and services suitable for the environment, either in terms of products or production methods. The concept has not been extended to issues of agriculture and PGRFA, except for traditional commodities. Given the underlying ecological and social motives in support of TK-based products, there is no reason why TK-based PGFRA for medical purposes should not be included in these efforts.

Differential tariffs for these products can be installed in two different ways. The first method is based upon unilateral GSP schemes by countries. The second can take place within the Harmonized System, allowing for separate tariff lines for such products. Within the HS two possibilities exist: either the tariff lines are negotiated by reforming existing classifications within the World Customs Organization (WCO) multilaterally, or they can be unilaterally determined by members, using digits 6–8 of the schedules, as for example adopted by Canada specifically for handicraft goods produced with TK, and by Papua New Guinea for rice varieties of particular significance for the indigenous population. Both are emanations of special and differential treatment for developing countries.

In this vein, we see four avenues by which SDT for TK-based products could be specified. The first one is GSP, the second one tariff classification, and the third and procedural one limitation to the use of trade remedies against TK-based products. The fourth solution would be to introduce a labelling requirement for TK-based products.

Recourse to the General System of Preferences (GSP)

Since the introduction of Part IV to the GATT 1947 (Article XXXVI ff.), the multilateral trading system recognizes the possibility of treating products from developing countries on a preferential basis. While the founding provisions remain relatively vague, it is the 1979 Enabling Clause adopted at the end of the Tokyo Round that granted authority to members, in particular industrialized members, to operate preferential import regimes for products from DCs. This Clause was necessary as it operates as a justification for what otherwise would amount to a violation of the Most Favoured Nation (MFN) treatment. Many industrialized countries have introduced, based upon this authority, unilateral schemes of preferences. Importantly, the criteria and products chosen are not defined in WTO law. In the same way as conditions, they remain a matter of national legislation and therefore vary from country to country. Critical voices of the concept find that developing WTO members, which are not under a GSP scheme, have undertaken more trade liberalization efforts than their peers under a GSP scheme (most recently, Brenton, 2003, p. 624). Secondly, it is said that GSP prevents groups in developing countries lobbying for exports to influence government, since the exports have low tariffs in any case and thus trade liberalization efforts are thwarted by the competing import lobbyists (Chambovey, 2003). While these objections addressing the problem of structural adjustment may be generally relevant, we do not think that they should impede GSP schemes that seek to support policies of long-term conservation through use of genetic resources and thus biodiversity as part of the global commons.

Nothing in the GATT Agreement would therefore imede a country from introducing criteria specifically to the benefit of TK-related products under its GSP, as long as the same conditions apply to all the beneficiaries alike. In the light of generally high tariffs in agriculture, GSP is an important port of entry for special rules relating to such products and the enhancement of their market access in industrialized countries.

In an effort to encourage these products, it is possible to define the products either in terms of substance and quality, or in terms of their production methods based upon the use of traditional knowledge. GSP schemes could therefore list crops that are of particular importance from the point of view of protecting biodiversity and a broad food base. Alternatively or in combination, they could also define products on the basis of the production and process methods by which they are produced. It is therefore possible to allocate low or zero tariffs for products based upon the conservation and use of traditional methods of production.

In defining conditions for GSP applicable to TK-based products, regulations would need to define products derived from crops that are essential for long-term and global food security. We suggest that this would need to entail products that under regular tariffs cannot compete and for which, consequently, no sufficient incentives for human investment in preserving these crops exist. Likewise, criteria should be developed to encapsulate relevant methods of production on the basis of traditional knowledge. The work undertaken to document such knowledge in international or national registries, developed in the context of active or defensive protection (see Chapter 5), could be referred to. Granting GSP to such methods of production is even more meaningful, since they often are less efficient than industrial methods. They need a competitive advantage if they are to fulfil their function of making a contribution to preserving global biodiversity.

Adjusting tariff schedules and the Harmonized System

The second approach consists of adjusting tariff classification through the introduction of special categories for TK-based products, both in terms of substance and of process and production methods. This can be done unilaterally or by way of multilateral negotiations.

As Vaughan describes for environmental goods, the 2002 revision of the HS customs codes 'set a precedent by explicitly referring to the importance of environmental and social criteria' (2003, p. 7). What he observed as a move towards more holistic product categories bears the opportunity to propagate the creation of a custom code for traditionally produced goods and/or goods of special significance for biodiversity (similar to environmental goods as proposed by Araya, 2003; Hamway *et al.*, 2003; and Vaughan, 2003).

While current schedules do not distinguish TK-based products as a category, the structure of the HS would not bar efforts to introduce product differentiation on this basis within the current system. The following table of examples shows that product differentiations within the same type of product already exist within the HS (Table 8.1). Modes of production, as well as of storage, can be observed. There are such distinctions as frozen/non-frozen foods, cheeses made with raw or fresh milk. The table brings together different product categories, which could be used as a basis for the possible development of TK-related differentiations of products.

UNILATERAL ADJUSTMENTS

Members are free to unilaterally introduce new subcategories in digits 6–8 reflecting TK products. These categories, however, would basically need to apply in law to all products qualifying due to MFN obligations. It would also be possible to insert specific categories exclusively applicable to developing countries, using authority under the enabling clause. However, since a large majority of TK-based products originate in DCs and LDCs, MFN would essentially benefit developing countries, and general categories for TK products would be preferable. Certain limits need to be respected in our view. A low or zero tariff rate can apply as long as this does not impair or nullify the benefits accorded under the present schedules to conventional products. If the introduction of TK-based category leads to the nullification and impairment of third party interests (e.g. upsetting the conditions of competition for exports of conventional products) and to a complaint, the GATT recognizes under Article XXIII GATT 1994 and Article 26 DSU the legitimate expectations of third country members and requires either under a non-violation or violation complaint, the member introducing the change to withdraw the measure or to grant compensation at least as a temporary measure under Article 22 DSU (Cottier and Nadakavukaren-Schefer, 1997).[106] Beyond this point, it will become a matter of negotiations.

In this context, it is important to note that within their negotiated tariff schedules, WTO members have considerable leeway in the tariff treatment of their products. The Appellate Body (AB) in the *EC-LAN* case refuted the Panel's approach based upon the protection of legitimate expectations. The AB thus granted substantial discretion to members in the practical operation of their schedules. On the question of whether the USA could expect the EC to classify LAN (local areas network) equipment as (lower-taxed) computer instead of (higher-taxed) telecommunications equipment, the AB said the following:

> On the basis of the erroneous legal reasoning developed and the selective evidence considered, the Panel was not justified in coming to the conclusion that the United States was entitled to legitimate expectations that LAN equipment would be

[106] See also on non-violation/violation complaints, Cottier and Nadakavukaren-Schefer (2000).

> accorded tariff treatment as ADP machines [computer equipment] in the European Communities.[107]

While the rejection by the AB of the doctrine of legitimate expectations resulted, in our view, in confounding subjective with objective expectations, it nevertheless failed to confirm the result reached by the panel. The AB thus confirms that members of the WTO enjoy considerable leeway in structuring and interpreting their tariff schedule classifications under prevailing textual modes of interpretation in the WTO.

Under current practices, members thus enjoy some leeway as to tariff classification, and they are entitled to introduce subcategories to this effect for TK-based products. Two countries at least have introduced a special tariff category in digits 6–8 for TK-based products. In order to recognize the TK of its indigenous population, Canada has unilaterally adapted digits 6–8 under the HS tariff headings to promote the marketing of such TK-based products. Canada recognizes:

> The following handicraft goods, having forms or decorations that are traditionally used by the indigenous people or representing any national, territorial or religious symbols of the geographical region where produced, having acquired their essential characteristics by the handiwork of individual craftsmen using tools held by hand or tools not powered by machines other than those powered by hand or foot, being non-utilitarian and not copies or imitations of handicraft goods of any country other than the country in which they originate, and not produced in large quantities by sophisticated tools or by moulding: puppets, musical instruments (other than guitars, viols, harpsichords or copies of antique instruments), gourds and calabashes, incense burners...[108]

Canada imposes the following condition:

> Under this Act, the Governor in Council may amend the list of goods in this tariff item. Goods may be classified under this tariff item on production of a certificate in duplicate in the prescribed form with the information required to be provided with the form, and signed by a representative of the government of the country of origin or any other authorized person in the country of origin recognized by the Minister of National Revenue as competent for that purpose.[109]

The differentiation of products whose forms and decorations are traditionally used or reflecting national, territorial or religious production methods in the Canadian tariff, demonstrates a production process (PPM)-oriented tariff category for TK-based products. In contrast to the following example of a substance-based tariff differentiation for TK-based products in the Papua New Guinean Tariff, the production process-based differentiation raises problems of evidence, since the end-product cannot be distinguished from a conventionally produced one. For this very reason, the Canadian government introduced a certification requirement, which could be adopted for the creation of a multilaterally recognized tariff differentiation for PPM-based TK-holding products under the HS. It should also be noted that scholarship on TK does not recognize religious practices or equipment as TK.[110]

Papua New Guinea recognizes, as perhaps the only rice-producing country in the

---

[107] *European Communities-Customs Classification of Certain Computer Equipment,* Appellate Body Report, WT/DS62/AB/R, WT/DS67/AB/R, WT/DS68/AB/R, 5 June 1998, para. 94, see also para. 75.

[108] Canada Customs and Revenue Agency, http://www.ccra-adrc.gc.ca/customs/general/publications/tariff2003/t2003–1/chap99ne.pdf

[109] Canada Customs and Revenue Agency, http://www.ccra-adrc.gc.ca/customs/general/publications/tariff2003/t2003–1/chap99ne.pdf

[110] See WIPO/GRTKF/IC/3/9 (Report of third session of the Intergovernmental Committee on Intellectual Property and Genetic Resources, Traditional Knowledge and Folklore), Traditional Knowledge – Operational Terms and Definitions.

Table 8.1. Enhancing market access. (Table by Gloria Pasadilla, WTI.)

		USA		EC		Japan
HS Code	**Potato**		*HS Code*		*Hs Code*	
07019010	Yellow (Solano) potatoes, fresh or chilled	0.5¢/kg	07019010	6.3		
07101000	Frozen potato vegetable	14%	07101000	15		8.5
07129030	Potatoes, whether or not cut or sliced but not further prepared	2.3¢/kg	07129030	13.3	071290050	12.8
	Tomato					
0702000	Tomatoes, fresh or chilled	3.17¢/kg*		annex	070200000	3
07108040	Tomatoes, frozen	2.37¢/kg*	07108070	15		–
07129074	Dried tomato (In powder)	8.70%	07129030	13.3		–
	Onions and garlic					
07031030	Pearl onions not over 16 mm in diameter, fresh or chilled	0.96¢/kg	07031030	10		
07111000	Onions, provisionally preserved, not for immediate consumption	5.10%	07111000	7.5	071110000	9
07122020	Dried onion powder or flour	29.80%	07122000	13.3	071220000	9
07032000	Garlic, fresh or chilled	0.43¢/kg	07032000	10	070320000	3
07129040	Dried garlic	29.80%	07122000	13.3		–
	Cabbages, cauliflower, broccoli					
07041020	Cauliflower and headed broccoli, if entered during the period from June 5 to October 15, inclusive, in any year, fresh or chilled	2.50%	see [1]	11.4 MIN 1.37 €/100 kg/net	07041000	3
07041040	—Not reduced in size	10%				
07041060	—Cut, sliced or otherwise reduced in size	14%				
07049020	Cabbage, fresh or chilled	0.54¢/kg	07049010	12.5 MIN 0.4 €/100 kg/net	070511000	3
07051120	Head lettuce, if entered in the period from June 1 to October 31, inclusive, in any year	0.4¢/kg	see [2]	11.37 MIN 1.67 €/100 kg/br		–
07051140	—Other	3.7¢/kg				
	Carrots, turnips, salad beets, radishes					
07061005	Carrots, fresh or chilled – reduced in size	14.90%	07061000	14.2	07061000	3
07061010	Carrots – under 10 cm in length	1.4¢/kg				
07061020	Carrots – other size	0.7¢/kg				

07061040	—Turnips	0%	07061000	14.2	07061000	3
07069020	—Radishes	2.7				
07069030	—Beets and horseradish	1.9	07069030	12.5		
07129010	Dried carrots – whole, cut, sliced, broken or in powder, but not further prepared	1.3	07129050	13.3	071290040\[i]	9
	Cucumbers					
070700	Cucumbers, including gherkins, fresh or chilled	4.255¢/kg*		annex	070700000	3
07114000	Cucumbers, including gherkins, provisionally preserved	7.7	07114000	12.5	071140000	9
	Asparagus					
07092010	Fresh, chilled, not reduced in size, entered during the period from September 15 to November 15, inclusive, in any year, and transported to the USA by air	5	07092000	11.2	070920000	3
07092090	—Other	21.3				
	Mushrooms and truffles					
07095100	Mushrooms, fresh or chilled	8.8¢/kg + 20%	see [3]	7.275	070951020	4.3
07095200	—Truffles	0	07095200	6.7	070952000	3
07108020	Mushrooms, frozen	5.7¢/kg + 8%	07108061	15		
07119040	Mushrooms, provisionally preserved	5.7¢/kg on drained weight + 8%	07119040	10 + 199 €/100 kg/net eda		
07123010	Mushrooms, air dried or sun dried	1.3¢/kg + 1.8%	07123000	13.3	071230010	12.8
07123020	—Other	1.9¢/kg + 2.6%				
07123040	Truffles, dried	0				
	Olives					
07099035	Olives, fresh or chilled	8.8¢/kg	07108010 [4]	15.8		

world, among the categories for rice multilaterally recognized (husked, milled, brown), the so-called 'special variety rice (Basmati)'.[111] This substance-based category, which possibly reflects a variety created by TK, might be a future option for recognizing varieties which have been TK-influenced or TK-created, as opposed to genetically modified varieties of the same plant product (genetic modification is also not yet recognized in the HS as a special tariff category).

While differential treatment of specific TK-based products is possible, questions arise under current law as to the distinctions made on the basis of production methods. Under prevailing views, product classification is limited by the requirements of treating like products alike under Article III GATT 1994 and not to impair tariff concessions under Article II of that agreement. Thus, members are not entitled to treat a product produced by a traditional method differently from one produced industrially, if the end products are 'like'. As it stands, the Harmonized System does not recognize process and production methods (PPMs) as criteria for granting a product lower tariff treatment. The matter is closely related to the same and controversial problem under Article III GATT 1994, discussed below.

Since the difference between a TK-based product and an industrially produced one will often lie only in the production method, as the Canadian example shows, we propose that the HS system should be amended in the interest of legal security so as to allow explicitly the creation of a tariff sub-category for traditionally produced goods, and to require certification of the traditional method as a precondition for granting the product lower tariff treatment. Linking the tariff treatment to a certification process for TK-based products leads to issues of what standards of certification to use, and to ensure that the importing country does not abuse these standards as a technical barrier to impede trade in TK-based goods. If certification is used for creating a different tariff treatment of TK-based products, certification requirements must be examined in the context of the WTO Technical Barriers to Trade Agreement (TBT).

MULTILATERAL ADJUSTMENTS

TK-based categories, however, should ideally be introduced, in parallel with unilateral efforts, by revising the structure of the Harmonized System. This would be best for legal security. Based upon an assessment of relevant products, developing countries could suggest the introduction of special subcategories for which they seek zero or low tariff from importing countries. These categories could be defined in terms of substance (e.g. rice propagated and cultivated from traditional local varieties and landraces) or alternatively be defined by the production and process methods employed (e.g. [crop] conserved, propagated, planted and harvested by way of traditional methods of farming).

Alternatively, such changes could also be introduced in bilateral or sectoral tariff negotiations, using the 6–8 digit of the existing HS, and eventually be multilateralized. Since negotiated tariff schedules form part of the overall agreement in accordance with Article II.7 of the GATT 1994, they enjoy similar status to other rules of WTO law and would thus be immune from challenges made by other members of the WTO.

Limiting recourse to trade remedies against TK-based products

A complementary and procedural manner to secure market access and enhance TK-based products could consist of limiting the use of trade remedies and of strengthening corresponding disciplines. Otherwise, the protectionist vagaries of anti-dumping and

[111] Papua New Guinea, Tariff Heading HS 1006 for rice: 1006.10.00 Rice in the husk (paddy or rough) Free Tonne P&R; 1006.20.00 Husked (brown) rice Free Tonne P&R; 1006.30.10 Semi-milled or wholly milled rice, Free Tonne 'whether or not polished or glazed'; 1006.30.90 Basmati (special variety rice) 11% kg.

countervailing duties are likely to impair the potential of TK products despite low or zero tariffs. In fact, the very nature of according low or zero tariffs renders particular efforts to this effect necessary. The fact that TK-based products embody non-trade concerns, such as biodiversity and sustainability, may justify the limitation proposed on the use of trade remedies. Moreover, these products are likely to be produced in lower quantities and at higher costs than conventional crops. Risk of distortions remains low. It is submitted that countries should be prohibited from applying so-called 'trade remedies' (Jackson and Benke, 2003), which are otherwise available and lawful under the WTO Agreements (safeguards, countervailing duties and anti-dumping measures), or at least the remedial measures of safeguards to TK-based products. Appropriate rules, privileging these products, could build upon, and learn from, the following provisions.

Article 3(a) of the Annex to the WTO Agreement on Textiles and Clothing (ATC, suspended 31 December 2004) prohibited the application of safeguard measures under Article 6 ATC to TK-based textiles originating in developing countries. Upon expiry of the Agreement at the end of 2004, the provision no longer precludes the application of the general safeguard measure under Article XIX of GATT 1994, if it ever did so. The provisions demonstrate that the idea of restricting trade remedies for certain products is not a new one.

Article 15 of the Anti-Dumping Agreement (ADA) demands that 'special regard' be given when applying an AD duty to a developing country member. The scope of the concept of SDT for developing countries is not clear. It could, however, also provide a basis for further work on the matter. Article 15 ADA could be amended in view of specifying that – just as Article 3(a) Annex ATC did for safeguards in the textile sector – AD duties may not be applied to TK-based products from developing countries. Another option would be to adopt an authoritative interpretation under Article IX, 2 WTO (Marrakesh) Agreement, to Article 15 ADA, which would interpret the current provision in the light of exempting TK-based products from AD duties.

The Agreement on Subsidies and Countervailing Duties (SCM Agreement) contains a limitation on the application of a Countervailing Duty (CVD) to a developing country member calculated on the basis of either the value of the subsidy imposed by the developing member on its products as compared to the value of the product or on the basis of the volume of the subsidized imports on the total number of imports of the like product (Article 27.10 SCM Agreement). Article 27 SCM could be amended to specifically exempt TK-based products from CVD measures applied against imports from developing countries.

The non-application of trade remedies to TK-based product imports would introduce a passive form of protection for developing countries and, inversely, developing countries should have the possibility of applying special safeguard measures under the WTO Agreement on Safeguards (ASG) against the over-exploitation of their resources (PGRFA and PGM) at the basis of their TK-products, including medicines. Such a proposal was made outside the WTO at the CITES Tenth Meeting of the Conference of the Parties with Resolution 10.19 (Rev.CoP 12) for traditional medicines, however, in a slightly different context.[112]

At the WTO efforts to empower developing countries with safeguards were undertaken, so far to no avail. In the preparation leading up to the Cancún Ministerial, in the WTO Committee on Agriculture

[112] See Conference of the Parties Resolution 10.19 (Rev.CoP 12), Traditional Medicines, Harare (Zimbabwe) 9–20 June 1997 available at: http://www.cites.org/eng/resols/10/10_19.shtml (last accessed 27 February 2005), which says in Recommendation (d): '[C]onsider, where appropriate and with sufficient safeguards, the application of artificial propagation and, in certain circumstances, captive breeding, to meet the needs of traditional medicines where this would relieve pressure on wild populations of species and is in accordance with their national legislation...'.

(CoA) a new special safeguards mechanism (SSM) under the AoA was proposed for developing countries. The Cairns Group, however, attempted to devalue its impact by linking the proposed SSM to developing countries' tariff reduction commitments.[113] Brazil and others wished to terminate the existent Special Safeguard for developing countries (SSG) under Article 5 AoA and to replace it with the SSM (Special Safeguard Mechanism).[114] Focusing such proposals on TK might enhance their potential for acceptance in the future.

Labelling requirements

Product differentiation based upon TK either in GSP schemes or within tariff lines will require identification of eligible products. The problem encountered is that customs officials need to be in a position to recognize which products are eligible for lower or zero tariff treatment or even for exemptions from trade remedies under the respective schemes. It would not be sufficient to treat all products imported from a particular group of countries alike. Since the distinction often entails production methods, appropriate methods of designation and labelling need to be developed.

Labelling requirements exist for the purposes of identifying organically produced foodstuffs as well as geographical indications,[115] and products for which the control of quality either by private producers' associations or government could be further developed. Moreover, electronic data collections on traditional knowledge will be of assistance in defining whether a product should be recognized as a TK-based product (Chapter 5). Importing countries granting preferential treatment to TK-based products could operate systems for recognizing them, entailing inspection on the site in order to assess whether conditions are being met. Again, this is not a new task as agricultural products are already subject to health inspections under the SPS Agreement, and combinations could be envisaged. The same is true for declarations to be made under the TBT Agreement as to whether technical standards and norms are complied with. The task of inspection could also be delegated to specialized non-governmental organizations operating in the exporting countries concerned. A label indicating the qualities and properties of a TK-based product would thus allow both for swift inspections at borders as well as for the purposes of consumer information.

A simple solution could be to grant preferential tariffs to products from developing countries protected by geographical indications and appellations of origin, since these products pertain to a particular region and often entail traditional knowledge applied to their production. The combination of formal protection by intellectual property on the basis of an international registration to be discussed at the WTO (see Chapter 5) could serve as a clear definition for products eligible for low or zero tariff protection. Tariff reductions for such products would at the same time entail an incentive to develop GI protection and encourage production of eligible production. Overall, the scheme would support the goals of bringing about greater variety of competing products originating in developing countries, and thus of PGR effectively used.

---

[113] See International Center for Trade and Sustainable Development (ICTSD) (2003), *Doha Round Briefing Series*, Cancun Update, Vol. 2 Nr.2, available at: http://www.ictsd.org (last visited 17 September 2003).

[114] See Agriculture: A Blank Check Bitterly Contested, *Bridges*, July–August 2003, available at: http://www.ictsd.org (last accessed 17 September 2003).

[115] The EC set up a Committee for designations of origin, geographical indications and certificates of special character in 1992 (Commission Decision 93/53/EEC of 21 December 1992 setting up a Scientific Committee for designations of origin, geographical indications and certificates of specific character [Official Journal L 13 of 21.01.1993]; see also EC Commission External Trade, Why do Geographical Indications Matter to Us? Brussels, 30 July 2003, available at: http://europa.eu.int/comm/trade/issues/sectoral/intell_property/argu_en.htm (last accessed 26 January 2005).

8.10 TK and the TBT and SPS Agreements

TK-based products, even under low or zero tariffs, may still face important non-tariff barriers, mainly in the field of technical barriers to trade and phytosanitary measures that have been legitimately imposed by the importing country for food safety reasons. In the field of PGFRA, these barriers are mainly dealt with under the SPS Agreement, while other products, such as textiles and other manufactured TK-based items, may be subject to requirements of the TBT Agreement. Both agreements clearly address the quality of the product itself according to its impact on the domestic market. The more difficult question is whether these agreements also extend to measures that differentiate between products on the basis of their PPMs and can be used as a legitimate reason for restricting imports. We will begin by discussing the TBT Agreement, as this agreement contains an interesting reference to traditional technologies and their relations to international standards.

8.10.1 Agreement on Technical Barriers to Trade (TBT Agreement)

Product standards

The TBT Agreement primarily focuses on product standards, defining its proper and required quality. The use of international standards is encouraged, and national standards employed should not go beyond what is necessary to achieve legitimate policy goals, such as plant, animal and human health, sustainability or cost-effectiveness. The agreement essentially enshrines the principle of necessity and proportionality, should WTO members enact such regulations and standards, as long as these are necessary to achieve a defined policy goal.

TK-based products

The main problem may relate to the fact that they may not live up to safety standards, which usually are defined with production in industrialized countries. The Agreement does not address the problem relating to developing countries. These have to comply with the product standards prescribed. If a developing country felt that the standard is excessively strict, consultations and dispute settlement before the WTO could be used.

Special and differential treatment exists only with respect to standards adopted by developing countries taking effect in their own jurisdiction. Traditional technology may be exempted from strict, in particular international. standards.

Article 12.4 TBT Agreement says:

> developing country Members adopt certain technical regulations, standards or conformity assessment procedures aimed at preserving indigenous technology and production methods and processes compatible with their development needs. Members therefore recognize that developing country Members should not be expected to use international standards as a basis for their technical regulations or standards, including test methods, which are not appropriate to their development, financial and trade means.

While the approach clearly promotes the use by developing countries of indigenous TBTs, it does not go so far as to require industrialized countries to allow entry into their markets of products respecting such indigenous exporting country TBTs.

Clearly, the text of Article 12.4 TBT Agreement situates itself in the context of the (often financial/technical) difficulty of developing nations in conforming to international standards promoted by the TBT. But, probably due to industrialized country pressure (who pleaded to keep their high standards for environmental/societal reasons), the agreement did not go so far as to demand that developed, importing nations recognize indigenous standards and thus facilitate the importation of such goods. Thus, 12.4 TBT is limited to national or, through extensive interpretation, to regional jurisdiction but does not bind the WTO members to respecting the lower TK-based standards. Consequently, Article 12.4 TBT is not effective at facilitating market access for manufactured TK (textiles, food products, medicines), through recognition by industrialized countries of 'lower' technical standards for TK-produced goods. To the contrary, industrialized members are entitled to impose technical barriers on such imports.

The trade-promoting effect of Article 12.4 TBT consists of enabling trade in TK-products between developing countries with lower than international standards, and through mutual recognition of indigenous standards South–South trade is fostered. The option for developing countries (Article 12.4 TBT) to maintain their lower-than-international technical standards may receive more attention in the light of Cancún's failure to open trade flows between North and South. With South–South trade likely to increase, the formation of new RTAs by developing country groups will grow in proportion to each setback in reaching multilateral tariff reductions (Forero, 2003). South–South trade does not, however, resolve the problem of market access to industrialized countries.

The other possible opening for indigenous TBTs/SPSs is that both preambles speak – without mentioning indigenous standards – of the difficulty of the developing countries in conforming to international standards. Yet this does not amount to an implicit obligation for developed WTO members to recognize different standards, which are not necessarily lower standards. Other than 12.4 TBT Agreement and the Preambles in both the TBT and SPS Agreements, there is no obligation in WTO law to respect standards commensurate with TK-based technologies.

Developing countries therefore could seek introduction of a measure of recognition for TK-based products into the TBT Agreement to the extent that existing product standards impair exportation. In this context, Article 12.4 could be amended in the sense that TK-based products should be taken into account in defining international and national product standards. Existing international and national standards should be reviewed as to whether they unduly impair the importation of TK-based products in the light of the interest in promoting conservation through use of a large variety of PGRs cultivated in developing countries. Such reforms of the TBT should take care not to jeopardize the product safety for consumers. The task is limited to avoiding the potentially protectionist effects of existing standards. Dispute settlement already offers the possibility of challenging such standards in terms of necessity and proportionality (see Cottier 2000; Pauwelyn 2001; Oesch 2003).

Another, albeit difficult, approach is to seek Mutual Recognition Agreements (MRAs), which take into account such needs while preserving product security at the same time. Article 6(3) TBT Agreement encourages such agreements and thus exempts them from MFN obligations. Practical experience demonstrates that these agreements are difficult to achieve, in particular in relation to PGRFA, where countries insist on imposing their own food and health standards. A particular and limited focus on TK-based products may facilitate the achievement of results.

Product Process Methods (PPMs)

In international environmental and labour law, it is generally recognized that production and processing methods are an essential element for the evaluation of a product. Thus, the question in this section is whether the TBT Agreement authorizes quantitative restrictions when environmentally unfriendly production methods, or those that disregard worker safety or even those that disregard traditional knowledge, have been used (Jackson, 1997, pp. 224, 235–238, 244–245). The TBT Agreement, in addition to product standards, takes into account PPMs as it says in the definition of 'technical regulation' in Annex 1 of the TBT Agreement: '[T]echnical regulation ... document which lays down product characteristics or their related processes and production methods...'.

It remains somewhat unclear how far PPMs are covered in the TBT Agreement (Meinheit, 1995, p. 190). The main problem relates to the fact that such measures often exert extraterritorial effect as they define PPMs as if products were to be processed and produced, instead of simply imported into the regulating country. States have notified PPM-related standards, which affect protection abroad. In the lists of notifications that WTO members can make to the TBT, the Netherlands for example made a notification relating to its higher standards for indigenous protected animal and plant species and products from such animals and plants.[116]

There is a widely shared view that the provisions of the TBT Agreement should generally be interpreted narrowly (Meinheit, 1995, p. 190). Consequently, PPMs should only then justify import restrictions when these processes influence the end characteristics of a product (e.g. when they leave traces of pesticides, hormones, nitrogen and heavy metals) (Meinheit, 1995, p. 192). If they do not translate into the quality of the product, they should not be able to justify trade restrictions under the TBT Agreement. But there are others who see Article 2.2 of the TBT Agreement as a departure from the *Tuna–Dolphin* rulings under GATT (which said that production methods do not qualify for the consideration of likeness for the obligation of national treatment) and to allow 'differentiation based on *how* a product is made, as opposed to the final product itself' (Trebilcock and Howse, 1999, p. 143). However, because it is more costly to examine in retrospect the characteristics of an end product and possibly to reject that product altogether, it is more efficient to lay down and recognize in advance the production and processing standards for products (Meinheit, 1995, p. 191; Trebilcock and Howse, 1999, p. 143). This is also what Article 2.8 TBT Agreement seems to underscore when stating that: '[W]herever appropriate, Members shall specify technical regulations based on product requirements in terms of performance rather than design or descriptive characteristics'.

Today, it thus remains unclear and unsettled how the TBT Agreement's reference to PPMs should be interpreted and why it should be interpreted narrowly. One possible explanation is that it should match with GATT Article III, the national treatment obligation for 'like' products, which constructs likeness as relating to the product and does not take into account 'differences in the country of export or its manufacturing environment' (Jackson, 1997, p. 236, and references therein). This is so because, as Jackson says: 'this is what comparative advantage is all about: differences in environments of production, including the environment of government regulation' (1997, p. 236). At the same time, it is evident that efficient protection of global commons requires a review of traditional perceptions. The TBT Agreement is not conclusive on this point, and clarification may be required in future negotiations. We return to this controversial issue after looking into the SPS Agreement.

8.10.2 The Agreement on the Application of Sanitary and Phytosanitary Measures (SPS Agreement)

The SPS Agreement is a special agreement, which grew out of the insights that the TBT Agreement was not fully suitable for regulating food standards, both for plants and animals. It is of particular importance in the context of TK-based products, as many of them will fall under the scope of the SPS Agreement. The Agreement builds upon the adoption of existing international standards. To the extent that members chose to rely upon their own and higher domestic standards, they are required to provide scientific evidence showing risks. As to risk management, members essentially remain free to adopt appropriate measures to address the risk on a level they chose.

Product standards

TK-based products therefore are essentially subject to international foods standards. It is a matter of further examination to what extent TK-based products do not live up to these standards, which again are mainly framed from a perspective of developed countries and their conditions and standards of production. The matter deserves discussion in the relevant organizations, in particular the Codex Alimentarius, a joint body of the WHO/FAO upon which the SPS Agreement essentially builds.

[116] See Notification of the Netherlands, WTO Doc. G/TBT/Notif.00/344, 7 August 2000.

At this stage, there is no provision equivalent to 12.4 TBT in the SPS Agreement, which would address the problem of standards of indigenous technology and PPMs for sanitary and phytosanitary purposes applicable in trade between developing countries. Equally, no similar SDT provision as to lowering international standards for domestic purposes, taking into account TK products, can be found in the SPS Agreement. It is unclear to what extent the absence of such a provision amounts to a practical problem.

A problem, however, exists in the absence of SDT provisions relating to the importation of TK-based products into developed markets. Here, it would be suitable to examine whether international standards sufficiently take into account the interest in promoting trade in TK-based products and whether they partly impair such trade. Again, it will only be a matter of addressing and identifying excessive and thus protectionist standards. Preservation of health in importing countries cannot give way to importation of potentially risky TK products as populations in importing countries have not been exposed to them and may not be immune to detrimental effects on health. Developed countries, such as the USA, are concerned that the SPS Agreement may lead *de facto* to a 'downward harmonization' because a member may sacrifice the health of its consumers in order to prevent a trade dispute (Silverglade, 1998). At the international level, the FAO/WHO Codex Alimentarius, which has been mandated pursuant to Article 3.4 SPS Agreement to establish international standards, is already concerned with 'cultural practices' that lead to 'food borne diseases', and one of its recommendations is to 'identify cultural practices that lead to increasing hazards associated with foods and development of programs to modify these behaviours within the context of the prevailing culture and social conditions'.[117] The question is one of singling out in the light of preserving valuable TK from TK that is potentially detrimental to human health. An example of TK that would be hazardous to human health may be the expression of a traditional culture which believes that washing one's hands will lead to rheumatic disease (Horton, 1998, footnote 84 to p. 153).

The Codex Trust Fund, a US$40 million joint FAO/WHO fundraising activity for increasing developing country participation in the standard-setting Codex Alimentarius Commission (CAC), which also formulates the SPS standards, supplies an important framework for developing countries. It is propagated that the WTO might join this initiative to enable developing countries to join in the rule-formulation, since substantively the SPS Agreement subscribes to Codex' rules.[118] Enhanced participation of developing countries in international standard-setting may facilitate taking into account specific interests relating to TK-based products and their value in preserving long-term food security by means of offering a wider range of uses for PGRs.

Product Process Methods (PPMs)

The SPS Agreement allows for a quantitative restriction based on PPMs when production methods in the exporting country are necessary to protect human, animal and plant life in the importing country (Meinheit, 1995, pp. 191–192). One commentator on this working paper has noted that the Annex A Definitions to the SPS Agreement exclude PPMs-based SPS measures.[119] Import restrictions based upon SPS meas-

[117] Expert Committee on Food Safety, Food and Agriculture Organization/World Health Organization, The Role of Food Safety in Health and Development (1984) WHO Technical Report 705, cited in: Linda R. Horton (1998) Food from developing countries: steps to improve compliance. *Food and Drug Law Journal* 53, 152.

[118] See Codex Trust Fund, 14 February 2003, available at: http://www.who.int/foodsafety/codex/trustfund/en/.

[119] See Lothar Ehring, oral comments to this Paper, presented on 29 April 2003, at the WTI, Berne, Switzerland; see also Annex A to the SPS Agreement, Definitions of Sanitary or Phytosanitary Measures: 'Sanitary or phytosanitary measures include all relevant laws, decrees, regulations, requirements and procedures including, inter alia, *end product criteria*; processes and production methods; testing, inspection, certification and approval procedures; quarantine treatments including relevant requirements associated with the transport of

ures need to be based upon the quality of the product. Production methods *per se* do not cause the risks addressed by the Agreement. They need to be related to the quality of the product. It is in this sense that Article 5.2 SPS Agreement requires PPMs to be taken into account. At the same time, it does not allow privilege of them either. In contrast to Article 12.4 of the TBT Agreement, the SPS Agreement does not contain a similar recognition of PPMs specific to developing countries. As a result, TK-based practices may amount to important trade barriers which cannot be overcome without changing TK practices, if need be in order to meet hygienic standards. For developing countries, aiming at enhanced market access for their TK-based products, the only realistic avenue to date is to examine whether existing international and national standards are excessive and unnecessarily exclude TK-based products for protectionist purposes. Such measures can be challenged in WTO dispute settlements (see Cottier, 2000; Pauwelyn, 2000; Oesch, 2003). In reforming the SPS Agreement, the inclusion of at least a 12.4 TBT Agreement-type provision for TK-based products should be envisaged, or perhaps a further-reaching amendment of the SPS standards to require some sort of special recognition by developed countries of developing country standards for TK-based products.

8.11 The Need for Joint Standard-setting Operations

Product standards under both the TBT and the SPS Agreements need to serve the purposes of product safety and promoting trade, in particular of environmentally friendly products. The threefold purpose of product regulation in the future could be expanded to include trade-promotion for TK-based products and products based on TK-related PPMs. Authorities protecting consumer health, trade and the promotion of the global commons should cooperate with a view to achieving optimal coherence. First efforts to this effect have been undertaken.

The CBD has recognized that especially in the framework of the Cartagena Protocol it is crucial the CBD and the WTO SPS and TBT Agreements work together in establishing import standards, which might have an impact on the biological diversity, especially of local communities.[120] In this context, the CBD has applied for an observer status in the WTO Committee on Technical Barriers to Trade. In his application for observer status, the executive secretary of the CBD stresses the close independence of trade, health and the environment. He notes that the Cartagena Biosafety Protocol, which regulates the transboundary movements, transit and handling/use of GMOs, sets out procedures for notification and decision-making on import and export of GMOs. It includes prior informed consent procedures, risk assessment and management, and socio-economic considerations. It entails taking into account the 'value of biological diversity to indigenous and local communities' (Article 26 Biosafety Protocol).[121] Moreover, the parties of the Biosafety Protocol and the members of the WTO are 'encouraged to cooperate on research and information exchange on any socio-economic impacts of GMOs, especially on indigenous and local communities'.[122] Another feature of increased cooperation between the WTO (specifically SPS and TBT Agreement) with the Biosafety Protocol with potential impact on

Note 119 continued

animals or plants, or with the materials necessary for their survival during transport; provisions on relevant statistical methods, sampling procedures and methods of risk assessment; and packaging and labeling requirements directly related to food safety' (emphasis added).

[120] Request for observer status by the Convention on Biological Diversity, WTO Doc. G/TBT/W/177, 15 July 2002 [hereinafter WTO Doc. G/TBT/W/177].

[121] Ibid.

[122] Ibid.

TK is the concept of the Biosafety Clearing House (BCH). The provisions on the BCH require that all domestic implementation regulation of the Protocol, bi- and multilateral agreements on the issue as well as the risk assessments and environmental reviews regarding GMOs, be transmitted to it. It serves the purpose of ensuring compliance and facilitates the exchange of information.[123] The clearing house mechanism may thus also contribute to the policy coordination and the interface of different policy goals.

Appropriate institutional structures allowing the interface of different regulatory purposes of standards need to begin at home. The different agencies affected need to cooperate in preparing for standard-setting in international organizations. Delegations to different standard-setting bodies should be in a position to take into account all relevant concerns of health, environment and trade. Effective domestic coordination will also facilitate enhanced cooperation and coherence on the level of international organizations. Members of the secretariat of these organizations should reciprocally follow the work on the basis of observer status and ensure that work is sufficiently informed about their concerns and that these are taken into account. The elaboration of food standards in the FAO Codex Alimentarius Commission therefore should also entail representation of the WTO and UNEP secretariat. It was also suggested that the WTO take a precursor role in merging the standard-setting activities of all these organizations (see Marceau and Trachtman, 2002, p. 840). While the allocation of the leading house remains an important issue, we submit that procedures of cooperation and interaction ultimately are more important and may also assist in overcoming traditional rivalries between different international organizations as well as domestic government agencies. This also allows specialized organizations to retain the initiative and the end of centralizing standard-setting with the WTO in a large number of different regulatory areas.

8.12 Non-discrimination Principles of GATT 1994 and Traditional Knowledge

8.12.1 Introduction

Problems and difficulties encountered in privileging market access for TK-based products throughout are partly rooted in the principles of non-discrimination underlying the multilateral trading system. Most-favoured nation treatment and national treatment provisions as well as related principles of multilateral liberalization amount to the WTO's constitutional precepts. The GATT framers installed non-discrimination to achieve a level playing field between domestic and imported products, which would eventually lead to realizing the best allocation of resources. Non-discrimination is crucial in assuring equal market access rights, competition and efficiency. The goals of validating TK may create tensions with non-discrimination, because TK is about long-term, equitable rather than short-term, economically efficient distribution of production factors. In the present context of discussing TK, we need to consider some of the downsides of the non-discrimination obligations, as they render considerations of equity and ecological considerations more difficult. The concerns for distributional justice and conservation of plant genetic resources require a review of traditional definitions of what are called like products in WTO law. While the law allows and requires distinguishing between different products of a different quality, it does not allow one to differentiate like products on the basis of differing and diverging production methods (PPMs). The problem was already encountered in discussing the implications of the TBT and SPS Agreement. We now turn to the issue under the General Agreement on Tariffs and Trade, which enshrines the principles of non-discrimination in its Articles I and III. So far, developing

[123] Ibid.

countries were opposed to considering PPMs in defining like and unlike products, because in their perspective industrialized countries were abusing PPMs in the form environmental and labour regulations as protectionist tools with extraterritorial effects, to impair developing countries' market access rights. With the advent of increasing awareness about TK, their resilient attitude may change and lend them an argument in favour of improving market access for their own products. If developing countries manage to trade off recognizing PPMs in the context of TK against industrialized countries' interests in the same instrument for labour and environmental purposes, a negotiating position for narrowing the like product definition by distinguishing different PPMs could be brought about. Thus, under such an amended Article III GATT 1994, a hitherto 'like' product would become 'unlike'.

8.12.2 Legal issues

Article I GATT obliges a member to grant unconditionally and immediately all privileges granted for the importation of a product originating in a member state to like products from any other member of the WTO. Article III applies the principle in relation to the treatment of domestic products. Upon customs clearance, all foreign products are entitled to obtain treatment no less favourable than domestic like products. In the context of TK, principles of non-discrimination pose the following problems.

First, under MFN, regulations applicable to preferential treatment of TK-based products would not only have to be extended to like TK products from other members, developed and developing alike. They also apply to all conventional products to the extent that they are to be considered 'like'. Regional integration under Article XXIV GATT may offer further privileges to products from selected countries. Other exceptions may be possible if TK products could be distinguished on the basis of process and production methods rather than on their proper quality. Whether or not this is possible needs to be examined in detail under the following section.

Secondly, national treatment obligations operate only to the benefit of foreign products. 'Less favourable treatment' of domestic products cannot be invoked. A WTO member would therefore be in a position to treat imported TK-based products more favourably, e.g. by requiring less stringent rules or by imposing lower taxes, than like domestic products. A member may do so on the basis of the product quality or on the basis of PPMs. A developed country therefore could adopt measures supporting the marketing of foreign TK-based products. This is important with a view to supporting the marketing of TK-based products in industrialized markets. However, could a member inversely impose more stringent requirements to imported like products and distinguish treatment on the basis of product requirements or production and process methods? This constellation is relevant for developing countries that seek to favour the use of TK-based home products *vis-à-vis* imported 'like' products. It is in this context that the controversy in PPMs in relation to Article III GATT becomes relevant. Should this be considered the pursuit of a legitimate policy goal or should this be prevented and ruled out as protectionist?

Whether or not regulations specific to TK-based products and different from those applicable to conventional products would be GATT-consistent depends upon the assessment as to whether TK and conventional products are to be considered like products under GATT. While there is ample jurisprudence under Article III on likeness, precedents relating to Article I are much scarcer. Likeness does not mean the same thing in all provisions. They need to be interpreted in context, and the Appellate Body called this the 'accordion approach'. In *Japan–Alcohol*, the Appellate Body said, referring to the criteria developed in the *Border Tax Adjustment* report that:

> No one approach to exercising judgement will be appropriate for all cases. The

> criteria in Border Tax Adjustments should be examined, but there can be no one precise and absolute definition of what is 'like'. The concept of 'likeness' is a relative one that evokes the image of an accordion. The accordion of 'likeness' stretches and squeezes in different places as different provisions of the WTO Agreement are applied. The width of the accordion in any one of those places must be determined by the particular provision in which the term 'like' is encountered as well as by the context and the circumstances that prevail in any given case to which that provision may apply...[124]

We therefore need to analyse likeness in MFN and the principle of National Treatment separately (see also Cottier and Mavroidis, 2000). The analysis is relevant for products that basically find themselves in a competitive relationship. To the extent that this is not the case, since they strongly differ in terms of quality or price, and do not form part of the same relevant market, the following principles do not apply.

8.13 The Impact of MFN

Likeness under Article I GATT has, so far, been read in a rather strict and narrow sense. In *Spain–Roasted Coffee*, it was held that different varieties of coffee are considered to be alike, in particular since they were blended, and thus should be subject to the same level of imposed tariffs. The case essentially relied upon basic product qualities and discarded arguments relating to consumer preferences or slightly varying physical qualities of the products involved.[125] Whether or not differentiation based upon PPMs would be admissible under current law has not been addressed. An answer cannot be given in isolation from jurisprudence relating to Article III GATT 1994, but certainly the accordion principle also applies here. At present it would not seem possible to privilege TK-based products over conventional products from other member states, unless strong reasons can be found in the context of Article III and the exceptions under Article XX GATT, in particular those in paragraph (g) relating to the protection of non-renewable resources. The prime avenue to be pursued therefore relies upon explicit differential treatment in treaty law. Privileges granted to TK-based products in terms of tariff and non-tariff treatment *vis-à-vis* conventional products should therefore be clarified in future negotiations. Avenues in the context of GSP and the HS were indicated before. Explicit admittance of PPMs for defining the likeness of products in amending the Agreement or adopting an interpretative understanding is yet another additional way to proceed. To what extent this could also be achieved in treaty interpretation much depends upon future developments relating to Article III GATT 1994.

8.14 The Impact of National Treatment

We stated above that national treatment does not exclude privileging imported products over domestic products. The principle of reverse discrimination (*discrimination à rebours*) offers the legal possibility of privileging TK-based products imported from developing countries over domestic like products. Legislation could envisage distinguishing products on the basis of product differentiation, but also on the basis of PPMs. This is an important finding, albeit difficult to realize politically where foreign and domestic products compete on the market. Foreign TK-based products thus not only could be accorded low or zero tariffs with a view to facilitating importation, they could also be given differential treatment regarding domestic taxation or requirements for marketing the product. International law therefore does not inhibit supporting TK-based products.

Problems arise in respect of the inverse constellation. Does Article III GATT allow

[124] *Japan-Taxes on Alcoholic Beverages,* Appellate Body Report, 4 October 1996, WT/DS8/AB/R, WT/DS10/AB/R, WT/DS11/AB/R, p. 21.

[125] See *Spain-Tariff Treatment of Unroasted Coffee,* GATT 1947 Panel Report, 11 June 1981, paras 4.6–4.9.

privileging *domestic* TK-based products over imported, conventional and competing products? LDCs may contemplate applying preferential tax rates or facilitate marketing through other regulatory measures with a view to developing local or national markets for TK-based products. The answer depends upon the assessment as to whether the domestic TK-based product and the imported product are 'like' products within the meaning of Article III GATT. As indicated above, jurisprudence adopted an approach examining the matter on a case-by-case basis, based upon the *Border Tax Adjustments* 1970 working party report and a number of criteria set forth therein:

> the interpretation of the term [likeness] should be examined on a case-by-case basis. This would allow a fair assessment in each case of the different elements that constitute a 'similar' product. Some criteria were suggested for determining, on a case-by-case basis, whether a product is 'similar': the product's end-uses in a given market; consumers' tastes and habits, which change from country to country, the product's properties, nature and quality.[126]

Likeness for the purpose of PGRFA will also depend on the distinguishing factors developed by GATT and WTO jurisprudence. Within Article III, different constellations need to be distinguished (see Cottier and Oesch, 2005). First, likeness in the context of taxation is defined narrowly in the first sentence of Article III(2). Secondly, likeness in the context of substitutable products is defined more widely in the second sentence of Article III(2), allowing the taking into account of non-protectionist regulatory purposes. Thirdly, likeness for general regulatory purposes under Article III(4) has witnessed enhanced flexibility in recent cases, allowing for example the taking into account of health concerns within the criteria of consumer preferences (Ehring, 2002). Much therefore depends on whether consumer preferences can be distinguished between imported conventional products and domestic TK-based products.

In our view, the purpose of Article III GATT is to avoid and prevent rent-seeking protectionism. It does not exclude product differentiation for the purpose of legitimate regulatory purposes, such as policies to protect biodiversity and conservation through use of traditional crops, to the extent that this translates into diverging consumer preferences. For the purpose of the present study, recent evolutions in the case law of the AB are particularly pertinent and will allow the drawing of some conclusions in the context of regulating traditional knowledge-based products.

8.14.1 The narrow definition of likeness for purposes of taxation

In *Japan–Alcoholic Beverages*, the Appellate Body laid down its reading of Article III(2) as follows:

> Article III:1 informs Article III:2 first sentence, by establishing that if imported products are taxed in excess of like products, then that tax measure is inconsistent with Article III. There is no specific invocation in this first sentence of the general principle in Article III:1 that admonishes Members of the WTO not to apply measures 'so as to afford protection'. (...) Read in their context and in the light of the overall object and purpose of the WTO Agreement, the words of the first sentence require an examination of the conformity of an internal tax measure with Article III by determining, first, whether the taxed imported and domestic product are 'like' and, second, whether the taxes applied to the imported products are 'in excess of' those applied to the like domestic products. If the imported and domestic products are 'like products', and if the taxes applied to the imported products are 'in excess of' those applied to the like domestic products, then the measure is inconsistent with Article III:2, first sentence.[127]

[126] *Border Tax Adjustments*, adopted 2 December 1970, L/3464, para. 18, available at: http://www.worldtradelaw.net

[127] *Japan-Taxes on Alcoholic Beverages*, Appellate Body Report, 4 October 1996, WT/DS8/AB/R, WT/DS10/AB/R, WT/DS11/AB/R, pp. 18–19.

A narrow reading of likeness was adopted in this context:

> Because the second sentence of Article III:2 provides for a separate and distinctive consideration of the protective aspect of a measure in examining its application to a broader category of products that are not 'like products' as contemplated by the first sentence, we agree with the Panel that the first sentence of Article III:2 must be construed narrowly so as not to condemn measures that its strict terms are not meant to condemn. Consequently we agree with the Panel also that the definition of 'like products' in Article III:2, first sentence, should be construed narrowly.[128]

WTO jurisprudence therefore is based upon a strict interpretation of likeness in the context of taxation. It would not seem to allow for differentiation of conventional and TK-based products within Article III(2) GATT 1994.

8.14.2 The broader definition for regulatory purposes

In *EC–Asbestos*, the

> EC contends that the Panel erred in its interpretation and application for the concept of 'like products', in particular, in excluding from its analysis consideration for the health risks associated with chrysotile asbestos fibres which cause cancer. According to the EC in this case, Article III calls for analysis of the health objective of the regulatory distinction made in the measure between products containing asbestos fibres, and all other products. The EC argued that under Article III:4, products should not be regarded as 'like' unless the regulatory distinction drawn between them 'entails [a] shift in the competitive opportunities' in favour of domestic products.[129]

The AB in *EC–Asbestos* took up the opinion of the EC and ruled that 'a determination of "likeness" under Article III:4 is, fundamentally, a determination about the nature and extent of a competitive relationship between and among products'.[130] Thus, at least for Article III:4, the AB defines likeness as a concept protecting the competitiveness of the marketplace, which leaves room for a broader approach to 'likeness' than the narrow definition in *Japan–Alcohol*.[131] The AB in *US–Cotton Yarn* held that 'two products are in a competitive relationship if they are commercially interchangeable, or if they offer alternative ways of satisfying the same consumer demand in the marketplace'.[132] The case understands itself as a further development from *Korea–Alcoholic Beverages*, and emphasizes that competition in the marketplace is a dynamic concept,[133] not exclusively determined by current consumer preferences, and that a static view is clearly erroneous.[134]

8.14.3 Assessment

The evolution of AB jurisprudence from *Japan–Alcohol* over *EC–Asbestos* to *US–Cotton Yarn* demonstrates the WTO's increasing reliance on consumer preferences and regulatory goals, shaping competitiveness and substitutability of products. Extensive interpretation argues that there is essentially no limitation as to the criteria which may be taken into account in establishing product differentiation under Article III:4 GATT 1994 (Howse

[128] *Japan-Taxes on Alcoholic Beverages,* Appellate Body Report, 4 October 1996, WT/DS8/AB/R, WT/DS10/AB/R, WT/DS11/AB/R, pp. 19–20.

[129] *European Communities-Measures Affecting Asbestos and Asbestos Containing Products,* Appellate Body Report, WT/DS135/AB/R, 21 March 2001, para. 86.

[130] Ibid., para. 99.

[131] See ibid., fn. 58 to para. 88 for an overview of all GATT/WTO cases defining 'like products', and paras 98–100, where the AB correlates Art. III:4 with the goal of ensuring competitive relationships.

[132] *US-Transitional Safeguard Measures on Combed Cotton Yarn from Pakistan,* Appellate Body Report, WT/DS/AB192/R, 8 October 2001, para. 91.

[133] See ibid.

[134] See ibid., para. 96.

and Türk, 2001). This evolution carries the potential that some day jurisprudence could recognize – stimulated by consumer demand and awareness – a TK-based product as a valid reason for treating the product unlike an industrially produced one.

For a country to be in the position to argue that it is favouring its own domestic and traditionally produced goods over imported industrialized ones, it will have to show that its consumers distinguish and favour such TK-based products with a view to supporting sustainable production and local identities.

To the extent that such evidence cannot be brought about and thus the concept of likeness applies, product differentiation could still be justified on the grounds of Article XX(g) GATT. According to this provision, measures can be justified to the extent that they serve the purpose of protecting non-renewable resources. Under such conditions, a country may even apply differential taxation to TK-based products, provided that the privilege does not amount to economic protectionism. Over the years, the Appellate Body has adopted a broad reading of Article XX(g), and there is no doubt that the purpose of protecting conservation through use of PGR could be justified as a measure in support of biodiversity.[135]

8.15 Product Process Methods (PPMs) in GATT 1994

8.15.1 The state of affairs

While TK-based products may differ in terms of competitive relationship from conventional products, they may also be like, but differ from them by the ways and processes of producing them. Does WTO law allow taking into account in the context of non-discrimination different methods of production? Typically, TK-based products entail a more cumbersome, less efficient means of production by using traditional crops and even landraces, to the benefit of protecting biodiversity. Is it possible to treat domestic products differently from more cumbersome, but ecologically valuable, modes of production under the WTO? We have already encountered the question in the context of the TBT and SPS Agreements and found that the answers are not clear. The matter is equally controversial under GATT 1994.

Article III GATT only recognizes those products as like when their end-use, physical characteristics and tariff categorization are similar. So-called production and process standards or measures (PPMs) have in GATT/WTO jurisprudence not been regarded as a criterion for constituting likeness or lack of likeness as first set out in the *Belgian Family Allowances* case under GATT 1947, where Belgium had decided to treat differently an exemption on a tax levy conditioned upon whether another GATT member's legal system had a similar, a dissimilar or no exemption from taxes for family allowances.[136] As these measures normally restrict market access and comparative advantages, mainly for protecting the environment or labour standards, product differentiation based upon them has mainly been opposed, in particular by DCs. Another reason why PPMs are refuted is the fact that they often entail extraterritorial effects in the sense that they oblige exporting countries to adopt legal standards of importing countries. The short answer therefore is that a PPM-based definition of TK-based products would not allow at this stage privileging domestic products *vis-à-vis* imported conventional products. It would be considered protectionist, unless

[135] *United States–Import Prohibition of Certain Shrimp and Shrimp Products,* Appellate Body Report, WT/DS58/AB/R, 12 April 1998, paras 128–130.

[136] See *Belgian Family Allowances,* Report adopted by the Contracting Parties on 7 November 1952, para. 3, available on: http://www.worldtradelaw.net, 'Belgian law would be irrelevant in this respect, and the Belgian legislation would have to be amended insofar as it introduced a discrimination between countries having a given system of family allowances and those which had a different system or no system at all, and made the granting of the exemption dependent on certain conditions.'

an exemption under Article XX GATT, in particular Article XX(g) GATT, could be found in the context of safeguarding exhaustible natural resources. Thus, US import restrictions for countries exporting tuna caught with nets killing dolphins and against third countries using this tuna for producing tuna-processed goods, were considered GATT-inconsistent, at the time (*Tuna I and II* both unadopted). However, production methods seeking to avoid accidental killings of dolphins were found to be potentially lawful, some years later, under Article XX(g) of GATT 1994 (*US–Shrimps Turtle*), justifying import restrictions under Article XI GATT.

In legal doctrine, production differentiation based upon PPMs, however, is partly affirmed within Article III GATT 1994. Howse and Regan argue that 'regulatory distinctions objectively related to actual non-protectionist policies are consistent with Article III, whether product- or process based' (Howse and Regan, 2000, p. 249). Instead of excluding PPMs from national treatment *ab initio*, the authors suggest addressing PPMs as a policy concern with respect to Article III. The authors also dismiss the danger that process-based measures are 'unilateral', 'extraterritorial' or 'coercive' and result in 'greater market segmentation', or 'impose unfair or distributively unjust' costs on producers in developing countries (Howse and Regan, 2000, p. 250).

Under existing rules, the matter ultimately has to be decided in dispute settlement. Relating to TK-based products, we recall that nothing impairs importing countries to voluntarily privilege imported products on the basis of PPMs as long as the measure is not applied in a discriminatory manner between importing countries. Thus, legislation could be enacted which allows for lower taxation or special marketing regulations for developing country products made in traditional modes. Whether or not it is possible to privilege domestic products over imported products currently depends upon the assessment as to whether the conditions of Article XX are complied with. We submit that this is possible provided that the PPMs serve the purpose of protecting biodiversity as one of the most important global commons and non-renewable natural resource. We also submit that the future is likely to see a more differentiated analysis of the like product analysis, allowing the taking into account of PPMs. The matter is at the heart of interfacing trade regulation with other policy concerns, in particular the environment, human rights, culture and gender. Product differentiation on the basis of PPMs will become a necessity in bringing about more coherence in treaty interpretation. Existing Multilateral Environmental Agreements (MEAs) show the way. Examples in point are the Basel Convention on the Control of Transboundary Movements of Hazardous Wastes and Their Disposal, 1989; International Tropical Timber Agreement (ITTA); International Commission for the Conservation of Atlantic Tunas (ICCAT); Convention on Biological Diversity (CBD); and Convention on International Trade in Endangered Species of Wild Fauna and Flora (CITES).[137] These agreements indicate that there has been an evolution in public international law towards recognizing the environmental, social and cultural implications on trade of the processes leading to a finished product. If the parties to an MEA match the membership of the WTO, the WTO adjudicating bodies would be held by rule 31(3)(c) of the Vienna Convention on the Law of Treaties (VCLT) to take into account as 'applicable law between the parties' the status of the discussion on PPMs in MEAs (Marceau, 1999; Marceau, 2001; Pauwelyn, 2001). If the membership of an MEA does not match, e.g. there are some Member States that are not part of the relevant MEA, then, instead of Article 31(3)(c) VCLT, it would be the more vague concept of dynamic treaty interpretation which might impose on the WTO a measure of recognition for PPMs. The 'PPMs in MEAs' discussion functions

[137] See Matrix on Trade Measures Pursuant to Selected MEAs, Note by the Secretariat, 14 June 2001, WTO Doc. WT/CTE/W/160/Rev.1.

as the external and additional reasoning for why the WTO should allow the integration of PPMs in the like product definition for purposes of non-discrimination within the national treatment obligation (Howse, 2000, p. 56).

8.15.2 Towards a new deal on PPMs

TK-based products offer a new perspective on the problem of PPMs, as they are primarily in the interest of developing countries. The debate, so far, has mainly been informed by interests of developed countries in protecting labour standards or environmentally friendly products and means of production, much to the detriment of market access for developing country products. Here, the constellation is inverse. Developing countries may be interested in protecting their own traditionally made products and achieving better market access for them abroad. This may provide grounds for a new deal on PPMs in future negotiations.

In addition to the evolution of case law, the issue of PPMs should be addressed and settled in multilateral negotiations. MEAs show that this is possible, at least on a sectoral basis. It is therefore conceivable to negotiate special provisions allowing for PPM product differentiation of TK-based products. Such an agreement would need to define relevant TK-based products. Again, we may refer to considerations made above on identifying and labelling such products. We may suggest that a special regime could be introduced for products qualifying for the protection as a geographical indication, as this category is generally supportive of broadening food security and the conservation of plant genetic resources through use. Such products may not only be granted privileged tariff treatment at the border, they may also be privileged domestically with a view to supporting the global commons of biodiversity, as well as the cultural diversity which they represent.

8.16 Conclusions

This chapter confirms that the promotion and protection of traditional knowledge reaches beyond the scope of intellectual property instruments, both in terms of TIP rights and GIs, as discussed in Chapter 5 (this volume; see also Panizzon and Cottier, 2005). Market access of TK-based products also extends beyond the removal of distorting export and domestic production subsidies, as discussed in Chapter 8, Part 1. This chapter proposes that market access for TK-based products and PPMs shall be addressed in other areas of trade policy, in terms of their competitive relationship to conventional products. A number of possibilities exist to privilege and support TK-based products.

High levels of tariff protection in agricultural products allow for meaningful product differentiation and privileges for TK-based products of developing countries. This can be brought about unilaterally in GSP or HS schemes, or by multilaterally extending the HS system. It is suggested that products qualifying for, and protected as, geographical indications could be made eligible for such treatment.

The main obstacles to TK-based products are likely to result from TBT and SPS measures. While the protection of health standards is imperative, enhanced involvement of developing countries and cooperation in standard-setting between different international organizations may assist in avoiding protectionist standards and in increased consideration of concerns for TK-based products.

A set of flanking and defensive policies could come in the shape of a prohibition for industrialized countries from using trade remedies such as countervailing and AD duties, as well as safeguards as far as indigenous products or traditional (artisan) process production methods are concerned. The now-defunct Agreement on Agriculture in its Annex, which expired by year-end 2004, has set the example for a revision of the ADA, CVD and ASG to that effect, by prohibiting the application of a safeguard to exports of handloom fabrics and other traditionally woven textile.

Since TK-based products are often defined by PPMs rather than by product quality, the problem of treating such PPMs in international trade regulation is at the heart of the matter. It is important to note that WTO law does not prevent importing countries from privileging TK-based products over domestic products. The principles of non-discrimination apply only to the benefit and protection of foreign and not domestic products. WTO rules, however, render the protection of domestic TK products more difficult. The law is not fully settled. While product differentiation is generally accepted as a matter of like product analysis, it is possible in terms of exceptions and existing international environmental agreements. The protection of plant genetic resources and TK-based products, being mainly in the interest of developing countries, may assist in making progress on the issue. Linking intellectual property and tariff policy amounts to a new and innovative approach with a view to supporting both biological and cultural diversity.

Trade policy measures, combined with new intellectual property rights over traditional knowledge, therefore offer the potential for substantially enhancing the protection of biodiversity including plant genetic resources, by means of facilitating market access for products, as well as PPMs, which represent conservation through use.

Bibliography

Part 1

Bope, L.B. (1991) *Afrikanische Gesellschaft im Wandel – Soziale Mobilität und Landflucht in Zaire.* Brandes & Apsel Verlag, Frankfurt, Germany.

Bridges (2003) Agriculture: a blank check bitterly contested. *Bridges* 7(6), 11–13 http://www.ictsd.org

Caritas International and CIOSE (2003) A background paper on the occasion of the 5th Ministerial Conference in Cancún, Mexico. http://www.agtrade.policy.org/page/resource/marketaccess.htm (accessed December 2005).

Carrol, R.C., Vandermeer, J.H. and Rosset, P. (1990) *Agroecology.* McGraw-Hill, New York.

Chang, H.-J. (2002) *Kicking away the Ladder. Development Strategy in Historical Perspective.* Anthem Press, London.

Desta, M.G. (2002) *The Law of International Trade in Agricultural Products. From GATT 1947 to the WTO Agreement on Agriculture.* Kluwer, The Hague.

Diaz-Bonilla, E., Diao, X. and Robinson, S. (2003) Thinking inside the boxes: protection and investments in the development and food security boxes. http://www.ifpri.org/pubs/confpapers/2003/diazbonilladiaorobinson.pdf (accessed December 2005).

Egger, U., Rieder, P. and Clemenz, D. (1992) *Internationale Agrarmärkte.* Verlag der Fachvereine an den schweizerischen Hochschulen und Techniken vdf, Zürich, Switzerland.

FAO (1996) *Report of the State of the World's Plant Genetic Resources for Food and Agriculture.* Prepared for the International Technical Conference on Plant Genetic Resources Leipzig, Germany, 17–23 June 1996.

FAO (2002) *FAO Papers on Selected Issues Relating to the WTO Negotiations on Agriculture.* FAO, Rome. http://www.fao.org/documents/show_cdr.asp?url_file=/DOCREP/005/Y3733E/Y3733E00.htm (accessed 9 January 2005).

Gerster, R. (2001) *Globalisierung und Gerechtigkeit.* h.e.p. verlag ag, Berne, Switzerland.

Groombridge, B. and Jenkins, M.D. (2002) *World Atlas of Biodiversity.* Prepared by the UNEP World Conservation Monitoring Centre. University of California Press, Berkeley, California.

Habte, M. (2001) *Economics of Export Subsidy.* Study carried out at the WTI, Berne, September 2001.

Hoekman, B.M. and Kostecki, M.M. (2001) *The Political Economy of the World Trading System – The WTO and Beyond,* 2nd edn. Oxford University Press, Oxford, UK.

International Institute for Sustainable Development (2003a) Doha Round Briefing Series – Agriculture. Vol. 1, No. 2, February 2003.
International Institute for Sustainable Development (2003b) Doha Round Briefing Series – Special and Differential Treatment. Vol. 2, No. 13, August 2003.
Knox, P. and Agnew, J. (1998) *The Geography of the World Economy*, 3rd edn. Arnold, London.
Mamaty, I. (2000) Special and differential treatment. In: FAO (ed.) *Multilateral Trade Negotiations on Agriculture: A Resource Manual*. FAO, Rome. http://www.fao.org/documents/show_cdr.asp?url_file=/DOCREP/003/X7353E/X7353E00.htm (accessed December 2005).
Mueller, D.C. (1989) *Public Choice II*. Cambridge University Press, Cambridge, UK.
Myers, N. and Kent, J. (2001) *Perverse Subsidies – How Tax Dollars can Undercut the Environment and the Economy*. Island Press, Washington, DC.
OECD (1993) *World Cereal Trade: What Role for Developing Countries*? OECD, Paris.
OECD (2000) *Agricultural Outlook 2000–2005*. OECD, Paris.
OECD (2001a) *Market Effects of Crop Support Measures*. OECD, Paris.
OECD (2001b) *Multifunctionality. Towards an Analytical Framework*. OECD, Paris.
Pearson, C.S. (2000) *Economics and the Global Environment*. Cambridge University Press, Cambridge, UK.
Rangnekar, D. (2003) Plant breeding, biodiversity loss and intellectual property rights. Economics discussion paper no. 00/5. Kingston University, UK, available at: http://www.kingston.ac.uk/research/0005.pdf (accessed December 2005).
Rieder, P. and Anwander Phan-Huy, S. (1994) *Grundlagen der Agrarmarktpolitik*. Verlag der Fachvereine an den schweizerischen Hochschulen und Techniken vdf, 4. Auflage, Zürich, Switzerland.
Schulze, E.-D., Beck, E. and Müller-Hohenstein, K. (2002) *Pflanzenökologie*. Spektrum Akademischer Verlag GmbH, Heidelberg/Berlin.
Sharma, R. (2000) Agriculture in the GATT: a historical account. In: FAO (ed.) *Multilateral Trade Negotiations on Agriculture: A Resource Manual*. FAO, Rome. http://www.fao.org/documents/show_cdr.asp?url_file=/DOCREP/003/X7351E/X7351E00.htm (accessed December 2005).
Shiva, V. (2002) Die WTO und die Landwirtschaft in den Entwicklungsländern. In: Mander, J. and Goldsmith, E. (eds) *Schwarzbuch Globalisierung*. Riemann Verlag, Munich, Germany. (Title of original: *The Case against the Global Economy* (2001) Earthscan, London).
Siebert, H. (1997) *Weltwirtschaft*. Lucius & Lucius Verlagsgesellschaft GmbH, Stuttgart, Germany.
Siebert, H. (2000) *Aussenwirtschaft*, 7th edn. Lucius & Lucius Verlagsgesellschaft GmbH, Stuttgart, Germany.
Steinberg, R.H. and Josling, T.E. (2003) When the Peace Ends: The Vulnerability of EC and US Agricultural Subsidies to WTO Legal Challenge. *Bridges* 7(8), 7–10.
Sterkenburg, J.J. (1987) Rural development and rural development policies. Cases from Africa and Asia. Nederlandse geografische studies. Geografisch instituut, rijksuniversiteit Utrecht, Netherlands.
Swingland, I.R. (2001) Definition of biodiversity. In: Levin Simon Asher (ed.) *Encyclopedia of Biodiversity*, Volume 1. Academic Press, New York.
Theissen, U. and Voigt, H. (1986) *Entwicklungsländer – Entwicklungshilfe*. Aulis Verlag, Köln, Germany.
Tisdell, C. (2003) Socioeconomic causes of loss of animal genetic diversity: analysis and assessment. *Ecological Economics* 45(3), 365–376.
Todaro, M.P. (2000) *Economic Development*. 7th edn. Addison Wesley Longman, Harlow, UK.
Tokarick, S., Sutton, B. and Yang, Y. (2002) How do industrial country agricultural policies affect developing countries? In: *World Economic Outlook (2002) Chapter II, Essays on Trade and Finance*. Available at: http://www.imf.org/external/pubs/ft/weo/2002/02/pdf/chapter2.pdf (accessed December 2005).
WTO (2004) *Agriculture Negotiations: The Issues, and where we are now* (revised 1 December 2004) http://www.wto.org/english/tratop_e/agric_e/negs_bkgrnd00_contents_e.htm (accessed February 2005).

Part 2

Araya, M. (2003) WTO negotiations on environmental goods and services: maximizing opportunities? In: *Global Environment & Trade Study (GETS).* Yale Center for Environmental Law and Policy.

Brenton, P. (2003) Integrating the least developed countries into the world trading system: the current impact of European Union preferences under 'Everything but Arms'. *Journal of World Trade* 37(3), 623–646.

Bridges (2003) Agriculture: a blank check bitterly contested. *Bridges* 7(6), 11–13. Available at: http://www.ictsd.org

Chamborey, D. (2003) Comments on the contribution by Prof. Thomas Cottier and Marion Panizzon, Differential Tariff Protection, Product Standards and Product Process Methods. In: *Rights to Plant Genetic Resources and Traditional Knowledge: Basic Issues and Perspectives, Second Expert Workshop of the Biodiversity Project*, 27–28 April 2003, Berne, Switzerland.

Chaytor, B. (2003) A primer on environmental goods and services: definitional challenges to the negotiation of further liberalisation. Study commissioned by the Royal Society for the Protection of Birds (RSPB), UK.

Cottier, T. (1998) The protection of genetic resources and traditional knowledge: towards more specific rights and obligations in World Trade Law. *Journal of International Economic Law* 1(4), 555–584.

Cottier, T. (2000) Risk management experience in WTO dispute settlement. In: Robertson, D. and Kellow, A. (eds) *Globalization and the Environment: Risk Assessment and the WTO.* Edward Elgar, Cheltenham, UK, pp. 41–62.

Cottier, T. and Mavroidis, P. C. (eds) (2000) *Regulatory Barriers and the Principle of Non-Discrimination in World Trade Law*, World Trade Forum Volume 2. University of Michigan Press, Ann Arbor, Michigan.

Cottier, T. and Nadakavukaren-Schefer, K. (1997) Non-violation complaints in WTO/GATT dispute settlement: past, present and future. In: Petersmann, E.-U. (ed.) *International Trade Law and the GATT/WTO Dispute Settlement System*, 11 Studies in Transnational Law. Kluwer Law International, The Hague.

Cottier, T. and Nadakavukaren-Schefer, K. (2000) Good faith and the protection of legitimate expectations in the WTO. In: Bronckers, M. and Quick, R. (eds) *New Directions in International Economic Law, Essays in Honour of John H. Jackson.* Kluwer Law International, The Hague.

Cottier, T. and Oesch, M. (2005) *International Trade Regulation.* Cameron May, London, and Stämpfli Publishers, Berne, Switzerland.

Cottier, T. and Panizzon, M. (2004) Legal perspectives on traditional knowledge: the case for intellectual property protection. *Journal of International Economic Law* 7(2), 371–400.

Ehring, L. (2002) De facto discrimination in World Trade Organization law – national and most-favoured nation treatment – or equal treatment? *Journal of World Trade* 36(5), 921–977.

Forero, J. (2003) Brazil pushes for South American trade pact. *The New York Times*, 17 September 2003. Available at: http://www.nytimes.com

Hamway, R., Hoffmann, U., Vikhlyaev, A. and Vossenaar, R. (2003) Liberalization of international trade for environmental goods and services. In: *Background Paper, UNCTAD/FIELD Project Building Capacity for Improved Policy Making and Negotiation on Key Trade and Environment Issues*, funded by the UK Department for International Development (DFID).

Hart, M. and Dymond, B. (2003) Special and Differential Treatment and the Doha 'Development' Round. *Journal of World Trade* 37(2), 395–415.

Hoekman, B. (2002) Strengthening global trade architecture for development: the post Doha agenda. *World Trade Review* 1(1), 23–45.

Hoekman, B., Michalopoulos, C. and Winters, A.L. (2004) Special and differential treatment for developing countries, towards a new approach in the WTO. Working Paper, Draft Version 4 April 2003.

Horton, L.R. (1998) Food from developing countries: steps to improve compliance. *Food and Drug Law Journal* 53(1), 139–169.

Howse, R. (2000) Adjudicative legitimacy and treaty interpretation in international trade law: the early years of WTO jurisprudence. In: Weiler, J.H.H. (ed.) *The EU, the WTO, and the NAFTA,*

Towards a Common Law of International Trade? Oxford University Press, Oxford, UK, pp. 35–69.
Howse, R. and Regan, D. (2000) The product/process distinction – an illusory basis for disciplining 'unilateralism' in trade policy. *European Journal of International Law* 11(2), 249–289.
Howse, R. and Türk, E. (2001) The WTO impact on internal regulations – a case study of the Canada–EC asbestos dispute. In: de Burca, G. and Scott, J. (eds) *The EU and the WTO, Legal and Constitutional Issues.* Hart Publishing, Oxford, UK, pp. 283–328.
International Center for Trade and Sustainable Development (ICTSD) (2003) Cancun Update. *Doha Round Briefing Series*, Vol. 2, No. 2.
Jackson, J.H. (1997) *The World Trading System*, 2nd edn. The MIT Press, Cambridge, Massachussetts.
Jackson, J.H. and Benke, R.T.J. (2003) The WTO cases on US 'trade remedies': opening a discussion. *Journal of International Economic Law* 6(1), 111.
Marceau, G. (1999) A call for coherence in international law, praises for the prohibition against 'clinical isolation', in WTO Dispute Settlement. *Journal of World Trade* 33(5), 87–152.
Marceau, G. (2001) Conflicts of norms and conflicts of jurisdiction, the relationship between the WTO Agreement, MEAs and other treaties. *Journal of World Trade* 35(6), 1081–1131.
Marceau, G. and Trachtman, J. (2002) The Technical Barriers to Trade Agreement, the Sanitary and Phytosanitary Measures Agreement and the General Agreement on Tariffs and Trade. *Journal of World Trade* 36(5), 811–881.
Meinheit, E. (1995) *Handelspolitik als Umweltpolitik im Agrarbereich? Eine ökonomische und GATT-rechtliche Analyse.* Wissenschaftsverlag Vauk, Kiel, Germany.
Oesch, M. (2003) *Standards of Review in WTO Dispute Resolution.* Oxford University Press, Oxford, UK.
Panizzon, M. and Cottier, T. (2005) Traditional knowledge and geographical indications: foundations, interests and negotiating positions. In: Petersman, E.U. (ed.) *Developing Countries in the Doha Round.* Robert Schuman Centre for Advanced Studies, European University Institute, San Domenico di Fresole, Italy.
Pauwelyn, J. (2000) Applying SPS in WTO disputes. In: Robertson, D. and Kellow, A. (eds) *Globalization and the Environment: Risk Assessment and the WTO.* Edward Elgar, Cheltenham, UK, pp. 63–80.
Pauwelyn, J. (2001) The role of public international law in the WTO: how far can we go? *American Journal of International Law* 95(3), 535–587.
Perkins, F. (2003) Africa's agricultural trade reform and development options. In: *The South African Institute of International Affairs, Trade Policy Briefing,* No. 1, June 2003.
Rosenberg, T. (2003) Why Mexico's small corn farmers go hungry. *The New York Times*, 3 March 2003.
Short, C. (2001) Making the development round a reality. In: Sampson, G.P. (ed.) *The Role of the World Trade Organization in Global Governance.* United Nations University Press, New York.
Silverglade, B.A. (1998) The impact of international trade agreements on US food safety and labelling standards. *Food and Drug Law Journal* 53, 537–542.
Sutherland, P. *et al.* (2005) The Future of the WTO. Addressing Institutional Challenges in the New Millennium. Report by the Consultative Board to the Director General Supachai Panitchpakdi, WTO, 2005.
The New York Times (2003) The Cancun Failure, Editorial. *The New York Times*, 16 September 2003. Available at: http://www.nytimes.com
Trebilcock, M.J. and Howse, R. (1999) *The Regulation of International Trade*, 2nd edn. Routledge, London.
Vaughan, S. (2003) Trade Preferences and Environmental Goods. *Carnegie Endowment for International Peace, Trade, Equity, and Development*, Issue 5, February 2003.

Index

acacias 97
access: to genetic resources and traditional knowledge
  and benefit-sharing
    Australia 103–104, 108
    in context of prior art 241–242
    in Convention on Biological Diversity 31, 61–62, 207, 290–291
    India 91–92
    involvement of grassroots stakeholders 295
    Jeevani case 138–139
    Multilateral System of Access and Benefit-Sharing 31, 66–67, 291–294
    necessity to identify information holders 299–300
    problem areas in international trade 295–296
  contracts 85–86, 89–90
  control by users of information 298–299
  financial mechanisms 87–88
    aims and functions 301–302
    elements 302–304
    models: assistance funds 304–309
    models: compensation funds 309–313
    models: Kyoto Protocol Flexible Mechanisms 313–317
    options for future mechanism 318–321
    relevant work within FAO 317–318
  by foreigners 91
  legislation
    effects on marketability 297–298
    laws not dependent on landrace identification 186
  open access 24
  scientific *vs* industrial use 88–89
  varied accessibility of traditional knowledge 160
access: to markets
  background 357–361
  effects of access legislation 297–298
  facilitation of trade through clearing houses *see* clearing house mechanisms
  impact of most-favoured nation principle 382
  impact of national treatment 382–385
  impact on traditional knowledge-based products 364
  non-tariff barriers
    Agreement on the Application of Sanitary and Phytosanitary Measures 377–379
    Agreement on Technical Barriers to Trade 375–377
    need for joint standard-setting operations 379–380
  special and differential treatment 364–366
    adjustment of tariff schedules 368–372
    General System of Preferences 366–367
    labelling requirements 374
    limiting recourse to trade remedies 372–374
  tariffication of trade barriers 361–364
  tariffs to facilitate exports xxx–xxxii
acquisitions: of patents 147
Africa
  agrarian structure 338
  OAU Model Legislation 82–84, 133–134, 214
Agreement on Agriculture (WTO) 72
  coverage 328–329
  domestic support *see* direct payments
  objectives 327–328
  reform programme 329–333
Agreement on Application of Sanitary and Phytosanitary Measures (WTO) 72–73, 377–379
Agreement on Safeguards (WTO) 373
Agreement on Subsidies and Countervailing Duties (WTO) 373

Agreement on Technical Barriers to Trade 375–377
Agreement on Textiles and Clothing (WTO) 373
Agreement on Trade-related Aspects of Intellectual Property Rights *see* TRIPS Agreement
agrobiodiversity *see* biodiversity
aid, food
  economics 341
  effects on agrobiodiversity 344, 346
AIDS: drug therapies 97–98
Andean Community
  regime on genetic resources and traditional knowledge 84–90
  sanctions for external infractions 298–299
Anti-Dumping Agreement (WTO) 373
Asia: agrarian structure 337
Australia
  Aboriginal intellectual property 98–102
  access and benefit-sharing 108
  connection between land rights and traditional knowledge 105–107
  Environment Protection and Biological Diversity Conservation Act 103–104
  legal system 102, 106–109
  megabiodiversity 96
  national and international legal obligations 102–103
  Native Title and Native Title (Amendment) Acts 104–106
  traditional knowledge and biopiracy 96–98

barley 181
bean 181–182
benefits
  access and benefit-sharing *see* access: to genetic resources
Berne Convention 76
biocollecting society (Drahos) 274
biodiversity
  in Australia 96
  causes of loss
    direct causes 11
    ex situ material 10
    genetic erosion 10, 338–339
    underlying economic causes 11–13, 27–30
    WWF root causes study 13–14, 15
  Convention on Biological Diversity *see* Convention on Biological Diversity
  deficiencies of international policies xxv–xxvi
  distribution and origin 6–7
  effect of export subsidies and food aid 340–346
  Farmer's Unit of Diversity Management 177–178
  impact of domestic support 350–351
  in situ on-farm conservation 164–165
  need to halt loss xxiv, 3–4
  obstacle to landrace protection 189
  promoted by subsistence farming 339
  reflected in language richness 15, 17–18
  registers and registration 137–138, 229
  sustainable use 44
  terminology 4–5
  and traditional knowledge 20–21
    patterns of loss 26
  value 7–10, 48–49
Biodiversity Act (Costa Rica) 244
Biodiversity Act (India) 91–92
Biodiversity Act (Venezuela) 244
biopiracy 92, 96–98, 134
bioprospecting 296
bioprospecting: agreements 105–106
biosafety 63, 248
biotechnology
  European Union Biotechnology Directive 93–94
  industry structure 147–151
Brazil 215, 296
breeders: rights *see* rights
Brundtland Report 36–37, 41, 43
bush food 97
by-products: of genetic resources 85

Canada: tariff schedules 369
Cartagena Protocol 63, 84–85, 379–380
cassava 184
CEFIC (European Chemical Industry Council) 258
CGIAR (Consultative Group on International Agricultural Research) 241–242
  access to collections 67
  guiding principles on intellectual property 65
circuits, integrated 78
civil liability 309–311
Clean Development Mechanism, Kyoto Protocol 315
clearing house mechanisms
  Convention on Biological Diversity 270–271
  elements and structure 275–277
  facilitation of trade in traditional knowledge 272–274
  models 273–275
  purpose and function 270–271
  WIPO online portal 271–272
climate: impact of change on food security 48
clones: distinctness and gene flow 183–184
Codex Alimentarius xxxi, 72, 380
collecting forms: for gene banks 195–197
  possible policy responses 197–198
collecting societies 266–269
commercialization 264
common heritage 58
communities, rights of *see* rights
competition 265
  adverse effect of intellectual property rights 142–143
  policies 145–147
  within seed industry 147–151
  sequential innovators 144–145

conocurovone 97
consent, prior informed 140, 242–244
conservation
in Convention on Biological Diversity 60–61
importance of *in situ* on-farm conservation 164
in International Treaty on Plant Genetic Resources for Food and Agriculture (2002) 65–66
contracts, access *see* access: to genetic resources
Convention on Biological Diversity
access and benefit-sharing regime 31, 61–62, 207, 290–291
application for observer status at WTO 379
broad notion of biodiversity concept 4
Cartagena Protocol 63
clearing house mechanism 270–271
concept of origin 165–166, 167, 168
concerns of indigenous peoples 21
and conservation 60–61
creation of incentives 31–34
definition of origins of genetic resources 6–7
financial commitments 62
institutional structure 62–63
integrative character 3–4
limitations xxv
prior informed consent 244
and state sovereignty 58–59
stimulus for international activity xxiii
sustainable use 41, 42–43
on traditional knowledge 16
copyright 75–76
collecting societies 267
inadequate protection for communal works 100, 101–102
Costa Rica 244
costs: internalization 12
countervailing duties 373
country of origin: definitions 165–166

data mining studies 193–195
possible policy responses 197–198
databases
multipurpose 258–259
need for resources on traditional knowledge 129, 241
SINGER 193–195, 228–229
WIPO online portal 271–272
dealing, exclusive 146
Declaration of Human Rights 45
descriptors, ethnobotanical 195–197
Desertification Convention 286–287
designs, industrial 77–78
developing countries: agriculture
economics of subsistence farming 333–335
rural–urban migration 335–337
special treatment and special products 331–332
structures 337–338
development, sustainable *see* sustainability
differentiation: of products
General System of Preferences 366–367
labelling requirements 374
tariffs 368–372
direct payments
conclusions 354–356
critique of current status 351–352
multifunctionality *vs* food security 352–354
support of agrobiodiversity 350–351
WTO regulations 348–350
disclosure 92, 126, 229–230, 242–244
distinctness: criterion of origin 169–170, 247–248
documentation *see* registration and documentation
Doha Declaration and Work Programme 325–327, 330–333
domain public payant 268
domestic support *see* direct payments
Drahos, P. 274
Duboisia 96–97
duties, countervailing 373

economics
of sustainability 30–31
underlying the loss of biodiversity 11–13, 27–30
ecosystems
destruction and degradation 11
ecosystem services 9
Emissions Trading, Kyoto Protocol 315–316
Environment Protection and Biodiversity Conservation Act (Australia) 103–104
equity
in Brundtland Report 43
definitions and general usage 36–37
and fairness 40
legal sources 35–36
and patent protection 119
traditional concepts 37–38
erosion, genetic 10, 338–339
European Chemical Industry Council (CEFIC) 258
European Patent Convention 93
patentability of living organisms 121–125
protection of traditional knowledge 127–130
protective criteria 126
European Union Biotechnology Directive 93–94
exclusive dealing 146
exclusive territories 145
exports
facilitated by tariffs xxx–xxxii
subsidies 340–343, 346

facilitators (Krattiger and Lesser) 274–275
fairness
equity as fairness 40
justice as fairness 39–40
see also equity
FAO (Food and Agriculture Organization) 317–318

Fourth International Technical Conference on Plant Genetic Resources, Leipzig, 1996 18, 48–49
farmers
  innovation 188, 189–190
    perverse incentives 191–192, 216, 220
    vs tradition 216–217
  position within Multilateral System 293
  rights *see* rights
  selection of landraces 175–177
    Farmer's Unit of Diversity Management (FUDM) 177–178
farming, subsistence 333–335
  promotes biodiversity 339
Flexible Mechanisms, Kyoto Protocol 313–317
food aid
  economics 341
  effects on agrobiodiversity 344, 346
food security 352–354
  and climate change 48
  current situation 46–47
  and individual motivations 23–24
  proposed strategies 49
  role of traditional knowledge 22
  and trade 47–48
  value of diversity 48–49
funds
  assistance 304–309
  compensation 309–313
  financial mechanisms for access to genetic resources *see* access

GATT (General Agreement on Tariffs and Trade) *see* World Trade Organization
gene banks
  collecting forms and ethnobotanical descriptors 195–197
  data mining studies 193–195
  possible policy responses 197–198
  and prior art 241–242
General System of Preferences 366–367
genetic erosion 10, 338–339
geographical indications xxx, 77, 231–232
  characteristics and scope 248–249
  further development 250–251
  and trademarks 249–250
  and traditional knowledge 251–253
germplasm: collecting forms and ethnobotanical descriptors 195–197
Global Environment Facility 305–307
grant backs 146
green box payments 348–350
Green Revolution 141–142

habitats, alteration of 11
handicrafts 369
health care: inclusion of traditional medicine 49–50
heritage, common 58, 88, 99
heritage, state 85
Hiddens, Les ('the Bush Tucker Man') 97
Honey Bee Network 139, 260

in-breeders: distinctness and gene flow 180–182
incentives
  Convention on Biological Diversity discussion 31–34
  Kyoto Protocol models 313–317
  perverse incentives 191–192, 216, 220
India
  benefit-sharing 138–139
  Biodiversity Act 91–92
  challenges to patents 135, 136
  local innovation 139
  Patents (Amendment) Act 92, 244
  Protection of Plant Varieties and Farmers' Rights Act 90–91, 133
  registration and documentation 260–261, 262–263
indications, geographical *see* geographical indications
indigenous peoples 16–17
  concerns about Convention on Biological Diversity 21
  intellectual property rights of Australian aborigines *see* intellectual property rights
  jurisprudence systems 20
  Peruvian development fund 89
Indigenous Peoples Rights Act (Philippines) 244
industries, research-based
  relationship to traditional knowledge xxiv–xxv
innovation
  by farmers
    challenge for *sui generis* protection 188
    perverse incentives 191–192, 216, 220
  local 139
  sequential 143–145
  vs tradition 216–217
intangible components: of genetic resources 85
integrated circuits 78
intellectual property rights
  Andean Community regime 86
  benefit-sharing *see* access: to genetic resources
  CGIAR principles 65
  as colonialist instrument 265–266
  commercialization 264
  and competition laws *see* competition
  definition 98–99
  double protection 82
  economic value 264–265
  effects on seed industry market structure 147–151
  forms of protection 75–80, 113–114
  function 114–115
  impacts on plant genetic resources and traditional knowledge 141–142
  of indigenous Australians 98–102
  levels of protection 120–121, 134–140
  market activation 265–266

intellectual property rights *continued*
protection of traditional knowledge xxviii
funding issues xxix
no true protection xxvi, 19–20, 24–25, 100, 134–137
objections 216–220
proposed solutions 137–139
reform and novel uses of existing rights xxviii–xxix
sui generis protection *see* varieties, plant: protection
WTO dispute settlement 71–72, 233
see also patents; rights; Traditional Intellectual Property Rights: proposed concept; TRIPS Agreement
International Oil Pollution Compensations Funds (IOPC) 311–313
International Rice Research Institute 242
International Treaty on Plant Genetic Resources for Food and Agriculture xxiii, xxv, 5, 16, 219
access and benefit-sharing 66–67
CGIAR collections 65, 67
concept of origin 166
conservation 65–66
fails to support rights of small farmers xxv
farmers' rights 285–286
multilateral system 208
Multilateral System of Access and Benefit-Sharing 31, 66–67, 291–294
revisions 64
sustainable use 41, 43
International Undertaking 284–285
International Union for the Protection of New Varieties of Plants *see* UPOV Convention
invention: definition 124
inventorship, joint 244–245
IUCN *see* World Conservation Union (IUCN)

Jeevani 138–139
Joint Implementation, Kyoto Protocol 314–315
joint inventorship 244–245

kava 137, 253
knowledge, traditional
adaptation of conditions for plant breeder rights 247–248
appropriation problems
analysis 140
examples 135–137
proposed solutions 137–139
and biodiversity 20–21, 26
and bioprospecting 296
common heritage not commodity 99–100
connection with land rights 105–107
considered as public domain 205, 206
in Convention on Biological Diversity 61
definition and concept 15–17, 158–159, 220–221
and existing patentability criteria 245–247
and geographical indications 251–253
as global public good 206–207
holders 18–20
intellectual property rights
do not afford protection xxvi, 19–20, 24–25, 159–160
impact 142
objections 216–220
interests: public *vs* private 161
and joint inventorship 244–245
legal status
international laws 207–214
recommendations and national laws 214–215
loss 23–27, 28
origin and ownership 161
Peruvian legislation 88–89
and prior art 117–118, 240–242
protected by farmers' rights 286–288
registration and documentation *see* registration and documentation
result of ecocultural learning process 17–18
and sustainable development 44–45
types 160
value
of indigenous knowledge in Australia 96–98
short-term and long-term 21–22
see also Traditional Intellectual Property Rights: proposed concept
Krattiger, A.F. 274–275
Kyoto Protocol 304–305
Flexible Mechanisms 313–317

labelling: for product differentiation 374
land rights 104–107
landraces 5
abandonment of old varieties 23–24
definition and concept 163–164
historical evolution of concept 174–180
distinctness and gene flow
clones 183–184
in-breeders 180–182
out-breeders 182–183
and food security 22
gene bank accessions
collecting forms and ethnobotanical descriptors 195–197
data mining for links with originating communities 193–195
possible policy responses 197–198
in situ on-farm conservation 164–165
origin
centres of diversity 165
country of origin: definitions 165–166
criterion of distinctness 169–170
criterion of time 171, 174–175
criterion of uniformity 170–171
difficulties of definition and interpretation 166–168
importance of 6–7

potential intellectual property policy responses
challenges of *sui generis* protection criteria 185–190
feasible situations 192–193
hypothetical case 190–191, 192
selection by farmers 175–177
Farmer's Unit of Diversity Management (FUDM) 177–178
focus on trait preservation 178–179
language
local names do not always reflect diversity 177–178
loss of diversity 25–26
richness reflects biodiversity 15, 17–18
Latin America
agrarian structure 337
laws to protect traditional knowledge 84–90, 215, 298–299
law, international
equity and fairness 35–41
European 93–94
goals of protective measures 34–35
WIPO model laws 210–213
see also Convention on Biological Diversity; International Treaty on Plant Genetic Resources for Food and Agriculture; UPOV Convention; World Trade Organization
law, national
Africa: OAU Model Legislation 82–84, 214
Andean Community 84–90
Peru 86–89, 215
Australia 102–109
Brazil 215
India 90–92, 214–215
Philippines 215
USA 95–96
WIPO model laws 214
learning, ecocultural 17–18
Leipzig Declaration 48–49
Lesser, W.H. 274–275
liability, civil 309–311
licensing 144, 145–146
likeness 359, 381–385

maize 176–177, 179, 183
chianquiahuitl 190–191, 192
markets
access for traditional knowledge products *see* access: to markets
creation 296–297
informal sector 335
local and global market failure 30
seed industry market structure and restructuring 147–151
Marrakesh Agreement 325–327
medicine, plant-derived 96–98
benefit-sharing 138–139
extent of use 8, 21–22
problems with patents 135, 137
medicine, traditional
definition 221
inclusion into primary health care 49–50
mergers: of companies 147
within seed industry 147–151
migration, rural–urban 335–337
millet 181, 182
Montreal Protocol: Multilateral Fund 307–308
most-favoured nation principle 75, 382
multifunctionality 352–353
Multilateral System of Access and Benefit-Sharing *see* International Treaty on Plant Genetic Resources for Food and Agriculture

Native Title and Native Title (Amendment) Acts (Australia) 104–106
needs, basic 45–46
neem 136
novelty 117, 125, 126–127
UPOV *vs* patent law 81

Organization of African Unity (OAU)
Model Legislation 82–84, 133–134, 214
origin *see* landraces
Our Common Future see Brundtland Report
out-breeders: distinctness and gene flow 182–183
ownership: communal *vs* individual 100–101, 161, 191

Papua New Guinea 369, 372
patents
application to traditional knowledge 117–118, 128–130
combination patents 224–225
competitive strategies 147, 149–151
contractual arrangements and licensing 144–146
European Patent Convention *see* European Patent Convention
European Union Biotechnology Directive 93–94
Indian regulations 92
joint inventorship 244–245
OAU Model Legislation 83
patentability of living organisms 78–79, 83, 86, 93, 95–96, 117, 121–125
problematical cases 135–137
protective criteria
disclosure 92, 126, 229–230, 242–244
inventive step 125–126, 245–246
novelty 81, 117, 125
prior art 117–118, 240–242
utility 125
rationale of patent protection 119–121
scope and limitation of protection 127–128
second-tier (petty) patents 246–247
TRIPS Agreement (1994) *see* TRIPS Agreement (1994)
payments, direct *see* direct payments
Peoples Plan of Action (NGO statement of FAO plant genetic resources conference, Leipzig, 1996) 18

Peru
  Draft Regulation on Access to Genetic Resources 87–88
  Fund for the Development of Indigenous Peoples 89
  Law introducing a Protection Regime for the Collective Knowledge of Indigenous Peoples derived from Biological Resources 88–89, 215
  Law on Conservation and Sustainable Use of Biodiversity 87
  registration system 259–260, 261–262
petty patents 246–247
Philippines 215, 244
Plant Variety Act (Thailand) 248
Plant Variety Protection Act (USA) 95–96
  effect on seed industry 147–148
pooling: of patents 147
potato 183–184
prices, restriction of 146
prior art xxvi, 117–118, 135, 240–242
prior informed consent 140, 242–244, 248, 262–263
products
  product process methods (PPMs) 372, 376–377, 378–379, 380–381, 385–387
  standards 375–376, 377–378
    need for joint standard-setting operations 379–380
products: differentiation
  General System of Preferences 366–367
  tariffs 368–372
property, intellectual *see* intellectual property rights
public domain 89
  considered to include all traditional knowledge 205, 206
  domain public payant 268
public good 13, 206–207
Public Interest Intellectual Property Advisors 228

reach-through royalties 146
registration and documentation
  Biozulua register, Venezuela 257–258
  definitions and terminology 256–257
  to differentiate farmer and breeder varieties 287–288
  as a marketing tool 263–264
  national examples 259–263
  questions relating to traditional knowledge 254–256
  registers of biodiversity 137–138, 228–230
  stakeholder positions 257–258
  types of resources 258–259
remedies, trade 372–374
resources
  biological *vs* genetic 4–5
  plant genetic resources
    access *see* access: to genetic resources
    by-products and intangible components 85
    characteristics 6
    definition 5
    impact of intellectual property rights 141–142
    low market value of new varieties 297
    see also landraces
rice 180–181, 242
rights
  of communities 83, 100–101, 217–218, 222, 287
  copyright and related rights *see* copyright
  of farmers
    African Model Legislation 83
    European Union 94
    India 90–91
    under the International Treaty on Plant Genetic Resources for Food and Agriculture 285–286
    under the International Undertaking 284–285
    a matter of contention 283–284
    means to protect traditional knowledge 286–288
    not recognized 85
    vs indigenous peoples 18–19
    vs plant breeders 81
  to genetic resources *vs* traditional knowledge 85–86
  land rights 104–107
  of landowners 104
  of plant breeders 130–132, 232–233
    adaptation to traditional knowledge 247–248
    African Model Legislation 84
    Andean Community 85
    European Patent Convention 122, 123–124
    exemptions 133–134
    India 90
    see also UPOV Convention
  of state to genetic resources 85
  see also intellectual property rights; patents; Traditional Intellectual Property Rights: proposed concept
Rio Declaration on Environment and Development 15–16
  sustainability 35–36
royalties, reach-through 146

second-tier patents 246–247
secrets, trade 79–80
security, food *see* food security
Seed Treaty *see* International Treaty on Plant Genetic Resources for Food and Agriculture
seeds
  industry market structure and restructuring 147–151, 206
  inter-community exchange and gene flow 181
  and perpetration of landrace traits 176–177, 178–179

primary objects of early trade xxiii
traditional supply systems 162
SINGER databases 193–195, 228–229
smokebush 97–98
sorghum 182
sovereignty 6–7
in Convention on Biological Diversity 60
limitations 57–58
over genetic resources 58–59
permanent sovereignty
formalization by UN (1962) 57
inequality between formal and informal sectors xxv
SPS Agreement *see* Agreement on Application of Sanitary and Phytosanitary Measures (WTO)
state, the
holder of property rights to genetic resources 85
subsidies, export
economics 340–342
effects on agrobiodiversity 342–343, 346
subsistence farming 333–335
promotes biodiversity 339
sui generis protection *see* varieties, plant: protection
sustainability
in Brundtland Report 36–37, 41, 43
definition 41–42
financial aspects 30–31
in Rio Declaration 35–36
sustainable development 43, 44–45
sustainable use 42–43, 60
sweet potato 184

tariffs *see* access: to markets
territories, exclusive 145
Thailand 248
time: criterion of origin 171, 174–175
trade
and benefit-sharing problem areas 295–296
facilitation *see* clearing house mechanisms
and food security 47–48
necessity of transparency 300
negative impacts of liberalization 27–30
support of products based on traditional knowledge xxix–xxx
tariffication of trade barriers *see* access: to markets
see also World Trade Organization
trademarks xxx, 76–77, 249–250
Traditional Intellectual Property Rights: proposed concept
content and scope of rights 222–227
duration of rights 227
fora for future negotiations 233
opposition and review procedures 230–231
registration requirements 228–230
relation to geographical indications 231–232
relation to plant breeders' rights 232–233
right holders 221–222
subjects covered 220–221
traditional knowledge *see* knowledge, traditional
traditional medicine *see* medicine, traditional
TRIPS Agreement (1994)
copyright and related rights 75–76
disclosure of prior art 243
discussions on protection of traditional knowledge 209–210, 296
enforcement 80
geographical indications 77, 231–232, 249, 250–251
industrial designs and integrated circuits 77–78
more suited to industrialized states 216
negotiations and establishment 73–74
NGO opposition 219–220
patentability of living organisms 121–122
patents 78–79
protective criteria 126
rights conferred by patent 123
scope and general provisions 74–75, 208–209
sui generis protection 132–134
trade secrets 79–80
trademarks 76–77
turmeric 135–136
tying: of sales 146, 149

uniformity: criterion of origin 170–171
United Nations
resolution on permanent sovereignty (1962) 57
UPOV Convention 79, 80–82
criteria for protection
adaptation to traditional knowledge 247–248
distinctness 169–170
position of landraces 186–190
definition of a plant variety 122
model for *sui generis* protection 133, 207–208
not ratified by India 90
rights conferred 123
scope and limitations of protection 131–132
USA
joint inventorship 245
patentability of plants and living organisms 117–118, 124–25
Plant Variety Protection Act *vs* patent protection 95–96
effect on seed industry 147–148
prior art: interpretation 135
protection of plant varieties 122
protective criteria for patents 126–127
Public Interest Intellectual Property Advisors 228
statement to TRIPS Council 296

varieties, plant: protection 122–123, 131–132
adaptation to traditional knowledge 247–248
new farmers' varieties 189–190

varieties, plant: protection *continued*
sui generis protection
examples 133–134
landraces: challenges 185–190
landraces: feasible situations 192–193
landraces: hypothetical case 190–191, 192
rationale 132–133
Vavilov, N.I. 165
Venezuela 241, 244, 257–258, 287

WIPO (World Intellectual Property Organization)
clearing house mechanism 271–272
definition of intellectual property 98–99
definition of traditional knowledge 220–221
Fact Finding Mission on Intellectual Property and Traditional Knowledge 20
model laws for protection of traditional knowledge 210–214
World Bank: structural adjustment programmes 336
World Conservation Union (IUCN) 27
World Conservation Strategy (1980) 43
World Health Organization (WHO)
definition of traditional medicine 221
promotion of traditional medicine 49–50
World Trade Organization
Agreement on Agriculture 72, 327–333
Agreement on Application of Sanitary and Phytosanitary Measures 72–73
Agreement on Safeguards 373
Agreement on Subsidies and Countervailing Duties 373
Agreement on Textiles and Clothing 373
Agreement on Trade-related Aspects of Intellectual Property Rights *see* TRIPS Agreement
Anti-Dumping Agreement 373
dispute settlement and enforcement 70–72, 233
GATT 1947 68–69, 325, 329, 363, 366–367
GATT 1994 343, 357, 362, 368, 372, 373, 377
likeness 381–385
non-discrimination principles 380–382
product process methods 385–387
Marrakesh Agreement 325–327
non-tariff trade barrier negotiations 69–70
regulation of domestic support *see* direct payments
tariffication *see* access: to markets
three major pillars 68
World Wide Fund for Nature 13